Bernhard Welz
Atomic Absorption Spectrometry

Distribution:

VCH Verlagsgesellschaft, P.O. Box 1260/1280, D-6940 Weinheim (Federal Republic of Germany)

USA and Canada: VCH Publishers, 303 N.W. 12th Avenue, Deerfield Beach FL 33442-1705 (USA)

ISBN 3-527-26193-1 (VCH Verlagsgesellschaft)
ISBN 0-89573-418-4 (VCH Publishers)

Bernhard Welz

Atomic Absorption
Spectrometry

Second, Completely Revised Edition

English Translation by Christopher Skegg

Dr. Bernhard Welz
Perkin-Elmer Bodenseewerk
Postfach 1120
D-7770 Überlingen

first edition 1976
reprint 1981
second edition 1985
First published in German under the title "Atom-Absorptions-Spektroskopie" by Verlag Chemie in 1972.
This second English edition was translated from the completely revised third German edition published under
the title "Atomabsorptionsspektrometrie" by Verlag Chemie in 1983.

Editorial Director: Dr. Hans F. Ebel
Production Manager: Dipl.-Ing. (FH) Hans Jörg Maier

Library of Congress Card No. 85-18391

Deutsche Bibliothek Cataloguing-in-Publication Data

Welz, Bernhard:
Atomic absorption spectrometry / Bernhard Welz.
Engl. transl. by Christopher Skegg. – 2., comp.
rev. ed. – Weinheim; Deerfield Beach, Fl.: VCH, 1985.
 Dt. Ausg. u.d.T.: Welz, Bernhard:
 Atomabsorptionsspektrometrie
 ISBN 3-527-26193-1 (Weinheim)
 ISBN 0-89573-418-4 (Deerfield Beach)

Composition and Printing: Zechnersche Buchdruckerei, D-6720 Speyer
Bookbinding: Wilhelm Osswald + Co., D-6730 Neustadt
Printed in the Federal Republic of Germany

Preface

In the nine years since the publication of the first edition of this monograph, atomic absorption spectrometry has undergone a remarkable development. This is perhaps not entirely true for flame AAS, which nowadays is established as a routine procedure in all branches of elemental analysis, but it is certainly the case with all other techniques of AAS. Even though flame AAS had already found acceptance in many standard methods due to its reliability in the mg/L range, it is only a few years ago that considerable doubt was cast on the ability of graphite furnace and hydride generation AAS to provide correct results at all in the μg/L and ng/L ranges.

The difficulties observed by many analysts using these techniques were due in part to the shortcomings of the instruments employed and in part to non-optimum application, since the significance of a number of parameters had not been recognized. In addition, the general problems of trace and nanotrace analysis had to be taken into consideration, since these newer techniques opened this concentration range to AAS.

These days, the causes of the majority of interferences and also the possibilities for their elimination are known. Even if all technical problems have not been completely solved, the way to their solution has been shown.

Thus, as well as the flame technique, the graphite furnace, hydride generation, and cold vapor techniques are nowadays of equal significance. The major field of application of these newer techniques is in trace, nanotrace and ultratrace analysis. Each of these techniques has its own atomizer, its own specific mechanisms of atomization and interference, and of course its own preferred field of application. In this second edition, these three techniques are thus treated separately whenever this appears expedient.

This made it necessary to substantially revise numerous chapters. Chapter 3 now deals only with atomizers, their historical development, and their specific characteristics for each technique. A new chapter 8 has been introduced in which the mechanisms of atomization and the interferences for each technique are discussed in detail. Additionally, typical interferences and their elimination are mentioned. A general discussion and classification of interferences is presented in chapter 7. Application of the Zeeman effect for background correction is also treated in detail in this chapter. This treatment includes the theoretical aspects of the method, the various configurations, and their advantages and disadvantages. In the chapters on individual elements and specific applications, the various techniques are, wherever applicable, weighed against each other.

A discussion on trace and nanotrace analysis has also been newly introduced, since the newer techniques of AAS are among the most sensitive methods for elemental analysis. Solids analysis is also treated since this has become possible with the graphite furnace technique. A section on environmental analysis has been included in the chapter on specific applications, and topical questions on the analysis of air, waste water and sewage sludge are addressed.

Among associated analytical methods, atomic emission spectrometry employing an inductively coupled argon plasma is discussed especially, since it is frequently regarded as a competitive technique to flame AAS. However, a broad treatment of this theme is outside the scope of this book.

Graphite furnace atomic emission spectrometry has also received attention even though, like atomic fluorescence spectrometry, it is rarely used in practice.

Finally, terms, nomenclature and units of measurement have been brought into line with the latest international standards – a fact reflected in the changed title of this monograph. Of particular help in this respect was my work on the committee of material testing within the German Institute for Standardization. This committee was chaired by Dr. HANS MASS-MANN, who, until his death, worked on the completion of DIN 51401 and who also made valuable suggestions for the second edition of this book – a fact greatly appreciated.

I should also like to thank those readers who wrote to me pointing out errors in the first edition; they have made valuable contributions to improving this work. I should particularly like to thank Sir ALAN WALSH who drew my attention to a number of errors and who proposed numerous improvements and more precise definitions.

The numerous new diagrams were prepared with the customary care by Mr. E. KLEBSAT-TEL, who receives my grateful thanks. I should also like to thank Mr. J. STORZ for designing the cover.

This book is the English-language version of its German forerunner "Atomabsorptions-spektrometrie" (formerly "Atom-Absorptions-Spektroskopie") which is now in its third edition. As for the first edition the translation has been very capably carried out by CHRISTO-PHER SKEGG to whom I extend my thanks.

Meersburg, May 1985   Bernhard Welz

Preface to the first Edition

It was very convenient that the translation of my book into the English language was undertaken just as I was completing the second German edition. Therefore, all the latest developments and publications could be incorporated directly into the English edition.

Usually, years go by between the publication of the original book and the completion of a translation which, therefore, typically does not represent the latest state. Here, however, the translation could be published about a year after the original German edition. This is of special importance for the rapidly growing field of furnace atomic absorption which was hardly known a few years ago when the first edition of my book was published. In the meantime it has found worldwide acceptance among analysts.

So, to all my friends and colleagues who have been involved in the translation and completion of this book, I would like to express my thanks for the time that they have spent and for all the effort that they have put into it so that it could be published so early. Last not least, I want to express my pleasure that my book on Atomic Absorption Spectroscopy has been accepted for translation into English. I hope that it will prove a stimulus to atomic absorption spectroscopy and will help analysts and spectroscopists in their daily work.

Meersburg, March 1976   Bernhard Welz

Contents

1.	**Introduction**	1
1.1.	History	1
1.2.	Atomic Spectra	3
1.3.	Selection of the Spectral Lines	5
1.4.	Thermal Excitation	7
1.5.	Absorption Coefficient	9
1.6.	Line Width	11
1.7.	Measuring the Absorption	12
1.8.	Instrumentation	14
2.	**Radiation Sources**	19
2.1.	Hollow Cathode Lamps	19
2.2.	Vapor Discharge Lamps	25
2.3.	Electrodeless Discharge Lamps	26
2.4.	Flames as Radiation Sources	28
2.5.	Continuum Sources	28
3.	**Atomizers**	31
3.1.	The Flame Technique	31
3.1.1.	Flame Types	31
3.1.2.	Nebulizers and Burners	38
3.1.3.	Special Sampling Techniques	42
3.2.	The Graphite Furnace Technique	48
3.2.1.	Graphite Furnaces	48
3.2.2.	Graphite Materials and Coating	55
3.2.3.	Purge Gas	58
3.2.4.	Temperature Programs and Heating Rate	60
3.2.5.	The Stabilized Temperature Furnace	64
3.2.6.	Automation	65
3.2.7.	Analysis of Solid Samples	67
3.3.	The Hydride Technique	69
3.3.1.	Methods of Hydride Generation	70
3.3.2.	Collecting the Hydride	72
3.3.3.	Atomization of the Hydride	73
3.3.4.	Automation	73
3.3.5.	Sample Volume and Volume of Measurement	74
3.4.	The Cold Vapor Technique	75
3.4.1.	Instrumental Developments	76
3.4.2.	Reduction and Liberation of Mercury	78

3.4.3. Amalgamation 79
3.5.   Miscellaneous Atomization Techniques 81

4.   Optics 83
4.1.   Spectral Bandpass 83
4.2.   Reciprocal Linear Dispersion 87
4.3.   Prisms and Gratings 89
4.4.   Resonance Detectors 91
4.5.   Multielement Instruments 92

5.   Analytical Measure and Readout 95
5.1.   Detectors 96
5.2.   Noise 97
5.3.   Analytical Measure 99
5.4.   Readout 101
5.5.   Automation 105

6.   Methods, Nomenclature and Techniques 107
6.1.   Important Terms, Quantities and Functions 107
6.2.   Calibration Techniques 115
6.2.1. Analytical Curve Technique 115
6.2.2. Bracketing Technique 116
6.2.3. Analyte Addition Technique 117
6.3.   Extraction, Enrichment and Separation 119
6.3.1. Solvent Extraction 120
6.3.2. Separation and Enrichment Techniques 121
6.4.   Problems of Trace Analysis 122

7.   Interferences in Atomic Absorption Spectrometry 129
7.1.   Spectral Interferences 129
7.1.1. Direct Overlapping of Spectral Lines 129
7.1.2. Molecular Band Overlap and Radiation Scattering on Particles 131
7.1.3. Background Correction with Continuum Sources 135
7.1.4. Application of the Zeeman Effect 140
7.2.   Non-Spectral Interferences 159
7.2.1. Classification of Non-Spectral Interferences 159
7.2.2. Elimination of Non-Spectral Interferences 161

8.   The Techniques of Atomic Absorption Spectrometry 165
8.1.   The Flame Technique 165
8.1.1. Atomization in Flames 165
8.1.2. Spectral Interferences 170

8.1.3. Transport Interferences 172
8.1.4. Solute-Volatilization Interferences 175
8.1.5. Vapor-Phase Interferences 179
8.1.6. Spatial-Distribution Interferences 185

8.2.   The Graphite Furnace Technique 186
8.2.1. Atomization in Graphite Furnaces 187
8.2.2. Spectral Interferences 199
8.2.3. Volatilization Interferences 208
8.2.4. Vapor-Phase Interferences 215
8.2.5. Analysis of Solid Samples 220

8.3.   The Hydride Technique 227
8.3.1. Atomization Mechanisms 227
8.3.2. Spectral Interferences 231
8.3.3. Kinetic Interferences 231
8.3.4. Oxidation State Influences 232
8.3.5. Chemical Interferences 233
8.3.6. Gas-Phase Interferences 240

8.4.   The Cold Vapor Technique 243
8.4.1. Systematic Errors 244
8.4.2. Chemical Interferences 248

9.   **Related Analytical Methods** 251
9.1.   Atomic Emission Spectrometry 251
9.1.1. Flame Atomic Emission Spectrometry 251
9.1.2. ICP Atomic Emission Spectrometry 254
9.1.3. Graphite Furnace AES 262
9.2.   Atomic Fluorescence Spectrometry 263

10.   **The Individual Elements** 267

10.1.   Aluminium 267
10.2.   Antimony 268
10.3.   Arsenic 269
10.4.   Barium 272
10.5.   Beryllium 273
10.6.   Bismuth 274
10.7.   Boron 275
10.8.   Cadmium 276
10.9.   Calcium 277
10.10. Cesium 279
10.11. Chromium 279
10.12. Cobalt 281

10.13. Copper 282

10.14. Gallium 283

10.15. Germanium 284

10.16. Gold 285

10.17. Hafnium 286

10.18. Indium 286

10.19. Iodine 287

10.20. Iridium 288

10.21. Iron 289

10.22. Lanthanum, Lanthanides 291

10.23. Lead 294

10.24. Lithium 295

10.25. Magnesium 297

10.26. Manganese 299

10.27. Mercury 300

10.28. Molybdenum 305

10.29. Nickel 306

10.30. Niobium 308

10.31. Non-Metals 309

10.32. Osmium 310

10.33. Palladium 311

10.34. Phosphorus 312

10.35. Platinum 314

10.36. Potassium 315

10.37. Rhenium 316

10.38. Rhodium 317

10.39. Rubidium 318

10.40. Ruthenium 319

10.41. Scandium 319

10.42. Selenium 320

10.43. Silicon 323

10.44. Silver 324

10.45. Sodium 325

10.46. Strontium 326

10.47. Sulfur 327

10.48. Tantalum 328

10.49. Technetium 329

10.50. Tellurium 329

10.51. Thallium 330

10.52. Tin 331

10.53. Titanium 334

10.54. Tungsten 335

10.55. Uranium 336

10.56. Vanadium 336

10.57. Yttrium 337

10.58. Zinc 338

10.59. Zirconium 339

11.   **Specific Applications** 341

11.1. Body Fluids and Tissues 341

11.2. Foodstuffs and Drinks 356

11.3. Soils, Fertilizers and Plants 361

11.4. Water 367

11.5. Environment 375

11.6. Rocks, Minerals and Ores 382

11.7. Metallurgy and Plating 393

11.8. Coal, Oil and Petrochemistry 408

11.9. Glass, Ceramics, Cement 415

11.10. Plastics, Textiles, Paper 418

11.11. Radioactive Materials, Pharmaceuticals, and Miscellaneous Industrial
    Products 419

Bibliography 425

Index 475

1. Introduction

Atomic Absorption Spectrometry (AAS) is the measurement of the absorption of optical radiation by atoms in the gaseous state.

1.1. History

The phenomenon of light absorption had already been investigated at the beginning of the eighteenth century, principally on crystals and liquids. It was observed that the original radiation intensity is resolved into three components: into reflected, transmitted and absorbed radiation. Using BOUGER's work [147], LAMBERT [723] found in 1760 that the amount of light passing through a layer of uniform thickness of a homogeneous medium is dependent upon the thickness d of this layer and that the ratio of the intensity of the transmitted light I_{tr} to the intensity of the incident light I_0 is independent of the radiant intensity.

$$I_{tr} = I_0 \cdot e^{-\varkappa' d} \tag{1.1}$$

The proportionality factor $\varkappa'$ is a measure of the power of attenuation of the layer and is termed the absorption coefficient.

If the irradiated medium is not a single substance but a solution of an absorbing substance in a non-absorbing medium, the absorption coefficient is proportional to the concentration c.

$$\varkappa' = k' \cdot c \tag{1.2}$$

LAMBERT's law underwent a thorough examination by BEER [95] and is nowadays mainly used in the form

$$A \equiv \log \frac{I_0}{I_{tr}} = k \cdot c \cdot d. \tag{1.3}$$

It states that the absorbance A (the logarithm of the reciprocal transmission) is proportional to the concentration of the absorbing substance and to the thickness of the absorbing layer.

The history of absorption spectroscopy is closely connected with the observation of sunlight [1248]. In 1802 WOLLASTONE discovered the black lines in the sun's spectrum which were later investigated in detail by FRAUNHOFER. In 1820 BREWSTER expressed the view that these Fraunhofer lines were caused by absorption processes in the sun's atmosphere. The underlying principles of this absorption were established by KIRCHHOFF and BUNSEN during their systematic examination of the line reversal in the spectra of alkali and alkaline-earth metals [648–651]. They conclusively demonstrated that the typical yellow line emitted by sodium salts in a flame is identical to the black D line of the sun's spectrum. The classical experimental arrangement is shown in figure 1.

The relationship between emission and absorption was formulated by KIRCHHOFF in his law, which is generally valid and states that any material that can emit radiation at a given wavelength will also absorb radiation of that wavelength.

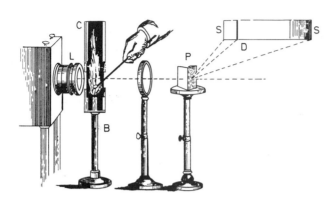

Figure 1. Experimental set-up of KIRCHHOFF and BUNSEN for investigating the line reversal in the sodium spectrum (according to [1248]). Radiation from a continuum source is focused by the lens **L** through the flame of a Bunsen burner **B** into which sodium chloride is introduced with a spatula. The radiation beam is dispersed by the prism **P** and observed on the screen **S**. The sodium **D** line appears as a black discontinuity in the otherwise continuous spectrum.

PLANCK (1900) established the quantum law of absorption and emission of radiation, according to which an atom can only absorb radiation of a well defined wavelength (frequency), i.e. it can only take up and release definite amounts of energy ε.

$$\varepsilon = h v = \frac{hc}{\lambda} \tag{1.4}$$

Characteristic values of ε and v or λ exist for each atomic species.

If continuous radiation falls, for example, on sodium or mercury vapor contained in a quartz vessel, spectral dispersion of the transmitted radiation produces the entire continuous spectrum of the radiation source with the exception of a narrow line seen as a black discontinuity. This "black" line is characteristic of the vaporized material and lies in the visible range at 589.2 nm for sodium and in the ultraviolet range at 253.6 nm for mercury. If the irradiated quartz vessel is observed at right angles to the beam path of the continuum radiation source, the atoms present in the vessel will emit only a single wavelength and this is found to be precisely that wavelength which appears as a black discontinuity in the spectrum of the transmitted radiation.

It has been shown that not only the same wavelength as is emitted is absorbed, but also that the same quantity of radiation is involved. This phenomenon is therefore termed "resonance fluorescence". On the basis of this and many other observations, BOHR proposed his atomic model in 1913, the fundamental principle of which is that atoms do not exist in random energy states, but only in certain fixed states which differ from each other by integral quantum numbers.

Upon absorbing a quantum of energy, an atom is transformed into a particular, energy-enriched state "containing" the radiation energy which has been taken up. Later (more precisely, after a period of 10^{-9} to 10^{-7} s) the atom can re-emit this energy—as a resonance fluorescence quantum—and thus return to the ground state.

Although KIRCHHOFF had already recognized the principle of atomic absorption in 1860 and the theoretical basis was steadily extended during the following decades, the practical significance of this method was not recognized for a long time.

Since the work of KIRCHHOFF, the principle of atomic absorption was mainly used by astronomers to determine the concentration of metals in the atmospheres of stars. Chemical analyses were only carried out very sporadically by this method; the determination of mercury vapor did, however, acquire a certain degree of importance [1354]. The actual year of birth of atomic absorption spectrometry was 1955. In that year, publications authored independently by WALSH [1282] and by ALKEMADE and MILATZ [33] [34] recommended atomic absorption spectrometry as a generally applicable method of analysis.

In the following years it was principally WALSH and his co-workers at C.S.I.R.O.*) who developed atomic absorption into a quantitative analytical technique of high sensitivity and selectivity. WALSH receives the credit for developing not only the theoretical basis, but also the practical applications and the instrumental principles.

1.2. Atomic Spectra

It has already been mentioned that only well defined amounts of energy can be absorbed and re-emitted by an atom, and that after absorbing a quantum of energy the atom is transformed into a particular, energy-enriched or "excited" state. Upon returning to the lower energy state, the atom usually releases the absorbed energy in the form of radiation.

Three different modes of radiative energy absorption or release can be distinguished:

If the atoms are excited thermally or electrically, then absorbed energy is released as an emission spectrum. If the excitation is by optical radiation, the atoms only absorb exactly defined amounts of energy (i. e. radiation of definite frequency) and an absorption spectrum can be observed. The energy absorbed in this manner is released in the form of a fluorescence spectrum.

If an atom could exist in only *one* excited state, then only one emission, absorption or fluorescence line would appear. As can be demonstrated experimentally, however, this is not the case. Thus the absorption spectrum of an atom is seen to consist of a sequence of numerous lines following one another, evidently according to a set plan. Towards the short wavelength end, the lines appear closer and closer together and decrease regularly in intensity until a "convergence limit" is reached; that is, a value beyond which there are no more lines and only continuous absorption is observed.

In the absorption spectrum of sodium, for example, it has been possible up to now to observe and measure 57 absorption lines (figure 2a).

The emission spectrum (figure 2b), which is produced on excitation by a glow discharge, a flame, or an electric arc, has many more lines and does not appear to exhibit the pronounced regularity of the absorption spectrum.

---

* C.S.I.R.O. = Commonwealth Scientific and Industrial Research Organization, Australia.

An atom can exist in many different states and many spectral transitions in absorption and emission are therefore possible. This situation is most clearly represented by an energy level diagram (refer to figure 3).

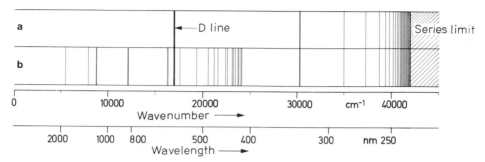

Figure 2. a) Absorption and b) emission spectra of sodium.—At the temperatures normally used in absorption measurements the majority ($\sim 99.9\%$) of the atoms formed are in the ground state. For the most part, only transitions that start from the ground state (main series) can occur in absorption. On thermal or electrical excitation of the sodium atoms, all possible excitation states can be reached from which the emission of the main and all secondary series is possible. The overlapping of these series leads to the observed multiplicity of lines in the emission spectrum.

Each spectral line can be regarded as the difference between two atomic states or terms $\tilde{\nu}$:

$$\frac{1}{\lambda} = \tilde{\nu}_k - \tilde{\nu}_j \tag{1.5}$$

The differences between two such terms are the energy differences between the excited atomic states multiplied by $1/h \cdot c$.

$$\tilde{\nu}_k - \tilde{\nu}_j = \frac{1}{h \cdot c} (E_k - E_j) \tag{1.6}$$

The various states in an atom under study can be described with the aid of three quantum numbers; the principal quantum number n, the secondary quantum number L and the "inner" quantum number J. In table 1 the possible terms in an alkali metal spectrum are given as an example.

In order to analyze a spectrum it is necessary to know which of these terms can be combined through optical transitions in absorption (or emission). The information is provided by selection rules, which in the case of alkali metal absorption spectra, for example, require an increase of the secondary quantum number L by one unit. The principal quantum number n, on the other hand, can alter by any amount.

The energy level diagram for sodium is shown in figure 3. The lowest term here is $3\,\text{s}\,^2\text{S}\frac{1}{2}$ because the photoelectron is situated in the 3s orbital above the filled K ($n=1$) and L ($n=2$) shells.

Table 1. Possible terms in alkali metal spectra*)

	$L = 0\,(s)$	$L = 1\,(p)$	$L = 2\,(d)$	$L = 3\,(f)$
$n = 1$	$1s\ ^2S\frac{1}{2}$			
$n = 2$	$2s\ ^2S\frac{1}{2}$	$2p\ ^2P\frac{1}{2};\ ^2P\frac{3}{2}$		
$n = 3$	$3s\ ^2S\frac{1}{2}$	$3p\ ^2P\frac{1}{2};\ ^2P\frac{3}{2}$	$3d\ ^2D\frac{3}{2};\ ^2D\frac{5}{2}$	
$n = 4$	$4s\ ^2S\frac{1}{2}$	$4p\ ^2P\frac{1}{2};\ ^2P\frac{3}{2}$	$4d\ ^2D\frac{3}{2};\ ^2D\frac{5}{2}$	$4f\ ^2F\frac{5}{2};\ ^2F\frac{7}{2}$
n	$ns\ ^2S\frac{1}{2}$	$np\ ^2P\frac{1}{2};\ ^2P\frac{3}{2}$	$nd\ ^2D\frac{3}{2};\ ^2D\frac{5}{2}$	$nf\ ^2F\frac{5}{2};\ ^2F\frac{7}{2}$

At the temperature of normal sodium vapor in usual flames, nearly all the atoms are in the ground state (see section 1.4), i.e. the photoelectron corresponds to the $3s\ ^2S\frac{1}{2}$ term. For this reason and as a result of the selection rules, all lines occurring in the absorption spectrum arise from transitions from this term to P terms.

$$3s\ ^2S\tfrac{1}{2} \dashrightarrow 3p\ ^2P\tfrac{1}{2},\tfrac{3}{2} \quad (589.593\ nm/588.996\ nm)$$
$$3s\ ^2S\tfrac{1}{2} \dashrightarrow 4p\ ^2P\tfrac{1}{2},\tfrac{3}{2} \quad (330.294\ nm/330.234\ nm)$$
$$3s\ ^2S\tfrac{1}{2} \dashrightarrow np\ ^2P\tfrac{1}{2},\tfrac{3}{2}$$

If, on the other hand, sodium vapor is excited thermally (e.g. in a hot flame) or electrically, terms with arbitrary n and L can be attained and the emission spectrum can contain all lines which are possible starting from these terms in compliance with various selection rules. A complete emission series can be ascribed to each secondary quantum number. Since these emission series overlap, the regularity of the individual series is obscured.

Hence it is obvious that the emission spectrum at these temperatures has many more lines than the absorption spectrum. This applies particularly to elements having several photo-electrons. While the situation is still fairly simple for sodium, atoms having complicated structures give rise to highly complex emission spectra.

Atomic fluorescence is the reversal of the absorption process; i.e. the fluorescence spectrum is produced by release of radiation energy taken up during the absorption process.

Fluorescence spectra are generally simple. They correspond to the transition of atoms excited by absorption of radiation energy to atoms in lower-lying excited states or in the ground state. Fluorescence lines generally have the same or lower frequencies (longer wavelengths) than the corresponding absorption lines. Relevant details will be discussed in a later section.

1.3. Selection of the Spectral Lines

From the energy level diagram of sodium (figure 3), it is clear that in absorption practically only those lines can occur which originate from the ground term of the neutral atom. These

---

* Details will be found in standard textbooks of physical chemistry.

lines are referred to as resonance lines since they can come into resonance with optical radiation of suitable frequency. Their radiation can be absorbed by atoms in the ground state.

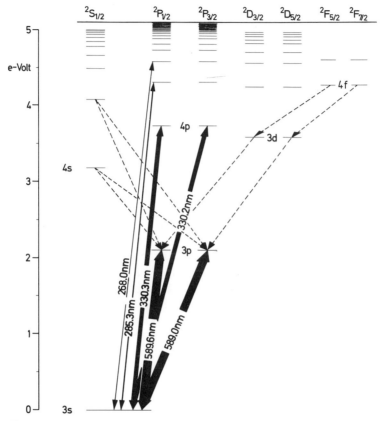

Figure 3. Energy level diagram for sodium.—The continuous lines with double arrows represent transitions for the main series and occur both in absorption and emission. The broken lines are for the transition of secondary series and at the temperatures of interest occur only in emission. Thick transition lines indicate strong spectral lines.

The most intense and easily excited line should be the "last line", corresponding to the transition from the ground state to the first excited state. The name "last line" comes from the fact that the transition to the first excited state requires the least energy, so that these lines are the absorption lines of lowest frequency and therefore appear at the red (long wavelength) end of the absorption spectrum.

Such a clear-cut situation, however, is encountered only with alkali and alkaline-earth metals owing to their very simple atomic structures. With various other elements, lines corresponding to a transition of the atom from the ground state into a higher excited state can also be more intense.

The atoms of the transition elements frequently exhibit a fairly large number of low-lying excited states, better described as ground states of slightly raised energy, whose populations depend upon the temperature. This problem will be discussed in detail in a later chapter. Which lines have the greater intensity can only be decided by experiment.

ALLAN [37] [38], DAVID [270], and MOSSOTTI and FASSEL [885] examined the absorption spectra of a large number of transition elements and lanthanides by means of an emission spectrograph in front of which was mounted a flame containing high concentrations of the respective metals. In this way the intensities of the absorption lines could be measured and the most sensitive lines determined.

In principle all elements can be determined by atomic absorption spectrometry since the atoms of any element can be excited and are therefore capable of absorption. At present the limitations of the method lie practically only in the field of instrumentation.

Thus measurements below 190–200 nm in the so-called vacuum UV range become difficult owing to incipient absorption by oxygen and especially by the hot flame gases. This range, for example, contains resonance lines of selenium (189.1 nm and 196.1 nm) and arsenic (189.0 nm and 193.7 nm), which can still be determined by a good atomic absorption spectrometer, and also the resonance lines of all gases and the typical non-metals. With somewhat modified instruments and a shielded flame [654] [669] or with a graphite furnace [779] [780] even such elements as iodine (183.0 nm), sulfur (180.7 nm) and phosphorus (178.3 nm and 178.8 nm) can still be satisfactorily determined.

Apart from the remaining non-metals, all metals and metalloids can be determined by atomic absorption spectrometry. The elements cerium and thorium are exceptions which up to the present have evaded determination by this method. Occasional references to direct determinations of cerium at 522.4 nm and 569.7 nm respectively are doubtful and the detection limits quoted are not very encouraging [582]. Recently, however, L'VOV and PELIEVA [2587] have found that the resonance line at 567.0 nm gives somewhat improved sensitivity, so that it might after all be possible to determine this element [2577].

Artificial and/or strongly radioactive elements have not yet been examined for obvious reasons.

1.4. Thermal Excitation

The differences between the thermal and the optical excitation process should first be appreciated in order that the principal differences in the observed emission, absorption, or fluorescence spectra be understood.

The ratio of the number of atoms N_j in an excited state j to the number of atoms N_0 in the ground state when the temperature is not too low is given by:

$$\frac{N_j}{N_0} = \frac{P_j}{P_0} \cdot e^{-E_j/kT}, \tag{1.7}$$

where P_j and P_0 are the statistical weights of the excited and ground states, E_j the energy of excitation, k the BOLTZMANN constant, and T the absolute temperature. Since the wavelength of the resonance line is inversely proportional to its energy of excitation (see equation 1.4), the number of excited atoms increases exponentially with increasing wavelength (see figure 4).

In equation (1.7) the exponent is inversely proportional to the absolute temperature, which means that the increase in the relative number N_j/N_0 of excited atoms with increasing temperature is exponential. WALSH [1282] has calculated the ratio N_j/N_0 for a number of elements at various temperatures (table 2). It is found that N_j is always small compared to N_0; i. e. the number of atoms in the excited state can be ignored relative to the number of atoms in the ground state at temperatures below 3000 K and wavelengths less than 500 nm.

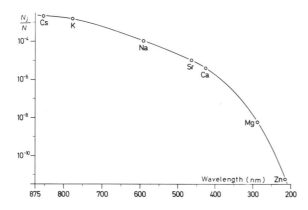

Figure 4. For a given temperature the number of excited atoms N_j increases exponentially with increasing wavelength. The values in this curve were calculated for 2500 K.

For absorption, it can therefore be assumed that the number of atoms in the ground state is virtually identical to the total number of atoms formed. This means that the number of atoms in the ground state is independent of the energy of excitation E_j and of the temperature, provided that the latter factor does not influence the total number N of atoms present, for example through chemical effects in the flame caused by changes in the gas ratios.

This influence of temperature on the total number of atoms, i.e. on the effectiveness of atomization, will be discussed at length in a later chapter. It has nothing to do with the statement in equation (1.7), however. If the total number N of atoms formed alters, then N_0 and N_j are both altered to the same extent. In practice this leads to the same percentage attenuation of both the emission and the absorption lines, the ratio N_j/N_0 remaining unaffected.

Table 2. Temperature and wavelength dependence of the ratio N_j/N_0—according to WALSH [1282]

Element	Resonance line nm	$\dfrac{P_j}{P_0}$	N_j/N_0		
			2000 K	3000 K	4000 K
Zn	213.9	3	$7 \cdot 10^{-15}$	$6 \cdot 10^{-10}$	$1 \cdot 10^{-7}$
Ca	422.7	3	$1 \cdot 10^{-7}$	$4 \cdot 10^{-5}$	$6 \cdot 10^{-4}$
Na	589.0	2	$1 \cdot 10^{-5}$	$6 \cdot 10^{-4}$	$4 \cdot 10^{-3}$
Cs	852.1	2	$4 \cdot 10^{-4}$	$7 \cdot 10^{-3}$	$3 \cdot 10^{-2}$

1.5. Absorption Coefficient

Free atoms in the ground state are able to absorb radiation energy of exactly defined frequency (light quantum $h\nu$) with concomitant transformation into an excited state. The amount of energy absorbed per unit time and volume is proportional to the number N of free atoms per unit volume, the radiation energy $h \cdot \nu_{jk}$, and the spectral radiation intensity S_ν at the resonance frequency:

$$E_{abs} = B_{jk} \cdot N S_\nu \cdot h \nu_{jk} \,. \qquad (1.8)$$

The proportionality factor B_{jk} is the EINSTEIN probability coefficient of absorption for the transition $j \rightarrow k$. The product $B_{jk} \cdot S_\nu$ is an expression for the fraction of all the atoms present in the ground state that can absorb a photon of energy $h \cdot \nu_{jk}$ per unit time.
In unit time a radiation unit of $c \cdot S_\nu$ (c = speed of light), or $c \cdot S_\nu / h\nu$ photons respectively, passes through a unit volume. The fraction of the photons that is absorbed by atoms in the ground state is proportional to the total number N of free atoms and to the "effective cross-section" of an atom, the so-called absorption coefficient $\varkappa_{jk}$. The total amount of energy absorbed per unit volume can then be expressed as the product of the number of absorbed photons and their energy.

$$E_{abs} = \varkappa_{jk} \cdot N \cdot c \cdot S_\nu. \qquad (1.9)$$

By equating the energies in equations (1.8) and (1.9), the absorption coefficient can be expressed as

$$\varkappa_{jk} = \frac{h \cdot \nu}{c} \cdot B_{jk} \,. \qquad (1.10)$$

An atom can also be regarded as an oscillating electrical dipole where the electrons circling round the nucleus represent the oscillators in equilibrium with the radiation. In the electromagnetic field of an optical radiation beam these can be excited to a motion of higher frequency.
According to the laws of electrodynamics, the total amount of energy absorbed by such a harmonic oscillator per unit time can be expressed as

$$E_{abs} = f \frac{\pi e^2}{m_e} \cdot S_\nu \,, \qquad (1.11)$$

where e is the charge and m_e the mass of an electron, and f is a dimensionless factor, the so-called oscillator strength, which merely represents the effective number of classical electron oscillators corresponding to transition $j \rightarrow k$ for the absorption effect of an atom. Expressed simply, f is the average number of electrons per atom which can be excited by the incident radiation ν_{jk}. By equating the energies in equations (1.8) and (1.11), and considering that equation (1.11) is valid for 1 atom, an expression for the absorption coefficient is obtained

$$\varkappa_{jk} = \frac{\pi e^2}{m_e \cdot c} \cdot f_{jk} .$$

(1.12)

The absorption coefficient $\varkappa$ has the dimensions of an area (as expected for an effective cross-section) and is a measure of the quantity of radiation at frequency v that can be absorbed by an atom.

In practice it is more convenient to use the absorption coefficient κ (dimension:length^{-1}) referring to the unit of volume rather than the absorption coefficient $\varkappa$ referring to an atom.

$$\kappa_{jk} = N \cdot \varkappa_{jk}$$

(1.13)

Equation (1.12) then takes the form:

$$\kappa_{jk} = \frac{\pi e^2}{m_e \cdot c} \cdot N \cdot f_{jk}$$

(1.14)

κ_{jk}, the probability of radiation absorption per unit volume, is now a measurable quantity. The other side of the equation is made up of an unequivocally calculable constant and two unknowns, namely the total number of atoms per unit volume N and the oscillator strength f. If N can be measured, f can be calculated. On the other hand, if f is known, the absolute number N of free atoms can be calculated. This procedure, originally proposed by WALSH [1282], has been successfully used by various authors for the calculation of f [777] [778] [1074] [1269] [1270].

A number of typical values for the oscillator strength f are presented in table 3.

Table 3. Absolute oscillator strengths f for spectral transitions—according to L'VOV [778]

Element	Resonance Line nm	Transition	f
Ag	328.1	5 $^2S^{1/2}$–5 $^2P^{3/2}$	0.31
Be	234.9	2 $^1S_0$ –2 $^1P_1$	0.62
Bi	306.8	6 $^4S^{3/2}$–7 $^4P^{1/2}$	0.077
Cd	228.8	5 $^1S_0$ –5 $^1P_1$	1.3
Ga	287.4	4 $^3P^{1/2}$–4 $^2D^{3/2}$	0.19
In	303.9	5 $^2P^{1/2}$–5 $^2D^{3/2}$	0.27
Pb	283.3	6 $^3P_0$ –7 $^3P_1$	0.19
Sb	231.1	5 $^1S^{3/2}$–6 $^4P^{1/2}$	0.042
Sn	286.3	5 $^3P_0$ –6 $^3P_1$	0.11
Te	225.9	5 $^3P_2$ –6 $^5S_2$	0.0018
Tl	276.8	6 $^2P^{1/2}$–6 $^2D^{3/2}$	0.29
Zn	307.6	4 $^1S_0$ –4 $^3P_1$	0.00013

Thus in absorption, the excitation energy is provided by a radiation quantum and no temperature factor is present. The absorption coefficient is determined by the product of the total number of atoms present per unit volume and the oscillator strength of the resonance line.

1.6. Line Width

In the previous section frequent mention was made of the frequency v_{jk} of the resonance line and also the radiation intensity S_{jk} at the resonance frequency. Nothing, however, was said about the dimensions. Rather it was assumed that the spectral radiation intensity S_v is constant in a finite frequency interval Δv and that the frequency v_{jk} lies within this interval. The expression for the absorption coefficient in equation (1.12) represents, for a spectral line of finite width, an integral over the line width (or the frequency interval):

$$\varkappa_{jk} = \int \varkappa_v \cdot d v \tag{1.15}$$

WALSH showed that owing to the probability distribution over each energy level and the HEISENBERG uncertainty principle, the natural width of a resonance line is in the order of 10^{-5} nm. This natural line width is influenced by a variety of factors, above all by the disordered thermal motion of the atoms and by various collisions of the atoms, so that the actual line width is somewhat broader.

The former influence, disordered thermal motion of atoms, causes the line profile to have the form of a MAXWELL distribution of the atomic velocities and this can be expressed as a GAUSS function. This so-called DOPPLER effect is given by

$$D = \frac{v}{c} \sqrt{\frac{2 R T}{M}} . \tag{1.16}$$

The DOPPLER line width D is proportional to the square root of the absolute temperature T, and inversely proportional to the square root of the atomic mass M.

A second effect, known as pressure or collision broadening, comes from the fact that the energy levels which are decisive for the emission or absorption of radiation quanta are somewhat indefinite due to the collision of atoms. For this reason, radiation quanta of slightly differing frequency are emitted and also absorbed. Depending upon the type of particles involved in the collision, the effect is also slightly different. If the interaction involves electrically charged particles it is referred to as the STARK effect, while collisions with uncharged particles lead to the VAN DER WAALS effect; collisions between atoms of the same kind give rise to the resonance (or HOLTSMARK) broadening effect. Since it is very difficult in practice to differentiate between the three effects, they are frequently designated collectively as LORENZ broadening.

Finally it should be pointed out that DOPPLER and LORENZ broadening often occur simultaneously, and both profiles then combine to give the VOIGT profile, which can only be expressed by relatively complicated functions. In figure 5 the profile of a resonance line is

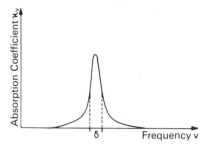

Figure 5. Profile of a resonance line.—The width at half height δ is frequently used as a measure for the line width.

shown schematically. Additionally, line broadening is caused by the hyperfine structure exhibited by many resonance lines due to nuclear spin. Furthermore, the resonance lines of elements that possess several stable isotopes undergo isotope shift. The same number of lines can be observed as there are isotopes. Except for the very light and very heavy elements, this isotope shift is so small that it cannot be resolved by normal spectrographic equipment and merely leads to a further line broadening. The significance of the peak width at half height, or half-intensity width, of emission and absorption lines for the absorption process will be discussed in more detail in the next chapter.

The half-intensity width of resonance lines increases with increasing pressure and with increasing temperature. Various authors [664] [665] [778] [953] [999] [1160] have measured or calculated the half-intensity widths of the resonance lines of a large number of elements. The values are in the order of 0.0005 nm to 0.005 nm.

1.7. Measuring the Absorption

Since atoms are only able to absorb radiation within a very narrow frequency interval, certain demands must be placed upon the radiation source. Although continuum radiation sources afford a high total illumination intensity, the illumination intensity in the interval of interest of about 0.0005 nm to 0.005 nm is nevertheless too weak. For this reason WALSH recommended that the radiation source used for absorption measurements should emit the spectrum of the element to be determined. With such an arrangement the required resonance line merely has to be separated from other spectral lines of the same element by means of a monochromator (see figure 6).

Up to now we have considered only the absorption of radiation per unit volume, a quantity which is difficult to measure in practice. If, on the other hand, the usual quantity employed in normal absorption measurements, the radiant flux Φ, is used for measurements in atomic absorption the BEER-LAMBERT law (equations 1.1 to 1.3) can be used in the form

$$\Phi_{tr} = \Phi_0 \cdot e^{-\varkappa_v N l}, \tag{1.17}$$

where Φ_0 and Φ_{tr} are the radiant fluxes prior to and after passing through the absorbing layer of length l, respectively. $\varkappa_v$ is the spectral atomic absorption coefficient and N the total number of free atoms.

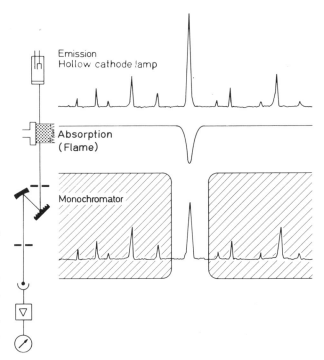

Figure 6. Measurements in absorption.—The spectrum of the element under study is emitted from a hollow cathode lamp. In the flame a portion of the resonance line, corresponding to the concentration of this element, is absorbed. Lines that do not occur in absorption are not attenuated. After dispersion of the radiation in a monochromator, the resonance line is separated by the exit slit and all other lines are masked. The detector 'sees' only the resonance line, whose attenuation is then displayed.

After rearrangement, equation (1.17) is derived from analogy to equation (1.3) in the form

$$A \equiv \log \frac{\Phi_0}{\Phi_{tr}} = 2.303 \cdot \varkappa_v N l, \tag{1.18}$$

where A, the absorbance, is directly proportional to N, the total number of free atoms present. The definitive quantity of measurement A, the absorbance, is the logarithm of the reciprocal transmittance τ,

$$A = \lg \frac{1}{\tau} = \lg \frac{\Phi_0}{\Phi_{tr}}. \tag{1.19}$$

The relationship between transmittance and the absorptance α is given by

$$\alpha = \frac{\Phi_0 - \Phi_{tr}}{\Phi_0} = 1 - \tau. \tag{1.20}$$

The total number N of free atoms present admittedly cannot be determined from equation (1.18), but this is not necessary for routine measurements. Like many other spectrometric methods, atomic absorption spectrometry is a relative method since a linear relationship exists between the concentration of free atoms in the measurement (sample) beam and their absorbance.

At higher absorbance values a deviation from the BEER-LAMBERT law is observed relatively frequently in practice. This deviation can be seen as a curvature of the analytical curve (absorbance versus concentration) towards the concentration axis. Various authors [11] [1006] [1073] have examined whether this effect is due to resonance broadening. It could be shown, however, that this is not the case [30]. Non-linearity of the analytical curve is occasionally caused by a hyperfine structure of the spectral line. Most spectral lines have such a fine structure due to isotope shifts and nuclear spin, which are, however, normally blanketed by the above broadening effects. Only with very light or very heavy elements do the individual components cause line separation. In these cases non-linearity due to the hyperfine structure has indeed been demonstrated [409] [1348].

Most frequently the non-linearity of the analytical curve is caused by radiation that cannot be absorbed [862], this can be stray light in the accepted sense, but more usually it is "background radiation" from the radiation source [1147]. For this reason this phenomenon is fully explained in the next chapter in connection with radiation sources for atomic absorption spectrometry.

It has already been mentioned that quantitative absorption of incident radiation by atoms can only take place if the half-intensity width of the emission line from the radiation source is considerably smaller than the half-intensity width of the absorption line. If this is not the case, the resonance line will be flanked on both sides by non-absorbable residual radiation which acts as radiation background. Such broadening of the resonance lines emitted from spectral line sources occurs, for example, with a high gas pressure in the radiation source. The self-absorption that usually takes place at the same time is a further source of interference in absorption measurements.

In addition to these effects, the total width of the emission lines also has a considerable influence on the occurrence of spectral interferences in atomic absorption spectrometry. FASSEL [342] in particular has drawn attention to this fact and has reported effective line widths (including the wings) in the order of 0.1 nm. This problem will be further described in the context of spectral interferences.

1.8. Instrumentation

The general construction of an atomic absorption spectrometer is simple and is shown schematically in figure 7. The most important components are a radiation source, which emits the spectrum of the analyte element; an atomizer, such as a flame, in which the atoms of the sample to be analyzed are formed; a monochromator for the spectral dispersion of the radiation with an exit slit for selection of the resonance line; a detector permitting measurement of radiation intensity, followed by an amplifier and a readout device that presents a reading.

Such a system, as shown in figure 7a, has a crucial disadvantage. Upon closer scrutiny, an atomic absorption spectrometer built on these lines is seen to be nothing other than a flame emission spectrometer with a radiation source irradiating the flame. In atomic absorption, however, the flame should ideally be an absorption cell in which the sample is atomized and only atoms in the ground state are produced. In contrast, the flame used in flame emission should produce as many excited atoms as possible, thus ensuring maximum feasible intensity of the emission spectrum produced by the element to be determined.

Although the number of excited atoms at normal flame temperatures is always considerably lower than the number of atoms in the ground state (see section 1.4), many practical examples are known in which the emission of the flame interferes with absorption measurements in this type of instrument. This is hardly surprising when it is considered that a "true sample" usually contains a multitude of elements which all display their own emission spectra in the flame (compare [176]).

Figure 7. Schematic construction of an atomic absorption spectrometer. — **1** Radiation source, **2** Flame, **3** Monochromator, **4** Detector, **5** Electrical measuring system with readout device, **a)** Single-beam direct current (DC) instrument; **b)** Single-beam alternating current (AC) instrument; the alternating radiation can be produced either by the use of pulsed current to the lamp, or mechanically by a rotating chopper situated between the lamp and the flame. **c)** Double-beam alternating current (AC) instrument; the radiation from the hollow cathode lamp is sent alternately through the flame and around the flame by a rotating mirror chopper. The two beams are recombined by a semitransparent mirror behind the flame.

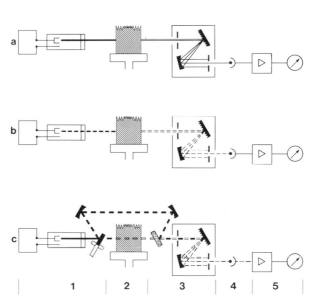

If it proves impossible to separate the resonance line of the analyte element from an emission line of another element, i. e. to eliminate the interfering line by means of the exit slit of the monochromator, then "spectral interferences" take place [7] [176] [605]. In such a case the quantity measured is not the ratio Φ_0/Φ_{tr}, from which the absorbance in calculated (see equation 1.18), but rather $\Phi_0/(\Phi_{tr}+\Phi_E)$, where Φ_E is the radiant intensity of an emission line of frequency v' which is not removed by the monochromator. The absorbance obtained from this ratio is always smaller than that expected. In highly unfavorable cases $(\Phi_{tr}+\Phi_E)$ can be even greater than Φ_0, resulting in an apparently negative absorbance.

To eliminate this interference due to flame emission, virtually all atomic absorption spectrometers nowadays operate with a chopped or pulsed radiation (AC) system as shown in figure 7b instead of direct current (DC) as depicted in figure 7a. The radiation beam is modulated either mechanically or electrically at a fixed frequency and the amplifier electronics are tuned to the same frequency (selective amplifier). In this AC system only the radiation from the primary source having the modulation frequency is amplified, while the emission from the flame, which is not modulated, is neglected. Accordingly, spectral interferences caused by emission of sample atoms in the flame are practically impossible in such a system.

A further refinement of this principle is the double-beam AC system shown in figure 7 c [604]. Here, the beam from the primary radiation source is divided by a rotating mirror-chopper into a sample beam (crossing through the flame, Φ_{tr}) and a reference beam (by-passing the flame, Φ_0) and then recombined behind the flame. The electronics of this system are designed to yield the ratio of the two beams with the sample beam as the denominator and the reference beam as the numerator. Since both beams come from the same radiation source, pass through the same monochromator, are received by the same detector and amplified by the same electronics, any variations in the emission of the primary source, detector sensitivity, or amplification appear in both the numerator and the denominator and therefore cancel out. The stability of this system is thus clearly better than that of a single-beam system [1273], and will only be influenced by variations in the atomizer, which nevertheless can be dominant in individual cases as will be shown later.

As will be discussed in section 7.1.4, the inverse Zeeman effect provides the requirements for an ideal double-beam system. A single-beam optical system is employed and the atomizer is mounted within a high magnetic field. Within this magnetic field the terms of the atoms split into different energy levels due to the magnetic moment of the electrons. This also causes the absorption lines of the atoms to split into various components; a π component, which is at the original wavelength and polarized in a plane parallel to the magnetic field, and two or more σ components that are slightly shifted to higher and lower wavelengths and polarized in a plane perpendicular to the magnetic field.

When a modulated magnetic field is applied and the modulation frequency is synchronized with that of the radiation source, and a polarizer is mounted in the optical path to eliminate the radiation polarized parallel to the magnetic field, then an optimum double-beam system results. Measurements are performed with a *single* radiation source and a *single* detector; the optical path and the profile of the emission line from the radiation source are identical in both measurement phases. In the phase when the magnetic field is off normal atomic absorption and possible background attenuation are measured, i. e. Φ_{tr}. In the phase when the magnetic field is on the atomic absorption is eliminated and only Φ_0 and possible background attenuation measured. The net atomic absorption is obtained by ratioing the two signals. Even variations in the atomizer are eliminated with this system. The Zeeman effect has become of major significance in the graphite furnace technique, as will be discussed in depth in a later section.

A further variation is the dual-channel system in which two radiation sources are used to irradiate the same atomizer (e. g. a flame) and are then handled independently by two monochromators, two detectors, and two electronic units. As well as permitting the simultaneous determination of two elements, this system affords a means of working with an "internal standard". Hence adverse influences of the nebulizer-burner system on the stability of the analyses can be reduced. The dual-channel system is a special form of the (not yet commercially available) multichannel system which will be discussed elsewhere.

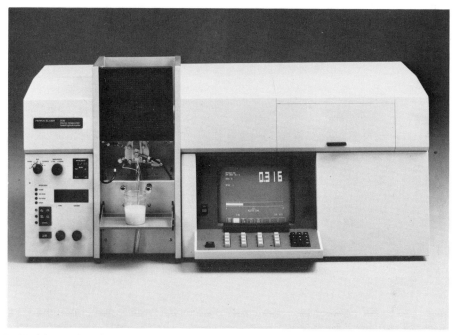

Figure 8. Model 3030 Atomic Absorption Spectrometer (Perkin-Elmer Corporation); a modern double-beam instrument with video display.

In the following chapters the principal components of an atomic absorption spectrometer and their significance for analyses will be discussed in full.

2. Radiation Sources

Very many of the advantages of atomic absorption spectrometry can be directly or indirectly traced to the narrow half-intensity widths of the resonance lines, i.e. the absorption of an element takes place within a very limited spectral range. This advantage becomes very noticeable if the radiation sources used for excitation emit the spectrum of the analyte element in spectral lines that are narrower than the absorption lines. Hollow cathode lamps and electrodeless discharge lamps are particularly suitable as radiation sources. BUTLER and BRINK [194] and SULLIVAN [2945] have published reviews on radiation sources for atomic absorption and atomic fluorescence in which the various lamp types and their function are described.

In spectroscopy, the term "light" is often used instead of radiation. According to the recommendations of the International Commission of Illumination (CIE), the traditional term light should be reserved for that portion of the electromagnetic spectrum that is capable of detection by the human eye (approximately 400 nm to 750 nm).

2.1. Hollow Cathode Lamps

In modern practice, hollow cathode lamps are used principally for excitation because they generally meet the requirements for performance and operating convenience. The technology of this type of lamp is not new since it was first described by PASCHEN [955] in 1916. After the introduction of atomic absorption, WALSH and co-workers [589] [1075] modified and simplified the construction. With the introduction of the Intensitron® hollow cathode lamps at the beginning of the nineteen-seventies it was possible to optimize the performance with respect to intensity and purity of the radiation.

A hollow cathode lamp consists of a glass cylinder filled with an inert gas (neon or argon) under a pressure of several hundred pascal (a few Torr) into which an anode and a cathode have been fused (figure 9). The cathode is generally in the form of a hollow cylinder and is either made of the analyte metal or filled with it. The anode is in the form of a thick wire, usually of tungsten or nickel.

Figure 9. Schematic construction of a hollow cathode lamp.

If a voltage of several hundred volts is applied across the electrodes, a glow discharge takes place. If the cathode consists of two parallel electrodes or of a hollow cylinder, under suitable conditions the discharge occurs almost completely within the cathode, where two processes then take place. A stream of positive ions, generated from the carrier gas by the discharge, strikes the surface of the cathode and atoms of the cathode material are released by the collisions. These atoms pass into the region of the intense discharge where they meet a concentrated stream of gas ions and excited noble gas atoms and are hence excited to radiate their spectral lines. Since the majority of the radiation is emitted from within the cathode, the resultant beam is relatively well bundled.

The manufacture of hollow cathode lamps is not entirely without problems. The nature and pressure of the fill gas (also termed carrier gas), the selection of the cathode material, and the applied voltage and current, for example, play an important part in the intensity and purity of the lamp spectrum and more importantly in the half-intensity width and form of the emitted lines. Since the quality of commercially available hollow cathode lamps is usually excellent, and several of the manufacturers give very detailed information on the optimum operating conditions, discussing these aspects in detail is irrelevant.

HUMAN [540] found that for hollow cathode lamps whose gas pressures were approaching zero, temperature had no noticeable effect on the line width. This means that the spectral lines emitted from hollow cathode lamps generally have an appreciably smaller half-intensity width than the absorption lines in the flame or graphite furnace since the latter are strongly broadened at atmospheric pressure and higher temperatures. Therefore, hollow cathode lamps represent ideal radiation sources for atomic absorption spectrometry.

Earlier hollow cathode lamps, however, possessed a number of disadvantages. The metal atoms displaced from inside the cathode by the gas ions formed a cloud of free atoms in the ground state in front of the cathode opening, especially when the cathode itself was relatively hot. This led to a reduction of the intensity of the emission owing to self-absorption and an alteration of the line profile because of resonance broadening (see section 1.6). This self-absorption could be particularly observed when higher currents were applied to metals of high vapor pressure [613] [1280].

The phenomenon of self-absorption or self-reversal, which is depicted in figure 10 on the example of copper, is worth further consideration; it occurs not only with spectral lamps of the most varying types, but may also be observed more often with emission and fluorescence measurements at higher atom concentrations.

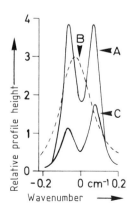

Figure 10. Profile of the copper line at 327.4 nm emitted from a hollow cathode lamp run at 25 mA.

A: Line profile with self-absorption before passage through the atomizer.

B: Absorption profile of a flame into which 10 mg/L copper solution is being sprayed.

C: Line profile after absorption of copper atoms in the flame (according to [2911]).

Normally, the emission intensity of a hollow cathode lamp can be increased by raising the applied current so that the number of metal atoms excited by gas ions is increased. With easily volatilized metals, the higher current and the resulting higher temperature of the cathode lead to increased vaporization of the cathode material, which, at the same time, increases the number of collisions with neutral and charged particles of higher temperature, hence leading to a clear broadening of the line. The cloud of neutral atoms in the ground state formed at the same time at the cathode opening is now in a position, owing to its relatively low temperature and the narrow half-intensity width of the absorption line, to absorb radiation selectively from the emission maximum while relatively little is absorbed from the line wings. This leads to a line profile with a very unsuitable frequency distribution for atomic absorption. At the position of greatest probability of absorption of the emission line there is a minimum (reversal dip), while on the wings, where the absorption probability falls off sharply, there are two maxima.

BRUCE and HANNAFORD [2114] found, however, that a cooler mantle of atoms in front of an emitting zone of higher temperature is not necessarily responsible for self-reversal. They showed that self-reversal can also occur when the emission and absorption lines have the same width. The only requirement is that there is a non-emitting, or only weakly emitting, layer in front of the main emission zone and that this layer has a sufficient optical density.

For the resonance line emitted from a calcium hollow cathode lamp, these authors found a half-intensity width of 0.0009 nm at a lamp current of 5 mA and a half-intensity width of 0.0015 nm at 15 mA. These values can be primarily assigned to DOPPLER broadening at the corresponding temperatures of 347 K and 429 K, respectively, in the hollow cathode and to self-reversal broadening.

This alteration of the line profile caused by self-absorption and self-reversal means that only part of the resonance lines emitted from the radiation source can be absorbed by the atoms in the atomizer. The absorption curve does not then approach 0% transmission asymptotically, but a value corresponding to the residual radiation. In practice this means a deviation from the BEER-LAMBERT law and thereby a non-linear analytical curve.

BRUCE and HANNAFORD [2114] calculated the influence of the final spectral line width emitted from a calcium hollow cathode lamp on the analytical curve when using a typical air/acetylene flame as atomizer. Assuming an infinitely sharp emission line and a single, non-shifted VOIGT absorption line, the error when using low lamp currents is about 10%. This result can be regarded as typical for resonance lines with little or no fine structure.

WAGENAAR and DE GALAN [3025] found that for resonance lines with hyperfine structure, the resulting analytical curve was non-linear and that at higher lamp currents this non-linearity was increased even further through self-reversal. Thus, for copper, the deviation from linearity is already 15% at an absorbance of 1.0 at a lamp current of 25 mA (refer to figure 10).

It would be appropriate at this point to mention several further causes of non-linearity of the analytical curve which are also due to the radiation source. All these effects have the same result; non-absorbable radiation passes through the monochromator and forms an additive member to Φ_{tr}, which for higher concentrations cannot approach zero but retains a finite value.

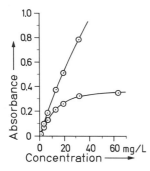

Figure 11. Analytical curves for nickel, measured with new and old hollow cathode lamps, respectively. The older lamp (lower curve) emits an intense hydrogen spectrum which leads to strong curvature.

As principal causes of the non-absorbable radiation, which passes through the monochromator within the transmitted spectral range, the following come into consideration: Non-specific background radiation; emission lines from the fill gas or the cathode support material, if this is not identical to the analyte element; and emission lines from the analyte element itself corresponding to higher transitions.

For a long time, obtaining high purity cathode materials free of hydrogen for a number of elements was a problem. With older lamps, hydrogen was often released during operation, which then under the influence of the glow discharge emitted an intense continuous spectrum. The proportion of this continuum to the total radiation increased on occasions to over 50% [851], thus greatly reducing the usability of such lamps (figure 11).

If the utmost consideration is not taken of the spectra of these foreign elements during selection of fill gases and support materials for cathodes of easily volatilized metals, it can easily occur that these materials have an emission line directly neighboring the resonance line of the analyte element, leading to effects similar to those of the hydrogen continuum. It should be noted, however, that in practice only two gases (neon and argon) come into consideration as fill gases for hollow cathode lamps and that not very many metals are suitable as supports for the cathode, so that on occasions the fact that lesser emission lines of the fill gas or support material lie within the set slit width of the monochromator cannot be avoided [810].

Finally, there are elements that have so many lines in their emission spectra that the resonance line cannot be separated from other non-absorbable lines [856]. This problem will be discussed further in connection with the spectral bandpasses of the monochromators used in atomic absorption spectrometry. When the reasons for the frequently observed deviation from the BEER-LAMBERT law became known and could largely be ascribed to hollow cathode lamps, considerable activity took place in the development of new lamps which led to significant improvements and to the elimination or reduction of many difficulties.

In 1959, WHITE [1316] showed that after several hundred operating hours, a normal cylindrical hollow cathode had changed its geometry, owing to metal transport from the hot inner side to the surface (figure 12). A cathode consisting of a hollow sphere with a relatively small opening was produced.

WHITE showed that with a ratio of the diameter of the opening to the internal diameter of the hollow sphere of 1:4, together with careful determination of the optimum gas pressure,

Figure 12. After a long period of operation of a cylindrical hollow cathode lamp **a**, the cathode alters its form owing to metal transport to the rim **b**. When the ratio of the diameter x of the opening to the diameter y of the hollow sphere is 1:4 virtually no further vaporization of metal from the cathode takes place.

x:y=1:4

virtually no more metal atoms could escape from inside the cathode and therefore no self-absorption or loss of metal could take place. This finding was verified by VOLLMER [1277] and has been used commercially since about 1967 [812].

Use of the WHITE cathode form, especially for easily melted and vaporized elements, together with complete electrical insulation of the cathode and anode with ceramic materials and mica disks (see figure 13) has limited the glow discharge practically to the inside of the cathode. A considerably increased radiant flux, as well as a much "purer" spectrum is obtained this way. This means that non-absorbable foreign lines and ion lines of the analyte element are both strongly reduced in their intensity in comparison to the resonance lines. Through this and the reduced self-absorption, many of the above described difficulties, which lead to strong curvature of the analytical curve, have been eliminated.

For a number of elements, particularly stable emission can be obtained if the metal is in the molten state in the cathode; a procedure only having lasting success in combination with the WHITE cathode form [813] [1275] [1276]. A further improvement in the radiant intensity can be obtained when the hollow cathode lamps are fed with short pulses of correspondingly higher current strength instead of direct or alternating current [1007]. With this procedure an improved signal-to-noise ratio can be achieved from commercially available hollow cathode lamps without significantly influencing their lifetime through the higher current strength in pulsed operation.

Figure 13. Schematic of an Intensitron®*) hollow cathode lamp.— **1** Hollow cathode, **2** insulation of ceramic, **3** restricted cathode opening, **4** anode, **5** mica disks.

---

* Intensitron® is a registered trade mark of the Perkin-Elmer Corporation, Norwalk, Conn., U.S.A.

In principle an atomic absorption analysis requires an individual radiation source for every element to be determined. Since this is frequently found to be a certain disadvantage, endeavors were made to produce multielement hollow cathode lamps at an early stage in the development of atomic absorption.

WALSH constructed a lamp with several cathodes of different metals within the same glass cylinder. The greatest difficulty with these lamps was correct alignment in the spectrometer. MASSMANN [828] and BUTLER and STRASHEIM [197] constructed multielement lamps whose cathodes consisted of rings of various metals pressed together. The rings were arranged in order of their volatilities.

The usual conception used nowadays for the construction of multielement lamps was developed by SEBENS *et al.* [1104] [1140]. Here, various metals in powder form are mixed in determined ratios, pressed and sintered. Using this method, virtually any desired metal combinations can be manufactured [364] [501] [811].

If a multielement lamp is to be used in AAS, however, various aspects must be taken into consideration. The first general requirement is that the intensity of the spectrum of each element in a multielement lamp is not markedly weaker than the comparable single-element lamp. Under no circumstances may the enormous advantage of atomic absorption spectrometry, namely the selectivity, be lost by the use of multielement lamps through the introduction of spectral interferences caused by unsuitable element combinations.

As previously explained, spectral interferences from atoms of concomitants in atomic absorption spectrometry are eliminated, in principle, by modulating the radiation source and tuning the amplifier to the same modulation frequency. Spectral interferences occur in emission spectral analyses if two emission lines of different elements cannot be separated in the monochromator and fall simultaneously on the detector.

By the use of unsuitable metal combinations in multielement lamps, resonance lines of various elements contained in the cathode can lie so close together that they cannot be separated in the monochromator. Since these lines are emitted from the same lamp, they have the same modulation frequency and are equally amplified by the detector. In practice it is no longer possible to assign a measured absorption to a specific element [574].

Great care must be exercised in the manufacture of multielement lamps to exclude spectral interferences, and it must always be ascertained when multielement lamps are purchased whether they have been specially manufactured for atomic absorption spectrometry.

While combination lamps with two or three elements can typically be used without hesitation, lamps with four or more elements are not recommended for all applications. With these multielement lamps the radiant intensity of the individual resonance lines is, in part, considerably lower than with the single-element lamps. This results in a somewhat poorer signal-to-noise ratio which can influence both the precision and the detection limit. Furthermore, the analytical curves occasionally exhibit stronger non-linearity owing to the multitude of lines of the lamps, so that the linear working range is shorter. Nevertheless, for a number of routine analyses, multielement lamps bring a clear advantage.

Various authors have described demountable hollow cathode lamps [431] [685] [686] [870] [1060] [1177]. Their use in routine work will not find great interest since their operation is not simple and requires experience of high vacuum techniques. For various research purposes, on the other hand, they may have certain importance. Developments in this direction do not, however, come within the scope of this book.

To eliminate self-absorption and to increase the radiant intensity, SULLIVAN and WALSH [1193] constructed a special type of hollow cathode in which the glow discharge on the cathode was used principally to atomize as much of the cathode material as possible. The cloud of atoms, which in normal hollow cathode lamps is responsible for the self-absorption, was then excited by a second discharge produced in front of the cathode by an additional electrode isolated from the first discharge. The authors reported a hundred-fold increase in the intensity of these lamps compared to that of normal hollow cathode lamps. The half-intensity width and profile of the resonance lines were said not to have altered and, above all, practically no more self-absorption or self-reversal was to have been observed. Accordingly, the first analytical curves published by SULLIVAN and WALSH showed good linearity up to high absorbance values. The findings of SULLIVAN were verified through work by CARTWRIGHT [213–215], in which attention was also drawn to the considerable improvements for such difficultly vaporized elements as Si, Ti, and V. For these elements, only lamps of low radiant intensity, which mostly gave unsatisfactory results, were available.

Although these high intensity lamps were initially greeted with great interest, it was soon discovered that they did not offer just advantages in routine operation. In particular, the second power supply for the additional discharge and the long warm-up period necessary for stabilization of the emission were found to be a nuisance. The higher price resulting from the complicated construction also bore no relation to the often observed short lifetime.

When it later became evident that the same results could be obtained from normal hollow cathode lamps through considerably simpler means, such as the WHITE cathode form and isolation of the space behind the cathode opening [812], the general interest in high intensity lamps disappeared rapidly. Today, high intensity lamps are practically only used for research purposes. In connection with atomic fluorescence spectrometry, on the other hand, they may gain a degree of importance since the obtainable sensitivity is directly dependent on the radiation intensity in this technique.

Recently, GOUGH and SULLIVAN [2311] suggested a further lamp design that emits a higher radiation intensity than conventional hollow cathode lamps. In one section of the lamp a volatile element is vaporized by means of an accurately controlled heater and then the vapor is excited in a discharge of low voltage but high current. This type of lamp is characterized by a high radiant intensity, very narrow spectral line width, and the absence of self-reversal.

2.2. Vapor Discharge Lamps

Very volatile metals such as mercury, thallium, zinc and the alkali metals were often previously determined using low pressure vapor discharge lamps since these were offered commercially at low prices and gave a high radiant intensity. Various work has been published in which these lamps are compared with hollow cathode lamps [810] [1074] [1151]. The disadvantage of these lamps under normal working conditions is that they emit strongly broadened lines owing to self-absorption and self-reversal of the high internal atom concentration and are therefore little suited to atomic absorption measurements [1240]. Vapor discharge lamps must therefore be operated with highly reduced currents to avoid excessive self-reversal in atomic absorption spectrometry. This can lead to an increasing instability of the lamp, however. All in all, low pressure gas discharge lamps are by no means so problem-free in use as modern hollow cathode lamps. Since excellent electrodeless discharge lamps for the volatile elements are nowadays commercially available, vapor discharge lamps are no longer of significance.

2.3. Electrodeless Discharge Lamps

Electrodeless discharge lamps are among the radiation sources used in atomic absorption and atomic fluorescence spectrometry exhibiting the highest radiant intensity and the narrowest line widths. A thorough investigation of this lamp type was made by BLOCH and BLOCH [2088] in 1935 and in subsequent years they were used in a variety of high resolution studies. Early in the nineteen-seventies electrodeless discharge lamps found increasing interest for atomic fluorescence spectrometry [25] [62] [169] [252] [263] [824]. For atomic fluorescence, line profiles and the spectral zone around the resonance lines are of less importance since the radiation emitted from the lamp does not fall on the detector. The radiant intensity is of principal interest since the sensitivity over a wide range is directly proportional to it.

The biggest advantage of electrodeless discharge lamps is the increased intensity of the radiation by several orders of magnitude compared to conventional hollow cathode lamps. Furthermore, they can be manufactured rather cheaply. They consist of a sealed quartz tube of several centimeters in length and about 5 10 mm in diameter, filled with a few milligrams of the analyte element (as pure metal, halide, or metal with added iodine) under an argon pressure of a few hundred pascal [263] [1355]. The tube is mounted within the coil of a high frequency generator (e.g. 2400 MHz) and excited by an output of a few Watts up to 200 Watts.

For some years, the manufacture of electrodeless discharge lamps was accompanied by various difficulties [25] [27] [516]. WEST and WINEFORDNER gave detailed instructions for the construction of good tubes [253] [255] [817] [1378].

Various factors, such as degassing and cleaning of the quartz tubes to be used, the tube dimensions, the filling pressure of the inert gas and, especially, the quantity and chemical nature of the filling material, all play an important part [105] [381] [567] [857].

HOARE [516] observed that small tubes, in particular, were only usable for a relatively short time. Good success could be obtained with tubes that were fused into a vacuum jacket for better thermal insulation. Particularly high stability could be achieved if the lamps were thermostatted [172]. HAARSMA and co-workers [454] published a critical review on the manufacture and operation of electrodeless lamps that can serve as an excellent guide for self-manufacture.

For some time, various views prevailed on the advantages of electrodeless discharge lamps in atomic absorption spectrometry since their higher radiant intensity does not influence the sensitivity; only the signal-to-noise ratio can occasionally be improved, leading to higher precision and to more suitable detection limits. From an early date a number of authors have, nevertheless, employed these lamp types in AAS with success [458] [1355]. Electrodeless discharge lamps are of great advantage especially for advancing into the vacuum UV range [654] [669] [670].

For the elements that can be determined in this range, no, or only unsatisfactory, radiation sources are available. Furthermore, the high radiant intensity is particularly important since higher energy losses due to the lower transparency of air, the flame and lenses, besides poorer reflection of mirrors, are to be expected.

The greatest disadvantages of early electrodeless discharge lamps for use in AAS were their relatively short life and the long warm-up periods usually required, during which they frequently drifted. Since 1973 lamps specially designed for atomic absorption spectrometry have been on the market, in which the known disadvantages are largely eradicated [74].

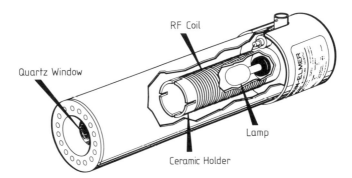

Figure 14. Schematic of an electrodeless discharge lamp. — The radiation source consists of a quartz sphere into which the element is fused under a filling gas pressure of a few hundred pascal. Excitation is by a high frequency field.

The principal difference from conventional lamps is that the analyte element is filled into a quartz tube, which is firmly joined to the high frequency coil and mounted into an insulated jacket (see figure 14). These lamps are very simple to operate, have a relatively short warm-up period and very good stability. A further advantage is that the lamps are operated with only 27 MHz so that the power supply unit is much simpler. BARNETT [74] reported a considerable improvement in the signal-to-noise ratio and thereby to the detection limit, and also in the linearity of analytical curves, particularly for arsenic and selenium.

Nowadays, electrodeless discharge lamps (EDL) are used routinely in atomic absorption spectrometry and have virtually replaced other radiation sources for a number of elements. This is exemplified by arsenic, where the sensitivity is improved by a factor of two and the detection limit by one order of magnitude when the equivalent EDL is used (see also under Arsenic). Electrodeless discharge lamps have also replaced vapor discharge lamps for cesium and rubidium, thereby permitting detection limits only previously possible by flame emission spectrometry to be attained [2061]. The determination of phosphorus has also become practicable since the introduction of an EDL [2061]. Electrodeless discharge lamps are now available for all volatile elements. The detection limits attained using these lamps are better by a factor of typically two to three compared to other radiation sources. Of further importance is the fact that while hollow cathode lamps for the volatile elements are often unstable and have short lifetimes, the equivalent electrodeless discharge lamps are very stable and have long lifetimes. In other words, hollow cathode lamps and electrodeless discharge lamps complement each other in an ideal manner.

2.4. Flames as Radiation Sources

In 1955 ALKEMADE and MILATZ [34] used a flame, into which high metal salt concentrations were sprayed, as a radiation source for atomic absorption spectrometry; various authors have followed this example [1134]. A flame as the primary radiation source is advantageous since it is cheap, universal and very flexible. Particularly for multielement analyses [197] [1178], it offers the advantage of admitting practically every desired element combination. A disadvantage of these radiation sources is that they are somewhat unstable and do not exhibit a particularly intense emission. A greater disadvantage, however, is that even under optimum conditions the half-intensity widths of the emission lines are the same as the absorption lines, while in practice they are mostly wider. They do not thus fulfil one of the basic requirements of an optimum radiation source; relatively non-linear analytical curves are the result. A variation is the use of an electric arc, such as the KRANZ arc [702], as the primary radiation source. These gas and wall stabilized arcs are said to have a high emission stability and a low background continuum. The results obtained with them can be compared with hollow cathode lamps [539].

2.5. Continuum Sources

Radiation sources that emit a continuous spectrum of sufficient brightness (hydrogen or deuterium lamps, high pressure xenon lamps [347] [820] [850], or halogen lamps [346]) would appear at first sight to be very attractive for a number of reasons. They show good stability, make multielement analyses possible and save costs, especially when many elements have to be determined [820]. Nevertheless, these apparent advantages are more than offset by two disadvantages which are based on the fact that the absorption only takes place within a very small frequency interval.

Over a width of 0.002 nm, the half-intensity width of an average resonance line, a continuum source has only a very low intensity compared to a spectral line source, even when the total intensity of such a continuum source is very large. Furthermore, such a radiation source places very high requirements on the resolution of the monochromator. With the best monochromators of commercially available atomic absorption spectrometers, a loss in sensitivity by a factor of around 100 cannot be avoided [807] [915]. This fact was recognized when the procedure was proposed for the first time [411], and the reason is that there are practically no monochromators available that have spectral bandpasses corresponding to the half-intensity width of an atomic emission line. If monochromators with larger spectral bandpasses have to be used, a considerably higher portion of non-absorbable radiation to either side of the resonance line will always fall on the detector. This leads to the above described reduction in sensitivity and to marked non-linearity of the analytical curve. Additionally, there is an increased risk of introducing spectral interferences when continuum sources are employed since the source itself is no longer selective. Despite these general disadvantages, FASSEL [348] has made a thorough study of these radiation sources and has found that under given prerequisites they can be quite useful.

ZANDER *et al* [3095] built an atomic absorption spectrometer in which they used either a 200 W xenon arc lamp or a 150 W Eimac® lamp as the radiation source. To achieve the required resolution they used a proprietry Echelle monochromator which they modified to permit wavelength modulation with a quartz refracting plate. These authors [3096] later showed that most types of spectral interference in AAS could be eliminated by the use of such a system.

O'HAVER *et al* [2701] found that using a 300 W Eimac® lamp with the same system, the radiant intensity within the narrow spectral range under consideration was comparable to that of a normal hollow cathode lamp.

The use of an Echelle monochromator together with a continuum source opens the possibility of simultaneous or rapid, sequential multielement analysis to atomic absorption spectrometry. This possibility is hardly attainable by other means. Nevertheless, many problems with this system still remain to be solved.

3. Atomizers

The emission spectrum of the analyte element emitted from the radiation source is passed through an "absorption cell" in which a portion of the incident radiation is absorbed by atoms produced by thermal dissociation, for example. Accordingly, the most important function of this absorption cell is to produce analyte element atoms in the ground state from the ions or molecules present in the sample. This is without doubt the most difficult and critical process within the whole atomic absorption procedure. The success or failure of a determination is virtually dependent upon the effectiveness of the atomization; the sensitivity of the determination is directly proportional to the degree of atomization of the analyte element in the sample. Ultimately, all known non-spectral interferences in atomic absorption spectrometry are nothing more than influences on the atomization, i.e. the total number of atoms that are formed.

The longest practiced, and therefore most widely propagated, method for transferring a sample into atoms in atomic absorption spectrometry is the spraying of a solution into a flame. In the decade from 1970, the graphite furnace technique, the hydride technique and the cold vapor technique all acquired considerable importance, especially for trace and ultratrace determinations. These techniques have become valuable supplements to the flame technique in AAS.

3.1. The Flame Technique

To atomize a sample in a flame, it is usually sprayed into the flame in form of a solution by means of a pneumatic nebulizer. Methods have also been described in which solid samples are introduced directly or as suspensions into the flame. A special case of solid sample introduction is the boat technique with its variants in which the solid (or dried) sample is introduced into the flame in a tantalum boat or nickel crucible. A further special case is the introduction of gaseous samples into the flame, as done in some forms of the hydride technique.

By using a nebulizer to spray the sample into the flame, a steady, time-independent signal is produced, whose height is proportional to the concentration of the analyte element, and which exists for as long as sample solution is aspirated and sprayed. For other types of sample introduction, such as the boat technique or the hydride technique, time-dependent signals are produced whose heights (or areas) are proportional to the absolute mass of the analyte element.

3.1.1. Flame Types

As will be discussed in detail in Section 8.1, the function of the flame is to convert the sample to the atomic state. Since the gases in the flame, and thus also the sample constituents, flow with a relatively high velocity, the time necessary for atomization of the sample can be expressed directly as the observation height in the flame above the top of the burner. It is thus desirable that atomization should occur as quickly as possible. Further, measurement of the absorbance should be performed at a position in the flame at which atomization is either complete or equilibrium has been reached.

Since in AAS the sample beam is passed through the flame, the latter should ideally be transparent and have no flame background, i. e. the flame should absorb either no or only a minimum of radiation and exhibit no emission. While emission of the flame does not cause false measurements, it can lead to increased noise in the signal. Generally, a flame should exhibit a high efficiency of atomization and avoid secondary reactions of the analyte element with concomitants in the sample or combustion products of the flame gases. To within certain limits the temperature of the flame is without great significance, more important are the oxidizing or reducing characteristics of the flame. These are determined by the partial pressures of the reactive products released during the combustion process.

The flame most well known and used in atomic absorption spectrometry is the air/acetylene flame. For many elements it offers a suitable environment and a temperature sufficient for atomization. In only a few cases (alkali metals) does noticeable ionization occur. The flame is completely transparent over a wide spectral range and only shows noticeable radiation absorption below 230 nm, increasing to about 65% at the wavelength of arsenic (193.7 nm) (figure 15). Furthermore, the emission of the air/acetylene flame is very low so that for many elements ideal conditions are given. Normally, this flame is operated stoichiometrically or weakly oxidizing; however, the ratio of the flame gases is variable over a wide range, thus further increasing its applicability. A number of noble metals such as gold, iridium, palladium, platinum and rhodium, for example, may be determined with the highest sensitivity and largely free of interferences in a strongly oxidizing air/acetylene flame.

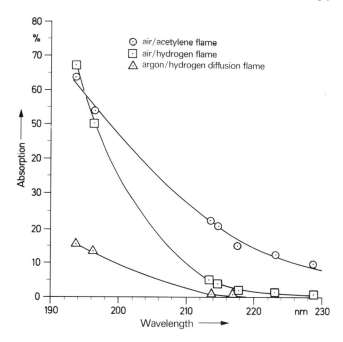

Figure 15. Absorption of various flame types in dependence on the wavelength.

While the alkaline-earth metals are determined most favorably in a slightly reducing flame (i. e. a slight excess of fuel gas), DAVID [271] found that molybdenum exhibited the most intense absorption in a highly luminous flame (larger excess of fuel gas). Similar flame conditions have also proved to be of value for chromium and tin. The disadvantages of these highly reducing flames are their relatively high emission and a noticeable absorption owing to unburnt carbon. The emission leads to increased noise, especially with low-energy radiation sources.

The temperature of the air/acetylene flame is nevertheless insufficient to dissociate a substantial number of principally oxidic bonds or to prevent their formation in the flame. RANN and HAMBLY [1018] were able to show in the case of molybdenum that oxidation frequently first takes place in the flame. They observed that in a stoichiometric air/acetylene flame, directly above the primary reaction zone there was in a narrow band a high concentration of molybdenum atoms that reacted quickly with the oxygen present.

The occurrence of a large number of interferences, which are often difficult to control, in the determination of molybdenum, tin and especially chromium in a fuel gas rich air/acetylene flame, indicates quite clearly that this flame is not particularly suitable for these elements. The occurrence of non-spectral interferences frequently indicates that certain chemical compounds are insufficiently dissociated in the flame or that the atoms formed react spontaneously with components in the flame. The frequent occurrence of such interferences should then be taken as an indication that the performance limit of a given flame has been reached. There are about 30 elements that cannot be determined at all using the air/acetylene flame. These are the elements that form very stable oxidic bonds.

For this reason therefore, attempts were made during the early years of atomic absorption spectrometry to determine those elements which form refractory oxides in a hot oxygen/acetylene flame [187] [281] [312] [349] [677] [885], frequently using organic solvents [40] [1146], or modified burners [47] [791], or even in an oxygen/cyanogen flame [1039] [2284]. The addition of complexing agents was also investigated [276]. It was shown that a fuel gas rich premixed oxygen/acetylene flame was suitable for the dissociation of several of these elements [345], but that its temperature [842] was not sufficient for all of them [241].

Certainly the most important development in flames was the introduction of the nitrous oxide/acetylene flame by WILLIS in 1965 [1333]. As a result of its low burning velocity, this hot flame offers a favorable chemical, thermal and optical environment for virtually all metals that give difficulties in the air/acetylene flame.

The usual nitrous oxide/acetylene flame is operated with a slight excess of fuel gas and exhibits a 2–4 mm high blue-white primary reaction zone above which is a characteristic 5–50 mm high red reducing zone. Above this is then a pale blue-violet secondary reaction zone in which the oxidation of the fuel gas takes place.

The red reduction zone is the one of analytical interest. Here the dissociation of the sample into atoms takes place, and no noticeable oxidation of the metal atoms can occur.

A large number of workers investigated this flame soon after its introduction [48] [150] [689] [794] [795] [976] [977] [1152] and found that it was largely free of interferences [547] [764] [1155]. MARKS and WELCHER [823] investigated the influence of the flame composition and instrument settings on interferences in the nitrous oxide/acetylene flame. They established that the height of observation and the oxidant/fuel gas ratio had the strongest influences, and that with the correct selection of these parameters, interferences disappeared or were at least reduced. To describe the flame conditions as exactly as possible, they introduced a parameter ϱ representing the molar oxidant/fuel gas ratio as a fraction of the stoichiometric ratio $3:1$ ($3\,N_2O + C_2H_2 \rightarrow 2\,CO + 3\,N_2 + H_2O$).

Various authors also established that a larger number of free atoms are produced in the nitrous oxide/acetylene flame than in all other usual flames [350] [575] [688]. These free atoms are exactly localized and strongly dependent on the stoichiometry, however, since the atomization is an equilibrium process [350]. RASMUSON and co-workers [1023] established that for every metal which has a monoxide with a dissociation energy of less than 6.5 eV, there is a flow relationship ϱ (nitrous oxide:acetylene) at which the atomization is complete (assuming complete vaporization of the sample).

Nevertheless, the nitrous oxide/acetylene flame has two disadvantages that must be taken into consideration. Firstly, many elements are more or less strongly ionized in the hot flame and thereby show reduced sensitivity. Secondly, the flame has a relatively strong emission [794]. While the ionization interferences can often be easily suppressed by adding an excess of another easily ionized element (see section 8.1.5), the emission can on occasions cause genuine difficulties [127].

The CN, CH and NH bands, which occur over a wide spectral range and are often very intense, can cause "emission noise" (see section 5.2) if they coincide with the resonance line of the analyte element. This noise can influence the precision of a determination, or in the case of weak radiation sources make a determination impossible.

Nevertheless, if these difficulties are understood, they can usually be avoided, and there is often no reason why the nitrous oxide/acetylene flame should not be used. The advantages and the success of this flame caused SLAVIN [1149] to state in a review, in 1969, that there were still a number of chemical interferences that could not be eliminated even by the use of the nitrous oxide/acetylene flame. Without doubt this flame has made a considerable contribution to the elimination of the difficulties encountered in dissociating stable oxidic compounds [1334].

FLEMING [378] described a system in which mixtures of air and nitrous oxide were used as oxidant with acetylene as the fuel gas. Such flames cover the temperature range between the air/acetylene flame and the nitrous oxide/acetylene flame. BUTLER and FULTON [2120] found that by utilizing the "variable" temperature of this flame, numerous interferences could be eliminated and that the flame had almost universal applicability. Flexibility and safety were said to be major advantages. BUTLER and FULTON [2120] also examined a nitrous oxide/propane-butane flame for special applications where freedom from interferences is more important than sensitivity.

A nitrous oxide/hydrogen flame was investigated independently by two reseach teams [260] [1336]. It was shown that for atomic absorption at least, this flame was inferior to the nitrous oxide/acetylene flame. LUECKE [765] reported a nitric oxide/acetylene flame which can give an increase in sensitivity of up to 60% for the elements barium, boron, strontium and zirconium. An excellent review on flames of higher temperature and their use has been published by WILLIS [1334].

ALLAN [43] found that the sensitivity for tin in an air/hydrogen flame was considerably better than in an air/acetylene or a nitrous oxide/acetylene flame. Other authors have verified this finding [206] [208]. More chemical interferences are, of course, to be expected in this flame. A thorough investigation of the air/hydrogen flame [1087] has shown that for the determination of a number of other elements it has distinct advantages, for example the lower ionization of the alkali metals in comparison to the air/acetylene flame. Furthermore, it has low absorption in the range 230 nm to 200 nm (see figure 15). However, more chemical interferences must be taken into account and the use of organic solvents, although possible, is critical [1219].

During his work on atomic fluorescence spectrometry, WINEFORDNER examined an argon/hydrogen/entrained air flame [1264] [1368], and DAGNALL has reported the low absorption of a nitrogen/hydrogen/entrained air flame [253] [254]. In both these flame types, hydrogen serves as the fuel gas while argon or nitrogen serves to nebulize the sample into the spray chamber. When the flame is ignited, the hydrogen, diluted with inert gas, burns in the surrounding air. This results in a flame with a unique profile, which is particularly noticeable if a three-slot burner head is used [139]. On the outside, where mixing with the surrounding air takes place, it has a temperature of about 850°C, while in the middle, the temperature only reaches 300–500°C, depending on the flame height [254]. This type of flame is termed a diffusion flame.

KAHN [617] investigated the applicability of the argon/hydrogen flame (this simplified name will be used in future) to atomic absorption spectrometry and found that for arsenic, selenium, cadmium and tin there are considerable improvements in the sensitivity and detection limits. In all probability the atomization takes place with the active participation of hydrogen [903] [1067] [1068], although the reaction mechanism is not completely clear. All authors, however, draw attention to the considerable interferences that occur in a flame of such low temperature; these can be both spectral and non-spectral interferences. This flame shows its greatest advantage at the start of the vacuum UV range because of its high transparency compared to other flame types (see figure 15). It is therefore particularly suitable for the determination of arsenic and selenium. In connection with hydride generation systems, the disadvantages are of no great importance because in this technique the analyte element is separated from the remaining matrix and conducted to the burner as a gas [799].

Also in connection with emission and atomic fluorescence measurements, WEST *et al* reported on separated air/acetylene [517] [518] [660] and nitrous oxide/acetylene [250] [658–662] flames. Here, the flames burn either in a quartz tube, or in an atmosphere of argon or nitrogen, so that the secondary reaction zone is separated from the primary. This separation and the shielding of the flame from atmospheric oxygen gives an intermediate zone with very low background radiation [668]. This is particularly interesting, because the highest atom concentration can be observed directly above the primary reaction zone of the flame. At this position in all hydrocarbon flames, the combustion of carbon monoxide takes place supported by the entry of atmospheric air. This is evidenced by the occurrence of particularly intense OH bands at 309 nm and 306 nm and the broad chemiluminescence continuum of the reaction $CO + O \rightarrow CO_2 + h\nu$ directly above the primary reaction zone. Through the separation of the primary and secondary reaction zones, the high atom concentration above the primary reaction zone is maintained, while the emission connected with the secondary zone is displaced to higher regions. A reduction in the emission by more than two orders of magnitude could be measured at the position of highest atom concentration in the separated flame. This has been shown to be particularly advantageous for emission and atomic fluorescence measurements.

KIRKBRIGHT *et al* [666] investigated the temperature relationships in normal and shielded nitrous oxide/acetylene flames and found, according to the observed height and stoichiometry, a decrease of 40–180°C in the intermediate zone of interest for atomic absorption. Despite this lower temperature, the authors found at least the same, if not better, sensitivity [659] for elements that form refractory oxides; that is to say, the flame temperature is not so decisive as a high partial pressure of reducing gases and low partial pressure of oxygen. Although the first applications of separated flames did not bring any great improvements to atomic absorption spectrometry [652] [657], they nevertheless showed interesting possibilities. RUBEŠKA [1066] established that no further interfering CaOH emission could be observed for the determination of barium in a shielded nitrous oxide/acetylene flame, even with a large excess of calcium.

KIRKBRIGHT *et al* [654] [669] were able to determine iodine, sulfur and phosphorus in the vacuum UV range successfully only by means of a shielded nitrous oxide/acetylene flame. The partial pressure of oxygen, which exhibits the strongest absorption in this range, is very small in a fuel gas rich, shielded nitrous oxide/acetylene flame. In principle, according to the reaction $2 N_2O + C_2H_2 \rightarrow 2 CO + H_2 + N_2$, no oxygen or water will be formed in such a flame. Additionally, the reaction $CN + O \rightarrow CO + \frac{1}{2} N_2$ can take place, so that operation is under strongly reducing conditions. The absorption of such a flame is also correspondingly low: At the wavelength of arsenic at 193.7 nm, a shielded nitrous oxide/acetylene flame only absorbs 5% of the radiation compared to about 65% with a non-shielded flame. Even at 178.3 nm, the resonance line of phosphorus, only 45% of the radiation is absorbed by a shielded flame (compared to nitrogen), so that it is still possible to work reasonably in this range.

The first practical applications of atomic absorption spectrometry by WALSH and his co-workers were carried out using air/coal gas flames. The low burning velocity of this flame gives a good absorption cell, but the low flame temperature is insufficient for complete atomization of many elements [7] [413]. As explained previously, the flame temperature has no noticeable influence on the absorption characteristics. For this reason the temperature of the flame should be raised as far as possible to avoid chemical interferences while still maintaining the good optical characteristics, but without introducing ionization too strongly. The air/coal gas flame, like the air/propane flame, is seldom used nowadays.

Flames using pure oxygen as the oxidant can only be employed with direct injection burners because of the high burning velocities. Owing to the high burning velocity and to the low efficiency of atomization, these flames lost importance, especially after the introduction of the nitrous oxide/acetylene flame, which has a comparable temperature but a considerably lower burning velocity.

The properties of various flame types used in AAS are compiled in table 4. However, the majority of the gas mixtures presented in the table are only of historical or scientific inter-

est. Nowadays the air/acetylene and the nitrous oxide/acetylene flames are used almost exclusively. Virtually all elements can be determined with good sensitivity and minimal interferences using these flames. Further, the hydride technique (refer to section 3.3) has made
superfluous the use of various flames with hydrogen as the fuel gas.

Table 4. Characteristics of various flame types*)

Gas mixture		Burning velocity (cm/s)	Temperature °C	Range**)	References
Oxidant	Fuel gas				
Argon + diffusion air	Hydrogen		400	(350–1000)	[254]
Air	Natural gas	55	1840	(1700–1900)	[842]
Air	Methane	70	1875		[842]
Air	Coal gas	55	1900		[746]
Air	Propane	80	1930		[842]
Air	Hydrogen	440	2045	(2000–2050)	[587]
Air	Acetylene	160	2300	(2125–2400)	[587] [936]
Nitrous oxide	Acetylene	180	2750	(2650–2800)	[936] [1377]
Oxygen	Hydrogen	1150	2660	(2550–2700)	[773]
Oxygen	Acetylene	2480	3100	(3060–3155)	[504]
Oxygen	Cyanogen	140	4500		[67] [2284]

  * A summary of further flames can be found in KNISELEY [676] and WILLIS [1334].
 ** Compare REIF [1029] and SNELLEMANN [1159].

Finally, a brief mention should be made of "electrical flames" or plasmas, which have been
investigated over some period of time by various workers [843] [1265] [1304] [1305] [1307].
This method of atomization has found hardly any practical importance in AAS up to the
present, because the equipment is expensive and the introduction of the sample is not without problems. In virtually all work, plasmas have been used for emission measurements
[300] [338] [667] because very high temperatures can be achieved.
In principle, plasmas should also be suitable in atomic absorption, especially for elements
that form stable compounds with flame gases, such as oxides. A plasma produces a chemically inert environment and the sample is so diluted that practically no interelemental effects can occur when the sample remains long enough in the plasma.
This was verified by MAGYAR and AESCHBACH [2606] during investigations on an inductively coupled plasma (ICP) for use in atomic absorption spectrometry. They found that
ICP-AAS, like flame AAS, exhibits high selectivity with hardly any chemical interferences.
The greatest advantage was found to be the virtually complete dissociation of the sample in
the hot, inert environment. Hardly any ionization of the atoms was observed. Disadvantages were found to be low sensitivity and poor detection limits. These disadvantages can
largely be ascribed to the shorter pathlength of the region of observation, which is about ten
times shorter in a plasma torch than in a flame. The authors found ICP-AAS useful in the
analysis of complex compounds that can only be dissociated with difficulty in a flame.

3.1.2. Nebulizers and Burners

Nowadays, premix burners are used almost exclusively because the laminar flame offers excellent conditions for determinations with minimum interference. Sample solution is aspirated by a pneumatic nebulizer and sprayed as an aerosol (or mist) into the spray chamber.
The sample aerosol is mixed thoroughly with the fuel gas and auxiliary oxidant before it reaches the burner slot, above which the flame is burning. Depending on the burner head, the flame is generally 5 cm to 10 cm long and a few millimeters wide. Normally, the radiation beam from the primary radiation source is passed through the entire length of the flame.
Premix burners already were used during the early work on AAS in Australia [237] [1328]. Over the years the design has been improved, especially with respect to durability, stability and effectiveness [190] [1139]. A typical premix burner is depicted in figure 16.

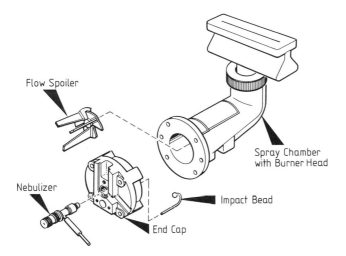

Figure 16. Premix burner optionally permitting use of a flow spoiler or an impact bead (Perkin-Elmer).

To permit use of the flame types mentioned in the preceding section (3.1.1), premix burners normally have interchangeable burner heads in which the slot widths are adapted to the various burning velocities. Proprietary burner heads are offered with slots around 5 to 10 cm long and 0.5 mm (air/acetylene) to 1.5 mm (air/propane) wide. Burner heads made of titanium are of especial advantage because of their high chemical resistance. A burner head much used at one time with three parallel slots 11 cm long and 0.45 mm wide was recommended by BOLING [139]. This three-slot burner head has better flame optics [140] [737], leading to an improved signal-to-noise ratio. A number of elements, such as calcium, chromium and molybdenum, can be determined with improved sensitivity [1172]. AGEMIAN *et al* [2004], however, point out that improved sensitivity can only be obtained for those elements that form oxides with high dissociation energies. For other elements, a loss in sensitivity often occurs as a result of the greater "dilution" in the larger gas volume. The greatest advantage of this burner head, however, is that relatively high total salt concentrations can be burnt without the slots becoming clogged. With normal slot burners, up to about 5% total salt content can be used, whereas solids contents of up to 30% or even undiluted blood serum can often be burnt for longer periods with the three slot burner. This burner head is suitable for the air/acetylene and air/propane flames, and also diffusion flames such as the argon/hydrogen flame.

An interesting variant is the burner head proposed by BUTLER [191] in which the slot is replaced by five rows of holes. For a number of easily atomized elements using an air/propane flame, an improvement in the sensitivity by a factor of ten was reported. ALDOUS *et al* [26] used a burner in which the slot was replaced by a row of capillaries. They found that this design gave lower emission and high stability. The greatest advantage seen by the authors was the ability to use extreme gas mixtures without the risk of a flashback of the flame. RAMSEY [2772] proposed a burner head in which the slot is likewise replaced by a row of capillaries, but adapted to the conical form of the radiation beam. The linearity of the analytical curve could be improved markedly with such a burner head.

SCHMIDT and SANSONI [1093] reported a burner which had cooling directly at the slot with water thermostatted at 30°C. They found that this burner gave a considerably improved signal-to-noise ratio. This burner design showed a very low blockage tendency when highly concentrated salt solutions were being nebulized, and subsequent spraying of water produced a high degree of self cleaning.

For the hot nitrous oxide/acetylene flame, specially designed burner heads are required [792] [912], which are, however, also suitable for other gas mixtures. Because of the somewhat higher burning velocity, a narrower slot must be selected to prevent flashback of the flame. Various endeavors have been made to avoid the deposition of pyrocarbon observed with this flame, but with only moderate success up to the present (compare [479]). It appears of advantage the hotter such a burner head can be run.

HELL [497] and VENGHIATTIS [1267] independently attempted to improve on the low use of the aspirated solution caused by separation of the larger droplets by heating the spray chamber. Desolvation of the solution droplets in the flame is thereby largely saved. For a large number of elements, an increase in the sensitivity by a factor of around ten was found. However, this procedure can only be used with dilute aqueous solutions.

In the decade from 1960, next to premix burners, direct injection burners were examined and used. These burners, known from flame emission spectrometry, consist of two concentric tubes through which fuel gas and oxidant are fed separately to the jet (see figure 17). The fuel gas and oxidant are actually mixed in the flame and thus considerable (optical and acoustic) turbulence takes place. The sample solution is aspirated through a central vertical capillary in the burner by the drop in pressure of the oxidant at the jet and is directly nebu-

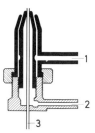

Figure 17. Direct injection burner. — **1** Fuel gas line, **2** oxidant line, **3** sample capillary.

lized into the flame. These burners are often referred to as total consumption burners because all the solution nebulized passes into the flame.

The most important advantages of the direct injection burner are its simple construction,
which enables it to be manufactured cheaply, its safety (a flashback of the flame is impossible), and the possibility of using virtually all imaginable gas mixtures—however, the corresponding burning velocities must be accounted for in the design. Complete burner units are
therefore offered for the various gas mixtures.

The fact that the relatively large volume of flammable gas in the spray chamber is a potential source of danger is regarded as the greatest disadvantage of the premix burner. Modern
burner designs, however, are largely protected against flashback of the flame. Premix burners also have the disadvantage that only flames of relatively low burning velocity can be
used; this excludes, for example, the use of gas mixtures employing pure oxygen.

In the following, both burner designs will be examined with respect to their atomizing properties: Both burner types operate with a pneumatic nebulizer system and have comparable
aspiration rates, which lie between 2 and 10 mL/min. In general, the direct injection burners have a slightly lower rate of aspiration. The rate of aspiration of the direct injection
burner is determined by the pressure and flowrate of the oxidant. An alteration in the pressure and flowrate of the fuel gas also causes an alteration in the aspiration rate. This is a
disadvantage of the direct injection burner since independent optimization of the nebulizer
and flame is not possible. Premix burners, on the other hand, usually have independently
adjustable nebulizer systems which operate with constant oxidant pressure.

Pneumatic nebulizers produce a sample aerosol with a wide variety of droplet sizes. There
is a maximum droplet volume up to which a given flame is just able to vaporize a given solvent within a determined time. Owing to the varying droplet sizes, desolvation takes place
at various positions in the flame; very small droplets are desolvated even before or shortly
after entering the flame, while large droplets are desolvated much later. This is naturally
valid as well for all the other processes in the flame through to atomization (refer to section
8.1.1).

Since both burner types operate with comparable nebulizer systems, the fineness of the
aerosol produced and the maximum droplet size are also comparable. While in the direct
injection burner the sample is nebulized directly into the flame, in premix burners the aerosol must travel a determined distance and pass contact surfaces (spoilers or impact beads)
before it reaches the flame. The larger droplets condense on the spoilers and run out
through a drain in the spray chamber. Although a large part of the sample solution is
wasted in this way (only 5–10% of the solution reaches the flame), the maximum droplet
size is considerably reduced. KERBYSON [639] carried out interesting comparative measurements in which the aerosol produced by a direct injection burner was first examined directly and then with a spray chamber inserted. The maximum droplet volume measured at
5 cm above the burner (no flame) was found to be approximately 7 nL for the direct injection
burner and about 0.05 nL for the burner with spray chamber.

Furthermore, the exit speed of the gases from the jet of a direct injection burner is greater
than from a laminar burner by a factor of about 50. It is then to be expected that the larger
droplets, which pass through the flame of a direct injection burner within a few milliseconds, do not have the opportunity to be quantitatively atomized. GIBSON [412] found that

with direct injection burners even water droplets could pass through the whole flame without being completely vaporized. This finding has been verified by other authors [282] [1344].

If the concentration of atoms in the turbulent flame of a direct injection burner is observed at any given height in the flame, only relatively small alterations can be detected; i. e. the atomization takes place throughout a wide area of the flame. In the same way, solvent droplets, salt crystals, etc., are to be found at every height over the whole length of the flame. The turbulent flame from a direct injection burner has little structure. In the laminar flame of a premix burner, on the other hand, a quite clearly formed zone structure is to be found, since the reduced droplet size makes optimum atomization within a defined flame zone possible [241] [460] [1018].

Various authors have compared the advantages and disadvantages of both burner designs [7] [682] [1145], especially with respect to the occurrence of chemical interferences [14] [1384]. The most common cause for the occurrence of interferences can be traced to the influence of concomitants in the sample on the degree of conversion of the aerosol droplets into free atoms of the element to be determined [25]. Other workers found that physical factors, in particular, influence the rates of volatilization and atomization. Decreasing droplet size of the aerosol, higher flame temperatures and longer residence in the flame reduce or eliminate a large number of interferences [511] [683] [1374]. This has also been verified by the discovery that interferences are less pronounced in higher flame zones than directly over the burner slot [35] [391]. Precise investigations on known interferences, such as the influence of phosphate [1155] and similar anions on calcium, have shown that they can be almost completely eliminated by the choice of suitable flames and burners [344] [368] [884] [976].

As mentioned at the beginning, the requirements placed on the flame are that it should be optically transparent and that the sample should be quantitatively atomized if possible. The latter requirement is well fulfilled by a laminar premix burner, since it produces a considerably finer aerosol and has a lower burning velocity than a direct injection burner. The temperatures achieved by both types are comparable. The transparency of a laminar burner is practically ideal, as demonstrated by means of schlieren photographs [140] [1145], while a turbulent flame is optically non-uniform, even when radiation scattering on non-volatilized particles is disregarded.

The optical noise of a flame results from the varying optical properties of the flame. These fluctuations are caused by convection of the gases and by turbulence. The influence of flame noise on the precision is considerably greater in atomic absorption spectrometry than in flame emission spectrometry or atomic fluorescence spectrometry, since in the latter two the flame is either not irradiated or the beam passing through the flame is not taken into consideration.

In a large number of investigations on various flames and burners, LEBEDEV [737] found that premix burners, especially with multislot burner heads, produced minimum optical noise. In such flames, completely stable middle zones can be observed. Direct injection burners, on the other hand, display the strongest optical noise, especially when organic solvents are sprayed.

For the above mentioned reasons, direct injection burners have practically disappeared from AAS nowadays.

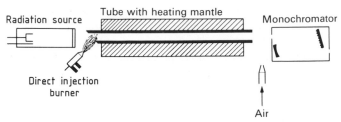

Figure 18. Long tube burner according to FUWA and VALLEE [399] in the heated form of MOLDAN [872].

In 1963, FUWA and VALLEE [399] [2284] introduced a long tube burner in which the flame of a direct injection burner is directed at the opening of a tube of ceramic material mounted in the beam path of an atomic absorption spectrometer (figure 18). In this way the atoms are forced to remain longer in the beam path, with the result that the sensitivity is increased. This is only valid however, if the atoms observed have a sufficiently long lifetime. According to equation (1.18) (page 13), the absorbance should increase proportionally to the length of the tube. To within a given degree this is also correct [685], but the length of the tube is always limited by the lifetime of the atoms [1236]. A number of difficulties, such as carry-over and increased chemical interferences, are certainly caused by the lower temperature in the tube [1186], but these can be largely removed by heating it [872] [1070] [1071]. Various fields of application have been treated in detail by AGAZZI [20], CHAKRABARTI [224], FUWA [398] [399], KOIRTYOHANN [685], RAMAKRISHNA [1015], RUBEŠKA [1064] [1071] and ŠTUPAR [1186] [1189].

3.1.3. Special Sampling Techniques

It has often been found unsatisfactory that the conventional pneumatic nebulizer, like the direct injection burner, produces droplets of greatly varying size and that the maximum droplet volume is considerably above the size that can be atomized in the flame. The effectiveness of this nebulizer is around 10%; i.e. about 90% of the aspirated sample solution is wasted in one way or another. This would appear to be a real starting point for an improvement in the sensitivity of atomic absorption analyses. Various authors have therefore described ultrasonic nebulizers which often produce a very even and fine aerosol. In practice, however, these nebulizers are not so easy to operate and the effective improvement in the sensitivity is not very great owing to their relatively low aspiration rates [515] [1163] [1187] [1188] [1307].

ISAAQ and MORGENTHALER [2413] [2414] [2415] reported an improved design in which a thermostatted heating chamber and a cooler to predry the sample aerosol are inserted between the ultrasonic nebulizer and the burner. Very high efficiency was achieved with this system; 86% of the aspirated sample reached the burner under removal of 72% of the solvent. Although sensitivities and detection limits are around a power of ten better than with a conventional pneumatic nebulizer, this system has little chance of success because it is too slow and complicated.

HARRISON and JULIANO [473], and also KASHIKI and OSHIMA [626], attempted to directly analyze solid samples by introducing them into the flame as suspensions. Water-insoluble tin compounds could be maintained in suspension without a stabilizer as long as required for the measurement [473], while aluminium oxide catalysts could be held in suspension in methanol with a special stirrer [626]. The results were dependent upon the physical nature of the samples, however showed few interferences, and by standardizing against similar suspensions, good agreement with results determined by digestion methods could be obtained.

WILLIS [3070] sprayed suspensions of geological samples directly into the flame and determined elements such as cobalt, copper, lead, manganese, nickel and zinc. He found that only particles smaller than 12 μm contributed substantially to the measured absorbance and that the efficiency of atomization increased rapidly with decreasing particle size.

FRY and DENTON [2273] used a Babington nebulizer to nebulize highly viscous liquids and suspensions. They determined copper and zinc in urine, whole blood, sea-water, condensed milk and tomato sauce.

FULLER [2281] determined copper, iron, lead and manganese in aqueous suspensions of titanium dioxide pigments. Since the particle size of these materials is highly consistent (10.0 ± 0.3 μm), good repeatability could be obtained. However, he found that when suspensions were aspirated continuously the nebulizer quickly clogged, so he changed to the direct *flame injection technique* proposed by SEBASTIANI *et al* [2835].

In this technique, a fixed sample volume of typically 50 μL or 100 μL is injected into the nebulizer rather than aspirating the sample solution (or suspension) continuously. The technique underwent a thorough examination by MANNING [2613], INGLIS and NICHOLLS [2408], and especially by BERNDT and JACKWERTH [2076] [2077]. Subsequently the technique was automated [2077] [2083] to improve the precision and operating convenience, and to increase the sample throughput (see figure 19). Using this auto sampling system, various elements were determined routinely in high purity aluminium [2077], the electrolytes Na, K, Ca and Mg, and also Li in 5–20 μL serum [2079], and iron, copper and zinc in less than 500 μL serum [2080].

MALLOY *et al* [2610] established that the efficiency of a pneumatic nebulizer changes when small, discrete sample volumes are nebulized. If an impact bead is mounted in front of the nebulizer venturi, the efficiency of aerosol production increases due to the improved separation of the droplets. This leads to a reduction of chemical interferences and of interferences resulting from incomplete volatilization. In the opinion of these workers, the opposite effect can occur without the impact bead. In the present author's experience, interferences caused by varying physical properties of sample and reference solutions can be reduced by the direct injection method.

SHABUSHNIG and HIEFTJE [2842] developed a dispenser that produces uniform droplets with a volume of less than 4 nL. Sample volumes of, say, 40 nL can be dispensed with a precision of 1.5%. Recently, CRESSER [2174] published a good review on discrete sample nebulization in AAS in which the advantages and disadvantages are discussed.

COUDERT and VERGNAUD [240] and also GOVINDARAJU [435] used a transport screw to feed solid, pulverized samples directly into the flame or into the flame gases. For various industrial products and geological samples, remarkably precise results were obtained.

Figure 19. Model AS-3 Autosampler for the flame injection technique (Perkin-Elmer). An aliquot of the sample solution is dispensed into a small funnel connected to the nebulizer aspirating capillary.

One of the continuous aims in atomic absorption spectrometry is to improve the sensitivity and the detection limits. Accordingly, many attempts have been made principally to improve the effectiveness of the nebulizer/burner system. Remarkable success has thereby been achieved with a very simple means of atomization, the so-called *sampling boat technique* [616], which acts on the two weakest points of the nebulizer/burner system. As already explained, only about 10% of the aspirated sample solution reaches the flame, while the other 90% is wasted. A further disadvantage of spraying the sample solution into the flame is that the transport of the sample to the flame is limited by the aspiration rate of the nebulizer. Both disadvantages are eliminated by the boat technique. A small amount of sample (maximum 1 mL) is introduced into a small tantalum boat, the solution is dried (for example, near the flame of an atomic absorption spectrometer or in a muffle furnace [527]) and then the boat is introduced into the flame of the spectrometer. The sample situated in the boat is then rapidly (within a few seconds) and quantitatively atomized. This results in a tall, narrow ballistic signal that can be registered on a strip chart recorder. Since the temperatures that can be achieved with the boat technique are only around 1200 °C, this procedure is only suitable for easily atomized elements such as arsenic, bismuth, cadmium, lead, mercury, selenium, silver, tellurium, thallium and zinc. For these elements, the boat technique brings an improvement in the detection limits by factors of 20–50 times (table 5).

Table 5. Absolute and relative detection limits with the boat technique and the DELVES system

Element	Absolute detection limits (g)		Relative detection limits (µg/L)		
	Boat technique	DELVES system	Boat technique (1 mL sample)	DELVES system (0.1 mL sample)	Flame AAS
Ag	2×10^{-10}	1×10^{-10}	0.2	1	1
As	2×10^{-8}	2×10^{-8}	20	200	140
Bi	3×10^{-9}	2×10^{-9}	3	20	20
Cd	1×10^{-10}	5×10^{-12}	0.1	0.05	0.5
Hg	2×10^{-8}	1×10^{-8}	20	100	170
Pb	1×10^{-9}	1×10^{-10}	1	1	10
Se	1×10^{-8}	1×10^{-7}	10	1000	70
Te	1×10^{-8}	3×10^{-8}	10	300	20
Tl	1×10^{-9}	1×10^{-9}	1	10	10
Zn	3×10^{-11}	5×10^{-12}	0.03	0.05	0.8

This technique rapidly became very popular, because elements of interest for toxicology and environmental control could be determined with increased sensitivity, and biological samples such as urine and blood could be analyzed directly [618]. CHENG and AGNEW [2155] determined tellurium in homogenized liver samples and FAVRETTO-GABRIELLI *et al* [2237] determined lead in dried, pulverized mussels in good agreement with digestion techniques. The boat technique has also been successfully applied to the analysis of water [186], milk [724] and other foodstuffs [527], and also to geological samples [329] [766] [767]. BEATY [92] found working with the boat to be particularly simple if only 10 µL sample volume were used, since the drying stage could then be omitted and at the same time better precision and increased lifetime of the tantalum boat obtained.

A little later, DELVES [292] advanced a modification of the boat technique, which he had developed specially for the determination of lead in blood. The long tantalum boat was replaced with a small round nickel cup, and to increase the sensitivity an open tube (of nickel or, better, quartz) was mounted over this. The radiation beam passed through the tube and the analyte element was atomized into the tube. This tube increased the residence period of the atoms in the beam path (figure 20). The blood samples were dried in the nickel cup on a hot-plate and chemically treated before introduction into the flame. Many authors have checked [359] and modified [85] [318] [357] [1056] this procedure for the determination of lead in small blood samples and found that it was especially suitable for screening tests. Later, the DELVES technique was also applied to the determination of cadmium in blood [218] [319] [592] and other biological materials [750]. DELVES [2197] found that the addition of primary ammonium phosphate, as in the graphite furnace technique, greatly facilitates the determination of cadmium in blood. By adding this matrix modifier, the non-specific signal generated by sodium chloride disappears almost completely and the determination can be performed without background correction.

Figure 20. DELVES system. The sample is introduced into the flame of a three-slot burner in a nickel cup. The atoms pass through a hole into the quartz tube which is mounted above the burner directly in the beam path of the atomic absorption spectrometer.

The determination of lead in urine [2421], silver in blood [2396] and thallium in biological materials [1120] have also been reported. The method has been used for the analysis of milk [456] and paints [502] [2537] [2670], including suspensions and solid samples.

Instead of a nickel cup and an air/acetylene flame, WARD et al [3032] and also KAHL et al [2446] used a molybdenum cup and a nitrous oxide/acetylene flame. These authors found that the sensitivity for volatile elements could hardly be improved, but chemical interferences were virtually no longer present. They were also able to determine a number of elements of lower volatility, such as cobalt, copper, chromium, manganese, nickel and tin, with detection limits in the lower nanogram range.

A further modification of the DELVES system has been described by CERNIK and SAYERS [219]; the blood sample is absorbed on a filter paper, and after drying, a piece is punched out and placed into the nickel cup for introduction into the flame. CERNIK reported the simplicity and reliability of this method [217], and that the signal is largely independent of alterations in the cup material. Other authors have been able to verify this finding [591] [593].

WATLING [3036] employed a slotted quartz tube with the boat technique and was thereby able to increase the sensitivity markedly for easily atomized elements.

It is important to realize that with the boat technique or with the DELVES modification, considerable non-specific signals can occur during atomization, due to the formation of smoke,

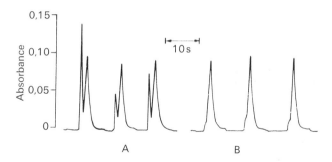

Figure 21. Determination of lead in blood with the DELVES system.— **A**: Without background correction a poorly reproduced signal appears before the lead signal. **B**: With background correction only the element-specific lead signal is obtained.

if special sample preparation steps [85] [592] or extraction of the analyte element with an organic solvent [1056] are not carried out before the determination. The use of a background corrector to improve the accuracy and precision is therefore recommended for this procedure (figure 21).

LAU *et al* [2536] mounted a cooled quartz tube in an air/acetylene flame that acted as an "atom trap" since a number of elements condense on it. After a sufficiently long accumulation period the cooling is turned off and the condensed elements volatilize. Depending on the accumulation period, a considerable increase in the sensitivity can be achieved.

NEWTON and DAVIS [2691] deposited cadmium electrolytically onto a tungsten wire loop, which they then introduced into a flame and additionally heated electrically. BERNDT and MESSERSCHMIDT [2082] used a platinum loop. They introduced microliter quantities of sample onto the loop, dried it near the flame (like in the boat technique) and then inserted it rapidly into the flame. To aid atomization the loop was additionally heated electrically. With this technique it was possible to attain detection limits of a few micrograms per liter, but considerable matrix interferences occur, even in simple samples [3063], so that reliable results can only be obtained after a solvent extraction [2075].

For elements that form gaseous covalent hydrides, such as antimony, arsenic, bismuth, germanium, selenium, tellurium and tin, the possibility exists of introducing the sample in gaseous form into the burner instead of spraying it as a solution. HOLAK [524] proposed this technique initially for arsenic. He produced nascent hydrogen by adding zinc to the sample solution acidified with hydrochloric acid and collected the arsine in a trap cooled in liquid nitrogen. At the end of the reaction he warmed the trap and conducted the arsine with a stream of nitrogen into the flame. DALTON and MALANOSKI [266] simplified this procedure by conducting the arsine together with the hydrogen produced directly into an argon/hydrogen flame. FERNANDEZ and MANNING [362] [799] optimized this technique and introduced a commercial system. Later, other reducing agents were recommended by POLLOCK and WEST [993] [994], and by SCHMIDT and ROYER [1092], for the generation of hydrogen to extend this method to other elements. FERNANDEZ [358] made a thorough examination of the reduction with sodium borohydride and obtained the optimum parameters with respect to acid concentration and reaction period of the reducing agent for seven elements.

In the meantime, this technique has been further developed and greatly refined. It is discussed in detail in section 3.3. Nowadays the hydride is rarely conducted into a flame; atomization is mostly performed in a heated quartz tube.

3.2. The Graphite Furnace Technique

In order to get around the disadvantages of the nebulizer-burner system, to design an absorption cell that permitted physical measurements to be performed more easily, as well as to achieve better detection limits, electrically heated atomizers were introduced at an early stage in the development of atomic absorption spectrometry.

3.2.1. Graphite Furnaces

In 1959, L'vov [775-777] developed a graphite tube furnace based on the work of KING [641]; it was 10 cm long and internally lined with tantalum foil. The sample was placed onto a graphite electrode, dried, and then introduced with the electrode through a hole into the graphite tube. The tube was heated by resistance heating and the sample was atomized by a direct current arc. In an improved version [778], a short tube about 5 cm long made of pyrolytic graphite was used and the graphite tube itself acted as the counter electrode. In this way, the arc which atomizes the sample is formed the moment the electrode is introduced into the graphite tube (figure 22). The whole arrangement was mounted in a sealed argon chamber having quartz windows to allow the radiation to pass through (figure 23), and it could be operated at increased or reduced pressures. With this atomizer, L'vov was able to obtain detection limits between 10^{-10} g and 10^{-14} g, values that are several orders of magnitude better than with flame atomic absorption.

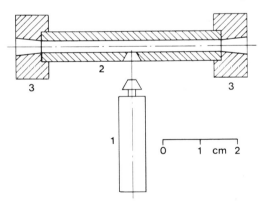

Figure 22. Graphite furnace designed by L'vov [778].—**1** Movable graphite electrode, **2** graphite tube, **3** contacts.

MASSMANN [832] [833] constructed a simplified graphite furnace (figure 24) which also consisted of a 5 cm long graphite tube but which was heated by passing a high current (500 A) at low voltage (10 V) through the tube. This resistance heating permitted a very fine gradation of the temperature to be made and thereby the selection of the optimum temperature conditions for the atomization of every individual element. MASSMANN's furnace was not mounted in a closed argon chamber like that of L'vov and therefore had to be continuously flushed with an inert gas stream to prevent the ingress of atmospheric oxygen. The sample (maximum 0.5 mg solid or 50 µL solution) was introduced into the graphite tube through a small hole in the tube wall.

Figure 23. L'VOV's argon chamber and graphite furnace with ten electrodes (by kind permission of B. V. L'VOV).

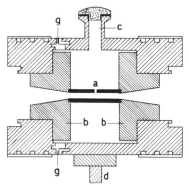

Figure 24. Graphite furnace designed by MASSMANN [832].—**a** Graphite tube, **b** steel flanges, **c** sample introduction port, **d** mount, **g** plastic insulator.

With this system MASSMANN achieved detection limits that were typically ten times poorer than those quoted by L'VOV. The reason for this can be found in the forced inert gas stream through the tube which reduces the residence time of the atoms. Also, the heating rate of resistance heating is slower than that of atomization in an arc, so that the atomization time is longer.

For a large number of elements, MASSMANN [829] [830] observed background absorption to varying degrees in his graphite furnace which he eliminated by using a dual-channel spectrometer. He also observed that the signal height was dependent on the nature and quantity of the matrix [831] and that this dependence was in part considerable.

WOODRIFF constructed a continuously heated graphite furnace into which the sample was sprayed by means of a pneumatic [1351] or an ultrasonic [1353] nebulizer. He found that the graphite tube was attacked by aqueous solutions, so he used absolute methanol solutions exclusively for his investigations. The characteristic concentrations obtained with this procedure were one to two orders of magnitude better than those attainable in the flame; the detection limits, on the other hand, were not greatly improved because of the relatively high noise level. Later, WOODRIFF introduced the evaporated or collected sample into the heated graphite tube in a small graphite pan and found detection limits in the order of 10^{-10} g for fifteen elements [1122] [1352]. KOIRTYOHANN [684] found that the furnace described by WOODRIFF was relatively free from spectral and non-spectral interferences and that it offered a very stable cell for fundamental studies. Because of its size and complicated construction, however, it is hardly suitable for routine analyses.

WEST proposed a very simplified version of a graphite furnace [1311]. A carbon rod, 1–2 mm in diameter and 2 cm long, is clamped between two electrodes in a glass cylinder through which argon flows (figure 25). By slightly altering the glass jacket, the same arrangement can also be used for atomic fluorescence measurements. The sample (e.g. 1 µL) is placed onto the carbon rod by means of a micropipet or syringe. The rod is then brought to incandescence by resistance heating (100 A, 5 V). A determination takes 5–10 seconds and the next sample can be introduced after about two minutes. In a further version, the glass jacket is abandoned and the carbon rod is merely flushed with argon from beneath [1310]. Because of its ease of access, this open version is particularly simple to operate.

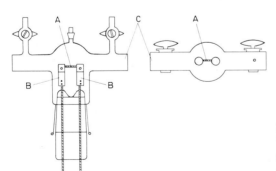

Figure 25. Carbon rod atomizer according to WEST [1311] for graphite furnace atomic absorption. **A** Carbon rod, **B** contacts, **C** glass jacket.

With this system, WEST found detection limits of 10^{-10} g for silver and magnesium. Sample introduction is rather critical and requires some experience with this procedure. Furthermore, a large number of interferences have been found [51] [52] which can be traced to the steep temperature gradient between the carbon rod and the environment. This could be confirmed by experiments in which the radiation beam is focussed directly above the glowing carbon rod [52]. A clear increase in the interferences and a decrease of the sensitivity was observed with increasing distance from the rod.

A similar principle was proposed by MONTASER *et al* [2675], who used a braid of 1.5 mm to 2 mm plaited from graphite fibers. Advantages shown by this system include a low current consumption and the ability to handle somewhat larger sample volumes, since the braid absorbs liquids better. However, it was not possible to obtain graphite braids of particularly high purity, and the lifetime was relatively short, even at reduced temperatures. The detection limits attained were hardly better than with the flame technique [2673], so that this system is more suitable for microanalyses rather than nanotrace determinations.

Virtually all proprietary graphite furnaces are based on the MASSMANN principle, i.e. they are tube furnaces heated by resistance heating. They differ, however, in the tube dimensions, in programmability, in flexibility and in operating convenience. Two such furnaces, the HGA-500® Graphite Furnace and the CRA-90 'Mini-Massmann'® Furnace are depicted in figures 26 and 27, respectively. Apart from a graphite tube, the CRA-90 furnace can also be fitted with a small graphite cup, which serves mainly for the determination of solid samples.

Figure 26. HGA-500 Graphite Furnace with optical temperature sensor (Perkin-Elmer).

®HGA is a registered trademark of the Perkin-Elmer Corporation.
®CRA-90 is a registered trademark of Varian Associates.

KOIRTYOHANN [684] compared the two systems and found that the 'Mini-Massmann' furnace had lower background absorption and higher absolute sensitivity, but exhibited the strongest matrix effects and was limited to a smaller sample volume. The HGA furnace had the better relative sensitivity and the highest precision, but was subject to greater background absorption problems. MORROW and MCELHANEY [882] found that the small cross-section of the 'Mini-Massmann' vignetted the radiation beam, leading to increased noise. Furthermore they had difficulty in pipetting volumes of a few microliters and found a poorer relative sensitivity. For these reasons, they used a tube of greater diameter with which they did not have these difficulties.

Figure 27. Model CRA-90 carbon rod atomizer in the 'Mini-Massmann' version (by kind permission of Varian Associates).

Examples of relative detection limits attainable with a good graphite furnace are compiled in table 6 and compared to those of the flame technique.

The individual components of a graphite furnace and their significance will now be discussed in more detail. As will be shown later, the height of the absorbance signal measured in a graphite furnace is dependent upon the density of the atom cloud. The density is de-

Table 6. Examples of relative detection limits (in µg/L) that can be attained with the graphite furnace and flame techniques. The values for the graphite furnace technique are referred to sample aliquots of 100 µL

Element	Detection limit µg/L	
	Graphite furnace (100 µL sample solution)	Flame
Ag	0.005	1
Al	0.01	30
As	0.2	20
Au	0.1	6
B	15	1 000
Ba	0.04	10
Be	0.03	2
Bi	0.1	20
Ca	0.05	1
Cd	0.003	0.5
Co	0.02	6
Cr	0.01	2
Cu	0.02	1
Fe	0.02	5
Hg	2	200
K	0.002	1
Li	0.2	0.5
Mg	0.004	0.1
Mn	0.01	1
Mo	0.02	30
Na	0.01	0.2
Ni	0.2	4
P	30	50 000
Pb	0.05	10
Pt	0.2	40
Sb	0.1	30
Se	0.5	100
Si	0.1	50
Sn	0.1	20
Te	0.1	20
Ti	0.5	50
Tl	0.1	10
V	0.2	40
Zn	0.001	1

pendent upon the atomization velocity (i.e. the rate of generation of atoms from the sample) and the residence time of the atoms in the graphite tube (i.e. the rate of loss of atoms from the absorption volume) [2936].

If the loss of atoms through a forced gas stream is temporarily neglected, the most important factor is the loss of atoms from the tube due to diffusion. According to equation (3.1), the relative mass loss of atomic vapor

$$\frac{\mathrm{d}M}{M} = -\frac{8D}{l^2} d\tau \tag{3.1}$$

is inversely proportional to the square of the tube length and independent of its diameter (D = diffusion coefficient). This means that the length of the tube plays a decisive part in the residence time of the atoms in the radiation beam. From equation (3.1) by differentiation the mean residence time τ of the atoms in the absorption volume is directly proportional to the square of the length l of the graphite tube

$$\tau = \frac{l^2}{8D}. \tag{3.2}$$

L'vov calculated the diffusion of mercury from a 5 cm cell at 1500 K as an example, and found a mean residence time of 1.8 seconds. A value of 0.5 seconds was obtained from measurements, this is a shorter residence time.

Thus, to achieve the longest possible residence time and thereby optimum sensitivity, the graphite tube should be as long as possible. However, the tube length cannot be increased at random, because with increasing length the current consumption also increases, so that tubes more than 10 cm long cannot be used. As shown experimentally, the optimum length is between 3 cm and 5 cm [8] [882] [1294]; although, in principle, the optimum tube length is different for every element and every temperature.

Not only the tube length, but also the tube diameter plays a certain role. Admittedly it does not influence the diffusion and thereby the residence time of the atoms in the tube, but since the volume increases with the square of the diameter, the cloud of atoms will be correspondingly diluted. In other words, the absolute sensitivity in a graphite tube furnace is inversely proportional to the tube diameter. Therefore, a long narrow tube is required for a good *absolute* sensitivity.

However, since a larger graphite tube can normally contain a larger sample volume, the *relative* sensitivity (i. e. the sensitivity in concentration units) is independent of the tube diameter in the first approximation. The same proportionality is valid for the tube length as for the absolute sensitivity, i. e. the requirement for a long tube.

The fact that the tube is mounted in the radiation beam of an atomic absorption spectrometer is a further detail that must on no account be forgotten when the tube dimensions are under consideration. The sample beam should pass as unhindered as possible through the graphite tube since pronounced shadowing, resulting from a tube cross-section that is too narrow, can lead to increased noise and other difficulties [834]. However, the tube cross-section also must not be too large since the atom cloud is then poorly utilized. The optimum tube dimensions should conform to the geometry of the radiation beam. A parallel radiation beam should be encircled as exactly as possible, while with a beam that has a focal point in the sample compartment, the tube length *l* and the diameter *d* should obey the mathematical interrelationship shown in figure 28 [1303],

$$d = b + l \cdot \tan \alpha. \tag{3.3}$$

Figure 28. Optimum dimensions for a graphite tube. — For an AA spectrometer having a sample radiation beam focus point in the sample compartment, optimum length and diameter of the graphite tube are calculated according to $d = b + l \cdot \tan \alpha$.

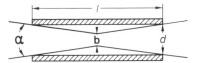

FRIGIERI and TRUCCO [2270] found that with normal graphite furnaces only a portion of the atoms in the tube were available for absorption because the radiation beam is focussed. They therefore made a tube adapted to the geometry of the radiation beam and obtained a two- to fivefold improvement in the sensitivity for a number of elements.

3.2.2. Graphite Materials and Coating

As well as the tube shape, the tube material and its surface characteristics play a decisive role. This will be discussed in detail in a later section, whereby the influence of the tube surface on the atomization mechanism and on interferences caused by reaction of carbon with the sample will be considered. Early on in the development of graphite furnaces it was realized, however, that hot graphite is very permeable to metal atoms.

Through experiments with radioactive tracers, L'VOV and KHARTSYZOV [781] were able to show that the diffusion of metal atoms through the walls of the graphite tube (in dependence on the temperature) could be greater than from the open tube ends. In the initial experiments, the inner surface of the graphite tube was covered with tantalum foil which immediately prevented these losses. Later, L'VOV found that tubes made of pyrolytic graphite (produced by pyrolyzing hydrocarbons, e.g. methane, at about 2000°C), or tubes that were coated with pyrolytic graphite, had the same properties. With these pyrocoated tubes, doubled sensitivity could be obtained for a number of elements because of the prevention of diffusion losses.

CLYBURN *et al* [2161] were the first to perform pyrocoating directly in the graphite furnace by passing a methane/inert gas mixture through the graphite tube at a temperature slightly above 2000°C. A dense, hard, impermeable and oxidation-resistant pyrographite layer is deposited on the tube surface. Initially, considerable problems were experienced with the durability of the pyrolytic graphite layer [2051], so MORROW and McELHANEY [883] mixed 10% methane with the inert purge gas to obtain a continuous renewal of the pyrographite layer and thereby a marked increase in the tube lifetime.

MANNING and SLAVIN [2617] [2618] treated pyrocoated tubes with molybdenum and were able to obtain reproducible signals, which they attributed to a sealing of cracks and flaws in the pyrographite layer. ORTNER and KANTUSCHER [2712] were the first to propose the impregnating of the graphite tubes with metal salts. They found that tubes impregnated with sodium tungstate gave the highest sensitivity and also the most reproducible tube lifetimes. RUNNELS *et al* [2813] used a layer of either lanthanum or zirconium to prevent contact of the sample with graphite and thereby the formation of carbides. THOMPSON *et al* [2970] like-

A

B

Figure 29. Graphite surfaces at 2000× enlargement. **A**: Uncoated (type RWO-HD, Ringsdorff). **B**: Tube coated with pyrolytic graphite deposited by the pyrolysis of methane at 2200°C.

wise proposed coating with lanthanum, while ZATKA [3097] used tantalum in hydrofluoric acid solution and HAVEZOV *et al* [2353] used zirconium. The generally proposed mechanism is that at increased temperature the metal salts react with the carbon to form carbides and these have a similar density and inertness as a layer of pyrolytically deposited graphite.

MANNING and EDIGER [2615] improved the procedure proposed by CLYBURN *et al* [2161] to permit direct *in situ* pyrocoating, and VÖLLKOPF *et al* [3022] have shown that under optimum conditions, tubes with excellent characteristics can be obtained. It now appears clear that a layer of pyrolytically deposited carbon gives the best tube characteristics. Impregnating with metal salts or coating with metal carbides can in general at best achieve the same quality as a good pyrocoating; an improvement is only possible when the pyrocoating has flaws.

The main use for pyrocoated tubes is the improved sensitivity and detection limits obtained for the determination of elements that react with carbon to form refractory carbides, such as molybdenum, titanium, vanadium [2933] and many more (compare table 6).

There are other elements, such as phosphorus, for which under given conditions better sensitivities can be obtained with uncoated graphite tubes [2933]. There is also a third group of elements for which equal sensitivities can be obtained with both tube types. In such cases, the choice of tube material is influenced by the sample matrix. Pyrocoated tubes exhibit a much greater resistance against oxidizing reagents and acids, etc.

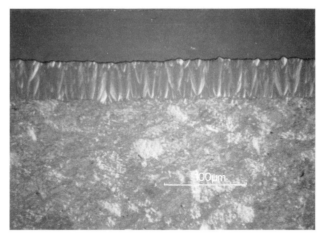

Figure 30. Section of a pyrolytically coated graphite tube showing a perfectly adhering layer (by kind permission of Ringsdorff, Fed. Rep. of Germany).

In figure 29 the difference in surface characteristics between an uncoated tube and a tube coated with pyrolytic graphite deposited by the pyrolysis of methane can be seen. Additionally, figure 30 shows how densely an optimum layer of pyrolytic graphite covers the substrate material.

3.2.3. Purge Gas

The inert purge gas plays a significant role in graphite furnace performance. The primary function of the purge gas is to prevent atmospheric oxygen from contacting the hot graphite components and causing their combustion. Since metal atoms react rapidly with oxygen at high temperatures, an inert gas atmosphere is a prerequirement. Next to this protective function, the inert purge gas transports concomitants vaporized during thermal pretreatment steps out of the radiation beam so that they cannot condense on cooler parts of the furnace.

In many furnace designs these aspects are hardly taken into consideration or are disregarded altogether. Yet a well controlled gas stream through the graphite tube can contribute much to the elimination of background attenuation (see figure 31). In figure 32 a graphite furnace is depicted in which the purge gas stream is divided logically by the graphite tube; an external protective gas stream sheaths the tube, while an internal purge gas stream flows through the tube. The purge gas stream is conducted from the ends of the tube to the middle and thus prevents vaporized sample constituents from condensing at the cooler tube ends. To permit reliable control of the gas stream and to exclude external influences, the furnace is sealed at both ends with removable quartz windows.

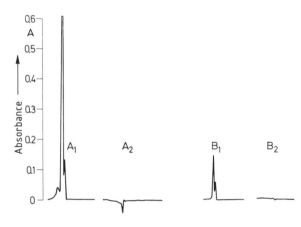

Figure 31. Influence of gas stream symmetry on the background signal generated by 20 μL urine in the graphite furnace. **A**: With an asymmetrical gas stream a high background signal (A$_1$) is generated that cannot be corrected with a continuum source background corrector (A$_2$). **B**: The background signal (B$_1$) is much smaller when a symmetrical gas stream is used and can easily be corrected (B$_2$). Wavelength: 283.3 nm.

This system permits the purge gas flowrate to be reduced during atomization, or even to be stopped altogether, without atmospheric oxygen entering the furnace. The longest possible residence time of the atoms in the radiation beam can thus be achieved, and consequently the highest possible sensitivity. On the other hand, if the concentration of analyte element in the sample is higher and best sensitivity is not required, this system permits a gas stream with predetermined flowrate to pass through the tube during atomization, so that the sensitivity can be preselected (see figure 33).

STURGEON and CHAKRABARTI [2936] found that a considerable loss in sensitivity occurred for difficultly atomized elements when the windows were removed from the furnace, since the atomic vapor could diffuse more easily to the cooler tube ends and condense. This ef-

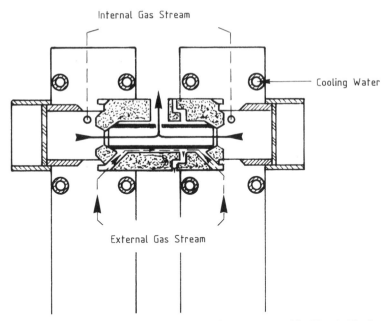

Figure 32. Internal and external gas streams in the HGA-500 Graphite Furnace (Perkin-Elmer). The internal purge gas enters from the tube ends and exits through the sample introduction hole. Volatilized sample constituents are thus transported from the tube by the shortest possible path. The external protective gas stream flows continuously around the graphite tube, even when the internal purge gas stream is shut off, and thus effectively prevents the ingress of atmospheric oxygen.

Figure 33. Sensitivity for 5.0 ng and 0.5 ng lead at various purge gas flowrates during atomization.

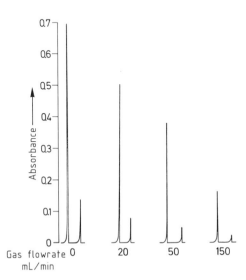

fect is not observed for easily atomized elements. SPERLING [2899] found that for cadmium, closing the furnace with only one window permitted better control over the generation of the atomic vapor. Although the sensitivity decreases compared to an open system, the reproducibility, even for complex matrices, is better.

VAN DEN BROEK *et al* [2998] carried out temperature measurements in graphite furnaces and found that under "gas stop" conditions the gas temperature followed the wall temperature closely to within a few degrees, albeit with a certain time delay. A flowing gas stream through the tube always causes a temperature difference, although this is relatively low when the gas enters from the tube ends. Gas entry from the middle of the tube is especially unfavorable because the cold gas meets the cloud of atoms, leading to considerable interferences.

High purity argon (e. g. 99.996%) is usually used as the purge gas. For reasons of economy, nitrogen is occasionally employed as the purge gas, but it should be borne in mind that at the temperatures prevailing in a graphite furnace, nitrogen can no longer be considered inert. MANNING and FERNANDEZ [802] reported that the determination of aluminium in argon was more sensitive by a factor of three than in nitrogen. This could be attributed to the formation of a cyanide. CERNIK [216] found better precision for the determination of lead in argon than in nitrogen. L'VOV and PELIEVA [2588] reported the observation of CN bands and the spectra of monocyanides of alkaline and alkaline-earth metals when these were atomized into a nitrogen atmosphere, clearly indicating that at high temperatures nitrogen reacts with the graphite tube material.

3.2.4. Temperature Programs and Heating Rate

In a graphite furnace the sample, either as a liquid, a solution or even a solid, is subjected to a series of stepwise or ramped temperature increases so that the analyte element is freed from as many of the concomitants as possible before it is atomized by a final, rapid increase in temperature. The first marked difference to atomization in the flame is that a portion of the concomitants (e. g. the solvent) is removed before atomization of the analyte element, so that this portion cannot cause interferences as in the flame.

As will be discussed later, for a determination in a graphite furnace, the better the separation of the concomitants from the analyte element before atomization, the freer the determination will be from interferences. This is valid for both non-spectral interferences and background attenuation. The efficiency of the separation depends on the volatilities of the analyte element and the concomitants. The separation will cause fewer problems the lower the volatility of the analyte element and the higher the volatility of the concomitants. Possibilities of controlling these problems will be discussed in the section on interferences and their elimination.

It is worth mentioning that especially for complex samples, programming flexibility of the graphite furnace is often decisive. For vaporization of the solvent (the drying step), experience has shown that raising the temperature rapidly to just below the boiling point, then increasing the temperature slowly to just above the boiling point and thereafter maintaining this temperature for 10 to 20 seconds is the best method. With unknown samples, the drying step should be carefully observed so that spattering of the sample in the tube can be avoided. Further thermal pretreatment steps are performed in a similar manner. To save time, temperature ranges in which the sample does not noticeably change can be covered rapidly. To remove larger quantities of salts or oils, or to char or ash biological or other organic materials, slow temperature ramping to the boiling or decomposition point is required for success. Thereafter an isothermal phase should follow to guarantee complete removal or decomposition. Frequently, a series of temperature ramps and isothermal phases is required to remove all concomitants successfully. The upper temperature limit for thermal pretreatment is governed by the volatility of the analyte element.

For atomization of the analyte element a rapid increase in temperature is usually selected since the density of the atom cloud, and thus the sensitivity, is greater the faster atomization occurs. The temperature necessary for atomization should not be greatly exceeded, however, since at higher temperatures expansion of the gas and losses through diffusion from the absorption volume increase. An "infinitely" fast rate of temperature increase to just above the atomization temperature would be the ideal case.

A technique using capacitor discharge proposed by MANTHEI [2622] in 1974 approaches closest to the ideal case. L'VOV [2576] and also CHAKRABARTI *et al* [2141] later examined this proposal in more detail. The technique did not come into general use, however, since the apparatus is complicated and there are difficulties with the signal handling.

LUNDGREN *et al* [771] developed a system in which the graphite tube is heated rapidly to a preselected temperature by passing a high current through the tube and then maintaining this temperature constant to within ± 10 °C. An infrared detector was used to measure the radiation emitted from the graphite tube and thus control the temperature.

A similar system was introduced commercially in 1977 [3064]. The graphite tube is heated at a rate of more than 2000 °C/s and the temperature is controlled by a silicon photodiode previously calibrated to the desired maximum temperature. As soon as this temperature is reached, maximum power heating is switched over to voltage-controlled heating and the

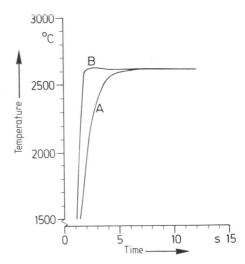

Figure 34. Temperature profiles for different modes of heating. **A**: Conventional voltage-controlled heating. **B**: Ultrafast maximum power heating with optical temperature control.

temperature is maintained constant. The temperature is usually chosen to lie just above the optimum atomization temperature. The temperature profile for maximum power heating compared to conventional voltage-controlled heating is shown in figure 34.

The main advantages of very fast heating rates are the ability to lower the atomization temperature by several hundred degrees (refer to figure 35) and the considerable increase in sensitivity for many difficultly atomized elements. Further, the separation of the analyte element from concomitants is facilitated in a number of cases, especially for easily atomized elements in non-volatile matrices [771] [3064].

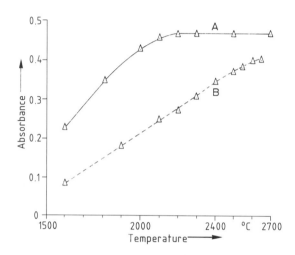

Figure 35. Atomization curves for copper (4 ng). With ultrafast heating (**A**) maximum sensitivity is attained at 2200°C, while with conventional voltage-controlled heating (**B**) it has not been reached at 2700°C.

In this connection the question of temperature accuracy as well as its measurement and control arises. MONTASER and CROUCH [2674] stressed that both current and voltage control can lead to temperature drift and changes. The main causes for a change of tube temperature can be found in boundary resistances and in alterations in the structure of graphite. A power control system should be able to compensate for resistance changes in the tube and thus provide good temperature repeatability.

Accurate measurements have shown that in practice resistance changes in a graphite tube over a wide temperature range can best be compensated by controlling the effective voltage across the tube (see figure 36). A resistance difference of 27% gives a linear deviation of only 4% over the temperature range of interest. This means that with correspondingly close tolerances for the resistance of a graphite tube, very good temperature constancy can be achieved.

Nevertheless, there are a number of good arguments for introducing real temperature control. In the first instance it is completely independent of the tube resistance, thus permitting the use of various graphite materials without influencing the temperature. Further, real temperature control would allow the use of any temperature/time program and make the maximum temperature independent of the heating rate. Naturally, for the very fast heating rate required for atomization the system must work without lag.

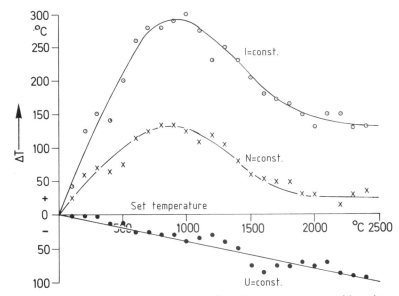

Figure 36. Deviation of the effective graphite tube temperature from the set temperature with various power control systems for a tube having a higher resistance. The temperature was set for a tube of 11.6 mΩ and measured with a tube of 14.7 mΩ (27% higher resistance). Voltage control provides the lowest deviation (-4%) over the entire temperature range.

Any form of temperature control can only be as good as the actual temperature measurement. Thus, only optical radiation measurement can be considered as a means for temperature control, since only optical systems react with sufficient speed [2674]. Unfortunately, up to the present there is no commercial system available that permits continuous monitoring over the entire temperature range from drying to atomization. A reliable optical measurement is usually only possible from 800°C. The measurement should also be performed inside the graphite tube, since otherwise it does not act as an ideal black body radiator; measurements on the outer tube wall exhibit considerable interferences due to alterations of the emission factor.

Temperature measurement (and control) via thermoelements or resistance wires, etc. is only possible in the lower temperature range. Such measurements are generally sluggish because heat transfer from the graphite tube to the temperature sensor always requires a certain time, so that these measurements are unsuitable for rapid increases in temperature. The temperature sensor should be removed from the graphite tube at the latest at 2000°C, because graphite begins to sublime at this temperature. The risk exists of carbide formation and thus an irreversible change in the metal or alloy used for the measurement. As long ago as 1932, KING [2467] observed that tungsten, which has a melting point of 3400°C, showed physical changes at 2300°C in a graphite furnace. At 2450°C tungsten rods melt, form balls, and at 2700°C are "soaked up" by the graphite. Other elements behave similarly and react with carbon to form carbides at temperatures markedly below their melting points.

3.2.5. The Stabilized Temperature Furnace

Even before the introduction of the first commercial graphite furnace, L'vov [8] had mentioned in his book that the furnace principle proposed by MASSMANN was by no means optimum. Due to the fact that the sample is dispensed into a cold tube which is then heated rapidly for atomization, the sample after vaporization is in an environment that is not in equilibrium with respect to either volume or time. This can lead to a host of interferences, as will be shown in detail in a later section. WOODRIFF has also constantly emphasized that far fewer chemical interferences occur in a continuously heated, isothermal furnace.

STURGEON and CHAKRABARTI [2934] found that about 60% of the atoms formed diffuse to the cooler tube ends and condense. SLAVIN *et al* [2883] discovered that at 2500°C there was a temperature gradient of 1000°C between the middle and the ends of a graphite tube. They made a modified tube with varying wall thickness over the length and thus obtained a temperature gradient of only 100°C.

However, even more important than temperature constancy over the length of the tube is equilibrium with respect to time. In both the furnace of L'vov and that of WOODRIFF, the sample is atomized into an environment of constant temperature. L'vov *et al* [2590] thus proposed that the sample should not be dispensed onto the tube wall, but onto a "platform" mounted in the tube (see figure 37). Since the graphite platform is only placed

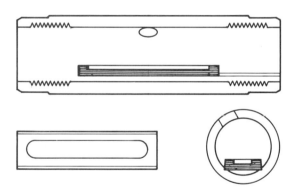

Figure 37. L'vov platform in a graphite tube. Side and end-on views of the tube and top view of the platform.

loosely in the tube, there is virtually no heat transfer through direct contact. This effect is even more marked when the platform is made of pyrolytic graphite, since this exhibits high anisotropy and permits virtually no heat conductance at right angles to the plane of the layers. If the graphite tube is heated rapidly in the atomization step, the temperature of the platform only follows this temperature rise sluggishly. Once the graphite tube and the gas phase have reached the final temperature and are largely in equilibrium, the platform is heated rapidly due to radiation and the hot gas. The sample is then atomized into an environment in which the temperature is no longer changing (see figure 38). With this arrangement, L'vov found significantly fewer chemical interferences. This finding was also verified by other workers [2880].

Later L'vov and PELIEVA [2585] introduced the sample on a tungsten wire into a preheated graphite tube. They found that this brought enhanced sensitivity for a number of difficultly

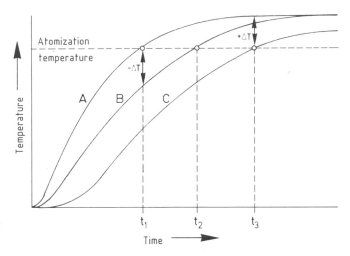

Figure 38. Heating profiles for the graphite tube wall (**A**), the inert gas (**B**), and the L'vov platform (**C**). For atomization off the wall the sample is volatilized at a time point (t_1) when the inert gas is still cooler ($-\Delta t$). Atomization from the L'vov platform is with substantial time delay (t_3) so that the inert gas has stabilized at a higher temperature ($+\Delta t$).

atomized elements. The freedom from interferences with this system is at least as good as atomization off the platform [2621]. The major disadvantage of the wire technique is the limited sample volume that can be used.

3.2.6. Automation

Automation has assumed great importance in graphite furnace atomic absorption spectrometry. In the flame technique using a nebulizer/burner system the operations are very simple and the procedure is quick; automation brings no advantage other than performing the work of the operator. With the graphite furnace technique, on the other hand, dispensing the sample is more difficult and requires greater care. Also, the times between each sample aliquot are rather long, making manual operation relatively unproductive. A well-proven commercial autosampler for the graphite furnace technique is depicted in figure 39.

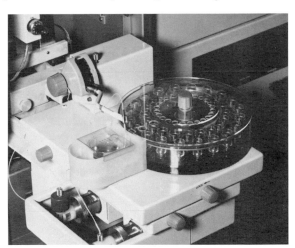

Figure 39. Model AS-40 Autosampler for the graphite furnace technique (Perkin-Elmer).

Of even greater importance, however, are the considerations of precision and accuracy with respect to sample dispensing. The place in the graphite tube onto which the sample is dispensed and the way in which the sample droplet is brought onto the graphite surface are of considerable significance. It is virtually impossible to keep these factors fully under control with manual pipetting. Further, there is the risk that particles around the edge of the sample introduction hole in the graphite tube can be introduced into the tube with the pipet tip, leading to contamination. The reproducibility that can be achieved with manual dispensing is usually in the order of several percent. With automatic dispensing, on the other hand, a reproducibility of better than 1% can be achieved [2920], as shown in figure 40.

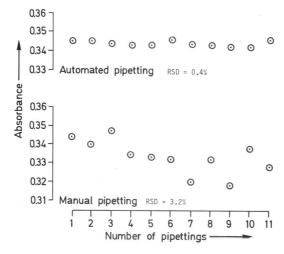

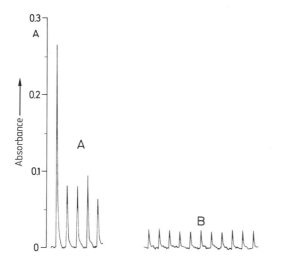

Figure 40. Comparison of the precision achieved for automatic dispensing and manual dispensing for the determination of lead (20 µL, 0.2 mg/L Pb).

Figure 41. Contamination of pipet tips with iron; repeated dispensing of 0.4 ng Fe (**A**) manually with a polypropylene pipet tip and (**B**) automatically with a PTFE pipet tip.

A further problem mentioned in many publications is contamination of the pipet tip. This cannot always be eliminated entirely, even after thorough leaching in acid and long rinsing [2074] [2316] [2897]. As shown in figure 41, a substantial blank reading is still obtained even after repeated pipetting when a pipet tip contaminated with iron is used. This leads to poor precision and false readings. This problem is virtually eliminated when an autosampler with a PTFE pipet tip is used [3046] [3047]. Employing the same type of autosampler, STOEPPLER et al [2920] found that the carryover for ^{89}Sr was 10^{-7}.

SCHULZE [2833] reported a modification of this autosampler to permit semicontinuous sampling of continuously flowing sample streams. This system is particularly useful for monitoring industrial plant and process systems.

MATOUŠEK [2633] proposed spraying sample solution using a nebulizer into the lightly heated graphite tube rather than employing an automated pipet. The advantages of this system are that the concentration sensitivity can be varied via the aspiration time and that virtually unlimited sample quantities can be deposited. The obvious disadvantages are that a much longer time is required to dispense larger sample quantities and that the sample requirement is in itself very large. As with all spray chamber nebulizers, only about 10% of the sample reaches the graphite tube, while the rest runs to waste. Added to this, with a nebulizer the transport interferences known from the flame technique are introduced to the graphite furnace technique.

3.2.7. Analysis of Solid Samples

It is a particular characteristic of the graphite furnace technique that solid samples can be analyzed relatively easily. Graphite furnaces without any modifications are usually employed. The sample is weighed into a small boat, which is then introduced into the graphite tube where the sample is tipped out. The boat is then removed and reweighed to determine the actual sample weight. Another possibility is to weigh a sample into a graphite boat or cup which is then left in the tube for atomization. Occasionally larger samples are placed directly with forceps through an enlarged sample introduction hole into the tube.

For their investigations, LANGMYHR et al [731] used a graphite furnace heated by a high frequency induction generator, which was apparently particularly suitable for the analysis of solids. While these authors introduced the sample in a small boat into the cold tube and then heated stepwise, other workers found it an advantage to introduce the sample into a furnace preheated to a constant temperature. ANDREWS and HEADRIDGE [2041] used an induction furnace with a crucible mounted on it into which metallurgical samples were placed. LUNDBERG and FRECH [2571] used a similar system for the analysis of metallurgical samples, but employed small, preheated graphite crucibles.

An interesting accessory for introducing powdered samples through the introduction hole into the graphite tube has been described by GROBENSKI et al [2320] (see figure 42). A glass capillary is fitted into a device similar to an injection syringe. The capillary is pressed repeatedly into the finely ground or pulverized sample until sufficient sample has entered the capillary. The accessory is then weighed, after which the capillary is introduced into the introduction hole in the tube, just like a micropipet, and the sample ejected by pressing down

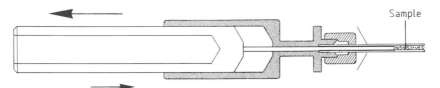

Figure 42. Powder sampler for introducing powdered samples into the graphite tube (Perkin-Elmer).

the plunger. The accessory is weighed again and the sample weight determined by difference. The dispensing step can even be partially mechanized, thus permitting convenient and
reliable insertion of the capillary into the graphite tube.

L'VOV [2575] has been particularly concerned with the disadvantages, such as weighing errors and inhomogenity, imposed by the limitation of sample weight to a few milligrams in
conventional graphite furnaces. He therefore suggested that solid samples should be atomized out of a cavity whose walls are permeable to gases.

In one version, L'VOV used a graphite capsule of 45 to 60 cubic millimeters volume, made
by boring out a graphite rod. About 40 mg of sample are mixed with graphite powder and
placed into the capsule, which is then sealed with a graphite plug. The capsule is placed between two graphite contacts and heated in the air/acetylene or nitrous oxide/acetylene
flame of a Méker burner (figure 43). To atomize the sample an electrical pulse is additionally applied to the capsule. The power required varies according to the analyte element,
ranging from 0.3 kW (Cd) to 2.6 kW (Ti). With this arrangement, up to about 50 elements
can be determined with detection limits of 10^{-5} to 10^{-7}%.

Figure 43. Graphite capsule proposed by
L'VOV for the direct analysis of solid samples. The capsule is located beneath the radiation beam of an atomic absorption spectrometer.

In the above version, the absorption of the atoms that diffuse out of the capsule is measured in the flame above the capsule. L'VOV constructed a second version comprising a double-walled graphite tube through which the radiation beam passes in the usual manner (figure 44). The outer tube is coated with pyrolytic graphite so that it is gastight, while the inner tube is made of porous graphite. The gap between the tubes has a volume of about 150 cubic millimeters and can accommodate approximately 100 mg sample mixed with graphite powder.

Since the mass of the double-walled tube is substantially greater and the tube is not heated in a flame, higher electrical power is required for atomization. This is between 1.0 kW (Cd) and 3.7 kW (Ni). The difficult-to-atomize elements, such as titanium, vanadium, and molybdenum, cannot be determined with this furnace. For the easily atomized elements, on the other hand, detection limits of 10^{-7} to $10^{-10}\%$ can be attained. This is at least two orders of magnitude better than with the capsule in the flame.

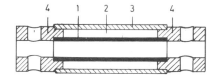

Figure 44. Double-walled graphite tube constructed by L'VOV for direct solids analysis [2575]. **1**: Inner tube of porous graphite; **2**: Space for the sample; **3**: Outer tube of pyrolytic graphite; **4**: Pyrolytic graphite contacts.

A further, interesting advantage of this system is that atoms can diffuse far more easily through the porous graphite than molecules. Far lower radiation scattering and molecular absorption are observed with this system than with an open system. NICHOLS et al [2692] made similar observations. They sealed biological samples into porous graphite capsules and inserted these into a graphite furnace heated at constant temperature. Background attenuation was two orders of magnitude lower than in an open system.

3.3. The Hydride Technique

The fact that arsenic and a number of the other representative elements of Groups IV, V and VI of the periodic table form volatile, covalent hydrides with "nascent" hydrogen has been known and utilized for more than 100 years (for example, in Marsh's Test and Gutzeit's Test for arsenic). The advantage of volatilization as a gaseous hydride lies clearly in the separation and enrichment of the analyte element and thus in a reduction or even complete elimination of interferences.

In the early 1950s a number of techniques were introduced for the determination of arsenic and other hydride-forming elements using colorimetric methods. The hydride was formed with zinc in acid solution and the gaseous reaction products were conducted into solutions containing ammonium molybdate or hydrazine sulfate, for example, which form characteristic, colored complexes with the hydride. Some of these techniques are still in use even today.

HOLAK [524], in 1969, was the first to apply hydride generation for the determination of arsenic using AAS. He generated hydrogen by adding zinc to the sample solution acidified with hydrochloric acid and collected the arsine in a trap cooled in liquid nitrogen. At the end of the reaction he warmed the trap and conducted the arsine with a stream of nitrogen into an argon/hydrogen diffusion flame to measure the atomic absorption. In the following years, numerous papers were published describing modifications and optimizations of the technique. Nevertheless, the technique only found wide acceptance within recent years after the first reliable accessories were introduced onto the market.

3.3.1. Methods of Hydride Generation

The most commonly used method in former times for the generation of nascent hydrogen, and thereby hydrides such as arsine, was the reaction of metals such as zinc with hydrochloric acid. It is therefore hardly surprising that this technique was used initially for AAS as well. The reaction vessels frequently comprised flasks fitted with dropping funnels that permitted zinc to be added to the acidified sample solution without having to open the system [799]. LICHTE and SKOGERBOE [2544] used a column packed with granulated zinc through which the sample solution was run.

In an automated system, GOULDEN and BROOKSBANK [2312] employed a suspension of aluminium powder in water as the reductant. In this system, however, the hydride had to be driven from a heated, packed column by a stream of inert gas. Other authors albeit were unable to obtain satisfactory results using this reductant [2751]. LANDSFORD *et al* [2535] found that for the determination of selenium using zinc as reductant, considerable interferences were caused by mercury, nitrates and, most particularly, by arsenic. They therefore proposed the use of tin(II) chloride, which they added to the sample solution acidified with 6 M hydrochloric acid; the selenium hydride had to be driven from solution by a stream of inert gas. POLLOCK and WEST [993] [994] employed a mixture of magnesium and titanium trichloride to generate the hydride by adding it to the sample solution acidified with sulfuric and hydrochloric acids.

Metal/acid reactions have a number of disadvantages, nevertheless, that played no small part in preventing the wider acceptance of the hydride technique. When zinc is used as the reductant, only antimony, arsenic and selenium can be determined. Further, granulated metals often cannot be obtained with the required degree of purity, so that it is necessary to work with significant, frequently varying, blank values. MCDANIEL *et al* [2651] established that using this reaction, only about 8 % of the hydride is released, while around 90 % is trapped in the precipitated zinc sludge or does not react. A yield as low as this is certainly unsatisfactory for trace determinations.

With the introduction of sodium borohydride as reductant a marked change occurred in the hydride technique. Using this reductant SCHMIDT and ROYER [1092] determined antimony, arsenic, bismuth and selenium, POLLOCK and WEST [994] determined germanium, and FERNANDEZ [358] optimized the conditions for these elements and also for tellurium and tin. THOMPSON and THOMERSON [2967] reported the successful determination of lead using sodium borohydride as reductant and thus increased to eight the number of metals that can be determined by this technique.

Initially this reductant was used in a similar manner to zinc, i. e. sodium borohydride pellets were dropped into the reaction flask containing the acidified sample solution. This mode of operation proved unsatisfactory, since poorly reproducible results were often obtained and contamination was also a problem, as with zinc. The reaction was also difficult to control because an alkaline zone formed around the borohydride pellet in which the processes were completely different than in an acid environment. MCDANIEL *et al* [2651] found that when using borohydride pellets in sample solutions acidified with 0.6 M hydrochloric acid, only about 10% of the hydride (for selenium) was released, while in 6 M hydrochloric acid solution under additional mixing with a stream of nitrogen, the yield could be increased to 40–60%. YAMAMOTO and KUMAMARU [3085] compared the use of zinc and sodium borohydride pellets for the determination of antimony, arsenic and selenium. They found that each reductant had certain specific interferences. Zinc was found to be preferable for the determination of arsenic, while sodium borohydride was preferred for the other two elements. The necessary reproducibility and control over the reaction was first achieved with the introduction of sodium borohydride solutions. These solutions can be stabilized by making them up in sodium hydroxide solution. The technique is also easier to automate since only solutions are involved, so that a higher sample throughput can be achieved. Mostly the borohydride solution is added to the sample [2168] [2489] [2857] [2893], or sometimes the sample to the borohydride solution [3037]. According to the type of reaction vessel used, to achieve good mixing and to drive out the hydride, either the solution is stirred with a magnetic stirrer [2489] or a stream of inert gas is passed through the solution [2168] [2857] [2893].

JACKWERTH *et al* [2431] showed that by using a reaction vessel with a conical bottom and by introducing the borohydride solution through a capillary into the bottom of the vessel, stirring is no longer required. Through the violent reaction of the alkaline reductant solution with the acidified sample solution and the conical form of the vessel, thorough, turbulent mixing takes place, guaranteeing a rapid and complete reaction. A commercial system using this form of reaction vessel is on the market (see figure 45) and functions reliably and without problems.

As well as borohydride, cyanoborohydride (BH_3CN^-) has been proposed as a reductant for the AAS hydride technique [2111]. The authors report a substantial improvement in yield of hydride and greater freedom from interferences in the presence of higher concentrations of

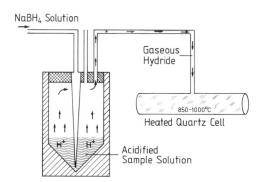

Figure 45. Schematic representation of a hydride system for the generation of gaseous hydrides.

cations such as nickel. The biggest disadvantage of this reductant, however, is the very slow reaction, which takes several minutes, making it necessary to collect the hydride, for example in a trap cooled in liquid nitrogen. This technique is thus suitable for interference studies, but can hardly be used for routine daily determinations.

3.3.2. Collecting the Hydride

Once the hydride has developed and been driven out of solution, it can be handled in a variety of ways. In his initial experiments, HOLAK [524] first collected arsine in a trap cooled in liquid nitrogen before warming to vaporize it for measurement. This procedure was later also used by a number of other workers since high sensitivity (measured in peak height) and a high degree of freedom from interferences could be achieved [2318] [2711]. This is especially true when an interference is caused by the hydride developing faster or slower from the sample solution than from the reference solution. Using radio tracers, McDANIEL *et al* [2651] found that the hydride can only be quantitatively trapped and then released for measurement without prior decomposition if the trap cooled in liquid nitrogen is filled with glass beads. Before freezing, the hydride should be dried. Calcium chloride is the most suitable drying agent since it develops a high heat of hydration that prevents hydride from being dissolved or adsorbed.

Since freezing of the hydride in a cold trap with subsequent vaporization through warming is a very time consuming procedure, FERNANDEZ and MANNING [362] developed a system in which the hydride is collected in a balloon. After 15 to 30 seconds collection time, the hydride and the hydrogen are conducted to the atomizer by a stream of inert gas. A disadvantage of this procedure is that the collection time for a number of elements must be maintained very exactly since their hydrides decompose very easily.

DALTON and MALANOSKI [266] therefore proposed conducting the hydride directly into a flame. They found that carrier gas was not even necessary with this procedure since the hydrogen liberated very effectively transports the hydride from the reaction vessel to the atomizer. Later investigations showed, however, that with an additional, slow inert gas steam through the sample solution the yield of hydride could be improved markedly [2651].

KANG and VALENTINE [2455] found that the analytical curves obtained with the direct "on-line" system were much more linear than when the hydride was collected in a balloon. They further established that peak area integration eliminates most of the disadvantages of the on-line method that are found with peak height measurements. The potential difficulties of a flow or direct system are to be found in the dependence of the reaction rate on the oxidation state of the analyte element and the possible influence of concomitants on the release of hydride from the sample solution.

Nevertheless, for routine determinations the direct (on-line) method offers the advantage of higher sample throughput and easy operation. Further, peak area integration frees this method from kinetic interferences. Trapping the hydride in a cold trap with subsequent rapid warming brings the highest yield [2651] and the best absolute sensitivity [2856], but the sensitivity of the direct system is nevertheless normally fully sufficient for general analytical work and the markedly higher speed is often decisive.

3.3.3. Atomization of the Hydride

Initially, the argon (or nitrogen)/hydrogen diffusion flame was used almost exclusively for atomization of the hydride. The auxiliary inert gas and the hydrogen liberated by the reaction can be conducted into such flames without significantly altering the combustion characteristics. If the hydride has been previously trapped by freezing out, even the hydrogen is removed and only the hydride with carrier gas is subsequently conducted into the flame. The temperature of the diffusion flame is fully adequate to atomize the hydrides and it has sufficient transparency in the far UV range to permit the determination of arsenic and selenium with a favorable signal-to-noise ratio.

Soon after the development of the hydride technique the use of electrically heated quartz tubes [2158] or quartz tubes heated in a flame [2967] for atomization of the hydride were proposed. Compared to a flame, a quartz tube offers the advantage of higher sensitivity and, especially for arsenic and selenium, negligible spectral background and thus an improved signal-to-noise ratio. DRINKWATER [2206] found that it was particularly important to prevent the hydrogen from igniting at the ends of the tube, since non-specific, difficult-to-correct signals are obtained. This finding has been fully corroborated in the present author's laboratory.

For their automated system, GOULDEN and BROOKSBANK [2312] used a quartz tube sealed at the ends with quartz windows and electrically heated to 1100°C. They soon established that the sensitivity could be increased substantially, and especially for antimony and arsenic, when oxygen or air was conducted into the gas stream. The best values, a more than threefold increase in the sensitivity, were obtained at an oxygen-to-hydrogen ratio of one to five. Under these conditions a fuel gas rich flame burnt in the quartz tube. Other workers took up the idea of a "flame in the tube" [2857] [2858] and introduced a small burner for the oxygen/hydrogen flame into the side entry tube.

DEDINA and RUBEŠKA [2188] found that in a cool, fuel gas rich oxygen/hydrogen flame burning in the side entry tube of a quartz T-tube, a very high degree of atomization (possibly 100%) for selenium was attained. They further showed that the atomization of selenium was not based on the thermal dissociation of the hydride, but was brought about by free radicals (H, OH) produced in the flame (refer to section 8.3.1).

3.3.4. Automation

Resulting from the high sensitivity of the procedure and also from the considerable significance that the determinable elements have for the environment, attempts were soon made to automate the hydride technique. GOULDEN and BROOKSBANK [2312] combined a 20-channel pump system with an atomic absorption spectrometer. They used a slurry of aluminium in water in the presence of tin(II) chloride in hydrochloric acid as the reductant for antimony, arsenic and selenium. PIERCE *et al* [2749] [2750] [2751] found that a solution of sodium borohydride as reductant was preferable to a slurry of aluminium for automated techniques. This was also substantiated by other authors [2767] [3019]. FISHMAN and SPENCER

[2251] developed an automated system for the digestion of water and sediment samples, and by combining this with an automated hydride system were able to achieve a throughput of 30 samples per hour.

3.3.5. Sample Volume and Volume of Measurement

The major difference between the hydride technique and other atomization techniques is certainly the separation of the analyte element, as its volatile hydride, from the sample matrix. The relatively small number of components that reach the atomizer thus make interferences rather unlikely. This is particularly true for spectral interferences caused by radiation scattering and molecular absorption, since they cannot be observed in a well designed system.

The hydride technique therefore has a decisive advantage over other techniques, although the virtual absence of interferences in the atomizer does not mean that the technique is completely without interferences. Possible interactions in the reaction vessel that can influence the formation and liberation of the hydride are discussed in detail in section 8.3.2.

However, it is necessary to mention another specific characteristic of the hydride technique since it can have great significance for practical work, namely the difference between the sample volume and the volume of measurement. The hydride technique is an absolute procedure, i.e. the signal measured in peak height, or occasionally peak area, is directly proportional to the absolute mass of the analyte element and not to its concentration in the solution. The volume of measurement plays a certain role in direct systems, especially if peak

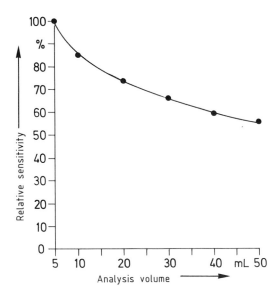

Figure 46. Dependence of the selenium signal on the volume of measurement in the hydride technique.

height is measured, but this is a secondary effect that is dependent upon the apparatus used and is never proportional to the degree of dilution (figure 46).

Most reaction vessels for the hydride technique are designed to accept a relatively large volume (e. g. 50 mL or 100 mL), but also require a minimum volume (e. g. 5–10 mL) to guarantee that the reaction takes place correctly. However, because of the high sensitivity of the hydride technique, sample volumes of 10 to 50 mL are rarely required; volumes of 1 mL or less are generally sufficient to permit working in the optimum measuring range.

In practice, approximately 10 mL acid are poured into the reaction vessel (a measuring cylinder is sufficient) and then the sample, say 0.5 mL, is pipetted accurately into the vessel. The volume of measurement can thus be substantially larger than the actual sample volume. As will be shown later, this opens up a multitude of possibilities for modifying the chemical environment for hydride generation so that interfering reactions can be controlled and suppressed over a wide range. Only in a few cases, such as the analysis of natural waters, is it necessary to use a large sample volume of 50 mL. In such a case the sample volume is then the same as the volume of measurement.

The relative determination limits that can be attained with the hydride technique are presented in table 7 and compared to those attainable with the graphite furnace technique.

Table 7. Determination limits of the hydride technique (50 mL volume of measurement) compared to the graphite furnace technique (0.1 mL sample volume)

Element	Determination limit µg/L	
	Hydride Technique	Graphite Furnace Technique
As	0.02	0.3
Bi	0.02	0.2
Sb	0.1	0.2
Se	0.02	1
Sn	0.5	0.2
Te	0.02	0.2

3.4. The Cold Vapor Technique

Mercury is the only metallic element that exists in the atomic state at ambient temperature. It also exhibits an appreciable vapor pressure. These unique properties led many workers to attempt the determination of mercury even at a very early stage in the development of elemental analysis. The relatively poor sensitivity exhibited by mercury in the flame and the ever increasing necessity of determining this element in very small traces were further stimulants to these endeavors. Since mercury has a vapor pressure of 0.0016 mbar at 20°C, corresponding to a concentration of approximately 14 mg per cubic meter in air, the possibility exists of determining it by AAS without the use of an atomizer at all. The element must merely be reduced to the metal from its compounds and transferred to the vapor phase.

3.4.1. Instrumental Developments

In 1939, before the actual rediscovery of atomic absorption by WALSH, WOODSON [1354] described an apparatus for the determination of mercury in air. In modified form, this apparatus was used by other workers. BRANDENBERGER [152] [153] developed a procedure for the electrolytic deposition of mercury on a copper spiral with subsequent vaporization into a quartz cell (figure 47) by electrically heating the spiral. Later, next to this dynamic method, in which the mercury is transported through the cell by a gentle gas stream, he developed a static procedure that was simpler to operate. In both cell types absolute quantities of 0.2 ng mercury can be determined. HINKLE and LEARNED [514] used a similar procedure except that they deposited mercury by chemical means onto a wire gauze which was then subsequently warmed.

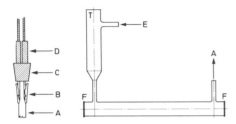

Figure 47. Apparatus for the determination of mercury according to BRANDENBERGER [152].—Left: Holder for copper spiral; **A** copper spiral, **B** crocodile clips, **C** rubber bung, **D** banana plugs. Right: Absorption cell; **T** funnel into which the copper spiral is inserted, **E** inlet for air, **A** outlet for air, **F** quartz windows.

The most successful technique for the determination of mercury was discovered by PoLUEKTOV and VITKUN [995] [996]. During their investigations on the determination of mercury by flame atomic absorption spectrometry, they discovered an unusually large increase in the absorbance, by one to two orders of magnitude, if tin(II) chloride was added to the sample being aspirated. This could be attributed to the reducing action of this reagent, which ensured that virtually all of the mercury being aspirated passed into the flame in the atomic state. Thereupon, they eliminated the nebulizer and flame, passed air through the sample after tin(II) chloride had been added and then conducted the air through a 30 cm quartz cell mounted in the radiation beam of an atomic absorption spectrometer. The detection limit obtained by this technique amounted to 0.5 ng mercury.

POLUEKTOV and co-workers were not the first to describe the reduction of mercury salts to metallic mercury with tin(II) chloride, but they were the first to use this reaction in combination with atomic absorption spectrometry. HATCH and OTT [481] extended this technique and employed it for the analysis of metals, rock and soil samples. Numerous authors have reported the use of this technique, occasionally with slight modifications, for the determination of mercury in most widely varying materials.

KAHN [611] described the first commercial system based on the procedure of POLUEKTOV and VITKUN (figure 48). Although the system is very simple and widely in use, a number of points must be carefully observed if false measurements are to be avoided.

The reasons for the observed systematic errors can all be traced, more or less, to the mobility of mercury and its compounds. Only those errors directly attributable to the apparatus

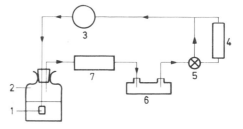

Figure 48. System for the determination of mercury by the cold vapor technique. — Free metallic mercury, released by reduction with tin(II) chloride in the reaction flask (**2**), is circulated by a pump (**3**) through the absorption cell (**6**). **1** Bubbler, **4** absorbant, **5** three-way valve, **7** drying agent.

will be mentioned in this section; a detailed treatise is presented in section 8.4.2. The reader is also referred to section 6.4 (Problems of Trace Analysis).

Many authors have dealt with the problem as to which material is the most suitable for vessels for storing solutions containing mercury. Normal laboratory glassware definitely has the worst characteristics. PTFE is better, but by no means satisfactory if it has not previously been fumed out in 65% nitric acid [2447]. In general, quartz, vitreous (glassy) carbon and FEP are the most suitable. Fuming out in 65% nitric acid reduces the readiness of the vessel to adsorb mercury [2448].

LITMAN *et al* [2551] also reported high rates of adsorption of mercury onto glass, PTFE, and especially polyethylene in the concentration range below 1 µg/L. They suspected that the losses could be attributed to a reduction to the metal. KOIRTYOHANN and KHALIL [2482] found using polypropylene vessels that the values measured were too low for reference solutions, but not for sample solutions if these contained an excess of oxidant remaining from digestion procedures. The cause was found to be di-t-butyl-methylphenol, which is added to polypropylene as an antioxidant and was found to act as a reducing agent. The problem could be eliminated by the addition of an oxidizing agent, such as potassium permanganate, to the reference solutions.

STUART [2926] has drawn attention to the fact that every component of the apparatus with which mercury vapor comes into contact after its reduction presents a potential problem. To avoid losses, the surfaces contacted by mercury should be as small as possible. Even with the greatest of care, some mercury will always remain in the system and can be released during a subsequent determination. Frequent blank measurements are thus essential, especially when the mercury concentration in the samples fluctuates markedly.

FRITZE and STUART [388] refer to possible losses of mercury in the tubing. TÖLG [2975] mentions especially red rubber tubes vulcanized with ammonium sulfides since mercury binds to the surfaces as mercury sulfide; in PVC tubes mercury is bound to non-saturated chlorine sites. TÖLG [2975] further mentions exchange reactions with mercury, for example through deposition on non-noble metal surfaces.

A particular problem is presented by the drying agent [388]. Calcium chloride is particularly unsuitable, since when moist it adsorbs most of the mercury [2926]. Magnesium perchlorate is much better since it only binds a few percent of the mercury. Concentrated sulfuric acid appears to be best, although a large dead volume in a wash bottle can cause problems [2503].

A detailed investigation has shown that water vapor does not absorb radiation at the mercury resonance line, so that a drying agent is really not required. It is only necessary to prevent larger solution droplets from being transported by the carrier gas [2926]. Frequently, slightly warming the absorption cell is sufficient to prevent the condensation of water.

Nowadays, the determination of mercury is performed in an apparatus essentially the same as that used for the hydride technique. When tin(II) chloride is used as reductant, certain modifications, such as the addition of a pump or switch-over to a higher gas stream to drive out the mercury, are required. These modifications are already included in many commercial instruments and changeover is simply effected with a switch.

In this connection the incompatibility of the reductants sodium borohydride and tin(II) chloride must be stressed. Use of both reagents in the same system without prior, thorough cleaning leads inevitably to the formation of precipitates and blockage of narrow bore tubing and valves. It is preferable to reserve a separate reaction assembly of the apparatus for each reductant.

3.4.2. Reduction and Liberation of Mercury

As mentioned above, initially tin(II) chloride was used almost exclusively for the reduction of mercury.

More recently, sodium borohydride finds increasing application as the reductant [2448] [2542] [2659] [2799]. The major differences between these two reductants are that sodium borohydride is by far the stronger reducing agent and that a great quantity of hydrogen is liberated when sodium borohydride reacts with acids. This hydrogen in fact transports the majority of the metallic mercury from the solution to the absorption cell, while for tin(II) chloride a stream of gas must always be bubbled through the solution to drive out the mercury. Interestingly, each of these reductants shows characteristic interferences that are not shown by the other.

In earlier instruments [481] [611] a closed system was normally employed. Air was pumped in a closed circuit through the reaction vessel and the absorption cell (refer to figure 48, page 77). After about a minute an equilibrium is established between air and the aqueous phase. The numerical value for this equilibrium constant is about 0.4 and is independent of most of the variables, such as mercury concentration, the volume ratio of the phases (in the range air to water 1:3 to 4:1), and the acid concentration (HCl and HNO_3). Solely sulfuric acid (higher H_2SO_4 concentrations increase the mercury concentration in the gas phase) and the temperature of the solution (higher temperatures raise the mercury concentration in the gas phase) have a marked influence [2482].

As well as the circulatory technique, in which a static signal is obtained after a certain lapse of time, there is the direct technique, in which air or another gas is bubbled through the reaction vessel and then conducted directly through the absorption cell. The dynamic signal thus generated is normally somewhat lower than the static signal obtained from the circulatory technique, but this disadvantage is balanced by the advantage of a higher sampling frequency.

When sodium borohydride is used as reductant only the direct technique is applicable because the large volume of hydrogen liberated would lead to a considerable increase in pressure in a closed system. Additionally, the reaction with sodium borohydride is much faster,

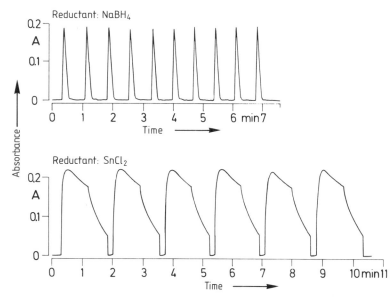

Figure 49. Determination of mercury by the cold vapor technique. Comparison between the direct technique using sodium borohydride and a closed system using tin(II) chloride. There is a break in the absorbance curve when the closed system is opened.

so that the peak height measured with the open system is almost as high as the signal obtained using tin(II) chloride in a closed system (figure 49).

A further major advantage of the rapid reaction of sodium borohydride and the open system is that mercury vapor is only in contact with tubes and other components of the system for a relatively short time, so that exchange reactions only play a minor role. In a closed system, in which mercury is enriched during circulation, losses through adsorption and carry-over from one sample to the next are a much greater problem [2659].

Important considerations affecting the efficiency of a system are reduction of the dead volume within the apparatus and diffusion of elementary mercury into the carrier gas [2355]. For example, KAISER *et al* [2448] reported times of three minutes for $NaBH_4$ and five minutes for $SnCl_2$ to drive 95% of the mercury into the gas phase. By using a well-designed reaction vessel with a conical bottom and by introducing the reductant into the bottom of the vessel, thorough and turbulent mixing of reductant and sample is ensured, so that the mercury is driven completely out of solution [2431]. The time required depends to some degree on the volume of the solution, but in this type of apparatus, one to one-and-a-half minutes are sufficient, as shown in figure 49 [3053].

3.4.3. Amalgamation

The necessity of determining mercury in the lowest concentrations led to the development of enrichment and separation techniques. Many of these techniques were based on the fact that mercury is an extremely noble metal and can therefore be deposited easily, either chemically or electrolytically, from solution, onto copper for example. It is only necessary to refer once again to the work of BRANDENBERGER [152–155] and HINKLE and LEARNED [514].

KAISER *et al* [2447] established that when the concentration of mercury is at or below 10 μg/L, the yield from these "static" procedures is poor, and cannot be improved through long electrolysis times of 10 hours or longer, or through stirring or ultrasonics. They therefore used a small column packed with copper gauze as the cathode and pumped the solution repeatedly through in a closed circuit. The copper gauze column acts rather like an ion exchange column. With this system, 50 pg mercury can be deposited quantitatively from 10 mL nitric acid solution in five minutes.

A technique that has found much wider use than separation and enrichment of mercury from solution is a combination of reduction and volatilization of the mercury with subsequent amalgamation from the gas phase. To avoid kinetic interferences, and to improve the sensitivity, mercury vapor is deposited on tin [621] or, more frequently, silver [2447] [2640] [2679] or gold [1247] [1262] [2447] [2448] [2636] [3053] [3087] (refer to section 8.4.2).

KAISER *et al* [2447] reported on a number of digestion apparatuses in which biological or non-volatile inorganic samples are digested in a stream of oxygen and volatilized mercury is collected by amalgamation on a gold gauze. Mercury vapor in air or in stack gases, etc. can also be collected and enriched using this procedure.

Most frequently, however, the amalgam technique is used quite simply to improve the detection limits of the cold vapor technique. Normally, the mercury is liberated slowly from solution over upward of one minute and thus generates a broad, low signal on the recorder. If instead the mercury vapor is collected by amalgamation, on a gold gauze for example, and then subsequently released by heating the gauze rapidly to 500–700 °C, a taller, narrower signal is obtained. As shown in figure 50, this enrichment brings an increase in sensitivity of more than one order of magnitude.

The detection limit attainable with this technique is below 0.1 ng absolute, corresponding to a relative detection limit of about 1 ng/L referred to 50 mL solution. A further improvement

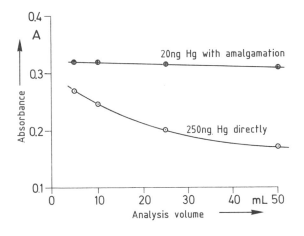

Figure 50. Enhancement of the sensitivity for mercury through amalgamation. The amalgam technique preconcentrates the mercury and makes the determination independent of the volume of measurement.

of this detection limit through optimization of all parameters [2355] or by combining the cold vapor technique with the graphite furnace technique [2855] is conceivable. Nevertheless, the omnipresent concentration of mercury in air is so high that for determinations in the lower nanogram range, significant blank values must be taken into consideration. It is therefore questionable whether detection limits below 0.1 ng can be utilized in practice [2447] [3053] (refer also to section 6.4).

3.5. Miscellaneous Atomization Techniques

The techniques thus far described—flame, graphite furnace, hydride-generation, and cold vapor—satisfy virtually all the requirements of routine analysis. Nevertheless, there are other techniques that have been, or still are, employed with some success for special applications.

In 1960, GATEHOUSE and WALSH [406] reported a demountable vaporization chamber operating on the principle of a hollow cathode lamp in which the sample was vaporized by a glow discharge in a cylinder open at both ends. Later, WALSH [1283] reported the determination of phosphorus in copper, and silicon in aluminium and steel by this method. GOLEB [424] [427] carried out isotope analyses in a similar vaporization chamber with good success. WALSH [1284] later described experiments on multielement determinations using this vaporization chamber as the atomizer. GOUGH *et al* [433] examined cathodic nebulization under reduced pressure in the glow discharge of an inert gas for the direct determination of a number of elements in alloys using atomic fluorescence spectrometry. The technique can also be used for AAS determinations [1286] [2310]. MCDONALD [2653] found that by using an internal standard, this technique could be employed for samples of widely varying composition and that good results were obtained.

LOFTIN *et al* [760] conducted air over graphite rods which were brought to incandescence in a high frequency field and then determined lead by atomic absorption in an attached heated quartz tube. MORRISON and TALMI [881] used a similar device for the direct analysis of solids. MARINKOVIC and VICKERS [822] investigated the possibility of atomization in a direct current arc and found that for difficultly atomized elements, such as boron and tungsten, the sensitivity was better than in a flame.

KANTOR *et al* [2456] vaporized solid samples in an electric arc and conducted the aerosol into a flame. They found an improvement in sensitivity of one to two orders of magnitude compared to the conventional flame technique.

LANGMYHR and THOMASSEN [731] heated a graphite tube by means of a high frequency induction generator and determined rubidium and cesium in standard silicate rocks directly from the solid sample. The temperature of 2200°C which could be achieved was relatively low and the rate of heating relatively slow, but in this case it appeared to offer certain advantages since the heat transfer time to the sample is small compared to the total heat-up time.

ROUSSELET *et al* [1062] investigated atomization by means of electron bombardment. In theory, this method offers two advantages; very high temperatures can be achieved (higher than 3400°C, the melting point of tungsten) and a partial pressure of electrons advantageous for the production of atoms in the ground state can be obtained. The sample, in the range 10–100 nL, is placed on a tungsten target mounted under the optical beam path and is heated under high vacuum by bombardment with accelerated electrons produced in a strong electric field (2 kV/cm). The analytical advantage is the use of very small sample amounts combined with a very good absolute sensitivity. In 1967, MOSSOTTI et al [886] reported the use of lasers for atomization, an idea that was later used by VULFSON et al [1278]. The main advantage with this procedure is likewise the small sample requirement (10–100 µg) and the excellent absolute detection limits (10^{-11} g for copper and silver, and 10^{-10} g for manganese). VULFSON analyzed various geological samples for trace elements by this procedure.

DONEGA and BURGESS [309] used an electrically heated boat of tantalum or tungsten foil in a closed chamber under reduced pressure to determine many elements with detection limits down to 10^{-12} g. This system is relatively simple, has a low current consumption, and has the advantage that it requires less than 0.1 second to heat up to 2200°C (so that the maximum atom cloud concentration can be formed).

This tantalum boat principle was later taken up by HWANG et al and introduced in a slightly modified form as a commercially available accessory. HWANG obtained detection limits of 10^{-13} g, but had problems with difficultly atomized elements because the maximum achievable temperature was under 2400°C. The use of hydrogen as purge gas was said to increase the lifetime of the tantalum boat and at the same time remove a number of interferences. Apart from the analysis of simple aqueous solutions and organic extracts, this system has also been used for the direct analysis of lead in diluted blood [551].

However, various authors have reported a strong dependence of the signal height upon the anion present [334] or upon the acid in whose presence the analysis takes place [827]. These interferences, in a similar manner to the carbon rod of WEST, are probably due to the large temperature gradient between the heated boat and the surrounding atmosphere.

NIXON et al [917] combined vaporization of a sample from a tantalum boat with plasma emission by conducting the vaporization products directly into the plasma. The detection limits were in the range 1 µg/L and less.

An interesting procedure for the direct atomization of solids—usually ores or rocks—was proposed by VENGHIATTIS [1266]. In this, the finely ground sample is thoroughly mixed with a gunpowder and the mixture is pressed into a pellet which is then ignited in the radiation beam of an atomic absorption spectrometer. The flame, so produced, has a similar temperature to an air/acetylene flame, so that about the same number of elements can be determined by this technique as in an air/acetylene flame. Since the dependence of this method on the matrix is greater than for conventional atomic absorption spectrometry, determinations must usually be carried out against preanalyzed standard rocks or ores, or the analyte addition technique must be employed. The achievable precision is then about 5–10%.

4. Optics

The spectral range of interest for atomic absorption spectrometry begins in the near infrared at 852.1 nm, the wavelength of cesium, and reaches down into the vacuum UV below 200 nm. At the present time, the instrumental limit for a non-flushed instrument and using a flame is 193.7 nm, the wavelength of arsenic. Atomic absorption spectrometry therefore covers much the same wavelength range that is of interest for atomic emission spectrometry or UV/VIS spectrometry. In principle it should then be possible to employ proven monochromators in atomic absorption. However, it has been shown that the requirements in AAS with respect to resolution and dispersion of the monochromator are different from those of other techniques.

4.1. Spectral Bandpass

As discussed in chapter 1, atoms can only absorb definite quantities of energy, i.e. they only absorb within a very narrow spectral range (refer to section 1.2, Atomic Spectra). One of the greatest advantages of atomic absorption spectrometry, namely its specificity, is based on the use of element-specific radiation sources that emit the spectrum of the analyte element in the form of very narrow spectral lines. While the quality of an instrument in other spectrometric techniques frequently depends on the resolution of the monochromator or on its spectral bandpass, i.e. the range of radiation that passes through the exit slit, these factors are not of primary importance in AAS.

If an alternating current instrument and an element specific radiation source emitting spectral lines are used, AAS is selective and virtually free of the spectral interferences caused by overlapping of atomic lines of different elements. The radiation source emits the spectrum of one (or several) elements and the unmodulated emission of the flame or graphite furnace is eliminated by means of selective amplification. The ability of atomic absorption spectrometry to differentiate between two elements is solely dependent on the half-intensity widths of the emission lines (~ 0.001–0.002 nm) emitted from the radiation source and of the absorption lines (~ 0.002–0.005 nm). In other words, values that lie on or beyond the limits of the resolving power of normal monochromators. The monochromator in an atomic absorption spectrometer has the sole task of separating the resonance line of the analyte element from other emission lines of the source. Experience has shown that this can be achieved for practically all elements with a bandpass of 0.2 nm. In principle, no advantages are gained if smaller slit widths than are required to separate other emission lines from the spectral source are employed. If larger slit widths are employed, AAS does not loose any of its selectivity or specificity, except in the case when multielement lamps are used and the resonance lines of two elements fall simultaneously on the detector. The disadvantages brought by slit widths that are too large in atomic absorption are lower sensitivity and an increasing non-linearity of the analytical curve. If a further non-absorbable emission line (see page 21) besides the resonance line passes through the exit slit and falls on the detector, this always "sees" the radiation of both lines. With increasing absorption of the radiation from the resonance line, the intensity of the second line remains unaltered so that the absorption curve does not approach 100% absorption (0% transmission) asymptotically as usual, but approaches a value corresponding to the percentage proportion of the second line. DE GALAN and SAMAEY [286] have shown that the causes for non-linearity of analytical curves are mostly trivial. Non-resolved multiplets (several resonance lines passing through the exit slit) and non-absorbable lines (apart from the resonance line, a non-absorbable line passes through the exit slit) were recognized as two main causes and this was verified by calculation.

The following figures illustrate this behavior and show the influence of the spectral bandpass on the sensitivity, the signal-to-noise ratio and the curvature of the analytical curve. Figure 51 shows the six resonance lines of silicon in the range 250 nm to 253 nm and their sensitivities in atomic absorption spectrometry. With a sufficiently small slit width every resonance line gives a linear analytical curve within the observed range.

In figure 52, the influence of the spectral bandpass of the monochromator on the analytical curve and on the signal-to-noise ratio for the 251.6 nm resonance line of silicon is shown. With decreasing spectral bandpass, the sensitivity clearly increases and the linearity of the analytical curve improves. If the optimum slit width is reached (i. e. the spectral bandpass at

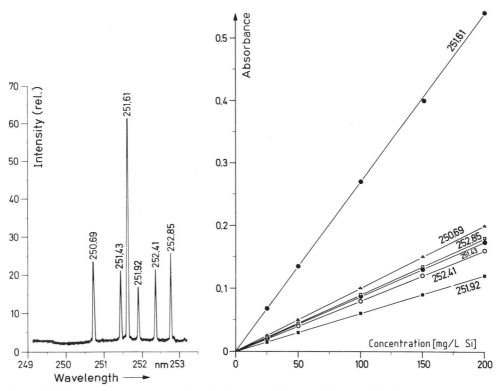

Figure 51. Various silicon resonance lines in the range 250 nm to 253 nm.—a) Emission spectrum of a Si hollow cathode lamp recorded at a spectral bandpass of 0.07 nm. b) Analytical curves of the six resonance lines measured at a spectral bandpass of 0.07 nm.

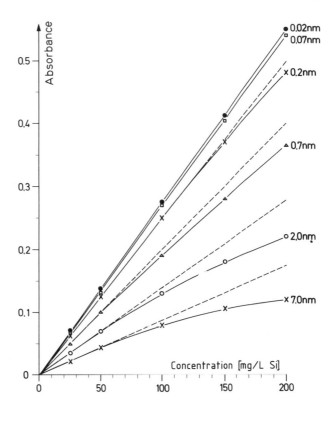

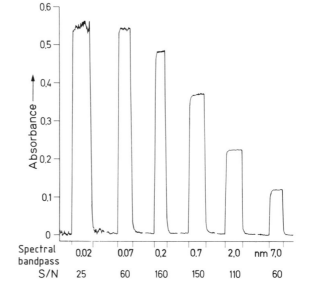

Figure 52. Influence of various spectral bandpasses on the determination of Si at the 251.6 nm resonance line.—a) The sensitivity decreases noticeably at bandpasses greater than 0.2 nm and the curvature of the analytical curve increases. b) The signal-to-noise ration (S/N) is optimum at a spectral bandpass of 0.2 nm and clearly decreases on both sides. (Reference solution 200 mg/L Si; nitrous oxide/acetylene flame.)

which only the resonance line reaches the detector)—in this case 0.2 nm—a further decrease of the slit width brings no noticeable advantages. As shown in figure 52 a, the linearity of the analytical curve and the sensitivity can nevertheless still be improved somewhat since in practice there is still a small amount of residual scattered radiation present which can be further eliminated by reducing the slit width.

Nevertheless, this slight improvement in the linearity is only obtained at the expense of an increase in the noise and a noticeably poorer signal-to-noise ratio, as clearly shown in figure 52 b. This effect will be further discussed in more detail.

In the case of silicon, the effect is caused by the influence of a non-resolved multiplet when spectral bandpasses greater than 0.2 nm are used. The non-linearity of the analytical curve is due to the various absorption coefficients $\varkappa_v$ of the individual resonance lines (see equation (1.18) page 13). This means that the absorption coefficient in equation (1.15) (page 11) is not constant over the integrated range if several resonance lines pass through the exit slit and fall on the detector.

To illustrate this further, the effect of the spectral bandpass on the analytical curve of antimony is shown in figure 53. Besides the resonance line of interest at 217.6 nm, an antimony hollow cathode lamp also emits an arc line at 217.9 nm and a spark line at 217.0 nm. Neither of these two lines is measurably absorbed by higher antimony concentrations, so that the radiation from them can be regarded as "non-absorbable".

It can be seen from figure 53 b that with smaller bandpasses of up to 0.2 nm, antimony behaves in a similar manner to silicon, i. e. decreasing the slit width brings no further advantage once the resonance line has been separated. With an increase in the spectral bandpass, there is a sudden decrease in the sensitivity and a strong increase in the curvature as soon as non-absorbable radiation falls on the detector. This is not surprising because the degree of the curvature is determined by the difference between the individual absorption coefficients. If two lines have exactly the same absorption coefficient (no such case is known), no curvature and no loss of sensitivity should take place. The strongest curvature quite clearly takes place when the second line is completely non-absorbable. In this case, the analytical curve will approach a value corresponding to the percentage transmission of the non-absorbable line asymptotically. DE GALAN and SAMAEY [286] calculated this value to be 55% transmission (absorbance 0.254) for antimony and also found this "limiting absorbance" experimentally for a 1000 mg/L Sb solution.

While it can be seen that for a number of elements a spectral bandpass of 0.2 nm is necessary to obtain good sensitivity and linearity of the analytical curve, for other elements the resonance line is more or less isolated so that larger slit widths can be used without disadvantage. It is obvious that in such cases the widest slit width at which the resonance line is just separated from other lines should be used since the signal-to-noise ratio is favorably influenced, as shown in figure 52 b.

If aberration and diffraction are negligible for a fairly wide slit, the spectral profile of the radiation passed by the monochromator will have the form of an isosceles triangle. The width at half height is termed the spectral bandpass. For wider slits, the calculated slit width and the measured spectral bandpass are equal. As the slit becomes progressively narrower, the relationship between slit width and spectral bandpass changes; the minimum spectral bandpass is limited by aberration.

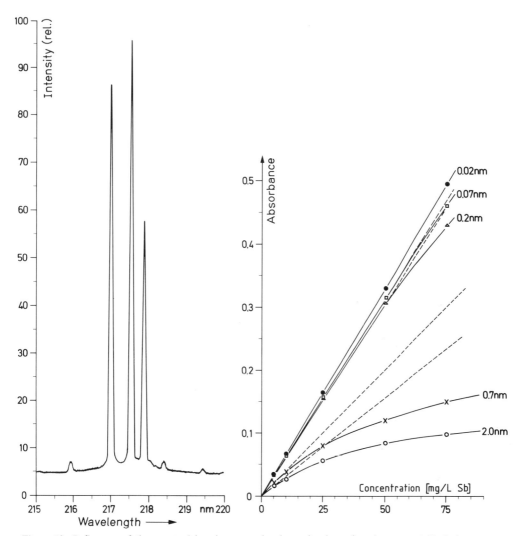

Figure 53. Influence of the spectral bandpass on the determination of antimony.—a) Emission spectrum of an Sb hollow cathode lamp recorded at a spectral bandpass of 0.07 nm. Next to the resonance line at 217.6 nm there is an arc line at 217.9 nm and a spark line at 217.0 nm. b) At increasing slit widths the sensitivity decreases dramatically as soon as non-absorbable lines pass through the exit slit.

4.2. Reciprocal Linear Dispersion

The size of the spectral bandpass should be so selected that only the resonance line of the analyte element falls on the detector and all other emission lines from the spectral radiation source are held back by the slit foils.

A further important dimension is the geometric slit width. This gives the effective mechanical widths of the entrance and exit slits in mm or μm for a given spectral bandpass. The geometric slit width determines the portion of the radiation emitted from the radiation source that really enters the monochromator and reaches the detector.

In an atomic absorption spectrometer, the image of the radiation source is formed on the entrance slit, i.e. a radiation beam of several mm diameter falls on the slit. It can be clearly seen from figure 54 that the geometric width of the entrance slit determines the amount of radiation that falls on the dispersing element and subsequently on the detector. With a wide entrance slit, a relatively large amount of radiant energy therefore falls on the detector; this means that the noise always present in the signal (the shot noise of the photon current for example) is relatively small compared to the signal. At the same time, lower amplification can be employed so that possible contributions of the electronics to the noise are reduced. For the analyst, low noise means a stable signal and hence good precision and low detection limits.

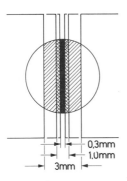

0,3mm
1,0mm
3mm

Figure 54. The image of the radiation source is formed on the entrance slit. Its geometric width determines the amount of radiation that falls on the dispersing element.

Two requirements have been placed on the monochromator of an atomic absorption spectrometer that are not directly in harmony with each other. Firstly, a spectral bandpass of 0.2 nm is required, and secondly the entrance slit should have as wide a geometric width as possible. However, the entrance and exit slits of a monochromator must have the same (or at least very similar) mechanical dimensions.

Spectral bandpass $\Delta\lambda_m$ and geometric slit width s of a monochromator are associated via the reciprocal linear dispersion $d\lambda/dx$, as shown in equation (4.1).

$$\Delta\lambda_m = s \cdot \frac{d\lambda}{dx}. \tag{4.1}$$

Reciprocal linear dispersion is expressed in nm/mm (often also shortened, incorrectly, to dispersion). A reciprocal linear dispersion of 2 nm/mm means that at a geometric slit width of 1 mm the spectral bandpass is 2 nm, or a geometric slit width of 0.1 mm is necessary to obtain the desired spectral bandpass of 0.2 nm.

From this consideration it is thus desirable to have as small a reciprocal linear dispersion as possible, i.e. to use a strongly dispersive element, because a relatively large amount of radiation then passes through the monochromator, even with small spectral bandpasses. On the other hand, because of the significance of the geometric slit width, the largest possible slit width just meeting the requirement for the isolation of the resonance line is always selected.

4.3. Prisms and Gratings

Prisms and gratings serve to disperse the radiation into individual wavelengths. A short examination will be made to determine which best meets the requirements of atomic absorption spectrometry.

The reciprocal linear dispersion of a prism is fixed by the dispersion of the prism material, i.e. by the wavelength dependence $dn/d\lambda$ of the refractive index. With a prism monochromator therefore, the reciprocal linear dispersion and thus the spectral bandpass at a fixed geometric slit width are wavelength dependent. In the first approximation, these dimensions become more unfavorable toward the long wavelength end exponentially (figure 55). In practice, this means that a variable slit system must be employed in which the geometric slit width is decreased with increasing wavelength so that a constant spectral bandpass is guaranteed. At the same time there is a continuous worsening of the transmission factor of the monochromator with increasing wavelength.

The spectral bandpass and the reciprocal linear dispersion of a grating monochromator are dependent on the grating constants, and these improve as the interval between two grating rulings becomes smaller, or as the number of lines per mm increases. Next to this, the order of the spectrum naturally plays a part, but in atomic absorption spectrometry this is not of practical interest since measurements are almost exclusively carried out in the first order.

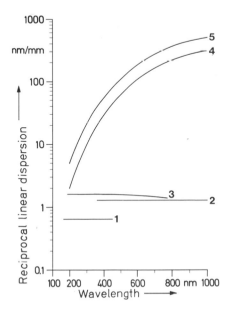

Figure 55. Reciprocal linear dispersion depending upon the wavelength in grating (**1–3**) and prism (**4–5**) monochromators.— **1** PERKIN-ELMER Models 4000 and 5000 in UV range (0.65 nm/mm), **2** PERKIN-ELMER Models 4000 and 5000 in VIS range (1.3 nm/mm), **3** PERKIN-ELMER Model 3030 (1.7 nm/mm), **4** and **5** Quartz prism monochromators.

Both dimensions are virtually independent of the wavelength, i.e. at a fixed geometric slit width, a grating monochromator has practically the same spectral bandpass and the same reciprocal linear dispersion over the entire wavelength range (figure 55). For atomic absorption spectrometry, which covers a range approaching 700 nm and for which the accuracy of the analyses as well as the detection limits is dependent upon the reciprocal linear dispersion of the monochromator, the grating monochromator offers considerable advantages over the prism. Furthermore, a good grating nowadays has a ruling density of 2000 to 3000 lines per mm (with a total of around 10^5 rulings), so that with the usual focal lengths employed a reciprocal linear dispersion of 1 nm/mm and less can easily be obtained. As clearly shown in figure 55, comparable reciprocal linear dispersions are achieved with normal prism monochromators only in the far UV region, while they are already several orders of magnitude poorer in the near UV.

Ruled gratings are used mainly in AAS. Holographic gratings exhibit far fewer irregularities and therefore lower stray radiation, but this is of secondary importance in AAS.

Radiation falling onto a grating is reflected and dispersed in a wavelength dependent arc to both the left and the right of the incident beam. Blazed gratings have a highly unsymmetrical, "saw tooth" profile that concentrates the radiation in one direction. This enhancement is only effective for a given grating angle, i.e. it is only optimum at a given wavelength setting. This is termed the blaze wavelength. The efficiency of a grating decreases the further the selected wavelength is from the blaze wavelength. The spectral efficiency profile thus exhibits a maximum at the blaze wavelength with a steep falloff toward shorter wavelengths and a more gradual decline toward longer wavelengths. Because of the great importance of the UV range, a blaze wavelength of relatively short wavelength is usually chosen, at the cost of a marked reduction in energy at the long wavelength end. Very efficient monochromators can be obtained by using two gratings in sequence. One grating can be blazed at a wavelength well into the UV (e.g. 210 nm) while the other can have a maximum in the visible range at, say, 500 nm. Gratings that are blazed at two wavelengths and thus have two maxima exhibit similar efficiency.

A number of technical solutions are known for the actual construction of the monochromator; the two most important, the simpler LITTROW monochromator and the more complicated EBERT, here in the modified CZERNY-TURNER form, are depicted in figure 56. Both monochromator types are used successfully in atomic absorption spectrometry.

Figure 56. Two frequently employed monochromator types.—**A** LITTROW monochromator and **B** CZERNY-TURNER monochromator.

In this connection, it should be briefly mentioned that filter monochromators (except perhaps for a few individual elements in the visible range) cannot be employed in atomic absorption spectrometry since the required resolution cannot be obtained, leading to the difficulties described in section 4.1.

4.4. Resonance Detectors

SULLIVAN and WALSH reported the use of resonance detectors as "monochromators" in atomic absorption spectrometers in various publications [1194] [1195] [1198] [1285].

Figure 57. Resonance detector.—After passing through the absorption zone, the spectral radiation emitted by the primary source falls on the atom cloud formed in the resonance detector. These atoms absorb radiation of the resonance line and re-emit it in the form of fluorescence radiation. Measurement takes place at right angles to the optical beam.

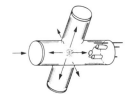

The mode of operation of these resonance detectors is depicted schematically in figure 57. After passing through the absorption zone, the spectral radiation emitted from the primary source falls on the resonance detector. The detector consists of a device for producing a cloud of neutral atoms in the ground state (similar to a hollow cathode lamp). Primary source and resonance detector must then be manufactured from the same metal. If the radiation which has passed through the absorption zone now falls on a cloud of atoms of the element to be determined, the resonance lines are absorbed while the remaining (non-absorbable) radiation passes through the atom cloud. The absorbed resonance radiation excites the metal atoms to fluorescence and this can then be measured by a detector at right angles to the direction of the incident radiation. The fluorescence signal is proportional to the intensity of the resonance radiation. Since the spectrum of the fluorescence radiation only consists of spectral lines that were absorbed by the cloud of atoms, resonance detectors can take over the function of the monochromator, which in atomic absorption spectrometry isolates the line to be measured.

Because of the absence of non-absorbable background radiation, it was expected that the analytical curves obtained with this type of "resonance monochromator" would be linear over a wider range than those obtained when the radiation of the primary source falls directly on the monochromator. This expectation was verified experimentally. A disadvantage of the resonance detector is that only a small portion of the fluorescence radiation emitted in all directions falls on the detector.

The absorption of resonance radiation by a cloud of neutral atoms can also be exploited in another manner to isolate resonance lines through so-called "selective modulation" [149] [193] [758] [1198]. Here, direct radiation from a hollow cathode lamp is passed through a discharge tube whose open cathode cylinder is made of the same metal as the cathode of the radiation source. The power supply to this discharge tube is modulated so that a pulsing atom cloud, which absorbs the resonance lines from the hollow cathode lamp at the modulation frequency, is formed. In this way, a beam consisting of modulated resonance lines and non-modulated non-resonance lines is produced. Only the resonance lines are amplified by a selective amplifier after the detector, while ion and carrier gas lines are eliminated.

Nevertheless, for this procedure a monochromator must be used since too much direct radiation would otherwise fall on the detector, leading to excessive noise. A simple monochromator without special requirements for the resolving power is sufficient; frequently filters can be employed. With this procedure of selective modulation, analytical curves which are linear over a wide range are likewise obtained.

SEBESTYEN [1105] showed that it is possible to obtain a similar effect in a normal atomic absorption spectrometer and with a normal hollow cathode lamp if short pulses of very high current strength are periodically superimposed on the operating direct current of the hollow cathode lamp. Thereby an intense atom cloud is periodically formed in front of the cathode which briefly absorbs the resonance radiation and thus modulates it. With suitable demodulation electronics tuned to this modulation procedure, analytical curves, which were linear into high absorbance ranges, could be obtained for iron and nickel, for example.

4.5. Multielement Instruments

It can be considered as one of the major disadvantages of atomic absorption spectrometry that it is a single element technique; simultaneous multielement determinations present not inconsiderable difficulties. Even rapid sequential multielement determinations are not easy, because at each change of element there are a relatively large number of parameters that must also be changed (refer to section 5.5).

Nevertheless, attempts were made early on in the development of AAS to perform multielement determinations and increase the sample throughput. MAVRODINEANU [884] described a multichannel instrument employing several hollow cathode lamps and several detectors. The use of resonance detectors has been reported in particular by WALSH et al (refer to figure 58) and operable instruments based on this principle have been built [1196] [1197] [1284]. MITCHELL et al [868] used a Vidicon detector in a multichannel atomic absorption spectrometer and were able to determine ten or more elements simultaneously. SILVESTER et al [1125] conducted the radiation from two multielement hollow cathode lamps through the flame or graphite furnace onto a three-way beam splitter. Each of the three beams was then directed to a monochromator with two exit slits and two detectors. Six elements could thus be determined. RAWSON [1027] also developed an instrument for determining six elements simultaneously. He conducted the radiation from four hollow cathode lamps via fiber optic bundles and a quartz radiation integrator to a monochromator with six exit slits. From each exit slit the radiation was again conducted by fiber optic bundles to six detectors. The major disadvantage of this system was the high radiation loss in the fiber optic bundles. BUSCH and MORRISON [188] have published a good review on multielement flame spectrometry.

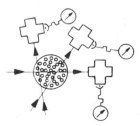

Figure 58. Multielement instrument according to SULLIVAN and WALSH [1198]. — By the use of several hollow cathode lamps which irradiate the same flame and whose radiation is captured by resonance detectors, various elements can be simultaneously determined in one sample.

It has already been mentioned in section 2.5 that continuum radiation sources are not particularly suitable for use in AAS. One of the reasons is the low radiant intensity within the narrow absorption profile of an atomic line. Another more important reason is the reintroduction of spectral interferences through overlapping atomic lines of different elements if the resolution is insufficient. It would appear that both of these problems have been solved in a simultaneous multielement instrument described by ZANDER *et al* [3095] [3096]. They used a very high intensity xenon arc lamp, an Echelle monochromator of high resolution, and wavelength modulation by means of an oscillating quartz plate. One advantage of this system is that the resolution of the Echelle monochromator is in the range of an atomic line [2464], so that spectral overlapping is hardly to be expected. At the same time, non-absorbable radiation is effectively masked, so that the attainable sensitivity is comparable with that of normal AAS.

An especially interesting procedure for simultaneous multielement determinations has been proposed by SCHARMANN and WIRZ [2823]. They utilized coherent forward scattering in a high magnetic field to separate resonance lines from the total radiation of a xenon arc lamp. A polarizer is mounted at each end of a graphite furnace with the planes of polarization at right angles, so that in principle no radiation can pass through. In the magnetic field around the furnace the energy levels of the atoms are split and thus polarized, so that a portion of the radiation at the resonance wavelength passes through the second polarizer. A normal monochromator is sufficient to disperse the resonance lines that are passed, and these can be directed onto a Vidicon detector. Simultaneous multielement determinations would be particularly expedient for the graphite furnace technique, since it is one of the most sensitive techniques in elemental analysis and permits more than 60 elements to be determined largely free of interferences. The fact that only one element per atomization cycle can be determined makes the technique relatively slow and thus less attractive for some types of application.

5. Analytical Measure and Readout

The demands made on the electrical measuring system and the readout in atomic absorption spectrometry often differ according to the nature of the application. If possible, an atomic absorption spectrometer should be able to meet all situations.

Three requirements are frequently demanded nowadays: Speed, high precision, and the lowest detection limits. The last two points can be combined under the requirement that the electrical measuring system must be capable of recognizing and displaying the smallest differences in the absorption. However, this is only possible and meaningful if the signals are steady and stable; the stability of the signals is primarily dependent upon the stability of the radiation source and the performance of the optics. This means that all components of an atomic absorption spectrometer must be well designed and matched to each other to obtain optimum performance. The radiation source must illuminate stably and supply maximum radiant energy. A flame must not interfere in the production of free atoms through unnecessary turbulence or irregularities, and the constancy of atom production and transport from the absorption volume is also of considerable significance for other types of atomizer. The monochromator must have a small reciprocal linear dispersion so that a high radiant flux reaches the detector. As previously explained, the relative noise of the signal will then be small. Additionally, lower voltages for the photomultiplier (i.e. less gain) can be used. A stable electrical signal is a prerequirement for speed, precision and good detection limits.

A stable signal can often be considerably expanded—atomic absorption spectrometers often allow an expansion of up to a hundred times—so that the readout accuracy can be improved many times over. In modern instruments it is possible to measure differences of 0.0001 absorbance units. The ability to expand the smallest absorbance or concentration readings permits the best detection limits to be attained. In addition, determinations of very high precision can be performed on samples containing high metal concentrations when the atomic absorption spectrometer allows segments of the reading range to be expanded (through zero, or baseline suppression). When high scale expansion is used, it is naturally necessary to dampen the noise (i.e. the fluctuation of the signal) by a corresponding increase in the response time. The lowest detection limits and the highest precision can only be achieved at the cost of speed.

The possibility of working with very short response times is for many applications equally as important as the ability to expand and dampen signals to clarify the smallest differences in absorption to obtain good detection limits and high precision. A disadvantage of high expansion and damping is the high substance requirements caused by the increased reading time. This can be of crucial importance with a conventional atomic absorption spectrometer with nebulizer and flame when only small sample amounts are available. Lower electronic damping in this case means lower sample consumption. With fast responding instruments, it is possible to carry out precise measurements with less than 0.1 mL sample solution for example, as shown in figure 59.

Under given conditions, even smaller volumes can be used [226]; BERNDT and JACKWERTH [2076] [2077] [2083] in particular have made a thorough examination of the injection method proposed by SEBASTIANI et al [2835] in which small, measured sample volumes are injected into the flame (refer to section 3.1.3).

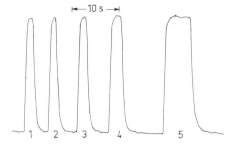

Figure 59. Aspiration times of only 1 second give sufficiently accurate signals with instruments having fast response times. The sample requirement is still below 0.1 mL per measurement. Aspiration times: Signals **1–3**, 1 second each; signal **4**, 2 seconds and signal **5**, 5 seconds.

Determinations with the graphite furnace technique, which has become a technique of major importance in recent years, can only be performed in spectrometers having very short response times. The signals generated by this technique often reach a maximum in less than a second and then return to the baseline within a second or two. This technique therefore requires instruments with very rapid response times having minimum time constants. This aspect will be treated in a later section.

In the following sections a number of important components of the electrical measuring system and the readout will be described with especial mention of their significance to the analyst.

5.1. Detectors

Photomultipliers are mainly used in atomic absorption spectrometry to convert optical radiation into an electrical signal. They consist of a vacuum photocell with an anode, a radiation sensitive electrode (photocathode), and a number of emission cathodes (dynodes) which have increasing positive potential with respect to the photocathode. Photomultipliers with a total of 10 electrodes are frequently used; in special cases the number of electrodes can be increased to 13.

A photoelectron released from the photocathode is attracted to the first dynode and falls upon it with a kinetic energy that is proportional to the voltage gradient. It then releases a number of secondary electrons which are accelerated and for their part release an even greater number of electrons, and so on, so that the effect is still further amplified. The amplification of a photomultiplier is dependent upon the applied high voltage. This can have values up to 1000 V to 1500 V.

The spectral range within which a photomultiplier can be used is dependent upon the radiation sensitive layer on the cathode and the window material of the tube. It is not easy to find a photomultiplier that has sufficient sensitivity over the entire spectral range of interest for atomic absorption spectrometry. With good photomultipliers radiant fluxes of 10^{-6} to 10^{-11} lm can be measured. The sensitivity is around 10–100 A/lm and the maximum current about 10 μA. If, at a given amplification, more radiation falls on the photomultiplier so that the 10 μA current is exceeded, the signal decreases rapidly. Reversible and irreversible alterations to the dynodes can then take place.

The dark current, which is the current flowing through the tube under the influence of the high voltage when no optical radiation falls upon the photocathode, is an important criterion for the quality of a photomultiplier. The so-called dark noise, the fluctuation of the dark current, can under circumstances be an important component of the detector noise; it increases with increasing voltage. Therefore, to obtain good separation of the signal from the dark noise, the highest possible radiant intensity on the detector and the choice of a suitable photomultiplier are important.

The quantum efficiency of the photocathode (spectral cathode sensitivity) is of special importance for a photomultiplier since in effect it gives how many photons are required to release an electron from the photocathode. A low quantum efficiency leads to considerable energy losses for the conversion of the photon current into an electrical current and thereby to an increase in the (shot) noise, which cannot be eliminated by higher amplification.

Other detector types have found little use in AAS. The resonance detectors proposed by SULLIVAN and WALSH [1194] [1195] [1198] [1285] have already been described in detail in section 4.4. For multielement instruments, Vidicon detectors [28] [868] [2025] [2239] [2823] have been used in addition to the classical spectrometer principle employing several exit slits with a photomultiplier for each slit [1027] [2015] [2574]. The greatest disadvantage of Vidicon detectors is their low responsivity in the far UV range below 250 nm, a range in which a number of important resonance lines are to be found.

The use of photodiode arrays [2164] may find interesting applications, particularly in connection with multielement instruments. Nevertheless, there is a great deal of basic work still to be done in this direction.

5.2. Noise

Reasons for the occurrence of increased noise have already been mentioned. Since noise, via the signal-to-noise ratio, has a direct influence on the achievable precision and on the determination or detection limit, further causes for the occurrence of noise will be referred to briefly. ALKEMADE *et al* [2029] [2097] have published comprehensive papers covering this topic.

The dark current was mentioned above. A second source of noise in the detector is the photon noise, caused by the statistical fluctuation of the photocurrent. Added to this, there is the lamp noise caused by fluctuations of the radiant intensity of the radiation source.

A substantial contribution to the noise can be made by a flame when it is the atomizer. The main cause is a fluctuation in the transmission of the radiation brought about by absorption in the flame or by radiation scattering. Furthermore, fluctuations in the absorption of the analyte element due to unsteady production and transport of the atoms are possible.

Emission noise caused by emission of the atomizer (flame, graphite tube) or of the constituents in the sample should also be mentioned. Although the emission from the atomizer is compensated in an alternating current system, it can nevertheless lead to increased photon noise when too much radiation falls on the photomultiplier.

The actual contribution to the noise made by the electrical measuring system should be so low in a well-designed instrument that its effect is negligible when compared to other sources of noise.

Fluctuations in the transmission of the flame are the dominant factor near to the detection limit [2098] [2545], although fluctuations in the emission of the radiation source and photon noise must also be taken into consideration [2545]. At high absorbance readings, on the other hand, only fluctuations in the absorption characteristics of the analyte element are of importance [2098] [2099]; this applies equally to single-beam and double-beam instruments [2546].

Fluctuations in the transmission of the radiation in graphite furnaces and quartz cells are considerably lower, so that noise in the region of the detection limit is much lower. The background absorption often observed in graphite furnaces leads to increased noise, however [2407]. When a continuum source is used for background correction, an increase in the noise by a factor of two to three must be taken into account.

In earlier years, emission noise was frequently observed when using weak radiation sources. Depending on the cause, increased noise could be observed after the flame had been ignited or when the sample was aspirated [176]. The effect was particularly pronounced when graphite furnaces were used with older atomic absorption spectrometers not designed for this technique [638]. Strong noise, or even a clearly marked shift of the baseline, caused by the direct radiation emission from the incandescent walls of the graphite tube, could be observed when the sample was atomized (refer to figure 60).

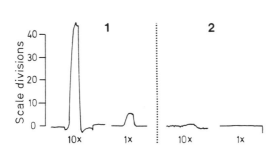

Figure 60. Influence of the direct radiation emission from the graphite tube on the baseline during atomization. — **1** Atomization at 2200°C at the wavelength of calcium (422.7 nm) produces a considerable offset of the baseline in an older instrument. **2** In an instrument whose optics and electronics have been matched to the requirements of the graphite furnace technique, the direct radiation emission plays no further part (422.7 nm, 2500°C) [638].

By means of a selective amplifier in circuit after the photomultiplier, element-specific signals are obtained in atomic absorption spectrometry since the amplifier only amplifies the signals having a determined modulation frequency. If a direct current amplifier was used instead, the emission produced within the same spectral range by other elements in the sample or by the flame would also be amplified and give rise to a signal. Especially intense signals would be obtained from broad emission bands, such as those emitted from glowing graphite, or also by molecular bands in the flame.

The selective amplifier eliminates this continuous emission, it "overlooks" the non-modulated components. Nevertheless, this radiation falls on the detector just the same as the modulated radiation and produces a photocurrent. The current leaving the photomultiplier therefore consists of a modulated component, which can be assigned to the radiation source, and a non-modulated component, which comes from the emission of either the flame or the graphite furnace. Since the high voltage applied to the photomultiplier is set according to the intensity of the modulated component (i.e. the lamp radiation), overloading of the photomultiplier can occur if the sum of both currents exceeds the maximum current, especially when the lamp intensity is somewhat weak. This overloading causes a loss in sensitivity of the photomultiplier, which is observed as a decrease in the signal. However, even when the detector is not overloaded, a high direct radiation increment causes fluctuations of the signal, leading to poorer detection limits and lower precision of the results.

In principle, various possibilities exist for the elimination or reduction of this emission noise. These are based either on a reduction of the direct radiation increment or on an increase in the modulated radiation increment. The increase in the modulated radiation increment can be most reliably achieved by using a more intense radiation source; occasionally an increase in the lamp current, or more seldom a decrease, can be of help. The reduction of the direct radiation increment, as far as it emanates from the flame, can be most reliably achieved by choosing another flame when this is possible; occasionally an alteration of the flame stoichiometry or the use of a smaller slit width can bring an improvement.

In the case of graphite furnaces, the direct radiation emission from the glowing graphite is typically several orders of magnitude more intense than the emission from a flame. For this reason the interference is more frequently noticed. In general, the same principles can be applied to eliminate the interference, namely the use of a more intense radiation source (e.g. a single element lamp instead of a multielement lamp, or an electrodeless discharge lamp instead of a hollow cathode lamp), or a reduction of the direct radiation increment.

The direct radiation emission from the graphite tube can be very simply and effectively removed by reducing the atomization temperature. The majority of elements can be quantitatively atomized with maximum sensitivity at temperatures between 2000 and 2500°C, or even lower when maximum power heating is used; a higher atomization temperature often brings only a small increase in the sensitivity but at the same time increases the emission noise considerably, so that as a whole, a clear deterioration of the signal-to-noise ratio is observed.

In contrast to the flame, the emission source in graphite furnaces is clearly optically localized; it is in the form of a ring around the cloud of atoms. Thus the possibility exists for largely masking this emission optically. The improvements that can be obtained with a modified optical system are shown in figure 60. In atomic absorption spectrometers of recent manufacture, the corresponding optical and electronic measures, such as the introduction of stops and diaphragms, reduction of the slit height, use of a suitable photomultiplier, etc. are normally taken into consideration.

At temperatures above 2500°C, components vaporizing from the graphite surface (e.g. C_2 radicals), or molecules or radicals formed with the purge gas (e.g. CN when nitrogen is used) begin to emit with increasing intensity. This direct radiation emission naturally cannot be masked—a further reason for avoiding atomization temperatures that are too high or nitrogen as the purge gas.

5.3. Analytical Measure

An atomic absorption spectrometer measures the attenuation of the radiant energy, or expressed more precisely, compares the transmitted flux Φ_{tr} after introduction of the sample with the incident flux Φ_0. The actual measure is thus the absorptance α or the transmittance τ. According to the BEER-LAMBERT law, the concentration is proportional to the logarithm of the reciprocal transmittance, i.e. the absorbance (refer to equation (1.18), page 13). Since it is generally much easier to work with linear functions rather than with logarithmic ones, the absorption signal must be converted to absorbance. This is performed automatically in all modern atomic absorption spectrometers, so that the readout is linear in absorbance.

The method in use for a long period was simple measurement of the voltage on a suitably calibrated voltmeter, either after the signal had been passed through a demodulator (single-beam instruments) or the ratio between the sample beam and the reference beam had been electronically formed (double-beam instruments). With this type of signal handling, all fluctuations are displayed on the meter in strengths dependent upon the selected time constant. After jumpwise alterations, such as occur when the sample is introduced, the meter only comes to the "middle" measurement reading asymptotically after several time constants have elapsed.

More recently, especially in connection with digital displays, true integration of the signals as a function of time has been increasingly applied. This means that a "clock" in the instrument must exactly define a precise integration time, since with time-independent changes of the absorbance the result is critically dependent upon the length of the integration period. In practice, the reproducibility of the integration time is very much more important than the absolute length. This problem is insignificant compared to the enormous advantage that the measured result is displayed steadily to within a known accuracy at the end of the selected integration period. It is neither necessary to observe the display to see whether a sufficiently accurate asymptotic limiting value has been reached, nor to estimate the mean value of a fluctuating display. The operator is largely relieved of the decision as to when and what should be read.

The accuracy of the analytical results is dependent upon the integration time in that it is inversely proportional to the square root of the time. Usually, the integrated signal is divided by the time unit at the end of the measurement so that the signal height is independent of the measurement period, and longer integration periods are expressed solely in a lower relative standard deviation. The integrated signal can also be displayed directly, so that longer integration times lead to correspondingly higher signals—nevertheless, the signal-to-noise ratio is not further improved.

While integration of the signal brings primarily a saving of time and some improvement in the precision for the flame technique, it can have a decisive influence on the accuracy of the results for the graphite furnace technique.

The electronic computation of the integral value has a totally different significance for time-dependent absorbance changes, such as obtained from discrete samples, than for continuous sampling (e.g. with a nebulizer). If a sample is completely atomized within a finite time, the integral of the absorbance over this time is proportional to the total number of analyte atoms. Under suitable conditions, kinetic effects occurring during atomization can be eliminated through integration [2576]. In such a case it would be pointless to divide the integral by the time, as in the flame technique.

Many authors have reported that for the graphite furnace technique, peak area integration brings an improvement in the precision and, especially, in the linearity of the analytical curves [751] [982] [2787] [2831] [2937] [2939]. An improvement in the sensitivity and the detection limit has also been reported [2787] [2937]. Nevertheless, signal handling techniques and the selected integration times are important. With noisier signals, integration times that are too long can lead to a substantial deterioration of the precision.

SIEMER and BALDWIN [2853] found that the influence of strong filtering on the integrated absorbance values was different if the filtering was performed before or after logarithmic conversion. Filtering falsifies the signal when it is performed before logarithmization since the signal is not proportional to the mass of analyte element. After logarithmization, no influence is evident since the signal is then proportional to the mass.

ERSPAMER and NIEMCZYK [2230] also reported signal falsification for time-dependent absorbance changes in atomic absorption instruments having slow signal handling. LUNDBERG and FRECH [2573] found that both the extent and the polarity of an interference can depend on the time constant of the instrument. They established that the minimum time constant required for distortion-free measurement of time-dependent absorbance changes is one tenth of the peak width at half height for the fastest signal.

5.4. Readout

Meters with calibrated scales were used mostly for readout of the analytical measure in earlier instruments. Nowadays, virtually all atomic absorption spectrometers have digital displays, and some even feature graphical presentation of the signals on the screen of a video display unit. A hard copy of the analytical results can be made on a chart recorder, a printer, or a plotter.

Meters had the advantage that they were fairly cheap. Also, fluctuations or changes in the reading could be followed easily. However, the readout accuracy of meters is restricted, thus limiting their use for precision and trace determinations.

Digital displays offer optimum convenience and accuracy, and also reduce subjective reading errors to a minimum. Normally the analytical result is presented in absorbance, but other converted quantities, such as concentration or mass, can be selected on most instruments.

For time-independent absorbance changes (e.g. nebulizer/burner), a choice is generally provided between quasi-continuous display of the momentary measure and display of the mean value calculated from the momentary measures taken over a selected time period. For time-dependent absorbance changes (e.g. graphite furnace technique), the maximum value (peak height) generated during the measurement, either at a given response time or after a series of short integrations, is displayed. A further possibility is display of integrated values calculated from the momentary measures taken over a selected time period (peak area).

Continuous display of the momentary measure permits changes of the readings with time to be easily observed and is thus particularly suitable for performing alignment or adjustment of the radiation source, nebulizer, burner or graphite furnace, etc. Selection of the most suitable response time or integration time for time-independent absorbance changes is also most conveniently performed in this operating mode.

Display of the mean value is most convenient for routine operation, since subjective reading errors are excluded. It should nevertheless be borne in mind that a single reading in this operating mode can give no indication of the precision of the determination. It is preferable to take a series of mean values so that the scatter can be estimated. As well as display of the mean values, many instruments nowadays also permit display of statistical values such as standard deviation and relative standard deviation (often termed coefficient of variation). These data provide a direct criterion for the precision of a determination.

For techniques generating time-dependent signals, display of the maximum value (peak height) has been used routinely for many years. Nevertheless, recent recognition of the influence of signal handling and readout techniques should start a trend in favor of the display of integrated values (peak areas).

It is essential at this point to warn of the errors that can occur when exclusively digital display of the readings, and especially the peak height, is used for time-dependent signals. Numerous examples are known where the peak height is influenced by various factors. This leads inevitably to measurement errors when a further control, such as recording of the analog signals, is not performed. The display of integrated readings is less susceptible in this respect, but only when a system meeting the requirements for an interference-free determination is used, such as a stabilized temperature furnace (refer to sections 3.2.5, 8.2.1 and 8.2.4). Even here, though, recording of the analog signal should not be completely neglected.

A printer to provide a hard copy of the readings is a logical extension to a digital display and is also an important step toward automation. A printed paper strip can be a valuable document if it contains several readings per sample, rather than just one value, and also the mean value and other statistical data. A sequential printer normally prints columns of figures identified by alphabetical symbols (see figure 61). A minicomputer, on the other hand, is capable of printing complete analytical reports (refer to figure 63).

In earlier years a chart recorder was the only means of obtaining any record of the measurements; even today it provides an important document for time-independent absorbance changes. A chart recording is virtually the only means of retaining the conditions prevailing during a determination; it provides information on aspiration time, signal height, on possible drift, fluctuation or other variations of the baseline and signal, on the signal-to-noise ratio, the response time, and on possible carryover or memory effects. A chart recording thus provides a permanent, reasonably clear record of the events taking place during a determination. In other words, on events that can be observed on a meter only at the actual moment of reading and, under circumstances, not at all on a digital display.

As mentioned above, the sole use of a digital display for time-dependent absorbance changes is very dubious, especially for maximum peak height readings. A tracing on a chart recorder is virtually indispensable for observing the signal form. Even when a method is used routinely, this additional control should not be forgone.

Recording of the time-dependent analog signals is particularly important when new analytical methods are being developed. As well as the corrected atomic absorption signals, it is very important to record the uncorrected signals or the background attenuation during the entire temperature program, including thermal pretreatment and heating out steps. Only in this way is it possible to optimize temperatures and times for a particular determination.

```
                          0.001
                          0.001
                          0.002
                          0.001   A V
                          43.30   C V
                          0.000   A Z
                          0.082
                          0.081
                          0.081
                          0.082   A V
                           0.71   C V
                          0.250   S 1
                          0.685   C
                          0.685   C
                          0.687   C
                          0.686   A V
                           0.17   C V
                          0.625   S 2
                          0.261
                          0.267
                          0.262
                  1       0.264   A V
                           1.22   C V
                          0.075
                          0.077
                          0.077
                  2       0.076   A V
                           1.51   C V
                          0.069
                          0.071
                          0.071
                  3       0.070   A V
                           1.64   C V
```

Figure 61. Typical printout of the readings with the average and relative standard deviation. AZ = zero; S1 and S2 = reference solutions; AV = average; CV = relative standard deviation.

Every recorder has a certain response time (the time required for the recorder pen to reach a stated percentage of full scale) and every analog signal has a certain time constant (deriving directly from the time constant of the electrical measuring system). The time constant is the time required for the voltage or current in the circuit to reach the fraction (1/e) of its final value. A multiple of the time constant is required so that the display approaches the momentary measure sufficiently close; after the sevenfold time constant the approach is 0.999.

In AAS the time-dependent signals frequently change very rapidly. This means that recorders with very short response times are necessary and that the shortest time constants must be employed. Thereby the noise is often increased markedly, leading to a negative influence on the precision. Normally recorders with a pen speed of less than 0.5 second for full scale (250 mm chart width) are used. However, a certain distortion of the signals is tacitly accepted [2936], even though this leads to increased curvature of the analytical curves and to a loss in sensitivity. As long as similar signals are compared, no further interferences are incurred. When such recorder tracings are compared with peak values determined digitally, however, differences are invariably found. This distortion of the recorder tracings, which is dependent on the speed of signal generation and on the time constant, means that a recorder is only of limited use for many research applications. For this reason, attempts were made to record distortion-free signals, especially in the development of the graphite furnace technique, and storing oscilloscopes were frequently employed [2612] [2936].

BARNETT and COOKSEY [2058] described the connection of a minicomputer to an atomic absorption spectrometer. With this system, 50 five-digit readings per second could be handled. If required, the signals could be digitally smoothed without introducing any distortion. Next to the maximum peak height and the integrated peak area, this system also enabled the appearance time of a signal (the first derivative becomes positive), the time point for peak maximum (first derivative goes through zero), and the end of the signal (first derivative becomes zero) to be determined. The use of highly resolved, undistorted signals for methods development, interference studies, and studies on atomization mechanisms is shown in figure 62.

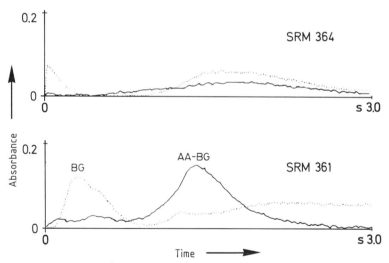

Figure 62. Typical tracings printed on a plotter for highly resolved graphite furnace signals. The determination of lead in two steel samples. BG = background; AA-BG = corrected signals.

A number of commercial atomic absorption spectrometers are already provided with built-in visual display units (video spectrometers). These instruments are capable of displaying highly resolved, distortion-free signals which can also be printed on a plotter, so that the use of a chart recorder is unnecessary. Instruments with video displays also offer a number of other capabilities, such as simultaneous display of single readings, mean values and statistical data. The simultaneous display of peak height and peak area, with and without background correction, in addition to the graphical presentation of the signal is also possible (figure 63). A data display of this type provides maximum information.

```
                                          AA          AA-BG         BG
     PEAK HEIGHT (ABSORBANCE)            0.156         0.155        0.007
     PEAK AREA (ABS-SECONDS)             0.089         0.091       -0.002

     INDIVIDUAL VALUES:
     01:    0.090    02:    0.091    03:    0.090    04:    0.091    05:    0.091




     STATISTICAL DATA:
     STANDARD DEVIATION:     0.000        COEF. OF VAR. (%):  0.53


                                                              TIME:  14:25
```

Figure 63. Typical printout of the results from an instrument with video display (Perkin-Elmer Model 3030). The printout includes peak height, peak area and individual readings, and statistical data. AA = uncorrected signal; AA-BG = corrected signal; BG = background signal.

5.5. Automation

Since atomic absorption spectrometry is a very rapid analytical technique, it is hardly surprising that in 1966, the year in which it received its greatest impetus and general recognition, the first automatic analyzer was introduced. GAUMER [408] constructed an instrument for the analysis of trace elements consisting of a commercial atomic absorption spectrometer to which an automatic sample changer, a digital evaluator [630] and a printer [1173] were connected. Soon afterwards, the first commercial instrument combination for automatic atomic absorption analyses was introduced [1138]. This consisted of an automatic dilutor, a sample changer for 200 vessels, a double-beam atomic absorption spectrometer with a digital readout and a printer. All stages of an atomic absorption determination were carried out automatically and the results were printed out directly in concentration units. Modern atomic absorption spectrometers are generally provided with facilities for operation with autosamplers and often also have the interface capabilities for direct on-line operation with an external computer. Digitalization of the signals in the spectrometer is an important factor for further signal handling. The tendency in AAS appears to be not so much toward interfacing to mainframe computers, but rather toward self-contained "information centers". The spectrometer and all peripheral instruments are controlled by a central processor that also provides printout of a complete analytical report.

Automation in flame AAS does not bring any substantial saving of time, since the technique is in itself very quick. A saving of time can only be expected when several elements are to be determined in the same sample solution. On the other hand, the situation is totally different for the graphite furnace technique. Here, sample dispensing is a difficult procedure and the time period between each dispensing is rather long. Automation of sample dispensing brings an improvement in the precision as well as facilitating operation, as already explained in section 3.2.6.

A definite disadvantage of AAS with respect to automation, especially for the fast flame technique, is the relatively long time required to change from one element to the next. Multielement lamps offer certain advantages if the corresponding combination of elements is to be determined, but a change of radiation source is in any case a quick and simple procedure. Setting the other instrumental parameters, such as wavelength, slit, gain, etc., and performing calibration are the most time-consuming procedures, and also the most difficult to mechanize and automate. Nevertheless, modern microprocessor technology and stepping motors for wavelength drive, etc. have made possible fully automatic element change [2166]. Despite the obvious advantages of sequential multielement determinations that such a system brings, a change of element is still the time-determining step. Thus, with such instruments, it is usual to determine one element in a series of samples, then change to the next element and rerun the series, and so on.

For simultaneous multielement determinations, which are of particular interest for techniques with time-dependent absorbance changes, a different instrumental approach is clearly required. This problem has already been discussed in section 4.5.

6. Methods, Nomenclature and Techniques

6.1. Important Terms, Quantities and Functions

The terms used are in accordance with the recommendations of the International Union of Pure and Applied Chemistry (IUPAC) and the International Organization for Standardization (ISO). (Refer to International Standard ISO 6955, "Analytical spectroscopic methods—Flame emission, atomic absorption, and atomic fluorescence—Vocabulary".)

Under given prerequisites, solid samples can be analyzed directly by AAS (refer to section 8.2.5), but the majority of samples are presented as liquids or solutions. A solution suitably made up from a test portion of the sample submitted for analysis is the **sample solution**.

Since atomic absorption spectrometry is a relative and not an absolute technique, a reference solution is required to permit quantitative determination of a given element. A reference solution is typically prepared from a **stock solution,** which is a solution of suitable composition containing the analyte element in an appropriately high, known concentration, frequently 1.000 g/L. Such a stock solution can be prepared by dissolving 1 g of the ultrapure metal or a corresponding weight of an ultrapure salt in water or a suitable acid and making up to one liter. Concentrated stock solutions are also offered commercially and must only be diluted to the given volume. With proper storage, stock solutions normally have a shelf-life of about a year.

Before the determination of the analyte element, a series of reference solutions of different concentration (**set of calibration solutions**) can be prepared by dilution of the stock solution. A **simple reference solution** is a solution containing a known concentration of the analyte element in the solvent. A **synthetic reference solution** is a solution containing a known concentration of the analyte element in the solvent, with the addition of chemicals used for preparation of the sample solution and other constituents influencing the determination in proportions similar to those in the sample. If spectrochemical buffers (refer to section 7.2.2) are added to the sample solution, it is absolutely essential that they are also added to the reference solutions in the same concentration. Ideally, a reference solution should be identical to the sample solution, except for the concentration of the analyte element. Depending on the actual concentration of the analyte element, a reference solution may only be stable for a day or two, so that in cases of doubt fresh reference solutions should be prepared daily. Reference solutions can also contain several elements in known concentration if these elements are to be determined in the same sample solution.

A **solvent blank**, consisting normally of the pure solvent, such as deionized water, is used for setting zero absorbance on the spectrometer. If larger quantities of acids or other chemicals are required to prepare the sample solution, they should be checked to see whether they contain the analyte element in measurable concentration. A **blank test solution** contains all the chemicals in the same concentrations as required for preparation of the sample solution, but it does not contain the analyte element. A **zero member compensation solution** or **matrix solution** contains all chemicals used in the preparation of the sample solution together with all constituents of the sample influencing the determination, especially spectrochemical buffers, in the same, or very similar, concentrations as in the sample solution. The analyte element is never added to this solution. Under certain circumstances the blank test solution or the solvent blank can be the matrix solution. The matrix solution serves to determine the point of intersection of the analytical curve with the concentration axis.

The relation of the measured absorbance A to the concentration c or the mass m of the analyte element is given by the **analytical function**

$$A = f(c) \tag{6.1}$$

or

$$A = f(m) \tag{6.2}$$

for the technique and the atomic absorption spectrometer employed. A graphical plot of the analytical function is the **analytical curve**.

In atomic absorption spectrometry, the relationship between the concentration c or the mass m of the analyte element and the analytical measure is determined from a set of calibration solutions and a matrix solution. This procedure, termed **calibration,** involves preparation and measurement of the reference solutions, and establishment of the analytical curve by plotting the measure (reading) against the concentration c or mass m of the reference solutions.

The slope S of the analytical curve is termed the **sensitivity**.

$$S = \frac{\partial A}{\partial c} \tag{6.3}$$

or

$$S = \frac{\partial A}{\partial m} \tag{6.4}$$

This relationship is only valid when the calibration function obeys the BEER-LAMBERT law (equation (1.3), page 1), independent of the concentration or mass. For non-linear calibration functions the sensitivity is a function of the concentration c or mass m.

To permit comparison of the sensitivity, i.e. the slope of the calibration curve for an element under given conditions, the terms **characteristic concentration** and **characteristic mass** are used. This is the concentration c_0 or mass m_0 of the analyte element corresponding to a net absorption of 1% or an absorbance of 0.0044. When integrated absorbance (peak area) is used for evaluation, characteristic concentration and characteristic mass are the concentration c_0 or the mass m_0 corresponding to 0.0044 A·s (absorbance seconds). In the past, the term "sensitivity" has been incorrectly used for these characteristic values. This is not in agreement with other spectroscopic methods and has led to confusion [284]. Likewise, the term "detection sensitivity" is nowhere defined and should therefore be avoided (the use of this and similar non-defined terms is not approved by IUPAC).

The measured result should naturally agree with the true value. Owing to various errors in the analytical procedure, however, the agreement is subject to limitation. The "correctness" of the measured result can be determined by comparison with a weighed standard or a reference material whose content of the analyte element has been established by several, independent analytical methods.

The **accuracy** relates to the closeness of agreement between the true value for an element in a sample and the mean obtained by repeating the analytical procedure a large number of times. The accuracy can be calculated from the difference between the true value and the measured values.

It is important to distinguish between accuracy and precision (see below). A series of determinations can have high precision even though the results are incorrect. This is indicated pictorially in figure 64 on the example of a target; the correct value is the center of the target.

The **precision** relates to the closeness of agreement, at a given level, between the results obtained by applying the analytical procedure repeatedly under given conditions. The precision can be determined by taking 2.83 times the standard deviation σ for 30 or more measurements, or the product $t\sigma$ if the number of measurements is less than 30, where t is STUDENT'S factor. In practice, it is usual to multiply the standard deviation by a factor of two or three. The IUPAC recommendation is that a factor of three should be chosen; the factor used should be stated to avoid ambiguity.

The **standard deviation** is given by

$$\sigma = \sqrt{\frac{\Sigma(x_j - \bar{x})^2}{n-1}} \tag{6.5}$$

where $\bar{x}$ is the mean value for all measurements, x_j is an individual measurement, and n is the number of measurements.

The **relative standard deviation** σ_r (RSD) can be calculated from the standard deviation.

$$\sigma_r = \frac{\sigma}{x} \tag{6.6}$$

Relative standard deviation is also frequently termed coefficient of variation (CV), although this name is discouraged by IUPAC.

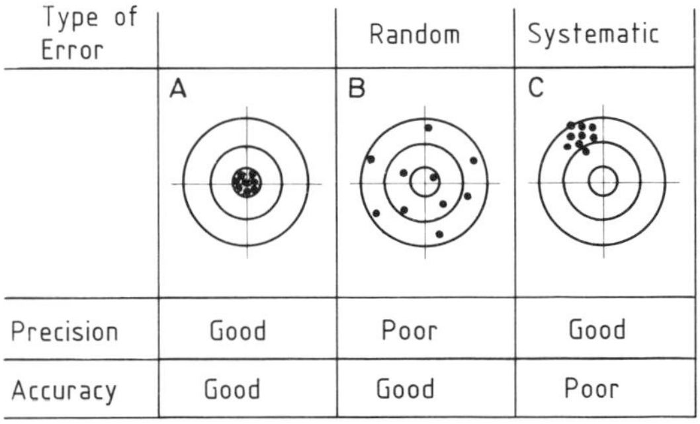

Type of Error		Random	Systematic
	A	B	C
Precision	Good	Poor	Good
Accuracy	Good	Good	Poor

Figure 64. Schematic representation of various types of error in the form of hits on a target.

The **detection limit** for a given analytical procedure is the concentration c_L or the mass m_L that can be detected with a stated statistical certainty. The detection limit is defined as

$$c_L = \frac{\partial c}{\partial A} \cdot k \cdot \sigma \qquad\qquad (6.7)$$

or

$$m_L = \frac{\partial m}{\partial A} \cdot k \cdot \sigma, \qquad\qquad (6.8)$$

where $(\partial c/\partial A)$ or $(\partial m/\partial A)$ is the reciprocal sensitivity of the procedure, and σ is the absolute standard deviation of measure A determined from measurements of the solvent blank. The factor k is usually taken as 2 or 3, depending on the required statistical certainty, and should be stated.

The detection limit is naturally dependent on the sensitivity, since for a given concentration or mass the sensitivity is a measure of the attainable signal height. The detection limit contains a second variable, namely the fluctuation of the background, generally termed **noise**. While the sensitivity is a largely natural constant, noise is essentially caused by the apparatus. Although there are many possible reasons for noise, under closer scrutiny clear contextual relations emerge. These have largely been treated in earlier sections and will only be mentioned here briefly.

A frequent cause for increased noise is low energy. This can originate in the radiation source if the radiant intensity is weak at the analyte wavelength. This was often the case with early hollow cathode lamps, as depicted in figure 65.

Figure 65. Influence of radiant intensity on the signal-to-noise ratio; left: earlier, low energy hollow cathode lamp; right: Intensitron hollow cathode lamp.

A low radiance monochromator can be a further cause of reduced energy. A poor reciprocal linear dispersion (high values in nm/mm) signifies a narrow entrance slit and hence only a fraction of the radiation emitted by the source can enter the monochromator.

Although the electrical measuring system can contribute to the noise, this is generally insignificant, so that the remaining major source is the flame. As demonstrated by BOLING [140] and SLAVIN [1145], a well-designed premix burner with a laminar flame produces very little noise, except at the start of the vacuum UV range. On the other hand, as the excellent researches of LEBEDEV [737] have shown, a direct injection burner with a turbulent flame causes maximum optical noise, which is naturally reflected in the signal (refer to section 3.1.2., page 39).

ROOS [1046] [1047] [1050] endeavoured to quantify the various causes for noise in AAS by assigning individual functions. He found that for the majority of elements the flame contributed the greatest amount of noise. The noise can comprise time-dependent changes in the atom concentration, alterations in the flame absorption (far UV), and emission noise (see page 97).

The other factor influencing the detection limit of an element is, as mentioned above, the signal height, or in effect the sensitivity. The sensitivity depends on the length (in cm) of the absorption volume through which the radiation beam passes and on the absorption coefficient κ. The degree of absorption, and thereby the sensitivity, is given by equation (1.14) (page 10). In simplified form this equation reduces to:

$$\kappa_v = \text{const.} \cdot N_v \cdot f_v \tag{6.9}$$

The only variable in this equation is N_v, the number of atoms available for absorption of radiation at frequency v. Thus, the factors influencing the dimension N_v should next be considered, initially ignoring the possibility of incomplete atomization.

Since the absorption coefficient κ_v for the transition of the line of frequency v is independent of the total number of atoms N per unit volume, but is dependent on N_v, the number of atoms available for radiation absorption of frequency v, it is interesting to compare the size of N_v in comparison to N.

It can generally be stated that atoms in the ground state are able to absorb radiation of exactly defined frequency, but as already indicated in section 1.3 (on page 6) this general statement requires modification since the ground state is not always clearly defined, especially for elements with complex atomic structure. Such atoms frequently exhibit a fairly large number of low-lying excited states whose populations at the prevailing flame temperatures are often higher than the actual ground state.

An interesting study clarifying this situation for tin has been published by CAPACHO-DEL-GADO and MANNING [208]. They found that for various flame types the sensitivity at individual resonance lines increased in the air/hydrogen flame (2050°C) compared to the air/acetylene flame (2300°C). The results are presented in table 8. Lines originating from the ground state undergo an increase by a factor of about 2.6 in the cooler air/hydrogen flame, while lines originating from metastable states (1692 K or 3428 K) undergo very little increase. This clearly indicates that a difference of 250°C in the flame temperature causes a shift in the populations of the ground state and metastable, excited states. L'vov *et al* [3102] have even developed a very sensitive and accurate technique for the measurement of gas temperatures based on the varying populations of energy levels in dependence upon the temperature.

The more complicated the electron shell of an element is, the greater is the number of metastable states. Especially in the hot nitrous oxide/acetylene flame, the probability that all of these states are populated to varying degrees is large. This inevitably means that the ratio N_v/N becomes smaller, resulting in a lower sensitivity at frequency v and a large number of resonance lines with comparable sensitivities. This is particularly striking for the lanthanides (see table 31). This behavior is indicated in table 9 on the example of titanium.

The ratio N_v/N is particularly unfavorable for those elements that must be determined at spectral lines emanating from poorly populated metastable states or from transitions of low probability. The resonance lines of mercury (184.9 nm) and phosphorus (177.5 nm), for ex-

Table 8. Absorbances measured at various Sn resonance lines in the air/hydrogen and air/acetylene flames [208]

Wavelength nm	Energy level (K)	Absorbance		$\dfrac{H_2}{C_2H_2}$
		Air/H_2	Air/C_2H_2	
224.605	0–44509	0.821	0.337	2.6
286.333	0–34914	0.492	0.193	2.7
254.655	0–39257	0.178	0.074	2.5
207.308	0–48222	0.013	0.005	2.6
				Mean 2.6
235.484	1692–44145	0.398	0.239	1.7
270.651	1692–38629	0.201	0.194	1.7
303.412	1692–34641	0.148	0.102	1.5
300.914	1692–34914	0.087	0.052	1.8
219.934	1692–47146	0.042	0.028	1.5
233.480	1692–44509	0.041	0.023	1.8
266.124	1692–39257	0.028	0.013	2.2
				Mean 1.7
283.999	3428–38629	0.108	0.119	0.9
242.949	3428–44576	0.105	0.109	1.0
226.891	3428–47488	0.065	0.075	0.9
317.505	3428–34914	0.053	0.055	1.0
220.965	3428–48670	0.023	0.021	1.0
248.339	3428–43683	0.014	0.009	(1.7)
228.664	3428–47146	0.006	0.005	1.2
				Mean 1.0

ample, lie in the vacuum UV range and are thus poorly amenable to AAS. Mercury must therefore be determined at the intercombination line at 253.7 nm, which has a lower sensitivity by a factor of around 100, while phosphorus must be determined at the 213.6 nm line [3103], emanating from a poorly populated metastable state (11361 K). The sensitivities of these elements are thus poor, and do not fall into line with the other elements in the respective groups of the periodic table.

Next to these "natural" phenomena influencing the sensitivity, the chemical environment is another important cause for an alteration in the total number of atoms N produced. In this respect, the formation of stable, difficult-to-dissociate compounds of the analyte element with constituents of the flame gases or carbides with the graphite tube material should be especially mentioned. Taken strictly, these phenomena are not interferences in the accepted sense, but they do affect the total number of atoms N produced and thus have a significant influence on the attainable sensitivity.

Table 9. Titanium resonance lines [214]

Wavelength nm	Lower energy level (K)	Observed intensity	Characteristic concentration (mg/L 1% absorption)
260.515	170	4	8.6
261.128	387	7	6.0
264.110	0	4	4.7
264.426	170	5	4.4
264.664	387	8	3.7
294.200	0	16	5.2
294.826	170	17	4.4
295.613	387	21	3.4
318.651	0	117	3.0
319.191	170	165	2.6
319.990	387	185	2.0
334.188 I	0 ⎫	230	3.7
334.188 II	4629 ⎭		
335.463	170	170	2.9
337.145	387	220	2.0
337.748	387 ⎫	93	7.0
337.758	170 ⎭		
363.520	387 ⎫	390	5.6
363.546	0 ⎭		
364.286	170	430	1.8
365.350	387	465	1.6
372.981	0	195	5.8
374.106	170	250	2.6
375.285	387	356	2.5
375.364	170	47	—
394.778	170	175	19.0
394.867	0	273	10.5
395.639	170	286	9.5
395.821	387	426	5.7
398.176	0	500 ⎫	6.2
398.248	0	100 ⎭	
398.976	170	490	4.7
399.864	387	640	4.3

Next to the detection limit, a term that has gained increasing significance in recent years is the **determination limit**. This is the lowest concentration c (the considerations are equally valid for the mass m) that can be determined with the prescribed precision for the particular procedure. HUBAUX and VOS [2397] have published a detailed treatise on these limits. Since for any determination only the measure y is obtained, it is necessary beforehand to determine the concentration limits c_L and c_D, corresponding to the measures y_L and y_D, respectively. Here it must be borne in mind that the measure for a given concentration does not assume a fixed value but is randomly distributed about a mean value, and that the analytical function on which the determination of the concentration is based is only an estimate of the true analytical function. The calculation of an unknown concentration c from the measure y via the analytical function thus leads to the determination of a confidence band for c.

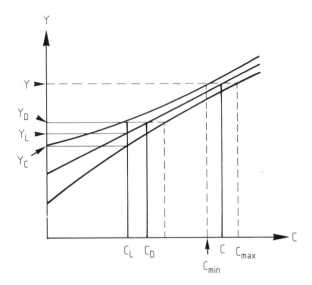

Figure 66. A calibration function and its confidence band in the region of the detection limit c_L and the determination limit c_D. Refer to the text for an explanation.

As depicted in figure 66, by taking the calibration function and its confidence band, the concentration range c_{min}–c_{max} and the mean concentration c for measure y can be obtained by interpolation. For a measured signal equal to y_c, the lower limit of concentration is zero. For signals equal to or lower than y_c it is not possible to distinguish whether the analyte element is present of not.

On the other hand, the lowest concentration that can be distinguished from the blank is fixed by the upper limit c_L of the concentration range determined by y_c; c_L is thus the detection limit. This is the lowest concentration at which qualitative detection is possible with reasonable certainty, but at which an accurate quantitative determination is not possible. For measures that are larger than y_c but lower than y_D, the lower confidence limit c_{min} is still below the detection limit c_L.

A quantitative determination is only possible when a value for y_D is reached for which the lower limit of the confidence band corresponds to the detection limit and can thus be distinguished from the blank. The mean concentration value c_D corresponding to the measure y_D is termed the determination limit. Only measures that are larger than y_D have sufficient precision to permit a quantitative determination.

It is also possible to define a usable concentration range within which the concentration c or mass m can be determined with a given precision. The lower limit of this range is the determination limit.

6.2. Calibration Techniques

Atomic absorption spectrometry is a relative and not an absolute method. This means that quantitative results can only be obtained by comparison with reference solutions or reference materials. This has the disadvantage that reference measurements must always be made. On the other hand, through choice of a suitable technique, or through selection of expedient reference materials or solutions, many interferences can be eliminated. Especially long term fluctuations and also changes from one sample series to the next or from day to day can be kept to a minimum by choice of a suitable calibration technique.

Different calibration techniques are available to meet the analytical requirements and the demands made with respect to accuracy, precision or speed. The advantages and disadvantages of individual techniques will be considered briefly. It is worth mentioning that the occurrence of errors can often be traced back to the choice of an unsuitable calibration technique [1133].

6.2.1. Analytical Curve Technique

The simplest and quickest calibration technique is the direct comparison of the sample solution with reference solutions. The calibration function is established by plotting the measures for a set of calibration solutions against the concentrations. The zero member has an absorbance of zero, while the highest member has an absorbance equivalent to the highest expected concentration of the sample solution (see figure 67). If the calibration function is linear, i.e. obeys the BEER-LAMBERT law, and the curve passes through the origin, then

$$c_i = \frac{A_i}{A_R} \cdot c_R \qquad (6.10)$$

or

$$m_i = \frac{A_i}{A_R} \cdot m_R, \qquad (6.11)$$

where c_i is the concentration or m_i is the mass of the analyte element in the sample solution, A_i is the absorbance of the sample solution, A_R is the absorbance of a reference solution, and c_R is the concentration or m_R is the mass of the analyte element in this reference solution.

Nowadays the reference curves are rarely plotted to permit deduction of the analyte concentration by interpolation. Spectrometers with digital displays permit entry of the concentration or mass of the analyte element in the reference solution. The spectrometer then determines the calibration function automatically and presents the analytical results in the desired units. If more than one reference solution is used to determine the calibration function, most instruments calculate the slope of the calibration curve according to the method of least squares.

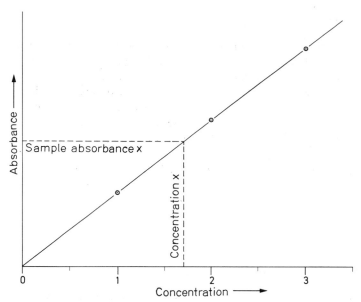

Figure 67. Analytical curve technique.—The measures for a set of calibration solutions are plotted against the concentrations. The concentration c_x of a sample is determined from its absorbance measure A_x by interpolation.

Modern microcomputer technology brings considerable advantages when non-linear calibration functions are encountered. While the results obtained from the calibration curves of such functions are relatively inexact, very precise algorithms permit calculation of these functions [2993]. With such instruments, it is only necessary to enter the concentration or mass of two or three reference solutions and presentation of the analytical results in the required units then takes place automatically.

6.2.2. Bracketing Technique

The bracketing technique is a variant of the analytical curve technique. The analytical function is usually determined using two reference solutions so selected that their concentrations or masses closely bracket the expected value for the analyte element in the sample solution. Thus, only a small section of the analytical curve is taken into consideration in this technique. The concentration c_i of the analyte in the sample solution is calculated according to

$$c_i = \frac{(A_i - A_1)(c_{R1} - c_{R2})}{A_2 - A_1} + c_{R1},$$ (6.12)

where c_{R1} and c_{R2} are the concentrations of the analyte element in reference solutions R1 and R2, respectively, and A_1 and A_2 are the corresponding absorbance values. Calculation of the mass m_i is performed correspondingly.

The advantage of the bracketing technique is the improved precision that can be obtained. Further, the technique can also be applied to non-linear sections of the analytical curve when the concentrations or masses of the reference solutions neighbor the sample solution closely enough.

ROOS [1050] emphasized that for every element there is nevertheless an optimum concentration range within which the relative error passes through a minimum (see figure 68). For many elements this range lies between 90 and 150 times the characteristic concentration or mass, or in the absorbance range 0.35 to 0.60. It is therefore advisable to adjust the concentration or mass of the analyte by suitably diluting solutions or to choose a suitable resonance wavelength so that measurements can be performed within this range. A deterioration of the precision must be expected when measurements are performed outside this range, especially for non-linear analytical functions.

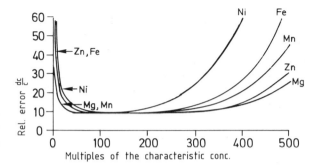

Figure 68. Relative error (normalized for a minimum value of 1.0) for a number of elements (according to [1050]).

6.2.3. Analyte Addition Technique

This technique consists in taking a number of replicate aliquot portions of the sample solution, adding to them increasingly known quantities of the analyte (usually as aliquots of a suitable reference solution) and diluting to the same volume. One aliquot portion is only diluted with solvent so that it contains zero added analyte. The series of solutions obtained thus serves as the set of calibration solutions. Deduction of the analyte concentration is performed by extrapolating the analytical curve to intercept the negative concentration axis (refer to figure 69).

The concentration c_i of the analyte element in the sample solution is given by

$$c_i = \frac{(A_i - A_0)(c_{R2} - c_{R1})}{A_2 - A_1}. \tag{6.13}$$

The mass m_i of the analyte element is calculated correspondingly. Since all calibration solutions have the same composition with the exception of their analyte contents, the influence of concomitants will be the same. When the absorbance values for each solution are plotted versus the added concentration of analyte, an analytical curve is obtained that intercepts the absorbance axis at a value greater than zero (figure 69).

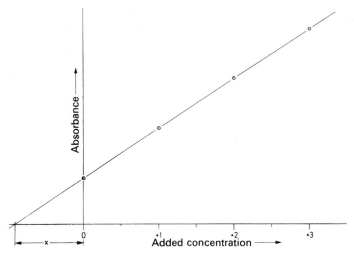

Figure 69. Analyte addition technique.—Reference solutions of varying concentration are added to replicate aliquot portions of the sample solution. By plotting the absorbance measures against the concentration of the added analyte and extrapolating to zero, the concentration c_x of the sample can be determined. The analyte addition technique serves principally for eliminating transport interferences.

The gradient of this analytical curve is specific for the analyte element in the given sample solution. If a series of samples of identical composition are to be determined, the analytical curve may be shifted in parallel until at zero concentration it passes through zero absorbance. This curve can then serve as the analytical curve for all the samples, as in the analytical curve technique.

If the sample composition varies from sample to sample, it may be necessary to employ the addition technique for every sample. The content of the analyte element in the diluted sample solution is deduced by extrapolating to zero concentration. It is then only necessary to take the dilution factor into account when calculating the content of the analyte in the original sample solution.

For the addition technique it is essential to realize that correct results can only be obtained when working in the linear range of the analytical curve. Extrapolating non-linear analytical curves to zero absorbance generally leads to analyte values that are too high for the sample solution. Dilutions and analyte additions should thus be so chosen that all measures are in the linear range. Most suitable is a measure of 0.1 to 0.2 absorbance for the least concentrated sample solution, and analyte additions that increase the measure by factors of about two, three and four.

FULLER [394] described a simplification of the addition technique for spectrometers having digital displays with polarity sign and automatic zero setting. Initially the absorbance measure of the sample solution is set to zero, then the solution with the highest analyte addition is set to the concentration of the reference solution used for the addition. Using further additions with reference solutions of differing concentration permits the linearity to be checked. Finally the blank test solution is measured and the displayed value (with negative sign) corresponds to the actual analyte concentration in the sample solution. RATZLAFF [2781] has developed a series of equations that permit optimization of the precision for the addition technique.

Since transport interferences and also to a large extent volatilization and spatial-distribution interferences, as well as vapor-phase interferences, influence the gradient of the analytical curve independent of the concentration, they can be quantitatively eliminated by the addition technique. A prerequisite is naturally that the concomitants do not depress the measure too much, since the precision deteriorates with decreasing sensitivity. Generally, strong interferences should first be reduced through the addition of spectrochemical buffers before they are finally eliminated through use of the analyte addition technique.

Interferences that depend on the concentration, such as ionization chemical interferences, cannot be eliminated by the addition technique. The effect of these interferences varies in dependence on the concentration or mass of the analyte element. The resulting curvature of the analytical curve is hard to control, making extrapolation to zero absorbance difficult and worsening the accuracy of the determination.

Spectral interferences **cannot** be eliminated by the analyte addition technique. The observed signal comprises two components, namely the specific absorption of the analyte element and the interfering absorption of the concomitants. Separation of these components through the addition of reference solutions is not possible.

The same is true for all other additive, systematic errors, such as residual blank values, since these increase or decrease all measures by the same amount. The analyte addition technique can only eliminate such interferences that alter the gradient of the analytical curve, but not such that cause a parallel shift of the curve.

6.3. Extraction, Enrichment and Separation

For a given task, insufficient sensitivity of the analytical method to permit detection or determination of the substance sought is a problem known to all branches of analysis. There are two solutions to this problem: Either a more sensitive technique is chosen, or the selected technique is retained and an attempt is made to bring the analyte content within the measurable range through enrichment (preconcentration), extraction, etc.

For reasons explained in section 6.4 on the problems of trace analysis, selection of the more sensitive technique is always the better solution. In trace analysis, every sample pretreatment step is a potential source of error and should thus be avoided if possible. Added to this, separation and enrichment steps are time-consuming and demand a degree of skill on the part of the operator.

Most enrichment techniques, on the other hand, also lead, to a large extent, to separation of the analyte from concomitants. In other words, an increase in the concentration can be accompanied by a decrease in interfering influences. Under this aspect, enrichment and separation techniques can become attractive, although they must be carried out with great care. Nevertheless, performed by an experienced analyst, such techniques can be a useful aid in trace analysis.

6.3.1. Solvent Extraction

Solvent extraction is a technique long known to analytical chemistry. For this reason, only the advantages for atomic absorption spectrometry will be mentioned briefly. Details of individual extraction procedures will be provided as relevant in later chapters for determinations that cannot be performed by direct methods. Further information can be obtained from the literature, such as the books authored by MORRISON and FREISER [880] and CRESSER [2173].

Solvent extraction is a process in which two immiscible liquids are brought into such close contact with each other that one or more elements in the one liquid phase are transferred to the other. Normally, one liquid phase is the aqueous sample solution while the other is an organic solvent immiscible with water. However, reversed phase extractions are also possible. Because of their ionic character, simple metal salts are generally much more soluble in aqueous media than in organic solvents. Metals must therefore be converted from the ionic form to a non-charged form to permit extraction with organic solvents. This is normally done by complexing with organic ligands.

Solvent extraction is used mostly in AAS when the concentration of the analyte element is too low to permit direct determination, or when the matrix is very complex and the resulting interferences cannot be controlled. Solvent extraction is thus employed for enrichment or separation and is used when other, simpler techniques are no longer satisfactory. A further application has also been reported by DELVES et al [294]. By selective extraction, they were able to determine 11 elements from only 1 mL of sample, so that it is thus possible to work with minimum sample volume.

For the flame technique solvent extraction offers a further, important advantage. Organic solvents in the flame often bring a substantial improvement in the sensitivity and detection limits. Thus, as well as the enrichment generally obtained through solvent extraction, a further signal increase is gained through use of an organic solvent. These advantages have led to the publication of many papers on the use of solvent extraction in AAS, despite the time-consuming and critical work involved.

The combination of ammonium pyrrolidine dithiocarbamate (APDC) as chelating agent and methyl isobutyl ketone (MIBK) as the organic solvent has found wide application [894]. APDC forms stable chelates with many metals over a wide pH range and MIBK exhibits ideal combustion properties in the flame. The use of other complexing agents and most organic solvents is naturally also possible. Solvents considered unsuitable are benzene, because it burns with a sooty flame, and highly chlorinated hydrocarbons, such as chloroform or carbon tetrachloride, because of their very poor combustion properties; but even these solvents could be used in principle. The use of highly volatile solvents is also not recommended because losses through evaporation make quantitative work difficult. Solvents having solidification points just below ambient temperature are equally unsuitable because they can solidify on the nebulizer orifice as a result of evaporation cooling.

Many extraction procedures have been optimized for the flame technique. With the more sensitive graphite furnace technique, on the other hand, it is not possible to perform such determinations directly. Nevertheless, many varying extraction procedures have been reported for this technique, mainly with the aim of eliminating the frequently observed interferences. However, it is very important that the metal concentration in the extract does not change with time, a frequent problem with many metal chelates that is aggravated by the time-consuming graphite furnace technique. Further, it is necessary to be able to dispense the solution easily and reproducibly, and the solution should not spread inside the graphite tube. For these reasons the use of organic solvents should be avoided.

Various authors have therefore proposed back extraction of the organic extract with a weakly acidified aqueous solution [2604]. In this way it is normally possible to obtain complete separation of the analyte element from the concomitants and thus obtain an almost pure aqueous solution that exhibits hardly any interferences. Nevertheless, it should be emphasized that such a double extraction procedure must be performed with extreme care, especially for trace determinations, and contains many sources of error. It is therefore quite possible that the determination performed on the final sample solution is free of interferences, but that the quantity of analyte element found bears no relationship with the quantity in the original sample. For ultratrace determinations with the graphite furnace technique, other procedures for elimination of interferences should be sought. In this respect, reference is made to the importance of thermal equilibrium at the time of atomization (refer to section 8.2).

6.3.2. Separation and Enrichment Techniques

For the analysis of high purity materials, the combination of wet chemical multielement enrichment and spectrochemical determination gives a high performance procedure for many applications. This procedure is particularly applicable when the blank value can be kept to a minimum, even after the sample pretreatment steps. The adsorption of trace elements onto acid-insoluble stationary phases has proved to be a valuable separation technique. As opposed to the use of acid-soluble phases, in this technique a concentrate is obtained that contains virtually only the trace elements and thus permits interference-free determinations.

Activated charcoal has proved to be a very suitable phase for this type of trace enrichment. A large number of trace elements can be enriched from various matrices by relatively simple manipulations. The basic scheme is essentially similar for most determinations: The material to be analyzed is taken into solution, the pH is adjusted to the required value, and a complexing agent is added to complex the trace elements. The solution is then filtered through a filter evenly coated with activated charcoal. The solution containing the "matrix elements" passes through the filter while the trace elements are retained by the charcoal. The filter is then dried and the charcoal carefully scraped into a small beaker. The complexes and the active surface of the charcoal are then destroyed by evaporating to dryness with concentrated nitric acid. The residue is taken up in a small, defined volume of acid, the charcoal separated by centrifugation, and the trace elements in the supernatant liquid are determined by AAS [563] [2077] [2081] [2423] [2425] [2426].

Similarly, silver halides [2078], copper sulfide [2428], or alkaline-earth nitrates [2430] can be used under given conditions as acid-insoluble stationary phases.

During chemical convertion of the matrix elements, an excess of reagent above the required stoichiometric quantity frequently causes a change in the characteristics of the separating system that can lead to a loss of some of the trace elements. These characteristics often become effective at the equivalence point. Excess reagent can cause the solubility product of trace compounds to be exceeded or a jumpwise shift in the distribution ratio, or charge reversal of precipitation products with concomitant charge-dependent adsorption of trace elements, or a sudden change of the NERNST potential with electrochemical reaction of the trace elements, etc. [2424]. Sources of error of this type can often be eliminated if it suffices when the matrix is largely, but not quantitatively, separated. Excellent examples of this technique include the determination of trace elements in ultrapure aluminium by partial dissolution of the matrix in the presence of mercury [519], and in ultrapure cadmium [2385] or high purity gallium [2427] by partial dissolution of the matrix.

A further technique bringing very effective separation and enrichment of the analyte element is electrolytic deposition on graphite. THOMASSEN *et al* [2961] separated a large number of elements from concentrated salt solutions by electrolytic deposition onto a graphite electrode. The electrode was subsequently ground and the graphite powder analyzed directly in a graphite furnace.

BATLEY and MATOUŠEK [2062] [2063] deposited cobalt, nickel and chromium under the addition of mercury(II) directly onto the inner surface of a pyrolytically coated graphite tube and then inserted the tube into a graphite furnace for determination of the respective element. They used a flowcell at a controlled potential for the deposition procedure with the graphite tube as the cathode. The deposition times were 15 minutes for cobalt and 10 minutes for nickel; the characteristic concentration for both elements was 0.02 µg/L. For chromium, the authors were able to separate Cr(III) and Cr(VI) by varying the potential and thus make a specific determination for each oxidation state.

VOLLAND *et al* [3021] also reported an electrolytic cell in which nanogram quantities of elements could be deposited galvanically in relatively short times from solutions with concentrations of less than 10 µg/L and thus be separated from the matrix elements. The cathode in the form of a tube made of ultrapure graphite could then be inserted directly into a graphite furnace for the determination of the deposited elements. The authors found that the determinations were impeded by the omnipresent concentrations of the respective elements.

HOSHINO *et al* [2394] [2395] described a variant of this procedure. They adsorbed the analyte element selectively onto a tungsten wire and then performed the determination in a graphite furnace. CZOBIK and MATOUŠEK [2183] examined this technique in more detail. At constant voltage it was possible to deposit elements such as cadmium, copper, lead, silver, or zinc within 30 to 300 seconds. No background attenuation could be observed during subsequent atomization.

A review on newer separation and enrichment techniques, including the pertinent literature, has been published by WILSON [3071].

6.4. Problems of Trace Analysis

In the following section the problems of determinations in the trace to nanotrace ranges will be discussed in general terms. Specific problems occurring during susceptible techniques or for critical elements are mentioned in the relevant paragraphs throughout the book.

With appropriate attention to laboratory technique, larger systematic errors are hardly to be feared when working in the mg/L range for solutions or μg/g range for solids. The situation becomes critical, though, for the μg/L range and below for solutions or the ng/g range for solids. The risk of systematic errors often increases exponentially with decreasing concentration.

The errors can lead to results that are either too low, due to loss of the analyte element during sample pretreatment, or too high, due to contamination with the analyte element. In AAS, systematic errors occur mostly during sample pretreatment and can rarely be ascribed to the analytical method itself. For this reason, direct methods are preferred for trace analysis since sample pretreatment steps can be avoided. However, solid samples can only be determined directly by the graphite furnace technique, so that a digestion is required as the starting step for the majority of determinations. The risk of losses due to volatilization is immediately obvious, especially for volatile elements or elements that form volatile compounds. Fusions are more prone due to the higher temperatures, but acid digestions are not without problems. The best safeguard against volatilization losses is to perform the digestion under pressure in an autoclave (see figure 70).

Figure 70. Autoclave for the digestion of difficult-to-dissolve samples at increased temperatures and pressure in a sealed PTFE beaker (Autoclave-3, Perkin-Elmer).

A particular source for systematic errors is the application of techniques suitable for higher concentrations to trace determinations without evaluation. This is particularly true for extraction techniques, since in the trace range the extraction need not be quantitative by any means.

Probably the most widespread cause of loss is adsorption onto materials that come into contact with the sample solution. This error can occur even at higher concentrations, but the effect is much more pronounced in the trace range. Neutral solutions of many metals are not stable and tend to hydrolyze. Silicic acid, on the other hand, is precipitated from acidic silicon solutions. The precipitates from dilute solutions often cannot be recognized and adhere to the walls of the container. Highly diluted solutions, even when acidified, are not stable over long periods, a fact that must be heeded for reference solutions. This effect has been impressively demonstrated by KAISER et al [2447] for mercury. But even with less mobile elements such as lead [2416], losses of up to 50% can occur within an hour when neutral aqueous solutions are stored in glass containers. Similar effects have also been reported for calcium and magnesium [2609], and can be expected to varying degrees for most elements.

The losses can frequently be controlled, at least for a few hours, by simply acidifying the diluted solutions with hydrochloric or nitric acid; some elements nevertheless require the addition of a complexing agent (such as potassium iodide for mercury) to prevent adsorption effects reliably. The proper choice of a suitable material for the container is of utmost importance; glass is unsuitable for virtually all elements in the trace range.

Since the adsorption is normally proportional to the adsorbing surface, the latter should be kept as small as possible. The use of filter papers is very critical since they have a very large surface area. The same is true for voluminous precipitates, which can adsorb the greater part of the trace elements.

A particular source of error in the graphite furnace technique is loss of analyte due to volatilization during thermal pretreatment in the graphite tube. This subject is treated fully in section 8.2.3, but is mentioned here to indicate that an element in a sample can, under given circumstances, be much more volatile than in an aqueous reference solution, since the chemical bonding and the ligands can have a substantial influence on the physical properties.

For example, BEHNE et al [2071] obtained a value of 22 ng/g chromium in brewer's yeast for a direct determination by the graphite furnace technique. After pressure digestion of the yeast in nitric acid, or also by neutron activation analysis, values of 150–160 ng/g chromium were measured. In this case it cannot be maintained that the direct method brings the more accurate results. In the graphite furnace technique it is therefore essential to avoid volatilization losses by carefully choosing the pretreatment temperature, and also through suitable digestion or matrix modification procedures.

A systematic error particularly widespread in the microtrace and nanotrace ranges is the contamination of the reagents, the laboratory ware, the instruments and the environment by the analyte element. In extreme cases the true content can be falsified by several orders of magnitude, so that the analytical result bears no relationship with the original concentration in the sample.

Reagents must therefore be chosen with extreme care for trace analysis. It is generally easier to maintain the purity of mineral acids than of salts, so that acid digestions are preferred to fusions. Further, the quantity of reagent used should be as small as possible, so that again, digestions in nitric acid, hydrochloric acid, hydrofluoric acid, etc. are especially favorable.

Frequently, however, even the quality of ultrapure analytical reagents is insufficient, because the sensitivity of the graphite furnace, hydride-generation and cold vapor techniques is simply too high. In figure 71, examples are presented of mercury blank values in proprie-

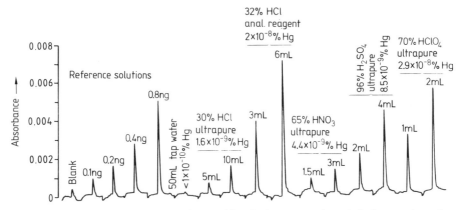

Figure 71. Blank values for mercury in various acids of ultrapure and analytical reagent grades, measured by the amalgam technique.

tary acids of varying purity determined with the amalgam technique. Hence the purification of reagents for use in trace analysis often cannot be avoided.

Acids are best purified by distillation at a temperature below the boiling point; a suitable apparatus is depicted in figure 72. Purified acids should only be permitted to come into contact with scrupulously clean containers and cannot be stored for any time, since they very quickly take up trace elements from the environment.

Ultrapure water [2940] is especially critical and should only be permitted to contact materials that have extremely low trace metal contents (PTFE, FEP, high impact polyethylene, or quartz); it should preferably not be stored at all. Since ultrapure water takes up trace elements very readily from the environment, it should be prepared fresh as required.

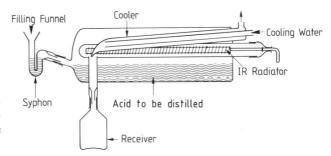

Figure 72. Apparatus for sub-boiling distillation (Kürner Analysentechnik, Federal Republic of Germany).

Substances used for matrix modification in the graphite furnace technique, such as ammonium salts, alkali metal and alkaline earth metal salts, can be purified by extraction. A solution of the salt at suitable pH is prepared and a universal complexing agent, such as APDC, is added. The solution is then repeatedly extracted with an organic solvent. In this way the trace element content can often be reduced by several orders of magnitude.

The importance of the laboratory ware in trace analysis has already been mentioned. Normal laboratory glassware is unsuitable for trace analysis. More suitable are materials with relatively inert surfaces, such as PTFE, FEP, high impact polyethylene, quartz (silica) and vitreous (glassy) carbon. However, it cannot be automatically assumed that these materials are not contaminated or that they remain uncontaminated for longer periods of time. It is therefore essential that vessels used for trace analysis are thoroughly cleaned before use. Experience has shown that washing or leaching is often insufficient, but that fuming out with nitric acid is mostly successful (see figure 73) [2074] [2897].

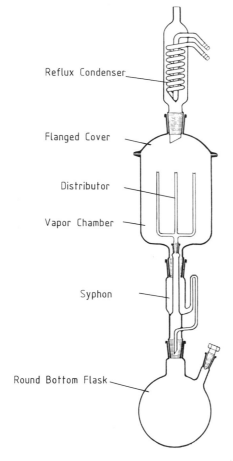

Reflux Condenser

Flanged Cover

Distributor

Vapor Chamber

Syphon

Round Bottom Flask

Figure 73. Apparatus for fuming out laboratory ware for use in trace analysis (Kürner Analysentechnik, Federal Republic of Germany).

As already mentioned in section 3.2.6, sample dispensing in the graphite furnace technique is especially critical. Pipet tips are frequently contaminated with trace elements and are difficult to clean [2074] [2897]. Proprietary yellow pipet tips, for example, are known to be contaminated with cadmium [2816]. An enormous improvement not only in the precision but also in the accuracy can be achieved through the use of an autosampler in which sample solution only comes into contact with materials such as PTFE (refer to figure 39, page 65) [3046] [3047]. The carryover from one sample to the next is also minimal in such a system [2920].

In trace analysis, automation has another major advantage, namely it replaces the analyst. Apart from the causes mentioned above, it is a well known fact that laboratory air and dust in the environment are potential causes of contamination. Thus it will often be necessary to perform trace determinations in a dust free area or even in a clean room. In such places the operator is the greatest source of contamination, hence the significance of automation.

As the concentration of the analyte element decreases, the systematic errors increase markedly. This is often hard to recognize, however, since the cause is to be found in the sample handling and pretreatment steps. The agreement of the analytical result with the true content in the original sample is the decisive question in trace analysis. It can only be answered by employing several independent analytical techniques, including independent digestion procedures, and by frequent inter-laboratory comparisons. The question of the accuracy can only be clarified when good agreement is found between the results obtained by various analytical techniques for the same sample [2977].

The quickest and simplest way to check the accuracy of measurements performed in a laboratory is to regularly analyze standard reference materials. Proof of the reliability of instrumental techniques is thus dependent on the availability of suitable reference materials [2976]. With the multitude of sample types presented for analysis and the enormously varying concentration ranges of the analyte elements (variations over many orders of magnitude), a satisfactory solution to this problem has not yet been found. Many national and international institutions are involved in seeking acceptable solutions, at least for the most pressing requirements. Among such institutions should be mentioned: The National Bureau of Standards (NBS), Gaithersburg/USA; the National Physical Laboratory (NPL), Teddington/UK; the Community Bureau of Reference (BCR), Brussels/Belgium; the International Atomic Energy Agency (IAEA), Vienna/Austria; and the Bundesanstalt für Materialprüfung (BAM), Berlin/Germany.

7. Interferences in Atomic Absorption Spectrometry

The presence of other constituents accompanying the analyte element in the sample can lead to interferences, which can cause systematic errors in the determination. The influence of the atomizing medium, such as the flame, graphite material, or quartz cell, or of the solvent is not regarded as an interference since sample and reference solutions are affected to equal degrees [2409]. An interference will cause an error in the analytical result only if the interference is not adequately accounted for in the evaluation procedure.

In spectrochemical analysis, interferences are generally divided into two classes: Spectral interferences and non-spectral interferences [2409]. Spectral interferences are due to the incomplete isolation of the radiation absorbed by the analyte element from other radiation or radiation absorption detected and processed by the electrical measuring system. With non-spectral interferences the analyte signal is affected directly.

7.1. Spectral Interferences

As already stated, spectral interferences are caused by the incomplete isolation of the radiation absorbed by the analyte element from other radiation or radiation absorption due to, or affected by, the interferent. Spectral interferences can arise from:
- Absorption of the source radiation by overlapping atomic lines or molecular bands of the concomitants.
- Scattering of the source radiation on non-volatilized particles from concomitants.
- The indirect influence of concomitants on the blank background absorption or scattering of the atomizer (e. g. the flame).
- By absorption of foreign radiation if, in addition to the analytical line, the source emits further radiation within the spectral bandpass of the monochromator. This effect is observed particularly when a continuum source is used.

In flame emission spectrometry it is often possible to observe spectral interferences that originate when thermal emission from the concomitants passes the monochromator or reaches the detector as stray light. Interferences of this nature do not arise in atomic absorption spectrometry. As a result of the alternating light system, in which the source radiation is modulated and the electrical measuring system is tuned to the same modulation frequency, all radiation emitted from the atomizer is eliminated since it is not modulated at the same frequency. This is a decisive advantage of AAS over all spectrochemical methods involving emission.

7.1.1. Direct Overlapping of Atomic Lines

At the temperatures prevailing in atomizers used in AAS, the number of absorption lines is much lower than the number of emission lines in the flame or from a high energy emission source. Further, since the absorption lines are relatively narrow, the probability of overlapping is relatively low. However, since a number of transition elements exhibit many spectral transitions, it is hardly surprising that spectral transitions of some elements do happen to have the same energy. A second element can thus absorb the radiation of the first; there is direct spectral overlapping. All spectral overlapping reported to date in the literature is presented in table 10.

The extent of the interference is dependent on the degree of overlapping of the emission line of the analyte element with the absorption line of the interfering element. Added to this, the absorption coefficient of the interfering element and the number of interfering atoms of concomitants in the optical beam play a part. As can be concluded from table 10, the interfering element must be present in large excess before a genuine interference occurs that can lead to incorrect measurements. The majority of these spectral interferences are of little analytical significance; the exceptions are between gallium and manganese [2030], and praseodymium and neodymium [2710].

Table 10. Spectral interferences in AAS caused by direct overlapping of analytical lines

Emission Line (nm)	Absorption Line Interfering Element (nm)	Signal Ratio
Al 308.215	V 308.211	200:1
Ca 422.673	Ge 422.657	—
Cd 228.802	As 228.812	—
Co 252.136	In 252.137	—
Cu 324.754	Eu 324.753	500:1
Fe 271.903	Pt 271.904	500:1
Ga 403.298	Mn 403.307	3:1
Hg 253.652	Co 253.649	8:1
Mn 403.307	Ga 403.298	—
Pr 492.495	Nd 429.453	—
Sb 217.023	Pb 216.999	10:1
Sb 231.147	Ni 231.097	—
Si 250.690	V 250.690	8:1
Zn 213.856	Fe 213.859	—

A spectral interference caused by direct overlap is best avoided by selecting an alternate analytical line. If this is not possible, then it is necessary to make an exact blank measure and subtract this from the analytical measure.

Next to direct overlapping of analytical lines, a second cause for interferences is the absorption of foreign radiation from the radiation source. Every source emits a multitude of spectral lines. If more than one of these lines lies within the bandpass of the monochromator, a non-linear analytical curve is the result (refer to chapter 4). Thus, as far as possible, it is usual to select a bandpass so that only the actual resonance line passes the exit slit of the monochromator. However, there are a number of cases where separation is no longer possible. For example, when an emission line exhibits hyperfine structure, the splitting of the individual components is far below the resolving power of monochromators used in AAS. In the case of multiplets, the individual lines have a wider separation, so that resolution is generally possible. However, it is sometimes preferable to forgo the better resolution in favor of an improved signal-to-noise ratio and to accept the more pronounced curvature of the analytical curve.

The risk of a spectral interference naturally grows with increasing number and width of the lines that pass through the exit slit of the monochromator. FASSEL [342] has also drawn attention to the fact that absorption lines can become relatively wide in the presence of high element concentrations. In the case of calcium, he was still able to measure 10% of the maximum intensity energy at a distance of 0.03 nm from the line center. The total width of an absorption line can on occasions increase to 0.1 nm; more than an order of magnitude wider than the assumed half-intensity width of 0.003-0.005 nm. PANDAY and GANGULY [2728] have reported overlapping that derives from broadening of absorption lines.

The indiscriminate use of multielement radiation sources brings a considerable risk for the occurrence of spectral interferences. Unsuitable elements may be combined in such sources if the manufacturer pays insufficient attention to this problem. In such cases a spectral line from a concomitant element may lie within the bandpass of the monochromator in addition to the resonance line. If this occurs the selectivity of AAS is no longer given and the analytical measure will be no longer element specific. Since the risk of this type of spectral interference increases with increasing bandpass, only the bandpass recommended by the manufacturer should be used. In cases of doubt the narrower bandpass should always be chosen.

Radiation sources that nominally emit the spectrum of only one element but *de facto* contain several elements pose an especial danger. This can happen if the manufacturer fails to use spectrally pure metal, if the cathode is made of an alloy or intermetallic compound for reasons of stability, or if a volatile metal is contained in a support cathode made of other metals. The second metal is not normally stated by the manufacturer, but it nevertheless emits a spectrum. Manufacturers of radiation sources do not, of course, normally combine unsuitable metals, but if a source is used that is not specifically offered for use in AAS, great caution is advised, especially at secondary resonance lines.

If a continuum source is employed instead of a spectral line source, then the door is thrown wide open to spectral interferences and the specificity of AAS is lost. Every element that can absorb within the bandpass of the monochromator will generate a signal; each signal depends only on the absorption coefficient and the concentration of atoms in the optical beam.

7.1.2. Molecular Band Overlap and Radiation Scattering on Particles

Although both effects—absorption of source radiation by overlapping molecular bands of concomitants and radiation scattering on non-volatilized particles—are completely different, they will be treated together since in practice it is often very difficult to distinguish between them. Also, the same measures are taken for their elimination. Both effects are combined under the general term non-specific or **background attenuation,** or sometimes non-specific radiation losses, or even background absorption, although scattering is naturally not an absorption phenomenon.

Background attenuation is caused by non-specific loss of radiation (i. e. radiation attenuation not caused by the specific analyte element) and always leads to an incorrectly high measure. This measure comprises the true absorption of the analyte element and the non-specific attenuation of the background. A separation of both signals is not possible.

The scattering of source radiation on non-volatilized particles of concomitants obeys RAYLEIGH's law of scattering. The scattering coefficient τ is given in the first approximation by:

$$\tau = \frac{J_t}{J_0} = 24\,\pi^3\,\frac{N \cdot v^2}{\lambda^4}\,. \tag{7.1}$$

Scattering is thus directly proportional to the number N of scattering particles per unit volume and to the square of the particle volume v, but indirectly proportional to the fourth power of the wavelength λ. Radiation scattering therefore occurs more strongly with increasing particle size (τ increases by a factor of 64 for a doubling of the particle radius) and decreasing wavelength (from 800 nm to 200 nm, τ increases by a factor of 256).

As experiments performed by KERBYSON [639] have shown, the maximum droplet volume for a direct injection burner is 100 fold greater than for a premix burner, so that the radiation scattering must be greater by a factor of approximately 10000. Radiation scattering rarely occurs when a well-designed premix burner is employed, provided that the total concentration of the solution aspirated does not exceed 5% and the wavelength is above 300 nm. At wavelengths below 250 nm certain interferences are to be expected when the solids content of the solution is higher, especially in cooler flames. BILLINGS [123] measured the radiation scattering caused by various salts in an air/propane flame. In figure 74 the values measured by BILLINGS for a 1% calcium solution are plotted versus the calculated curve. Nowadays it can be assumed that for a well-designed premix burner and an air/acetylene flame, virtually no unvolatilized particles pass into the optical beam; interferences

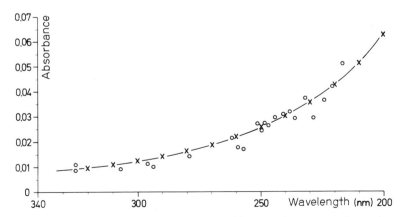

Figure 74. Non-specific radiation losses caused by scattering, measured at various wavelengths; 1% Ca solution in an air/propane flame: o values measured by BILLINGS [123]; × theoretical curve calculated from RAYLEIGH's law of scattering.

caused by radiation scattering are then hardly to be expected. For graphite furnaces, however, the situation is quite different. Particularly when organic materials are charred in an inert gas atmosphere, considerable quantities of particles can pass into the optical beam. Also, when volatilized inorganic materials pass into cooler zones they frequently condense and lead to radiation scattering [447].

The reasons for the occurrence of higher concentrations of molecules in the optical beam vary for the different types of atomizer. In a flame, higher concentrations of molecules are often formed through reaction of one of the concomitants with components of the flame gases. If an easily atomized element is determined in a flame of relatively low temperature in the presence of a concomitant that forms a difficultly dissociated oxide, hydroxide, cyanide, etc., the analyte element will be atomized, but not the concomitant. This is the most common cause for molecular absorption because of the buffer action of the flame gases, which are always present in large excess. The spectra of halides have also been observed in the flame when the sample contained high concentrations of halogen compounds.

In a graphite furnace no gas is present in sufficient concentration to have a buffer action, so the concomitants in the sample influence the atmosphere. The compounds having the greatest thermodynamic stability at the prevailing temperature will thus be formed. For many elements these are gaseous halides, or frequently monocyanides in the presence of nitrogen [2880]. GÜÇER and MASSMANN [447] have undertaken detailed studies on these types of molecular spectra in graphite furnaces. For example, they scanned the dissociation continua of numerous molecules that exhibit characteristic, broad bands (figure 75) and whose long wavelength limits are determined by the dissociation energy.

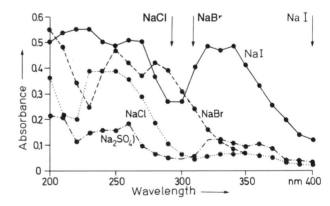

Figure 75. Dissociation continua of NaCl, NaBr, and NaI scanned in a graphite furnace.—Limiting wavelengths for photodissociation are indicated by arrows (according to [447]).

A spectral interference naturally only occurs when a molecular band of this type coincides with the analytical line of the analyte element. The effect is then virtually identical to that of two directly overlapping atomic lines. The total absorption thus comprises the specific absorption of the analyte element and the molecular absorption; separation of the two components is not possible.

Various procedures have been proposed to eliminate the non-specific attenuation caused by molecular absorption and radiation scattering on particles. For the flame technique, one of the most effective measures is to use a flame of higher temperature and strongly reducing properties, i. e. having a low partial pressure of oxygen. With a well-designed premix burner and an air/acetylene flame, particles in the optical beam are hardly likely to be a problem, as mentioned above. Similarly, the number of molecules are reduced drastically when the change is made to a nitrous oxide/acetylene flame. Background attenuation is hardly observed in this flame. However, a number of elements exhibit markedly reduced sensitivity in the hotter flame, so that it cannot always be used.

In the graphite furnace technique, matrix modification is a very effective means of at least considerably reducing background attenuation. A reagent is added to the sample in high concentration to make either the analyte element more stable or the concomitants more volatile, so that better separation is achieved before the atomization step. This procedure is treated more fully in section 8.2.2.

A further procedure that has been proposed is to measure a "blank sample". This sample should exhibit the same background attenuation as the true sample, but should not contain the analyte element. This procedure can rarely be performed in practice, however, because such blank samples are seldom available and attempts to prepare them synthetically mostly fail due to the high purity requirements placed on each constituent. Furthermore, the composition of the concomitants often varies from sample to sample and their true content is frequently unknown.

Most procedures originally proposed for the elimination of background attenuation were based on the fact that while atomic absorption occurs in a very narrow spectral range, molecular absorption and radiation scattering are both broad band phenomena. One such procedure is the reference element technique (also frequently termed "internal standard technique", although this name is disapproved by IUPAC). In this technique, the background attenuation of a spectral line of the reference element having a wavelength close to the analyte element is measured. The background attenuation is then subtracted from the analyte measure. A prerequisite for this technique is naturally that the reference element is not contained in the sample, and that under the given conditions it does not itself atomize and cause an absorption signal.

It is extremely important to realize that outside of a spectral range of only a few tenths of a nanometer, constant background attenuation cannot be guaranteed. According to RAY-LEIGH's law of scattering, the radiation scatter at 200 nm, for example, will change by about 20% at a distance of 10 nm. Changes in molecular absorption are even more rapid, depending on which part of the band the measurement is performed. A basic requirement for all procedures for the correction of background attenuation (frequently termed background correction procedures) is that correction should be performed "as close as possible" to the analyte absorption line.

For the reference element technique, the "next best" element must be chosen for correction of the background attenuation [1171], i. e. the element whose wavelength lies closest to that of the analyte. Frequently however, the next-lying resonance line is so far away that only a qualitative, or at best semiquantitative, estimation of the background attenuation at the analyte spectral line can be made.

WILLIS [1330] and SLAVIN [1140] therefore proposed correcting background attenuation by measuring the effect at a non-absorbable line and then subtracting the value obtained from the analyte measure. Non-absorbable lines are, for example, emission lines ending at a higher excited state, ion lines, or carrier gas lines; in general, emission lines that cannot be absorbed by the analyte element in the atomizer. Although such lines do not exhibit element-specific absorption, they can be attenuated by scattering, molecular absorption, etc. The advantage of using such lines for correction is that any attenuation measured can only be non-specific and that many more lines are available than resonance lines.

Nevertheless, even this procedure has the disadvantage that frequently there is no suitable non-absorbable line in the direct neighborhood of the resonance line, so that exact correction is not always possible. The precision of the correction procedure can be increased by measuring the background attenuation at several non-absorbable lines to either side of the resonance line, but this is time consuming and complicated.

Nowadays, instrumental techniques are used almost exclusively for the correction of background attenuation. A well-established technique for correction is to measure the radiation attenuation of a spectral line source (element specific radiation source) and a continuum source in rapid sequence. A further technique for very exact background correction introduced in recent years makes use of the Zeeman effect.

7.1.3. Background Correction with Continuum Sources

For the correction of background attenuation, KOIRTYOHANN [681] and L'VOV [778] used dual channel systems in which the radiation from a hollow cathode lamp was passed alternately with the radiation of a continuum source through the flame or graphite furnace, respectively. KAHN [608] described the first commercial system based on this principle that could be used with a proprietary atomic absorption spectrometer. By means of a rotating chopper with sector mirror, the radiation from the primary source and the radiation from a continuum source is passed alternately through the atomizer. Such a system is depicted schematically in figure 76. After passing the monochromator, both radiation beams fall on the same detector. The electrical measuring system forms the ratio from both radiant intensities. A deuterium lamp is used most frequently as the continuum source.

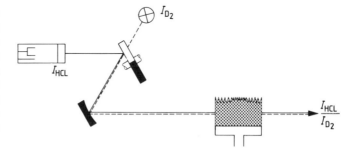

Figure 76. Schematic configuration of a deuterium background corrector. — By means of a rotating chopper with sector mirror, radiation from the primary source (I_{HCL}) is passed alternately with radiation from the deuterium lamp (I_{D_2}) through the atomizer.

The mode of function of such a background corrector is depicted in figure 77. The exit slit of the monochromator separates the resonance line from the emission spectrum of the element-specific (primary) radiation source, while from the continuum source a band of radiation equivalent to the bandpass leaves the exit slit. The half-intensity width of the spectral line from the primary source is about 0.002 nm; the width of the continuum corresponds to the selected bandpass, usually 0.2 nm or 0.7 nm. The intensity I_{ps} of the primary source is equalized to the intensity I_{cs} of the continuum source before the determination so that the ratio I_{cs}/I_{ps} is unity and no reading is presented on the display.

When normal radiation absorption by the analyte element in the atomizer occurs, I_{ps} is attenuated in proportion to the atom concentration. Naturally I_{cs} is also attenuated at the res-

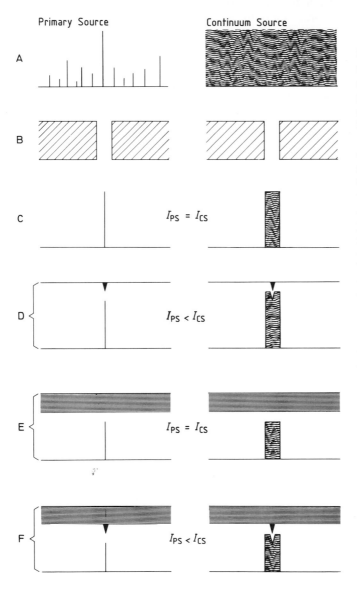

Figure 77. Mode of operation of a deuterium background corrector. — A The primary source emits a line spectrum while the deuterium lamp emits a continuum. B The exit slit of the monochromator isolates the resonance line from the spectrum of the primary source, with a half-intensity width of approximately 0.002 nm, and passes a band of radiation from the deuterium lamp equivalent to the selected bandpass (around 0.2 or 0.7 nm). C The radiant intensities of the two sources are equalized within the observed spectral range. D For normal atomic absorption by the analyte element, I_{PS} is attenuated by an amount equivalent to its concentration, while I_{CS}, in the first approximation, is not attenuated. E Broad band background attenuates the intensity of both sources to the same degree. F Atomic absorption by the analyte in addition to the background attenuates I_{PS} again by an amount equivalent to its concentration, while I_{CS}, in the first approximation, is not further attenuated.

onance wavelength, but since the half-intensity width of the resonance line is only about 0.003 nm while the continuum has a width of at least 0.2 nm, even for 100% absorption of the resonance radiation from the primary source, maximum 1.5% of the continuum would be absorbed. Thus for normal measurements, absorption of the continuum is under 1% and can be neglected. This is naturally the more so for bandpasses of 0.7 nm and above. Consequently, while I_{cs} is virtually unattenuated when it reaches the detector, I_{ps} is attenuated proportionally to the concentration of the analyte in the sample. The ratio I_{cs}/I_{ps} is greater than unity and a reading is obtained. In effect, I_{cs} serves as the reference beam in a double-beam system.

If non-specific attenuation of the radiation occurs, whether through scattering or molecular absorption, both radiation beams will be attenuated to the same extent, since background attenuation is usually a broad band phenomenon. It is nevertheless assumed that the background attenuation is constant over the observed spectral range of 0.2 nm or 0.7 nm.

Since radiation scattering and molecular absorption attenuate I_{cs} and I_{ps} to equal degrees, the ratio I_{cs}/I_{ps} remains unity; no reading is obtained, and thus the background is corrected. If element-specific absorption takes place in addition to background attenuation, I_{ps} will be further attenuated, leading to a normal reading for the analyte measure.

Background correction by comparing the attenuation of the spectral line source (hollow cathode lamp, electrodeless discharge lamp, etc.) with the attenuation of a continuum source has the decisive advantage that correction is at the resonance line, and not at a neighboring line as in other procedures. Genuine correction of the background with good precision and accuracy is thus possible, assuming that the background attenuation is continuous within the observed spectral range. Correction is also automatic and requires no further operating steps.

While background interferences in a well-designed premix burner are relatively seldom, as explained earlier, they are quite frequent in the graphite furnace technique. It is thus hardly surprising that with the increasing interest in this technique, background correction procedures were also further developed. Consequently single-beam spectrometers permitting the use of a background corrector [1292] and double-beam instruments with a background corrector in each channel to permit true double-beam operation were introduced onto the market. Further, the deuterium lamp has been complemented with a halogen lamp to permit correction over the entire spectral range of AAS. At this point it should be briefly mentioned that a hydrogen hollow cathode lamp is little suited as the continuum source because of its very low radiant energy, leading to relatively high noise during the measurement (refer to figure 78).

Nevertheless, background correction with a continuum source has its disadvantages and limitations. Alone due to the second radiation source and the altered signal handling, the noise is increased by a factor of two to three. Strong background attenuation brings a further increase in the noise. In general, it should be attempted to keep the background attenuation to below 0.5 absorbance; frequently values of over 0.7 to 0.8 absorbance are incompletely corrected, especially the fast, dynamic signals from the graphite furnace.

DE GALAN and MASSMANN in particular have repeatedly drawn attention to the fact that the uncritical use of a background corrector cannot guarantee the accuracy of the results; on the contrary, it can even cause errors. For samples of unknown composition, it is impor-

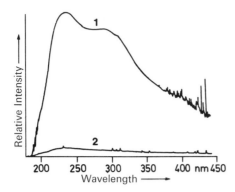

Figure 78. Emission of a deuterium arc lamp at an operating current of 800 mA (**1**) compared to the emission of a hydrogen hollow cathode lamp at an operating current of 40 mA (**2**).

tant to perform measurements both with and without background correction. In this way, important information on the nature and magnitude of the background attenuation can be won. This information would be lost if measurements were performed solely with background correction. Most modern atomic absorption spectrometers permit simultaneous measurement of the uncorrected or background signal and the corrected signal.

It has already been mentioned that the continuity of the background within the bandpass of the monochromator is a prerequirement for correct functioning of a continuum source background corrector. This is given for radiation scattering and dissociation continua. If molecules absorb radiation having a quantum energy higher than the dissociation energy, the final state is not a discrete energy state. It is then possible to observe a spectral continuum above a given limiting wavelength.

Background correctors employing continuum sources are incapable of correcting the background attenuation of electronic excitation spectra because these comprise many narrow lines. These spectra derive from electronic transitions within the molecule and the structure of the bands is determined by transitions from the rotational and vibrational levels of one electronic state to those of a different electronic state; i.e. transitions between discrete energy states of the molecule [2629].

For this type of electronic excitation spectrum, the actual background correction depends on the degree of overlap between the elemental spectral line and the individual molecular rotational lines [2365] [2630]. This demands powers of resolution far beyond the capabilities of the monochromators used in atomic absorption spectrometers.

As an example, a portion of the absorption spectrum of indium chloride scanned with a very high resolution spectral apparatus is depicted in figure 79. The resonance line of gold at 267.6 nm lies exactly in the middle between two rotational lines, so that the actual background attenuation is relatively low. If a continuum source background corrector is now used, it determines the mean absorbance over the observed spectral range, and this is naturally higher than the attenuation at the gold resonance line. This results in overcompensation when the readings are subtracted and thus a false analytical measure [2384].

MASSMANN *et al* [2630] reported similar interferences for the determination of bismuth (306.8 nm) and magnesium (285.2 nm) through absorption bands of the OH radical in an air/acetylene flame. These authors have also scanned the electronic excitation spectrum of

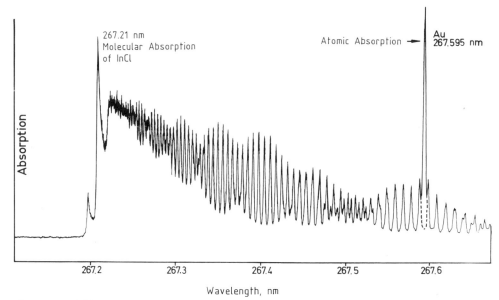

Figure 79. Highly resolved absorption spectrum of InCl in the neighborhood of the gold resonance line at 267.6 nm. The resonance line is exactly in the middle between two rotational lines of the InCl molecule (from [2629]).

PO in an air/acetylene flame and found coincidence with the resonance lines of iron (247.3 nm), palladium (244.8 nm and 247.6 nm) and ytterbium (246.4 nm).

MARKS *et al* [2624] mention a possible source of error when a continuum source is used due to absorption by concomitant atoms of the continuous radiation passed by the monochromator. This also results in overcompensation. MANNING [2614] reported this type of spectral interference for the determination of selenium in the presence of iron and VAJDA [2995] found a number of similar examples.

Although the limitations of continuum source background correction have been known for some years, there still appear to be analysts who have blind faith in its infallibility. Naturally a background corrector is an extremely useful device that frequently leads to the correct results. However, for samples of unknown composition, a check should always be made to see whether correction is complete or not.

Particularly for determinations in a graphite furnace it is important that the radiation beams from both sources coincide exactly within the atomizer. If this is not so, false results will be obtained when the sample vapor is not distributed homogeneously. This can be checked quite easily by holding a screen in the optical beam path; the display should not change markedly when the screen is moved backward and forward.

By simply diluting the sample, it is quite easy to check whether correction is complete. If the measure is disproportionally smaller, then non-corrected background was present. A further qualitative check can be made by performing the measurement with background correction at a neighboring, non-absorbable line. If a signal is obtained, correction must be incomplete since no element-specific absorption can take place at this line.

A good way of testing the presence of background attenuation due to fine structure, for example from electronic excitation spectra, is to perform several measurements at different bandpasses. If the extent of correction (not the absorbance) is dependent on the bandpass, non-continuous background is in all probability present and the risk of over or under correction exists.

7.1.4. Application of the Zeeman Effect

This effect, the splitting of spectral lines in a magnetic field, was discovered by the Dutch physicist ZEEMAN [3098] in 1897. Under closer scrutiny it is not the spectral lines that split, but the energy levels (terms) in the atoms. In figure 80, the splitting of the terms and the resulting splitting of the spectral lines is depicted schematically.

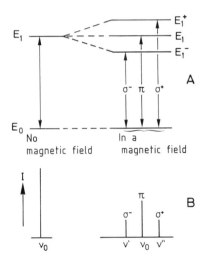

Figure 80. Normal Zeeman effect. **A**: Splitting of the terms; **B**: The resulting splitting of the spectral lines.

In the simplest case the terms, and thereby the spectral lines, split into three components; a central π component whose energy and frequency is unchanged with respect to the original level, and two components of slightly higher and lower energy, respectively. Designated σ^+ and σ^-, these components are shifted to the left and right of the original spectral line, i.e. to higher and lower frequencies. The extent of the shift depends on the applied magnetic flux density (also frequently called magnetic field strength). The distribution of the energy or intensity between the three components is according to $\sigma^+ : \pi : \sigma^- = 25:50:25$. The sum of the individual components equals 100, the intensity of the original line.

Simultaneously with the splitting of the spectral lines the radiation is also polarized. The plane of polarization varies according to the direction of the magnetic field and also ac-

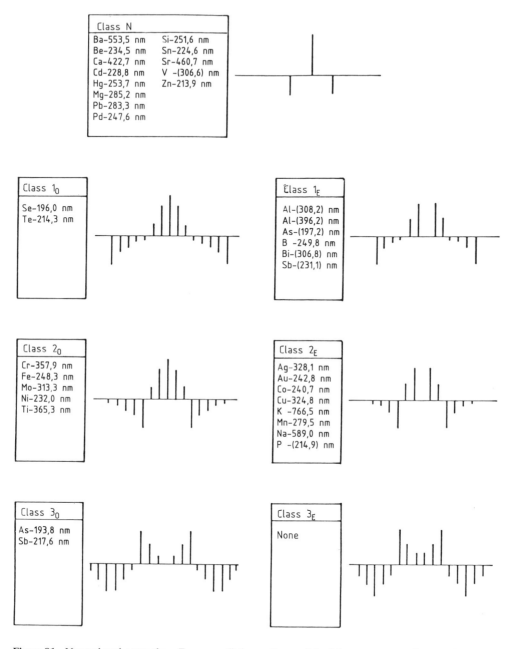

Figure 81. Normal and anomalous Zeeman splitting patterns of the 35 most common elements in AAS. Class N: Normal Zeeman splitting. Classes 1_O, 2_O, 3_O: Anomalous Zeeman splitting into an odd number of components. Classes 1_E, 2_E, 3_E: Anomalous Zeeman splitting into an even number of components. Secondary resonance lines are given in brackets.

cording to the direction of observation (in the direction of or at right angles to the magnetic field). The significance of this effect will be treated in detail later.

Splitting of the spectral lines into three components occurs only for singlet lines (terms with $S = 0$) and is designated the *normal Zeeman effect*. Singlet lines are, for example, the main resonance lines of Group IIA and IIB elements (Be, Mg, Ca, Sr, Ba, Zn, Cd, Hg). All other lines exhibit an *anomalous Zeeman effect* and split into more than three components. It is characteristic for the anomalous Zeeman effect that the π component also splits into several lines and thus no longer coincides exactly with the original resonance line. A further distinction must be made between splitting into an odd number of π components, where the original line is at least retained in one component, and splitting into an ever number of π components, so that the original line disappears completely. Zeeman splitting patterns for a number of common elements together with their resonance wavelengths are presented in figure 81. For elements exhibiting fine structure, due to isotope shifts for example (figure 82), a further variant is present since each isotope displays its own Zeeman splitting pattern.

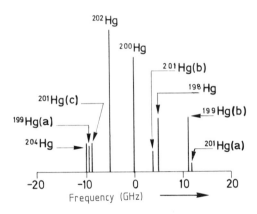

Figure 82. The individual isotope lines of the mercury line at 253.65 nm. Each of these lines exhibits its own Zeeman splitting pattern (according to [3089]).

As mentioned above, next to the Zeeman splitting of the resonance lines, the radiation is also polarized. The polarization varies according to the direction of observation. If the radiation is observed in a direction perpendicular to the magnetic field, or if the magnetic field is applied at right angles to the radiation beam, the π component is polarized in a direction parallel to the magnetic field, while the σ components are polarized in a plane perpendicular to the magnetic field (figure 83). This configuration is termed the **transverse Zeeman effect**. When the direction of observation is parallel to the magnetic field, or the magnet is parallel to the radiation beam, we speak of the **longitudinal Zeeman effect**. In this latter effect the π component is missing from the spectrum and only the circularly polarized σ^+ and σ^- components can be observed (figure 83). The longitudinal Zeeman effect is thus distinguished from the transverse effect not only by the form of polarization, but also in appearance since the π component is missing. The intensity distribution is moreover different, since both σ components exhibit half the intensity of the original spectral line.

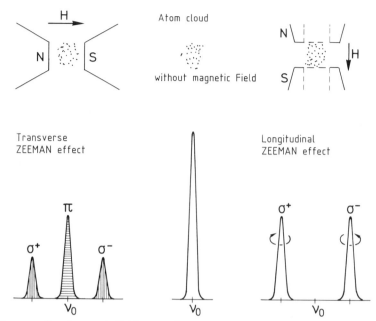

Figure 83. Transverse Zeeman effect (magnetic field perpendicular to the radiation beam) and longitudinal Zeeman effect (magnetic field parallel to the radiation beam).

As mentioned at the start of this section, it is in fact the energy levels (terms) in the atom and not the spectral lines that are split. This means that the magnetic field cannot be situated arbitrarily in the radiation beam, but must be applied to the cloud of atoms. For this there are two possibilities in AAS: The atom cloud in the primary radiation source and the atom cloud in the atomizer.

When the magnetic field is applied to the radiation source, we speak of the **direct Zeeman effect**. Here, the energy levels of the emitting atoms are split and splitting of the spectral lines can thus be observed, as well as varying polarization of the components. Applying the magnetic field to the atomizer is termed the **inverse Zeeman effect**. In this case the radiation source emits non-split spectral lines as usual in AAS. In the inverse Zeeman effect the energy levels of the absorbing atoms are split, and thus their absorbing characteristics totally changed. Since the radiation source emits the original resonance line, this radiation can only be absorbed by the π component of the analyte, but not the σ components. For the direct Zeeman effect, on the other hand, absorption can also be measured at the wavelengths of the σ components, since both π and σ components are emitted by the radiation source. Normally, of course, the analyte atoms only absorb the non-shifted π component.

A final difference, particularly noticeable in the manner of measurement, is whether a constant field or an alternating field is used to split the energy levels. A constant field, generated by a permanent magnet or a direct current electromagnet, splits the emission or absorption lines of the atoms permanently. To permit measurement of the various components, it is necessary to utilize their differing polarization, for example by placing a rotating polarizer in the radiation beam. An alternating magnetic field, generated by an alternating current electromagnet, only splits the spectral lines when the field is on. When the field is off, the lines are not split and normal AAS measurements are performed.

From the various ways of applying the Zeeman effect—transverse or longitudinal, direct or inverse, constant field or alternating field—there are eight possible configurations. Between these possibilities, there are considerable differences with respect to their realization, application, and manner in which the measurement is performed. A survey is presented in table 11.

Table 11. Various configurations for the application of the Zeeman effect in AAS and their particularities

Location of magnet	Orientation of magnet to radiation beam	Type of magnetic field	Particularities
At the radiation source (direct)	Parallel (longitudinal)	Constant	Rotating polarizer
		Alternating	No polarizer required
	Perpendicular (transverse)	Constant	Rotating polarizer
		Alternating	Fixed polarizer
At the atomizer (inverse)	Parallel (longitudinal)	Constant	Not applicable in AAS
		Alternating	No polarizer required
	Perpendicular (transverse)	Constant	Rotating polarizer
		Alternating	Fixed polarizer

PRUGGER and TORGE [2765] first proposed the application of the Zeeman effect in 1969 for the correction of background attenuation and for compensation of flame noise. They applied a magnetic field to the radiation source and introduced a rotating polarizer into the optical beam to permit measurement in rapid sequence of the π component polarized parallel to the magnetic field and the σ components polarized perpendicular to the magnetic field. While the π component is attenuated by both the analyte atoms and the background, the σ components are attenuated by the background alone. Subtraction of the two readings gives the true atomic absorption. The requirements for automatic background correction are given when the measurements are performed in rapid sequence (e. g. at 50 Hz).

PARKER and PEARL [2730] later showed that by applying the Zeeman effect to a single-beam instrument, a double-beam effect could be achieved. Since both components are emitted by the same radiation source, pass through the same atomizer and have the same optical path, by comparing the two signals in rapid sequence, drift phenomena of the radiation source can be corrected just as well as noise and radiation absorption of the flame.

Up to 1976, virtually all applications made use of the direct Zeeman effect, i. e. the magnetic field at the radiation source. The system proposed by PRUGGER and TORGE [2765] was also taken up by KOIZUMI and YASUDA [2483], and STEPHENS and RYAN [2914] [2915]. In this system, the emission lines from the radiation source are permanently split into the π and σ components. When the rotating polarizer is parallel to the magnetic field, the non-shifted π component is passed and both atomic absorption and non-specific attenuation are measured. When the polarizer is perpendicular to the magnetic field, the slightly shifted σ components are passed. These are shifted sufficiently to prevent overlapping with the absorption line and so only the background attenuation is measured. It is important to note that in the two measurement phases, two slightly different wavelengths are employed.

The use of a transverse alternating field at the radiation source is not mentioned in the literature. As with all systems employing an alternating field, measurements are taken in rapid sequence with field on and field off. All measurements in the "field-off" phase are conventional atomic absorption; the radiation source emits a resonance line at the original frequency and both atomic absorption and background attenuation are measured. In the "field-on" phase, Zeeman splitting of the source radiation into polarized π and σ components taken place. A rotating polarizer is not required for an alternating field, however; a fixed polarizer perpendicular to the magnetic field is all that is required to distinguish between the two phases. The π component is held back in the field-on phase, while the background attenuation is measured at the slightly wavelength-shifted σ components.

Since 1971, HADEISHI *et al* [2333] [2335] have undertaken extensive studies on various Zeeman systems. They started by examining a longitudinal constant field at the radiation source. In a longitudinal field the π components are missing, so that measurements at the original resonance line are not possible. HADEISHI used an electrodeless discharge lamp filled with the pure isotope ^{199}Hg (later with the isotope ^{198}Hg) as the radiation source. In a suitable magnetic field (about 0.7 Tesla), the isotope line splits in such a way that the σ^- component coincides with the center of the absorption line for natural mercury, while the σ^+ component is situated at the far, blue end of the absorption profile. The atomic absorption and background attenuation are measured with the σ^- component, while the background attenuation alone is measured with the σ^+ component. A rotating polarizer is used to distinguish between the polarized σ components. This technique is only possible for elements having suitable natural isotopes and exhibiting sufficient isotope shift. It is therefore limited to only a few elements.

A system with a longitudinal alternating field at the radiation source, on the other hand, can find general application since in the field-off phase the normal resonance line is emitted. In the field-on phase the shifted σ components are emitted, permitting measurement of the background attenuation, since the analyte atoms do not absorb. As the π component is missing in the longitudinal Zeeman effect, atomic absorption cannot occur during the field-on phase. A polarizer is therefore not required to separate the components. In this configuration the Zeeman shift is used for background correction.

For the inverse Zeeman effect with the magnetic field at the atomizer, the same considerations apply in principle as for the direct Zeeman effect. The major difference is that the radiation source emits the normal resonance line continuously, while the ability of the analyte atoms to absorb is governed by the magnetic field.

A transverse constant field has been described most frequently for the inverse Zeeman effect, probably because it is the simplest technical solution. KOIZUMI and YASUDA [2486] [2487] studied this configuration in detail, and also DAWSON *et al* [2187] and FERNANDEZ *et al* [2246] have examined the applicability of this system.

Due to the magnetic field at the atomizer the ability of the atoms to absorb is permanently changed. The absorption spectrum (which would naturally only be visible when a continuum source was used as the primary source) splits into a π component polarized parallel to the magnetic field and having the same frequency as the original absorption line, and into σ components that are polarized perpendicular to the magnetic field and are shifted to slightly higher and lower frequencies. If a rotating polarizer is introduced into the optical beam, the atomic absorption of the π component and the attenuation due to the non-polarized background are observed when the polarizer is parallel to the magnetic field. When the polarizer is in the phase perpendicular to the magnetic field, the π component is no longer passed, i. e. only the attenuation due to the non-polarized background is measured, but not the atomic absorption. The difference between the two measures gives the true atomic absorption.

It is important to note that, in contrast to the direct Zeeman effect (magnetic field at the radiation source), measurements are performed at the actual resonance line since this is emitted from the radiation source. The σ components, provided they are shifted far enough, do not affect the measurement since they do not overlap the radiation from the primary source.

In 1978, DE LOOS-VOLLEBREGT and DE GALAN [2192] [3101] proposed applying a transverse alternating field to the atomizer (graphite furnace) since this configuration offers a number of advantages [2194]. During the field-off phase the atomic absorption and background attenuation are measured as usual. During the field-on phase, a fixed polarizer orientated in a plane perpendicular to the magnetic field holds back the π component, so that only the non-polarized background is measured. FERNANDEZ *et al* [2244] examined the efficiency of this system in correcting high non-specific attenuation in the graphite furnace, while BRODIE and LIDDELL [2107] applied it to the flame technique.

Turning to the two possible configurations of a longitudinal magnetic field at the atomizer, the direct field system cannot be realized. The π component disappears permanently from the spectrum, so that the measurement of atomic absorption is impossible. The second configuration, longitudinal alternating field, was examined in 1975 by UCHIDA and HATTORI [2990], and a year later by OTRUBA *et al* [2716], in each case with a flame as the atomizer. In 1978, DE LOOS-VOLLEBREGT and DE GALAN [2192] proposed that this system was also optimal for the graphite furnace.

With a longitudinal alternating field at the atomizer, atomic absorption and background attenuation are measured during the field-off phase. During the field-on phase only the σ components absorb, so that with a sufficient shift no atomic absorption and only background attenuation is measured. In this configuration a polarizer is not required; utilization is made of the shift of the σ components and the extinction of the π component by the Zeeman effect.

It would be appropriate at this point to consider the instrumental aspects and the influence the Zeeman effect has on important components such as the radiation source and the atomizer.

With the direct Zeeman effect a strong magnetic field is applied to the radiation source. The standard, proprietary sources used in AAS are not especially suitable for this. Due to their dimensions, an air gap of several centimeters between the poles of the magnet is unavoidable. To obtain the required magnetic flux density of around 1 Tesla (10 kGauss) the magnet would have to be very large.

A greater problem than the dimensions of the magnet is the fact that it is difficult to start a glow discharge and to maintain stable emission in a strong magnetic field. Solely electrodeless discharge lamps run at reduced power appear to be suitable radiation sources. Mercury is the only element, however, that has sufficient vapor pressure under these conditions to emit stably. It is therefore hardly surprising that the initial investigations on the Zeeman effect were concerned with this element almost exclusively [2333–2335] [2483].

With other elements, even when they form volatile halides, the electrodeless discharge lamps must be heated to 250°C to 300°C to maintain the required vapor pressure. This is necessary to obtain stable emission when the source is run with 2 W power at 100 MHz.

Electrodeless discharge lamps, even when they are additionally heated, can only be manufactured for elements of relatively high vapor pressure, such as arsenic, cadmium, lead, selenium and zinc. This lamp type thus does not come into question for the majority of elements determinable by AAS. Consequently, for the direct Zeeman effect, it is necessary to develop special radiation sources that emit stably in a strong magnetic field.

For several elements, STEPHENS [2910] constructed capacitively coupled sources, run at a relatively low frequency of 2 MHz, that were magnetically stable. KOIZUMI and YASUDA [2485] designed a lamp in which the anode and the cathode were in the form of two parallel plates. This source was operated at 500 V under a high frequency field of 100 MHz. MURPHY and STEVENS [2678] attempted to run conventional hollow cathode lamps at 2.5 MHz, but found that the cathodes were rather quickly destroyed. Other attempts to operate hollow cathode lamps in strong magnetic fields were also not particularly successful [2911] [2912].

The application of the direct Zeeman effect to AAS founders on the inability to operate conventional hollow cathode lamps in strong magnetic fields. The direct Zeeman effect thus requires the development of special radiation sources; this is certainly possible for a few cases, but presents great difficulties for the majority of elements. It is hardly likely that radiation sources stable in high magnetic fields can be developed for all the elements determinable by AAS.

It can thus be seen as a great advantage of the inverse Zeeman effect, with the magnetic field at the atomizer, that the conventional radiation sources for AAS can still be used. The question now arises, however, as to any restrictions regarding the atomizer.

In most earlier work involving the Zeeman effect in AAS a flame was used as the atomizer. Under these circumstances it is hardly surprising that the direct Zeeman effect was used almost exclusively, since it is not easy to combine a magnet with a burner. Problems are caused by the dimensions of conventional burners—the flame has a length of 5–10 cm—and by the radiated heat. The radiated heat usually necessitates water-cooling of the magnet and a larger clearance between the magnetic poles and the flame. Resulting from the length of the flame and the large air gap, magnets must be bulky and unwieldy, and a compromise must generally be made regarding the attainable magnetic flux density.

Recently, various flame/Zeeman magnet combinations have been reported. A transverse constant field was used in two systems [2246] [3089] and a transverse alternating field in a third [2107]. In all cases, however, the magnetic flux densities attained were not as high as those used in graphite furnace systems [2246] [3089] and were hardly sufficient to cause optimum splitting [2107].

There are obviously considerable difficulties in achieving a sufficiently high magnetic flux density in the flame. This problem is offset by the fact that in principle there is no necessity whatsoever for applying the Zeeman effect to the flame technique. Background attenuation is usually low, and can be corrected with a continuum source background corrector. Even in the far UV at the wavelengths of arsenic and selenium the marked radiation absorption of the flame can be corrected by a deuterium background corrector, incidently bringing an improvement in the signal-to-noise ratio and the determination limits [2059].

Genuine problems with background attenuation and its correction occur in the graphite furnace technique. However, the combination of a Zeeman magnet with a graphite furnace appears to cause no great difficulties, as several publications indicate.

Graphite furnaces are generally small enough to allow the air gap between the poles, and thus the magnet, to assume reasonable dimensions. The graphite tube is not continuously heated, so the heat radiation is clearly lower and cooling of the magnet is not required. FERNANDEZ *et al* [2244] have demonstrated that a conventional graphite furnace can be placed in a Zeeman magnet without notable alteration.

A homogeneous magnetic field with a flux density in the order of 1 Tesla (10 kGauss) is required for satisfactory splitting and separation of the π and $\sigma^{\pm}$ Zeeman components. Depending on the air gap required to accommodate the atomizer or radiation source, a relatively massive magnet is required.

For a constant magnetic field, either a permanent magnet or a direct current electromagnet can be used. The latter offers the advantage that the magnetic flux density can be varied and thus adjusted to any special requirements of the determination. An alternating magnetic field is generated by an alternating current electromagnet. Since in this mode of operation measurements are performed with the magnetic field switched on and off in rapid sequence, it is necessary to have a low inertia system so that the magnetic field dies quickly when the current is switched off. The electrical measuring system must be tuned to the electromagnet such that one measurement is performed in the phase when the flux density is at its highest, while the next measurement is taken when the magnetism has completely died away.

To obtain a transverse magnetic field, the poles of the magnet are located to either side of the atomizer or radiation source, perpendicular to the radiation beam. For a longitudinal magnetic field, the poles of the magnet must be located in the direction of the optical beam. The poles are thus located in front of and behind the atomizer or radiation source and suitably perforated to permit the radiation to pass through. A longitudinal field can be more easily generated when the atomizer or radiation source is located in an air-core solenoid.

From the instrumental standpoint there is a major difference whether a constant field or an alternating field is used. In a constant magnetic field, the splitting of the absorption or emission line into a central π component and shifted σ components is permanent. To measure the intensity ratio of the components it is necessary to discriminate between them instrumentally. This can be done by inserting a double-refracting filter [2187] [2486] or a polarizing prism [2485] into the optical beam. To measure the components alternately, the filter or prism must be rotated in the optical beam so that the radiation polarized parallel to and then perpendicular to the magnetic field is passed in rapid sequence.

Both filters and prisms have the disadvantage that in the far UV (below 220 nm) they become increasingly opaque and thus cause greater losses of radiation. Due to polarization, only 50% of the radiation is in any case passed. A further difficulty is that the polarizer must be rotated at a relatively high speed (e. g. 50 Hz or 100 Hz), and due to its high mass a degree of imbalance can easily occur. A very important factor in the use of a constant field (whether at the radiation source or the atomizer) is the fact that the reflection of the π and σ components from the monochromator grating is unequal. Further, this preferential reflection of one of the polarized components can be strongly wavelength dependent. Instruments employing a constant field must therefore be fitted with an adjustable polarizer and the components must be equalized every time the wavelength is changed [3101].

These problems do not occur when an alternating magnetic field is used. The absorbance measured at maximum flux density is compared to the absorbance measured when the magnet is switched off. Optimum sensitivity is attained when the π components of the radiation are eliminated. With a transverse field this is achieved by placing a fixed polarizer in the optical beam such that only radiation polarized in a plane perpendicular to the magnetic field passes. Naturally also here half of the radiant energy must be sacrificed—in the field-on phase the π component, while in the field-off phase statistically half of the primary radiation polarized in all directions of space.

Losses of radiant energy are not a problem for a longitudinal alternating field since a polarizer is not required—the π component is missing in the field-on phase. This configuration is thus the most advantageous with respect to energy. The absence of a polarizer also makes the instrument simpler.

After considering these various instrumental aspects it is clear that a magnet at the atomizer is preferable, since the normal, reliable AAS radiation sources can be used. Constant magnetic fields are normally easier to generate, but then a rotating polarizer is required, as well as additional components to permit equalization of the intensities of the variously polarized components. Alternating magnetic fields do not present these problems and are therefore preferential [3101].

There are two important factors to be considered before introducing the Zeeman effect to atomic absorption spectrometry. Firstly, a Zeeman atomic absorption spectrometer should be an ideal double-beam instrument, and secondly it should offer the best possibility of accurate correction of background attenuation. When considering which system offers the best analytical performance, the basic principle is that it should have as many advantages with as few disadvantages as possible. The question is then, which instrumental principle provides the best correction of the background attenuation and the most stable double-beam system, while at the same time maintaining, as far as possible, the same sensitivity, detection limits and linearity as conventional AAS.

When considering the *double-beam principle,* a distinction must be made between the direct Zeeman effect (magnet at the radiation source) and the inverse Zeeman effect (magnet at the atomizer). With the direct Zeeman effect, the sample beam (π) and the reference beam ($\sigma^{\pm}$) are at different wavelengths and have different spectral line profiles. With the inverse Zeeman effect, both the sample beam and the reference beam are at exactly the same wavelength (ν_o) and have the same line profile (the emission line from the radiation source). The two optical beams are discriminated in a constant magnetic field (rotating polarizer) through their differing polarization, while in an alternating magnetic field (fixed polarizer) even this distinction in not present. Direct Zeeman AAS is thus a two-wavelength technique, while inverse Zeeman AAS is a true double-beam technique.

YASUDA *et al* [3089] showed that with direct Zeeman AAS a baseline drift occurred when the radiation source was switched on, but later disappeared. The baseline exhibits instability even after the radiation source has been operating for some time, however, due to variations in the degree of self-absorption of the π and σ components of the emission line. These self-absorption effects cannot be avoided for any radiation source. In contrast, inverse Zeeman AAS exhibits ideal linearity after the radiation source is switched on and is free from fluctuations. The double-beam principle is thus met in an ideal manner by the inverse Zeeman effect, since both beams originate from the same source, have exactly the same wavelength and line profile, follow the same optical path, pass through the sample and optical components identically, and fall on the same detector.

The *accuracy of background correction* is independent of the type of magnetic field, whether the transverse or longitudinal Zeeman effect is used, and whether an element exhibits the normal or anomalous Zeeman effect. Like the double-beam principle, the accuracy of correction depends solely on whether the magnetic field is at the radiation source or the atomizer. In the first case, the direct Zeeman effect, measurement of the total absorption (atomic absorption and background attenuation) is at the resonance line (π component), while measurement of the background attenuation is only via the σ components to either side of the resonance line. In the second case, the inverse Zeeman effect, both measurements are at the resonance line. The total absorption is measured in the field-off phase (or when the rotating polarizer is parallel to the field), while in the field-on phase (or perpendicular orientated polarizer) the π component is eliminated and only the background attenuation is measured. How far the σ components are shifted from the original wavelength is important when considering the accuracy of correction. With a magnetic flux density of 0.8 to 1.2 Tesla (8 to 12 kGauss) the wavelength shift is about 0.01 nm, i.e. relatively small. With a quasi continuous background, such as from radiation scattering on particles or the photodissociation of numerous molecules, both systems are equally good and provide excellent correction of the background attenuation. They are definitely superior to continuum source background correctors, since equalization of the radiant intensities is not required and there are no alignment problems.

With a structured background (rotational fine structure), however, measurement of the background attenuation at 0.01 nm from the resonance line can lead to errors if the background changes in this range (refer to figure 79, page 139). In this respect the inverse Zeeman effect is superior to the direct Zeeman effect. DAWSON *et al* [2187] were the first to point out that with the inverse Zeeman effect, correction of the background attenuation is at the same wavelength as the atomic absorption. KOIZUMI *et al* have also established that higher background radiation losses can be better corrected with the inverse Zeeman effect than with the direct Zeeman effect [2485] [2487]. DE LOOS-VOLLEBREGT and DE GALAN [2192] found that all Zeeman systems provide good background correction, provided that the background attenuation is continuous and that non-absorbable lines are absent. When the background has fine structure, however, the direct Zeeman effect has no real advantage over continuum source background correction. In this case, only inverse Zeeman AAS provides the correct results.

STEPHENS and MURPHY [2913], and also MASSMANN, have drawn attention to the fact that molecules can exhibit the Zeeman effect, so bringing a further potential source of error. However, in a magnetic field with a flux density of less than 1.5 Tesla, molecular absorption bands are hardly or not at all influenced [3089], so that inverse Zeeman AAS provides the best results.

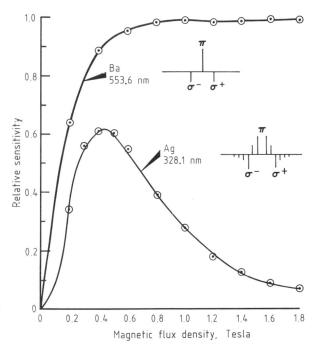

Figure 84. Relative sensitivities for barium (normal Zeeman effect) and silver (anomalous Zeeman effect) at various magnetic flux densities in a variable constant field.

The **sensitivity** of a determination by Zeeman AAS is independent of the location of the magnet; it only depends on how well the π and σ components are separated. With normal Zeeman splitting, the same sensitivity as with conventional AAS can be achieved with all magnet types, provided the magnetic flux density is sufficiently high to separate the σ components from the π component (about 0.8 Tesla for barium, as shown in figure 84). It is essential for the direct Zeeman effect that the $\sigma^{\pm}$ components of the radiation source do not overlap the wings of the absorption line, and for the inverse Zeeman effect that the $\sigma^{\pm}$ components of the analyte atoms do not overlap the emission line (refer to figure 85). Due to the lower temperature and pressure in the radiation source, the emission line is considerably narrower than the absorption line, so that separation of the components can often be achieved at a lower flux density with the inverse Zeeman effect. For the majority of elements, a magnetic flux density of about 1 Tesla is adequate to provide good separation.

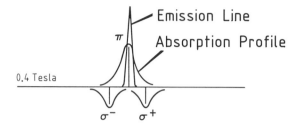

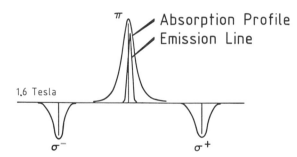

Figure 85. Line profiles for cadmium (normal Zeeman effect) at 228.8 nm in magnetic fields of 0.4 and 1.6 Tesla, respectively, with the magnet at the atomizer (inverse Zeeman effect).

For elements and wavelengths exhibiting anomalous Zeeman splitting, the attainable sensitivity is dependent on whether a constant field or an alternating field is applied. When a variable constant field is used, the sensitivity increases with increasing magnetic flux density, since the σ components are shifted further from the resonance line (refer to the curve for Ag in figure 84). With a further increase in flux density, a maximum is reached, after which the sensitivity decreases. This loss in sensitivity is explained by the fact that with anomalous splitting, several π components are present and these separate in the increasing magnetic field until they can no longer absorb. This effect is depicted in figure 86 on the example of chromium; the optimum flux density is about 0.4 Tesla. The π component is little broadened, so that there is good overlap between the absorption line and the emission line. However, the separation of the σ components is insufficient, so that they exhibit considerable overlap and thus absorption.

The maximum sensitivity for most elements is obtained at a flux density of 0.4 to 0.6 Tesla, in other words, at markedly lower values than for normal Zeeman splitting (refer to table 12). Elements having an even number of π components fare badly, because the original wavelength is virtually missing. At high flux densities a type of "line reversal" can occur,

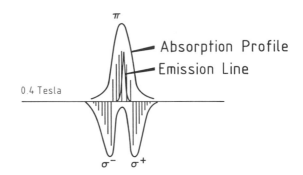

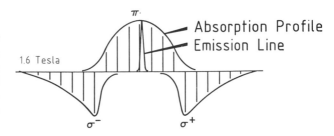

Figure 86. Line profiles for chromium (anomalous Zeeman effect, class 2_O) at 357.9 nm in magnetic fields of 0.4 and 1.6 Tesla, respectively, with the magnet at the atomizer (inverse Zeeman effect).

similar to self-absorption, leading to a substantial loss in sensitivity. An electromagnet with a variable constant field permits optimization of the magnetic flux density and thus maximum sensitivity for most elements. Nevertheless, it should be noted that this maximum sensitivity is clearly lower than that attainable with normal AAS (refer to table 13). A permanent magnet, or an electromagnet with fixed flux density, does not permit optimization and the attainable sensitivities are often less than half of those for normal AAS, depending on the magnetic flux density.

Table 12. Optimum magnetic flux densities for a number of elements using a direct current electromagnet

Element	Zeeman splitting	Optimum magnetic field (Tesla)
Ag	anomalous	0.5
Al	anomalous	1.0
As	anomalous	0.5
Ba	normal	0.9
Cd	normal	0.9
Cr	anomalous	0.4
Cu	anomalous	1.3
In	anomalous	1.6
Mo	anomalous	0.3
Pb	normal	0.8
Zn	normal	1.0

When an alternating magnetic field is applied, normal atomic absorption measurements are taken in the field-off phase and so the sensitivity is as for conventional AAS. In the field-on phase only the background attenuation is measured since the absorption line is not passed by the polarizer. The form of the π component is quite without significance, so that a magnetic field of sufficient flux density can be applied to shift the σ components away from the resonance line. Thus, with a magnetic field of suitable strength—a flux density of about 1 Tesla is generally sufficient—the full sensitivity of conventional AAS can be attained with an alternating magnetic field, even for elements exhibiting anomalous Zeeman splitting.

Table 13. Maximum attainable sensitivities for a number of elements using a variable constant field electromagnet compared to the normal sensitivities of AAS

Maximum attainable sensitivity %	Elements
90	Al, Cd, Si, Sn, Ti
80	Co, Ni, Pb
70	Bi, Fe, Pt, V
60	Ag, As, Cr, Mn
50	Se, Tl
40	Cu

With regard to the **detection limits** of Zeeman atomic absorption spectrometry, the same principles apply as for the sensitivity, since detection limit is a direct function of sensitivity. A second factor for the detection limits is noise. Discounting the strong fluctuations and instability of the radiation source that cannot be avoided with the direct Zeeman effect, no differences are otherwise to be expected for the various Zeeman configurations. It can be assumed that noise will be similar to, or perhaps slightly higher than, conventional AAS without background correction [2911]. Compared to background correction using a continuum source, noise will definitely be lower.

For the same sensitivity (normal Zeeman splitting and anomalous Zeeman splitting in a sufficiently high magnetic field), better detection limits can be expected than for normal AAS with continuum source background correction. Taking the improvement in background correction additionally into consideration, then for "real samples" substantially better determination limits will be attained. If a constant magnetic field is applied, especially with a permanent magnet of fixed flux density, these advantages cannot be used to the full because of the poorer sensitivity [2246].

Considerations on the **linearity of analytical curves** are particularly complex for Zeeman AAS. In conventional AAS the signal is dependent on a single absorption coefficient κ. This is not strictly proportional to the concentration of the analyte element, but nevertheless represents a monotonically increasing function of this concentration. This means that although analytical curves in conventional AAS may well be non-linear, they nevertheless rise continuously, so that a definite concentration can be ascribed to each absorbance reading.

In Zeeman AAS this is not implicitely so, since the measure is the difference between two absorption coefficients, κ_1^a and κ_2^a, according to the equation

$$\ln \frac{I_2}{I_1} = (\kappa_1^a - \kappa_2^a) + (\kappa_1^b - \kappa_2^b) + \ln \frac{I_2^0}{I_1^0}. \tag{7.2}$$

The final curvature of the analytical curve derives from the difference between two non-linear curves.

I_1^0 and I_2^0 are the radiant intensities of the radiation before passing through the absorbing layer and I_1 and I_2 are the radiant intensities after passing through this layer. κ_1^a and κ_2^a are the absorption coefficients for atomic absorption, while κ_1^b and κ_2^b are the absorption coefficients for the background. A detailed treatise on the appearance of analytical curves in Zeeman AAS has been published by DE LOOS-VOLLEBREGT and DE GALAN [2193].

The final profile of the analytical curve depends on the position of the magnet (direct or inverse Zeeman effect) and the type of magnetic field (constant or alternating), on the magnetic flux density and on the type of Zeeman splitting (normal or anomalous). The situation for the normal Zeeman effect is again the simplest. The same linearity as for conventional AAS can be achieved when the magnetic field is strong enough to shift the σ components out of the absorption region of the resonance line. This is valid for all forms of the Zeeman effect.

The same linearity as for conventional AAS can be achieved for elements and wavelengths exhibiting the anomalous Zeeman effect, provided an alternating field is applied with suffi-

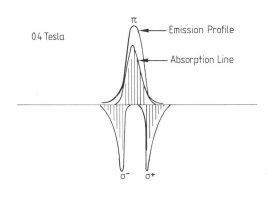

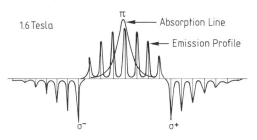

Figure 87. Line profiles for chromium (anomalous Zeeman effect) at 357.9 nm with magnetic fields of 0.4 and 1.6 Tesla, respectively, at the radiation source (direct Zeeman effect).

ciently high flux density to shift the σ components out of the absorption region. When a constant field is applied the results are very strongly dependent on the location of the magnet.

With the direct Zeeman effect (magnet at the radiation source), broadening and splitting of the emission line takes place with increasing flux density, while the absorption line remains unchanged (refer to figure 87). This broadening means that the emitted radiation becomes decreasingly absorbable, thus leading to increasing curvature of the analytical curve.

With the inverse Zeeman effect (magnet at the atomizer), the absorption line becomes broader as the magnetic flux density increases, while the emission line remains unchanged (refer to figure 86). Initially this leads to better linearity of the analytical curve, but only for relatively low concentrations. This must also be seen in connection with the somewhat lower sensitivity that is attained with this system. An overview of the sensitivity and linearity that can be achieved with various Zeeman systems in the usual concentration range is presented in figure 88 on the example of silver; measurements were performed at the 328.1 nm resonance line.

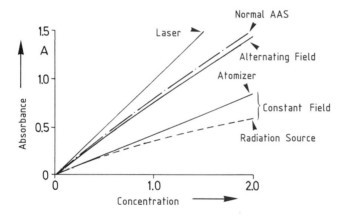

Figure 88. Sensitivity and linearity of analytical curves for silver at 328.1 nm (anomalous Zeeman splitting, class 2_E) in constant and alternating magnetic fields at the radiation source and the atomizer compared to normal AAS without a magnetic field. The magnetic flux density was 0.6 Tesla for all measurements (according to [2193]).

The deliberations so far have referred to the lower and middle concentration ranges. When higher concentrations are taken into consideration, an effect typical for Zeeman AAS is observed; namely **rollover of the analytical curve**. Again however, this effect depends on the type of magnetic field. It is a clear requirement for good linearity of the analytical curve with the anomalous Zeeman effect that the magnetic field must be strong enough to shift the σ components away from the original analytical line. The more successfully this is achieved, the higher will be the absorption coefficient κ_1^a, the lower κ_2^a, and thus $\kappa_1^a - \kappa_2^a$ will be greater and more linear. This requirement is best met with an alternating magnetic field. As shown in figure 89 for the absorption curves of the π and σ components, in a constant field κ_2^a becomes smaller with increasing field strength (better separation of the σ components), but κ_1^a also decreases after passing through a maximum due to broadening of the π component. With decreasing field strength, κ_1^a decreases while κ_2^a increases; just the opposite of the requirement for good linearity. Compromises must always be made when a constant field is applied.

Figure 89. Dependence of the π and σ absorption curves on the magnetic flux density in a constant field. For the anomalous Zeeman effect, the absorption of the σ components decreases with increasing field strength, since they are increasingly shifted away from the resonance line. The absorption of the π components also decreases, however, since they are shifted out of the line profile due to broadening.

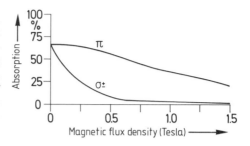

The final analytical curve in Zeeman AAS is obtained from the difference $\kappa_1^a - \kappa_2^a$. The ideal case is given for an alternating field of sufficient flux density, because then κ_1^a is identical with the normal absorption coefficient of AAS and κ_2^a is virtually zero. $\kappa_1^a - \kappa_2^a$ is then identical to the normal absorption coefficient. When an alternating field of insufficient strength is applied, or more especially a constant field, the following situation arises: κ_1^a increases initially faster than κ_2^a, but κ_1^a begins to flatten out earlier than κ_2^a, i.e. the rate of increase of κ_1^a compared to κ_2^a decreases with increasing concentration. As soon as the rate of increase of κ_1^a equals that of κ_2^a, the maximum of the Zeeman AAS analytical curve is reached. For higher concentrations the sensitivity decreases, i.e. the curve rolls over (figure 90). The biggest problem with rollover of the analytical curve is its ambiguity, since two concentration values could be assigned to every absorbance reading. The only reliable way of recognizing this interference is to dilute the sample.

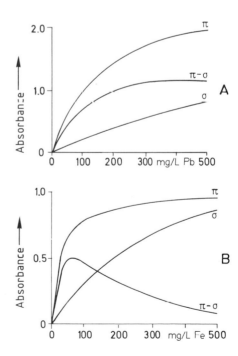

Figure 90. Examples of rollover of the analytical curve $(\pi - \sigma)$ with a constant magnetic field of 0.9 Tesla at the atomizer (inverse Zeeman effect).
A: Lead at 283.3 nm (normal Zeeman effect).
B: Iron at 248.3 nm (anomalous Zeeman effect, class 2_0). (According to [3089].)

In the graphite furnace technique, rollover of the analytical curve can be recognized from the signal form (figure 91). The signal recorded from the graphite furnace reflects the change of atom concentration with time. Initially the concentration rises, reaches a maximum and then decreases somewhat more slowly. For a higher element concentration, when the σ components are insufficiently separated, the signal rises at first proportional to the increasing atom concentration, then reaches the rollover concentration and begins to decrease. As the atom concentration decreases, the signal increases to a maximum once more and then falls toward zero with the decreasing atom concentration.

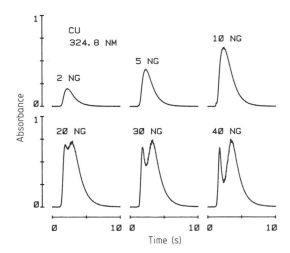

Figure 91. Effect of rollover of the analytical curve on the signal form in the graphite furnace technique. Determination of high copper concentrations in an alternating magnetic field of 0.8 Tesla (according to [2244]).

From all these considerations it is clear that the optimum Zeeman system has a magnet located at the atomizer (inverse Zeeman effect), operating with an alternating field at a flux density of about 1 Tesla. For practical intrumental reasons the transverse Zeeman effect is selected. This configuration offers an optimum double-beam system, high accuracy background correction, linearity and sensitivity similar to conventional AAS, detection limits better than with continuum source background correction, and fewer problems with rollover of the analytical curve. Such a system permits background attenuation up to absorbance 2 to be reliably corrected.

It seems logical to restrict the Zeeman effect to the graphite furnace technique, since the greatest problems with background attenuation are encounted here. The difficulties of applying the Zeeman effect to the flame technique have already been discussed in detail and, in the absence of suitable fields of application, the expenditure is hardly worthwhile.

The Zeeman effect is certainly superior to all other background correction techniques used at the present time. Nevertheless, the Zeeman effect should not be considered as the magic potion that can cure all problems of graphite furnace AAS. Caution is advised against thoughtless remarks, such as from HADEISHI [2333–2335], for example, that even an "inexperienced person such as a fisherman" can easily perform an analysis, since accurate background correction permits simplification of the chemical sample treatment. KOIZUMI and YASUDA [2484] also reported the determination of lead in blood and liver, and cadmium in urine directly in the graphite furnace without pretreatment, and even without drying or ashing. Such a simplification of the problems of trace analysis shows a lack of basic analytical knowledge and will in all probability lead to false results. Background attenuation is one thing, while sample treatment and non-spectral interferences and their elimination are quite another thing, and have nothing to do with the former.

7.2. Non-Spectral Interferences

Non-spectral interferences affect the analyte signal directly. Since atomic absorption spectrometry is a relative method (i. e. quantitative measurements can only be made by comparison with reference materials; usually reference solutions), any behavior of the sample that is different from the reference material constitutes an interference. Interferences that cannot be specified because their cause is unknown or of a complex nature are termed "effects". The matrix effect, for example, is a composite interference due to the concomitants in the sample. If a solvent other than water is used the influence is not termed an interference, since by definition the reference solutions are made up in the same solvent. Nevertheless it is possible to talk of a solvent effect. The same is true for all additives to the sample used to eliminate interferences or to achieve a certain effect. Their influence is not termed an interference since they are also added to the reference solutions.

7.2.1. Classification of Non-Spectral Interferences

These interferences are most conveniently classified according to the place, stage or process responsible for their occurrence; for example, transport, solute-volatilization, spatial-distribution, or vapor-phase interferences. The earlier classification into just physical and chemical interferences is discouraged because frequently both the physical and chemical properties of a concomitant combine in an interference [2409].

A physical interference is one caused by differing physical properties of the sample and reference material, such as varying viscosity, surface tension, or density of the solutions. These properties influence principally the nebulization of the sample and the transport of the aerosol into the flame and are thus nowadays preferentially termed **transport interferences.** This term encompasses all processes that affect the efficiency of transport of the sample into the actual atomizer. The interference can be positive as well as negative, i. e. can cause signal enhancement as well as depression. Transport interferences occur more or less predominantly in the flame technique due to the pneumatic nebulizer and premix burner. This type of interference is virtually unknown in the graphite furnace technique, unless a nebulizer is used to spray sample continuously into the graphite tube. In the hydride technique, the retardation of hydride liberation through foam or a larger sample volume can also be considered a transport interference; the same is also valid for the liberation of mercury in the cold vapor technique.

Taken strictly, a chemical interference is any formation of a compound that prevents quantitative atomization of an element. A chemical interference is then understood as a change in the number of free atoms formed per unit volume or time through chemical combination when samples and reference materials are compared under identical conditions (atomizer, solvent, etc.). This interference can be positive as well as negative, i.e. relative to the reference material the sample can form either more or fewer atoms. In principle, the occurrence of a chemical interference can have two causes: Either the conversion of the sample into atoms is not quantitative (for example, a difficultly melted or volatilized salt is formed, or the molecules do not dissociate), or the free atoms react spontaneously with other atoms or radicals in the environment to form new molecules or radicals and are thus not available sufficiently long to absorb. Since the causes for such interferences in the flame technique and graphite furnace technique can be markedly different, they are best distinguished according to the place and process responsible for their occurrence.

The term **condensed-phase interference** encompasses all processes from the formation of compounds during evaporation of the solvent, rearrangement reactions, through to complete volatilization of the analyte element as molecules or atoms. In the flame, this type of interference is generally termed a **solute-volatilization interference.** It is caused by a change in the rate of volatilization of the analyte element from aerosol particles due to the presence, or absence, of a concomitant. The interference is specific when the analyte and the interferent form a new phase of different thermal stability. Such interferences can also be non-specific when the analyte element is dispersed in an excess of a high melting point concomitant and therefore volatilizes slowly.

Solute-volatilization interferences occur in the graphite furnace especially when the analyte element is volatilized at a lower temperature in the presence of an interferent than in its absence. The risk then exists that losses of the analyte element will occur during thermal pretreatment. **Vapor-phase interferences** occur when the analyte element is not completely dissociated into atoms in the ground state. One cause can be that the analyte element forms a compound with a concomitant which is not dissociated to the same degree as the reference material. A further cause for vapor-phase interferences is that the atoms formed react rapidly with vaporized concomitants.

According to the definition of an interference made in the introductory remarks, the reaction of analyte atoms with components of the flame gases or with the purge gas in a graphite furnace (such as the formation of a monocyanide in the presence of nitrogen) is not an interference. A (positive or negative) influence by a concomitant on reactions of this type is nevertheless a vapor-phase interference. The reaction of an element with oxygen or hydroxyl radicals in the flame with the formation of a hard-to-dissociate monoxide or hydroxide is not an interference; a shift of the dissociation equilibrium by concomitants, however, is definitely an interference. Vapor-phase interferences occur frequently in Massmann-type graphite furnaces due to the analyte element volatilizing before the graphite tube or the inert gas have reached the final temperature. Since there is no thermal equilibrium in such furnaces and the gas phase is cooler than the tube surface, strong recombination of the analyte atoms with concomitants can be observed. This leads to a substantial depression of the signal.

The combustion products in flames are always present in large excess so that sample constituents have very little influence on the partial pressures of the individual products. The flame therefore has a strong buffer action. The inert gas atmosphere in a graphite furnace, on the other hand, has no buffer action worth mentioning. Here, the vaporized sample constituents influence the atmosphere, so that their effect can be more pronounced than in the flame.

A further interference observed in the vapor phase is the **ionization chemical interference**. This is mostly observed in flames of higher temperature. Again here, the ionization of atoms is in itself not an interference, but a shift of the ionization equilibrium through concomitants is. Only the latter leads to false measurements. Since ionization of the analyte element often leads to a considerable loss in sensitivity and the analytical curve exhibits noticeable curvature toward the absorbance axis, it is nevertheless necessary to eliminate this effect as far as possible (refer also to section 8.1.5).

All vapor-phase interferences are specific. They are easily recognized experimentally, since they occur when the analyte element and the interferent are volatilized separately in the same atomizer. This can be achieved, for example, by using a premix burner with twin nebulizers or by volatilizing analyte element and interferent from different locations in a graphite tube.

In the hydride technique, vapor-phase interferences can be caused when the hydrogen radicals, which promote atomization and are only present in low concentrations, are used preferentially by another gaseous hydride. A concomitant thus suppresses atomization of the analyte indirectly.

Spatial-distribution interferences can be observed when changes in the concentration of concomitants affect the mass flowrate or mass flow pattern of the analyte in a flame. They can be caused by changes in the volume and rise velocity of gases formed by combustion. In extreme cases they can influence the size and shape of the flame. When only the flame geometry is affected the interference is non-specific. Such interferences can become specific, however, if a concomitant causes a delay in the volatilization of particles. Such a delay reduces the time available for lateral diffusion of the analyte species before they reach the optical beam of the spectrometer.

The various interferences will be treated in detail in the next chapter (8. "The Techniques of Atomic Absorption Spectometry"), since each technique has characteristic interferences. Further information on interferences is presented as appropriate in chapter 10 with reference to the individual elements and in chapter 11 on specific applications.

7.2.2. Elimination of Non-Spectral Interferences

Procedures for the elimination or bypassing of non-spectral interferences will only be discussed in general terms in this section. Specific procedures will be handled in detail in chapters 8, 10 and 11.

In general, non-spectral interferences can be eliminated by making sample and reference solutions as similar to each other as possible. In the ideal case, a reference solution contains not only the same solvent, but also the same concomitants. In other words, it should be identical to the sample solution, except that the analyte element is present in a defined concentration. If this really is the case, then strictly speaking, no interferences will be observed since matrix effects influence the analyte element in both sample and reference solutions to equal degrees.

This ideal situation is rarely found in practice, however. In the first place, the exact composition of the sample is unknown and even if it were, often cannot be reproduced. Assuming that this condition could be met, the preparation of a reference solution matching the sample solution exactly requires reagents of the highest purity and also a good deal of time. If a large number of samples of varying composition is to be analyzed, it is clear that this procedure is quite impracticable.

Fortunately, however, interferences in AAS are rarely so pronounced that reference solutions must match sample solutions exactly. Frequently it is sufficient to use the same solvent and to match the major constituent of the sample. Especially in the flame technique, routine determinations can frequently be made directly against simple reference solutions. When the sample composition changes markedly or more complex matrix effects are evident, the **analyte addition technique** can be recommended. Since an individual analytical curve is in effect established for each sample (through addition of the reference solution), the ideal case described above is more or less met. Using this technique it should be possible to eliminate all non-spectral interferences, provided that they are independent of the analyte concentration.

Ionization is clearly dependent on the concentration, so that the addition technique cannot be used to eliminate this effect. In the hydride technique the oxidation state of the analyte element often is decisive for the sensitivity; this fact must be taken into full account if the addition technique is employed. In the graphite furnace technique the bonding form and the ligands often have a substantial influence on the thermal behavior, and thus the volatility, of an element. If the element in the addition has a bonding form different to that in the sample, its behavior can be totally different and the interference will not be eliminated. Ideally the addition technique can be applied to eliminate all non-specific interferences, such as transport interferences in the flame technique. If this technique is used to eliminate specific interferences, then great care must be taken to ensure that the influences on the analyte element are the same in both the sample and the added reference solutions.

The **reference element technique**, in which another element is added to the sample in known concentration, is, by definition, only suitable for the elimination of non-specific interferences, such as transport interferences. Since logically all specific interferences are specific to the analyte element, it cannot be expected that they can be eliminated by measuring another element. As already explained in section 7.1.2, this technique makes very high demands on the analytical apparatus and technique, and is thus little suited for AAS.

An important prerequisite for these techniques is that the signal depression caused by concomitants is not too severe. If the interferences are so large that the reference or addition technique curve is too shallow, suitable measurements often cannot be made. The loss in sensitivity also leads to a worsening of the determination limit.

A technique used widely for the elimination of interferences is the addition of **spectrochemical buffers** to sample and reference solutions. Depending on the type of interference, different buffers with varying modes of action are employed. *Nebulizing aids,* for example, can be used to bring the physical properties of the solutions, such as viscosity or surface tension, into line and thereby eliminate transport interferences. A further group of spectrochemical buffers *(volatilizers)* improves volatilization or atomization of the analyte element by converting it to a more suitable form.

Releasers are frequently used to prevent the analyte element from entering a thermally stable compound. This technique for the elimination of solute-volatilization and vapor-phase interferences is based on the knowledge that these interferences occur during desolvation, during initial precipitation from the saturated solution, or through rearrangement reactions in the solid particles. If a cation that forms a compound of lower solubility product with the interfering anion than with the analyte is added in excess to the sample solution, then according to the law of mass action the interfering ion will be bound almost quantitatively to the added cation. The analyte element can then be atomized free of interferences. This technique can be used very well for the elimination of very marked interferences, although it has the disadvantage that contamination cannot always be avoided when reagents are added in large excess. The best known example for the use of this technique is the addition of lanthanum when calcium is to be determined in the presence of sulfate, phosphate, aluminate, silicate, etc. in an air/acetylene flame.

An *ionization buffer* is added to suppress ionization. By adding an easily ionized element, such as cesium or potassium, the concentration of the free electrons in the absorption volume is increased substantially, thereby suppressing and stabilizing ionization of the analyte.

In the graphite furnace technique, **matrix modification** is a procedure for reducing or eliminating volatilization and vapor-phase interferences. A reagent is added to bring the physical and chemical properties of the analysis substances into line. The reagent frequently serves to convert either the analyte element into a less volatile form, so that higher thermal pretreatment temperatures can be applied, or the concomitants into a more volatile form. Either measure serves to bring about more effective separation of the analyte-element from concomitants during thermal pretreatment.

In the flame technique it should always be borne in mind that the dissociation of molecules into atoms is a temperature-dependent equilibrium reaction and that this reaction can be influenced by parallel reactions, both positively and negatively. Further, atomization in a flame is a dynamic process, i.e. time is a decisive factor. All processes from desolvation to release of free atoms from molecular bonds should occur within a few milliseconds, since this is the rise time from the burner slot to the observation height. One reason for the occurrence of interferences—or at least for a stronger occurrence—is that equilibrium between molecules and atoms has not been reached. Different interferences can be measured at different observation heights (refer to [31]). Vapor-phase interferences can sometimes be avoided or reduced by measuring in higher flame zones. Interferences caused by reaction of the analyte atoms with flame gas components, such as oxygen, can occasionally be avoided by measuring nearer to the burner slot [1048]. The greater the affinity of the analyte element for oxygen is, the earlier and more rapid is the decrease of atom concentration with the observation height.

A flame temperature that is too low or an unsuitable chemical environment often shift the temperature-dependent equilibrium between molecules and atoms to the left. It can be generally stated that with increasing flame temperature, solute-volatilization and vapor-phase interferences decrease markedly. This is particularly noticeable when a nitrous oxide/acetylene flame is used, because apart from the relatively high temperature (approx. 2750 °C), an ideal reducing environment for the dissociation of numerous compounds is present. The hotter oxygen/acetylene flame (approx. 3050°C), on the other hand, causes many interferences due to its strongly oxidizing characteristics.

From these observations on the causes of interferences and on their elimination, the view could be advanced that interferences are due to false operation, i.e. to the use of an unsuitable means of atomization.

This view is justified, especially when reference is made to the large number of publications concerned with the most widely varying interferences which occur with elements in the alkaline-earth group (see [1012]). These interferences can virtually be eliminated without exception through the use of a premixed nitrous oxide/acetylene flame.

Thus, many interferences can be eliminated by raising the flame temperature or by changing the chemical environment. Similar principles also apply to the graphite furnace technique. Solute-volatilization interferences can be eliminated through matrix modification, while vapor-phase interferences can be eliminated by atomizing into a thermally stabilized environment (refer to section 3.2.5).

8. The Techniques of Atomic Absorption Spectrometry

In this chapter, the flame, graphite furnace, hydride-generation and cold vapor techniques will be discussed in some detail, especially with respect to their particular characteristics, advantages, disadvantages, etc. Emphasis will be placed on practical analytical aspects, such as, for example, the interferences characteristic for each technique and their elimination. For a better understanding, the processes leading to atomization of the analyte element for each technique will also be treated.

8.1. The Flame Technique

The sample, usually in the form of a solution, is aspirated by means of a pneumatic nebulizer into a spray chamber from whence it passes into the flame. This generally used procedure is the basis of the following discussions. Other forms of sample introduction, such as described in section 3.1, are not within the scope of this chapter and the reader is referred to the original literature.

Non-spectral interferences occurring in the flame technique will be handled in the order in which they are met: transport interferences, solute-volatilization interferences, vapor-phase interferences, and spatial-distribution interferences.

8.1.1. Atomization in Flames

For atomization, a liquid or dissolved sample (e.g. the solution of a salt in water) is sprayed into a spray chamber and then passes in the form of a fine aerosol into the flame. First, desolvation takes place, i.e. the solvent evaporates. The rate of desolvation is dependent upon the solvent. The solid particles formed (e.g. salt crystals) can undergo various changes, depending on the flame temperature; organic substances burn, inorganic components react with each other or with the flame gases. Hydrated chlorides, formed by many elements from hydrochloric acid solution, can form oxides under the elimination of hydrogen chloride, or they can eliminate water with the formation of dehydrated chlorides. Similarly, oxides can be formed from carbonates or sulfates, and pyrophosphates from phosphates. Frequently, all subsequent processes, such as melting or volatilization of the solid particles, are dependent on the rearrangement reactions; in this context the thermal stability of a number of refractory oxides is mentioned (see [36]). The causes of numerous interferences are to be found at this stage. These thermal rearrangement reactions frequently depend on the reaction partners present in the sample. For example, a calcium ion will form a dehydrated chloride from hydrochloric acid solution and this chloride is readily volatilized; if, on the other hand, sulfate or phosphate ions are present in the solution, an oxide or a pyrophosphate will be formed that is much more slowly melted and volatilized, or only at higher temperatures. Consequently it is clear that temperature plays an important role in the discussion on non-spectral interferences.

Once the particles have volatilized to give gaseous molecules, the actual thermal dissoci-ation process of molecules into atoms can begin. This process is an equilibrium reaction that is often very complicated since numerous other similar reactions take place in parallel or join in.

In a flame there are a variety of combustion products besides the atoms under study, and these are, in some cases, in considerable excess (CO_2, CO, C, H_2O, O_2, O, H_2, H, OH, NO, N_2). The cleavage products of the solvent and other possible substances in the sample must also be taken into consideration. Atoms, radicals, ions, and so on are not capable of exist-ence under normal conditions (except the inert gases), and are only formed in the flame by temperature-dependent equilibrium reactions. If the same cleavage product is formed from two equilibrium reactions, both reactions influence each other according to the law of mass action through the partial pressures of the reaction products. For these discussions it is of interest if the anionic constituent of the metallic compound under analysis is also produced simultaneously in high concentration (high partial pressure) as a result of other equilibrium reactions in the flame. The thermal dissociation of the analyte element can then be strongly influenced. In principle, the flame can be regarded as a "solvent" in which a trace of metal atoms is to be found [580]. Besides the suppression of the dissociation equilibrium

$$MeX \rightleftharpoons Me + X$$

through a parallel equilibrium reaction in which X is also produced, the total number of metal atoms Me formed can also be influenced by a subsequent reaction, such as ionization (compare [1065])

$$Me \rightleftharpoons Me^+ + e^\ominus$$

or through the formation of a compound with another reaction partner

$$Me + Y \rightleftharpoons MeY .$$

A schematic representation of the most important processes possible in the flame is shown in figure 92.

Once it is known which physical and chemical processes take place (or can take place) on the way from solution droplets to free atoms, the factor of time must be taken into consider-ation. Every flame has a burning velocity characteristic for the gases used and their mixing ratio; a further important factor is the geometry of the burner slot. The sample aerosol neb-ulized into the flame will be transported with approximately the same velocity as the flame gases. Consequently the processes depicted schematically in figure 92 do not only take place in the given sequence but also at different heights in the flame.

It must also be taken into account that a flame is not an isothermal atomizer, but exhibits a temperature profile that has an additional influence on the various processes. Further, chemical rearrangement occurs in the flame, so that the partial pressures of various compo-nents also vary with height in the flame. These can influence the degree of atomization.

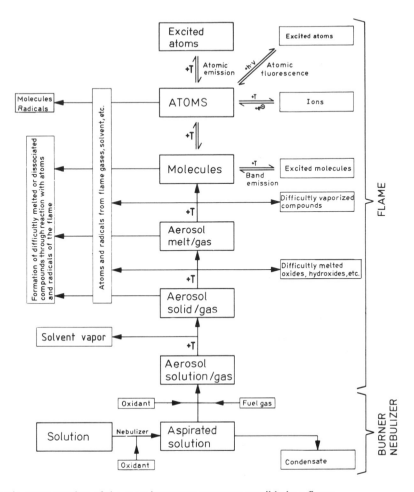

Figure 92. Schematic representation of the most important processes possible in a flame.

The time required to convert a droplet of solution into gaseous molecules depends on the size of the droplet and the flame temperature. The atomization itself also depends on the temperature and frequently on the partial pressure of one or more components (combustion products) of the flame. This can be influenced through the choice of a suitable flame; for example, the partial pressure of atomic oxygen is 200 times higher in an oxygen/acetylene flame than in an air/hydrogen flame. Secondary reactions can also be influenced by choice of a suitable flame. Ionization, on the other hand, is temperature dependent in the same manner as atomization. Accordingly, the temperature should be so chosen that atomization takes place rapidly but that ionization does not occur. It is also possible to influence the entire atomization process by prolonging the residence time of the sample in the flame, i.e. atomization will be more complete in flames of lower burning velocity. Even in a flame of relatively low burning velocity, such as the air/acetylene flame, a distance of 1 cm is crossed in less than 10 ms.

Despite the enormous advances that have been made in atomic absorption spectrometry since the nineteen-sixties, little is still known about the fate of metal atoms and the sample in the flame. Yet it is exactly the processes in the flame, which influence the efficiency of atomization, that are of decisive importance for an understanding and the correct estimation of non-spectral interferences.

In 1966, ROBINSON calculated that of the 10^{15} copper ions contained in 1 mL of an 0.1 mg/L copper solution, only about 10^{-3}% are atomized when the solution is nebulized into an air/acetylene flame. This can be attributed to insufficient nebulization, to insufficient thermal dissociation, and to the fact that the flame is not chemically inert, as should be expected from an ideal atomizer.

These observations are always referred to the total burner unit and are altogether too complex to permit a statement on the actual atomization process to be made. It is therefore certainly of no advantage to include nebulization of the sample in the considerations. The strongly increased interest in alternate atomization techniques with their considerably better sensitivities reawakened the question of the efficiency of atomization with and without flames during the nineteen-seventies.

DE GALAN and WINEFORDNER [285] calculated the portion of free atoms formed in an air/acetylene flame for 22 elements. They found that several elements, such as copper and sodium, were virtually quantitatively atomized. This calculation cannot be extended to all elements, however, since it requires a knowledge of the oscillator strength f (see section 1.5). The latter is not known for the majority of elements. WILLIS [1335] verified these findings through independent measurements and mentioned that with the previous calculations important points had been omitted that had led to degrees of atomization that were too low. KOIRTYOHANN [690] therefore attempted to determine the efficiency of atomization experimentally by setting the atomic absorption and molecular emission in a relationship to each other. He assumed that when using a premix burner and a hot flame, the portion of the sample reaching the flame is quantitatively volatilized; i.e. converted into atoms and molecules.

If the portion of free atoms is β and the portion of bound atoms γ, and the above assumption is true, then

$$\beta + \gamma = 1 \tag{8.1}$$

and

$$\frac{A_1}{A_2}\beta + \frac{E_1}{E_2}\gamma = 1, \tag{8.2}$$

where A_1 and A_2 are the absorbance values of the analyte atoms and E_1 and E_2 are the emission intensities of the molecules formed in fuel gas weak and fuel gas rich flames, respectively.

The relationships are especially easy to obtain when only one molecular species is formed in the flame, the monoxide for example. This is the case for a large number of the rare earth metals in the nitrous oxide/acetylene flame, and KOIRTYOHANN obtained values for these and other elements that were in good agreement with the calculated values. FASSEL *et al* [350] especially have drawn to attention that the free atoms of metals which form stable monoxides are localized to a high degree in nitrous oxide/acetylene flames. The total profile of the absorbance observed over the height of the flame alters with the stoichiometry of the flame and is in direct connection with the stability of the monoxide. This indicates the importance of this compound for the formation of free atoms. The β values are then only meaningful dimensions for many elements when they are referred to a particular point in the flame and to a determined stoichiometry.

RASMUSON *et al* [1023] found that in a nitrous oxide/acetylene flame there is a determined flow ratio ϱ at which atomization is complete when the metal forms a monoxide with a dissociation energy of less than 6 eV (on condition that the sample is completely volatilized). Elements belonging to this type are, for example, aluminium, beryllium, copper, iron, lithium, magnesium and sodium. SMYLY *et al* [1158], on the other hand, found that in a separated oxygen/hydrogen flame considerably worse values were obtained, which again shows that the chemical environment, and more especially the partial pressure of oxygen, plays a decisive role in the atomization processes.

L'VOV *et al* [2580] examined the composition and temperature of air/hydrogen, air/acetylene, and nitrous oxide/acetylene flames over a wide range of fuel-oxidant ratios. The results were used to determine the respective abilities of each flame and whether monoxides could be dissociated. It could be shown theoretically and also verified experimentally that virtually all elements determinable by AAS are dissociated almost quantitatively in a nitrous oxide/acetylene flame of suitable composition.

These authors also demonstrated the formation of refractory carbides for lithium and tin in the presence of carbon in the flame. This effect is also responsible for a number of anomalies that these elements exhibit in flames of lower temperature. It can be presumed that in flames such as a fuel gas rich air/acetylene flame, similar carbide formation takes place much more frequently than the thermodynamic data would suggest. The substantial differences in absorbance observed for elements in the iron group in carbon-rich flames, for example, might be attributable to carbide formation. The influence of iron, cobalt and nickel on the absorption of chromium in a reducing flame might also be due to the formation of refractory carbides.

RASMUSON *et al* [1023] were the first to indicate that several elements form stable monocyanides in reducing flames of higher temperature. L'VOV *et al* [2580] concluded from thermochemical data that alkali elements, such as potassium and sodium, and boron should belong to this class. The energy for the dissociation of the gaseous monocyanide BCN into B and CN is about 200 kJ/mol. The low sensitivity of boron when measured in the red reducing zone of a nitrous oxide/acetylene flame should be mainly attributable to this phenomenon.

HALLS [2342] [2343] established that the chemical environment and not the temperature has the greatest influence on the concentration of atoms in the flame. This view has gained wide acceptance in recent years and can be taken for granted in those cases where the flame gas composition influences atomization.

False values are often obtained from the theoretical dissociation equilibria for atomization. In such cases it is preferable to explain the atomization mechanism via a reduction through components of the flame. It is important to remember that the "chemical environment" in a flame is dependent on the components and these are present in far higher concentrations than any constituents of the sample.

Carbon monoxide and hydrogen do not have sufficient free energy to reduce most monoxides, and the free energy that is released during the oxidation of C_2 or H radicals is just sufficient to bring about the reduction of those metal oxides that are in any case easily atomized. CH radicals and C atoms should be able to reduce more stable oxides such as SiO_2. Calculation of the efficiency of atomization (β) for sodium and magnesium based on a reduction by H radicals gives values agreeing much more closely with the experimental values than the values calculated from the dissociation equilibria [2243]. The distribution of the individual components throughout the flame must also be taken into consideration. C_2 and CH radicals, for example, are only found in lower zones of the flame; H radicals, on the other hand, remain longer in the flame and can therefore intervene more effectively in the atomization process. HALLS [2342] found, for example, that the close relationship between the distribution profile of H radicals and indium atoms leads to the conclusion that H radicals are more or less forced to participate in the atomization of indium.

The dependence of the atomization of calcium on the fuel gas/oxidant ratio is well known. Also here HALLS [2342] indicated the high probability that H radicals participate in the reduction of CaO and in the atomization. The comparison between magnesium and aluminium is also interesting. Calculation of the ratio [M]:[MO] (where M = metal and MO = metal oxide) on the basis of thermal dissociation equilibria gives a value of 0.16 for magnesium and 0.016 for aluminium in an air/acetylene flame. This would indicate that magnesium is ten times more sensitive to determine than aluminium. In reality, the efficiency of atomization for magnesium with $\beta = 0.64$ is very high, while aluminium atoms can hardly be detected. Consequently flame components, such as H radicals, must also play a part, since the differences cannot be explained via thermal dissociation.

8.1.2. Spectral Interferences

As already discussed in section 7.1, spectral interferences in AAS caused by direct overlapping of the emission line from the primary radiation source and the absorption line of a concomitant element are seldom. In the presence of higher concentrations of concomitants, in the range of several grams per liter, substantial line broadening in the flame is nevertheless possible. FASSEL [342] in particular has drawn attention to the fact that the wings of the absorption profile can overlap the primary emission line, even when this is not expected on the basis of the half-intensity width and line separation. In the case of calcium, for example, he was still able to measure 10% of the maximum intensity at a distance of 0.03 nm from the line center. The width of an absorption line at the base can increase on occasions to 0.1 nm.

On the basis of such line broadening, PANDAY and GANGULY [2728] reported the absorption of grams/liter quantities of terbium at the 285.2 nm magnesium line and of chromium at the 290.9 nm osmium line. NORRIS and WEST [2698] utilized spectral overlapping between antimony and lead absorption lines to determine high lead concentrations in copper alloys.

Spectral interferences caused by radiation scattering on particles in the flame or by molecular absorption are relatively seldom when a well-designed premix burner is employed. The extent of the observed interferences depends among other things on the flame conditions and the height of observation. Through optimization of all parameters, background attenuation can usually be markedly reduced, although not eliminated in every case.

Radiation scattering hardly ever occurs in the flame of a premix burner, but it can be frequently observed in a direct injection burner. KERBYSON [639] established that the droplet volume of a direct injection burner is around 100 times greater than for a premix burner. Resulting from the quadratic relationship between droplet volume and radiation scattering, the later must be about 10000 times greater for a direct injection burner. Solely in isolated cases has radiation scattering been observed with premix burners using flames of low temperature, such as an air/propane flame [123].

The background attenuation observed in the flame of a premix burner is mostly due to absorption by molecules or radicals. These can be produced in great quantity when a volatile element is determined in a flame of lower temperature in the presence of a non-volatile matrix, or when a concomitant element present in higher concentrations forms a stable oxide or hydroxide radical with the flame gases. It is naturally a prerequirement that the analyte element and the molecules absorb in the same spectral range. A typical example is the CaOH band in an air/acetylene flame that has an absorption maximum virtually at the resonance line of barium (figure 93). CAPACHO-DELGARDO and SPRAGUE [210] found that a 1% calcium solution measured at the barium resonance line (with a barium hollow cathode lamp) gave an absorption of 50%. This effect cannot be observed in a nitrous oxide/acetylene flame.

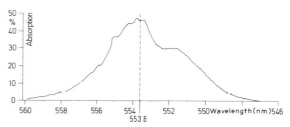

Figure 93. Absorption spectrum of CaOH radicals (1% Ca solution) in an air/acetylene flame in the spectral region around the barium resonance line at 553.6 nm.

A further typical example for molecular spectra in flames of lower temperature is the spectra of alkali halides; these can mostly cause interferences in the short wavelength range below 300 nm. For example, SIERTSEMA [2861] found an interference in the determination of iron in serum at the 248.3 nm line due to the spectrum of sodium chloride. The interference was markedly increased through HCl and trichloroacetic acid. However, it is likely that an unsuitable burner system was used, since other authors could not verify this interference.

FRY and DENTON [2274] investigated whole blood, condensed milk, and tomato sauce within the spectral range 190 nm to 300 nm and found no measurable background attenuation. Even the background signal for urine was hardly higher than the noise. For 5% solutions of sodium chloride and calcium chloride, these authors scanned the molecular spectra in an air/acetylene flame and found absorbance maxima of 0.02 A and 0.03 A, respectively, in the range 230 nm to 240 nm. Undiluted sea-water exhibited a maximum absorbance of

0.015 A. In a nitrous oxide/acetylene flame, molecular absorption could be reduced by a power of ten, so becoming more or less without significance. The sensitivity of a number of volatile elements is worse in the hotter flame, however, so that it cannot always be used— nevertheless, its use should always be investigated.

Spectral interferences caused through absorption by molecules or radicals are relatively seldom when a premix burner is used, but if present can normally best be eliminated with a continuum source background corrector. Since the background attenuation in flames is never very high, the working range of a continuum source background corrector will not be exceeded. Also, a static signal is generated in the flame, so that problems due to speed are not to be expected.

Nevertheless, MARKS et al [2624] have drawn attention to possible interferences in the determination of trace elements in metals and complex alloys when a continuum source background corrector is employed. It is possible that atoms of major constituents, which absorb within the selected bandpass, can absorb radiation from the continuum source. Consequently, the radiation of the continuum source, but not of the primary source, is attenuated, leading to overcompensation.

HÖHN and JACKWERTH [2384] reported a similar interference for the determination of traces of gold in indium. This interference can be attributed to molecular absorption by indium chloride, which exhibits fine structure; the individual lines have about the same width as the gold absorption line (refer to figure 79, page 139). For measurements with the element-specific primary source, only the background directly at the gold line is detected. Owing to the relatively wide bandpass of the monochromator in an atomic absorption spectrometer, a band of radiation around the analytical line will be attenuated from the continuum source. As seen in figure 79, the gold line at 267.6 nm does not coincide with an absorption line of the background. This means that the background attenuation measured with the primary source is less than with the continuum source. When the difference is taken, too large a value for the background is subtracted; the resulting measure is too small or even negative.

In such situations the use of a nitrous oxide/acetylene flame is recommended, since the interferences can be reduced significantly. It is also worth considering the use of a more sensitive technique, such as graphite furnace AAS. Here it is possible to work with substantially greater dilutions and hence with reduced interferences.

The application of the Zeeman effect for background correction in a number of special cases has been reported [3089]. However, interferences occurring in the flame can generally be eliminated by much simpler means. There is thus no real cause for introducing the Zeeman effect to the flame technique, since the disadvantages are greater than the advantages.

8.1.3. Transport Interferences

In flame AAS the concentration of the analyte element in solution and not the concentration of atoms in the flame must be determined. It has been shown that for numerous elements atomization is virtually quantitative when a suitable flame of appropriate stoichiometry is selected. The actual limiting factor is frequently the efficiency of solution transport into the flame. This depends on the rate of aspiration and the efficiency of nebulization. Under constant experimental conditions, these factors depend on physical characteristics, such as viscosity, surface tension, vapor pressure, and density of the solution.

Transport interferences are caused by alterations in the mass flow of aerosol through the horizontal cross-section of the flame at the observation height. An excellent mathematical treatment of the individual processes and effects has been published by RUBEŠKA and MUSIL [2811].

The change in the density of a solution brought about by inorganic salts has no influence on the rate of aspiration and only little influence on the droplet size of the aerosol. On the other hand, changes to the viscosity brought about by inorganic salts or free acids can be 10% or higher. If the viscosity of a solution is increased by adding increasingly greater quantities of a salt, it will be observed that, at a fixed nebulizer setting, less and less solution is aspirated and that the aerosol droplets become larger. This also means that a greater portion of the solution is deposited in the spray chamber, so that only a smaller portion reaches the flame. This effect becomes more and more noticeable for total salt contents above 1%; organic macromolecules, such as proteins and sugars, have a stronger influence than pure inorganic salts [683].

HÖHN and UMLAND [520] pointed out that a signal depression in the presence of increasing salt concentration must not necessarily be due to a reduced rate of aspiration. Rather more it can be due to increased embedding of the analyte element in the salt, which is not then dissociated.

Surface tension has virtually no influence on the rate of aspiration, but it does influence nebulization and has a decisive effect on the size of the droplets. The surface tension of aqueous solutions is little affected by inorganic salts, but organic materials have a marked influence. Surfactants, on the other hand, only have a limited effect since they do not have sufficient time during nebulization to migrate to the surface of the droplets.

In real samples, viscosity and surface tension naturally alter simultaneously. Sometimes the effect of a change of one physical property cancels the effect of another. A reduction in viscosity, for example, increases the rate of aspiration, but at the same time the mean droplet size gets bigger.

Changes in a sample brought about by varying salt or acid contents are rarely sufficient to cause transport interferences. The changes caused by greater differences in temperature cannot be neglected, however. Organic materials have the greatest influence, especially organic solvents. The use of organic solvents, either mixed with water or pure, to promote signal enhancement in flame emission spectrometry was proposed and checked by various authors [116] [133] [134]. ALLAN [40] carried out tests for AAS and found similar effects for the sensitivity of the determination. The enhancing effect of a number of solvents or water/solvent mixtures is presented in table 14.

Since most organic solvents have a lower viscosity and a lower specific gravity than water, they are more easily aspirated. The surface tension, which is often substantially lower, brings about finer nebulization, so that considerably more sample reaches the flame per unit time. These are the physical causes that lead to an enhancement of the signal; they are

Table 14. Enhancing effect of organic solvents on the signal for copper in a premix flame (according to ALLAN [40])

Solvent		Relative sensitivity*)
0.1 M hydrochloric acid		1.0
Methanol	40%	1.7
Ethanol	40%	1.7
Acetone	40%	2.0
Acetone	80%	3.5
Acetone	20% ⎱	2.35
+ Isobutanol	20% ⎰	
Ethyl pentyl ketone		2.8
Methyl isobutyl ketone		3.9
Ethyl acetate		5.1

* Referred to water = 1.0

typical for all solvents. Next to this is a chemical effect, which causes certain differences from element to element [683].

While the dissociation of water (which always takes place in the flame when aqueous solutions are nebulized) is a strong endothermic reaction that noticeably reduces the flame temperature, the combustion of an organic solvent is generally (except for CCl_4, $CHCl_3$, etc) an exothermic reaction which increases the flame temperature. A higher flame temperature can lead to better atomization of the analyte element, so that an enhancing effect is also to be expected. Finally, many metal atoms are presumed to be more easily released from organic compounds than from various inorganic ones, since organic molecules are thermally less stable than inorganic. This fact can also lead to an increase in the signal; this effect, just as the increased flame temperature, means that serious chemical interferences are less probable in organic solutions.

For a number of elements it could be shown that a linear relationship exists between the increase in the sensitivity and the logarithm of the product between viscosity and boiling point [740]. The physical nature of the effect is evident for the element under study.

Transport interferences are generally fairly easy to eliminate. The simplest way is to match the physical properties of sample and reference solutions, for example by using the same solvent, diluting the sample solution, or by adding the interferent to the reference solutions. If any of these measures are not possible, then the analyte addition technique is recommended, since this eliminates such transport interferences completely and reliably.

The reference element technique has also been proposed for the elimination of transport interferences [353] [1034] (see also section 7.2.2). Nevertheless this technique makes several demands; among others, a dual-channel spectrometer should be available, otherwise the procedure is too complicated. It must also be emphasized that this technique can only be employed for the elimination of transport interferences. The elimination of other interferences by this technique has no theoretical or practical basis.

Various authors have reported the use of sampling pumps with fixed liquid flowrates to avoid transport interferences [585] [1044]. The published results are surprisingly good, but such systems are generally too slow for routine operation. Also, variations in the rate of aspiration can be eliminated with a pump, but not the variations in nebulization and aerosol transport.

8.1.4. Solute-Volatilization Interferences

The solvent in the sample aerosol evaporates either in the spray chamber or directly after entering the flame. This process is usually without problems. The sample thus enters the flame as a solid.

The volatilization of solid particles in hot gases is one of the least researched processes in flame spectrometry and is thus based largely on conjecture. In the simplest case, volatilization is a simple physical process, but reaction of the analyte element with concomitants or with flame gases can also occur. Such reactions can influence the volatilization, and when this influence varies in the presence or absence of an interferent, a solute-volatilization interference is the result.

Solute-volatilization interferences can cause both enhancement and depression of the signal. They are generally specific and depend to a large extent on the properties of the compounds formed in the aerosol after desolvation. (For this reason they have also been termed chemical interferences.)

It can be assumed that in a premix, laminar flame volatilization of solid particles begins as soon as they enter the primary reaction zone. Volatilization requires a given period of time and this depends strongly on the size of the particles. In a flame, time can be translated directly into height. Consequently the occurrence of an interference can depend critically on the observation height.

A fully inert salt can also influence the volatilization of the analyte element. If the boiling point of the interfering salt is lower than the flame temperature, volatilization is determined by the heat transfer. The addition of a relatively volatile salt (such as an alkali halide) will retard heating of the particles and hence volatilization of less volatile elements. In this case, the occurrence of an interference depends on the observation height. The opposite effect can also occur when the salt volatilizes very rapidly so that the particles are disrupted; volatilization of the analyte element is then promoted. Salts that sublime (e.g. NH_4Cl) or decompose into gaseous products (e.g. NH_4NO_3) are very effective.

Frequently, however, volatilization is not a simple physical process, since the analyte element can react and form a less volatile compound. The most serious processes in this respect are the formation of stable oxides and the reduction of the element by flame gases to refractory metals or carbides. The extent of volatilization or of the formation of refractory compounds depends to a very high degree on the form in which the analyte element is precipitated during desolvation. The process of desolvation is thus of considerable importance and will be examined in greater detail.

Essentially, there are two factors that influence the form in which the analyte element will be present; namely the concomitant anions and the possible ligands. Since in every sample solution more than one anion species is generally present, the analyte will combine with an anion or cation to give the compound of lowest solubility during desolvation and crystallize out.

This salt can then undergo secondary reactions. If the salt is hydrated, water will be released; this process is often accompanied by hydrolysis. The salt can also thermally decompose and reactions with concomitants can take place with the formation of new phases of higher thermal stability. Heterogeneous reactions with flame gases can also occur, particularly the reduction to metal or the carbide. The most volatile salts are generally the halides (fluorides, chlorides). The oxides usually formed by decomposition of the salts of oxidizing acids are mostly less volatile. It should also be remembered that many halide salts contain water of crystallization, so that they can also form oxides during thermal decomposition. RUBEŠKA and MUSIL [2811] have shown that it is possible to determine the nature of the thermal decomposition from the flame profile.

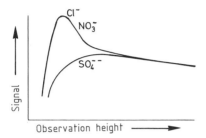

Figure 94. Flame profiles of various magnesium salts (according to [2811]).

The flame profiles for several magnesium salts are depicted in figure 94. Since magnesium chloride, like the nitrate, decomposes to give the oxide ($MgCl_2 \cdot H_2O \xrightarrow{581\,°C} MgO + 2\,HCl$), the flame profiles for these two salts are identical. Calcium chloride, on the other hand, loses its water of crystallization to form largely a dehydrated chloride, which is only partially hydrolyzed to the oxide. The flame profiles for the chloride and nitrate of calcium thus differ (figure 95).

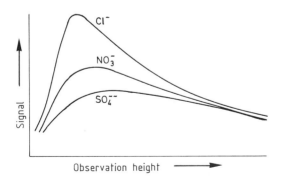

Figure 95. Flame profiles of various calcium salts (according to [2811]).

All elements that form stable oxides (i. e. the elements that must be determined in a nitrous oxide/acetylene flame) also form oxides from hydrochloric acid solution. The oxides of the elements in Groups III and IV of the periodic table, in particular, are very difficult to volatilize because they form large, three-dimensional, polymer structures. The sensitivity of these elements can be markedly increased if the formation of a metal-oxygen bond is prevented during crystallization of the final form of the analyte. SASTRI *et al* [2821] demonstrated this for titanium, zirconium, hafnium and molybdenum by nebulizing these elements as the acetylacetonates (containing a metal-oxygen bond) and the metallocenes (without this bond) into the flame. The same results were obtained for niobium and tantalum, by nebulizing the oxalate and a fluoro complex.

Many elements thus form an oxide, irrespective of the anions originally present in the solution. At the relatively high temperatures in the flame, oxides can react with one another and form phases of even higher thermal stability. Thereby the volatility of the analyte element is further reduced. In particular, spinel types ($MeO \cdot Me_2O_3$), or ilmenite and perowskite types ($MeO \cdot MeO_2$) form very stable lattices that do not melt at the temperature of an air/acetylene flame.

This type of interference is best eliminated by adding to the sample solution (and reference solution) an excess of another element which forms a compound with the interferent that is more stable than the compound with the analyte. The analyte element can then form another compound free of the influence of the interferent and thus volatilize without hindrance. Probably the best known element in this respect is lanthanum, which is frequently added to the alkaline earth elements to eliminate the interferences caused by aluminium, phosphate, etc.

It has been proposed that the mechanism is based on a lower free bonding energy of the added element with the interferent compared to the analyte element with the interferent. Other authors suggest that it could be a mass action effect.

A further procedure for eliminating such interferences involves complexing. The analyte cation is bound to a complex that is chemically resistant and little dissociated, so that salt formation with the interfering anion is prevented. The metal thus enters the flame within the "protective shell" of the complexing agent (frequently EDTA). At the flame temperature the complex is destroyed and the analyte released. This technique may be widely applied, even though it has been relatively seldom reported. It has the sole disadvantage that relatively concentrated acidic solutions frequently cannot be used.

Since interferences of this nature are based on the formation of a thermally stable phase, they naturally depend strongly on the flame temperature. The majority of the interferences thus disappear in the hot nitrous oxide/acetylene flame. The temperature is sufficient to volatilize the oxides. Further, the red reducing atmosphere in the intermediate zone of this flame attacks the oxide chemically.

An exception is the interference of iron on chromium, which has been interpreted as due to the formation of chromite, $FeCr_2O_4$. However, in an air/acetylene flame both iron and chromium oxides are readily reduced to the metal or carbide. The interference, which is only observed in a reducing flame, is thus probably due to the formation of mixed carbides. The addition of ammonium chloride makes chromium more volatile and eliminates the interference [1049]. This can be explained by the form in which chromium crystallizes out from solution. In the presence of a large excess of ammonium chloride, chromium crystallizes out of the hydrochloric acid solution as an ammonium complex. When heated, this complex forms chromium chloride, which has a sublimation point of $1300\,°C$. The same complex is also formed when potassium dichromate is heated with ammonium chloride. In the absence of ammonium chloride, chromium forms refractory oxides which are the cause of the interference.

The element that is bonded directly to the analyte can also strongly influence reduction in the condensed phase. A reduction of the analyte to metal or carbide may accelerate or retard its volatilization, depending on which form is the more volatile [2811]. The boiling point is often taken as a measure of the volatility. Comparison of the physical data shows that for elements such as Mg, Ca, Sr, Ba, Cr or Mn, the metal has the lowest boiling point, while for others it is the oxide, e.g. B_2O_3, SiO, TiO_2, V_2O_3, MoO_3, FeO_3, CoO or NiO, while for a very few it is even the carbide, e.g. BeC_2 or Al_4C_3. Consequently the adition of ligands that facilitate reduction will lower the sensitivity for those elements that are most volatile as their oxide.

L'vov and ORLOV [2583] showed that the extent of reduction of Cr, Fe, Co and Ni to carbides before volatilization in a reducing air/acetylene flame depends on the anion present. Chlorides are reduced the least, since they volatilize the fastest. Nitrates decompose easily to oxides, which are volatilized more slowly, so that reduction can take place in the flame gases. The greatest signal depression is observed for sulfates, since these are very stable and therefore volatilize the slowest. With chlorides, signal depression increases with increasing hydrochloric acid concentration in solution. This may be explained by partial hydrolysis in the droplets during desolvation or through thermal decomposition of the corresponding hydrated complex. The reduction to carbides in the condensed phase opens up the possibility for numerous mutual interferences, since the majority of the carbides are combinable with each other. Interferences for Cr, Mn, Fe, Co and Ni based on the formation of mixed carbides are more pronounced in reducing flames and for sulfates and nitrates. They disappear completely in oxidizing flames and in hydrochloric acid solutions [2811].

If a sample contains a compound that is consumed by the reducing constituents, the analyte element will volatilize more easily because the reducing constituents react preferentially with the concomitant. The heats of formation of the transition element carbides generally decrease in the order $Me(VI) > Me(V) > Me(IV)$, and aluminium carbide has the lowest heat of formation of all elements. Thus, in the presence of aluminium the reduction of most metals is retarded. If the metal oxide is more volatile than either the metal or the carbide, then signal enhancement will be observed. Aluminium therefore enhances the sensitivity for Mo, W, V, Si, Ti, and B [1306] [2162] [2727] [2808] [2812]. Aluminium also exhibits a marked enhancing effect for Cr, Fe, Co and Ni in a reducing air/acetylene flame. Aluminium is frequently used as a buffer instead of lanthanum in the determination of titanium and vanadium.

It can be assumed that aluminium oxide, which is produced when aluminium salts are calcined, decomposes according to the reaction

$$Al_2O_3 \xrightarrow{\ 2080\ K\ } Al + AlO + O_2$$

and that the oxygen released facilitates volatilization of the analyte element as its oxide. In effect, an "oxidizing microclimate" is maintained in the particles [2811].

Volatilization interferences should not be expected for aluminium and beryllium, since they are most volatile as their carbides and are easily reduced under the prevailing conditions; this expectation has been observed experimentally.

The platinum metals form relatively volatile oxides, but as the metals are very non-volatile. They all have boiling points above 3000°C and their heats of atomization are greater than 500 kJ/mol, except for palladium. All the platinum elements are easily reduced to metal, and in an air/acetylene flame are atomized very ineffectively due to their high atomization energies and their tendency to form agglomerates. Since they also form alloys among themselves, a second platinum element in the sample will increase the final particle size in which the analyte is bound as an alloy.

The mutual signal depression of the platinum metals in an air/acetylene flame is well known. Oxidizing spectrochemical buffers, such as lanthanum salts [1095] [3002] or a combination of cadmium, copper and alkali sulfates [571], suppress these interferences and enhance the sensitivity for the platinum metals. Next to their oxidizing action, these reagents also disperse the particles and hinder the formation of agglomerates.

8.1.5. Vapor-Phase Interferences

The radiant flux, which is converted in the spectrometer to the measured signal, carries the information on the spectroscopically active form of the analyte element in the axis of observation in the atomizer. A flame is a dynamic system, but it can nevertheless be assumed that the residence time of the sample in the axis of observation is long enough, in the first approximation, to permit equilibrium between free atoms and compounds (dissociation equilibrium), ions (ionization equilibrium) and excited atoms (excitation equilibrium) to be reached. These equilibria can be described via the law of mass action, the SAHA equation, and the BOLTZMANN distribution law. All the processes are temperature dependent, and since temperature also varies within the flame, the concentrations of individual components in the vapor phase will also change [1029].

The equilibrium and incomplete conversion of the analyte element into a spectroscopically active form (i.e. atoms in the ground state) cannot be considered as an interference as such. Solely the effect of a concomitant is an interference when it alters the equilibrium and thus the fraction of dissociated, ionized or excited element. This presupposes that the analyte element and the interferent are present in the vapor phase simultaneously and that they have a common partner, such as a common anion or free electron.

For these thermodynamic considerations, the mechanism responsible for equilibrium plays no part. Such a consideration is only possible, however, when equilibrium is reached in a time shorter than the residence time up to the point of measurement, i.e. shorter than the time required to reach the observation height. Since the velocity of the flame gases is around 10^3 cm/s, time available to reach equilibrium is only a few milliseconds. If this time period is too short for the equilibrium to be reached, the processes must be regarded as kinetic and not thermodynamic.

Alterations to the equilibrium can take place when a reaction is retarded or through a change in the reaction mechanism. A condition of equilibrium is reached very rapidly in the primary reaction zone of a flame. Once the sample leaves this zone, however, equilibrium is attained very much more slowly, since the number of radicals able to react and the number of electrons decrease sharply. An example here is the ionization equilibrium, which is rapidly attained in the primary reaction zone due to fast charge transfer. In higher zones, the much slower collisional ionization in the flame gases is responsible for changes in the equilibrium.

Vapor-phase interferences are easily recognized by nebulizing the analyte element and the interferent separately into the same flame using twin nebulizers. If a vapor-phase interference is present, the effect must be the same as when both are nebulized together. Vapor-phase interferences change the slope of the analytical curve by a determined factor, which can be more or less than one.

When considering vapor-phase interferences, it should not be forgotten that the flame gases are always present in large excess and act as buffers. Their composition is hardly affected by the sample. The composition of the flame gases depends on the gas mixture and on the design of the burner. If pure sample solution is aspirated, the composition of the flame gases can change as a result of the solvent, but not of the analyte element since its concentration is too low.

The compounds present in the vapor phase are essentially diatomic and triatomic molecules, such as monohydroxides ($NaOH$, $BaOH$, $BeOH$, etc.), cyanides [2586], or oxides (Cu_2O, etc.). The degree of dissociation can change substantially with temperature. It is known that the aspirated solvent can influence the flame temperature markedly; albeit, according to definition this is not an interference. However, this fact must be taken into consideration if the solvent is changed. It is especially significant for organic solvents with varying water contents. The influence of the remaining sample matrix on the flame temperature can be neglected.

In practice, a change of flame temperature can only be achieved by altering the type or composition of the burner gases. Simultaneously with a change of temperature, however, there is a change in the composition of the flame gases and this can lead to a shift in the equilibrium.

Dissociation equilibria frequently take place between components of the flame gases. Typical examples are the dissociation of oxides, hydroxides, cyanides or hydrides. The concentrations of O, OH, CN and H in a flame are determined by reactions between the natural components in the flame. The influence of the nebulized sample is negligible due to the buffer action of the flame gases.

This is also valid even when equilibrium has not be reached, especially in the primary reaction zone where higher concentrations of these radicals are present, so that the buffer effect is strengthened. In the primary reaction zone of a nitrous oxide/acetylene flame, the concentration of oxygen radicals is three orders of magnitude greater than the equilibrium concentration. Thus, a competition reaction for oxygen in the vapor phase as an explanation for the interference of aluminium on vanadium is clearly wrong [1023]. Similar considerations are also valid for all equilibrium reactions involving flame components, such as the dissociation of cyanides [2586], etc. The only requirement is that reactions between flame gas components should proceed rapidly enough, and this is usually the case.

It should be emphasized that under conditions of equilibrium, the mechanism of reduction of the analyte element to atoms is irrelevant. The way in which equilibrium is reached has no influence on the equilibrium itself [2343]. Under these conditions the dissociation equilibrium of oxides, hydroxides and other compounds with components of the flame gases will not be influenced by constituents of the sample. Consequently, vapor-phase interferences cannot occur under equilibrium conditions.

The situation for the dissociation of halides is quite different, in contrast, since halides are not normally components of the flame gases. Their concentration in the flame is dependent on their concentration in the aspirated sample solution. In general, the quantity and nature of the anion bonded to the analyte element is insignificant when the halide concentration in the solution exceeds 1 M. If the analyte element is present as the halide, volatilization is normally complete, but dissociation into atoms must by no means be complete. The existence of undissociated halides, especially in cooler flames, has been repeatedly demonstrated through the molecular spectra of monohalides. Since the dissociation constants of halides are temperature dependent, the strongest dissociation interferences are found in flames of lower temperature.

The energy of dissociation of halides increases with the decreasing atomic mass of the halogens. The determination of indium in the presence of halogenic acid is an example where the increase of the energy of dissociation in the order $HI < HBr < HCl$ agrees with an increase in the depression. Good agreement is also found for copper. Often, though, the interferences caused by halogenic acids are much more complex since they can also influence volatilization.

The atomization processes in the flame are frequently influenced by free radicals, thus leading to a deviation in the equilibrium. Theoretical reasonings on the efficiency of atomization based on the thermal dissociation of the monoxide without taking radicals into consideration are frequently wrong and do not agree with experimental findings [2342] [2343]. The absorbance signal for many elements increases with the fuel gas/oxidant ratio, although the temperature of a hydrocarbon flame changes very little. This signifies that atomization is not due to thermal dissociation alone.

It has been known for many years that oxides with energies of dissociation greater than 6.3 eV per bond (523 kJ/mol) behave as "refractory oxides". This energy value is suspiciously close to the bonding energies of CO and C_2, suggesting reduction mechanisms according to:

$$MO + CO \rightarrow M + CO_2$$

or

$$MO + C_2 \rightarrow M + CO + C .$$

Nevertheless, the change in entropy has not been taken into consideration in the above deliberations. A species can only effect reduction when the resulting change in free energy is negative, i.e. when the change of free energy of the reaction

$$2 CO + O_2 \rightarrow 2 CO_2$$

is smaller than the free energy for the formation of metal oxides. Numerous calculations have shown that reduction with CO or H_2 does not occur to any large extent, but that C_2 and H radicals are very efficient reductants [2343].

As indicated above, the reduction process is hardly influenced by constituents of the sample. Solely constituents that change the number of C_2 or H radicals can influence reduction in the vapor phase. In the first instance these are organic solvents; this explains their enhancing effect on the absorbance measure. According to definition, however, organic solvents are excluded as a source of interferences, so that any effect must be due to other organic materials in the sample.

If equilibrium in the flame is reached slowly, concomitants can have a catalytic effect in that either they accelerate atomization or they accelerate competing reactions and thus retard atomization. An example for the second possibility is the catalytic action of elements such as Cr, Ca, Sr, Ba, and compounds such as SO_2 and nitrogen oxides on the recombination of H radicals and their effect on tin.

If the catalytic effect of tin on the recombination of hydrogen atoms is via free tin atoms, then the atomization of tin should be proportional to the concentration of hydrogen radicals above the equilibrium concentration. This could explain the high degree of atomization of tin in cooler hydrogen diffusion flames, since these have a higher concentration of H radicals than air/hydrogen or air/acetylene flames. All elements and compounds that catalyse the recombination of hydrogen should depress the signal for tin. This supposition is in agreement with the observations made in flames and long tube burners [1068] [2672]. In long tube burners good agreement was found between the effectiveness of the catalyst and the depressive effect of various metals on tin [2807].

A further, relatively widely occurring interference in the vapor phase is the ionization chemical interference. Ionization occurs mainly in the primary reaction zone through charge transfer from such molecules as $C_2H_3^+$ or H_3O^+. In higher zones of the flame, mainly collisional ionization with relatively long times of relaxation is responsible for a shift in the equilibrium. All ionization reactions involving species produced in the primary reaction zone come to equilibrium slowly, so that the concentration of neutral analyte atoms increases with increasing observation height. Ionization equilibria can be shifted in the presence of electron acceptors. Cyanides can increase ionization by capturing free electrons [1023] according to

$$CN + e^{\ominus} \rightarrow CN^-.$$

In contrast, ions are produced in the primary reaction zone through reaction with flame gases, such as

$$CH + O \rightarrow CH^+ + O^-.$$

These ions suppress ionization of the analyte element.

The question whether ionization occurs or whether an ionization chemical interference is to be expected depends on the ionization energy of the analyte element, possible interfering concomitants, and the flame temperature. Ionization normally occurs in the hot nitrous oxide/acetylene flame for elements having an ionization energy of 7.5 eV or lower. In flames of lower temperature, ionization is practically limited to the alkali metals. Moreover, the extent of ionization for a given element and temperature is dependent on the concentration of the element. As shown in figure 96 on the example of barium, ionization is stronger at lower concentrations than at higher. This results in a concave analytical curve that can be explained by the more frequently observed rapid recombination of ions and electrons at higher concentrations.

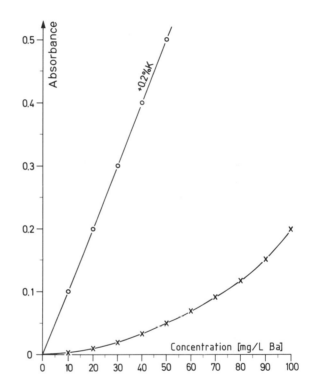

Figure 96. Interference caused by ionization for the determination of barium in a nitrous oxide/acetylene flame.—The lower curve (×) was obtained for measurement of aqueous solutions, while the upper curve (○) was obtained after addition of 0.2% potassium (as KCl) to suppress ionization.

Non-linearity of the analytical curve, caused by ionization, has two major disadvantages for practical analysis: Firstly, numerous reference measurements are required to establish the curve, and secondly, the loss of sensitivity in the lower range precludes trace analysis. For these reasons elimination of ionization is very desirable.

In principle there are two ways of suppressing ionization. The analyte element could be determined in a flame of lower temperature, for example. This is possible for the alkali metals; they are noticeably ionized in the air/acetylene flame, but hardly ionized in the cooler air/hydrogen flame (refer to table 15). This is not practicable for the majority of elements, however, since in a cooler flame they are either not determinable (lanthanides, etc.) or considerable solute-volatilization interferences must be taken into account (e.g. barium).

Table 15. Ionization of a number of elements in the air/acetylene and nitrous oxide/acetylene flames

Element	Concentration employed mg/L	% Ionization	
		Air/acetylene	N$_2$O/acetylene
Li	2	0	
Na	2	22	
K	5	30	
Rb	10	47	
Cs	20	85	
Be	2	—	0
Mg	1	0	6
Ca	5	3	43
Sr	5	13	84
Ba	30	—	88
Y	100		25
Lanthanides			35–80
Tm	50		57
Yb	15		20
Lu	1000		48
Ti	50		15
Zr	500		10
Hf	1000		10
V	50		10
U	5000		45
Al	100		10

The second way is to suppress ionization by shifting the ionization equilibrium

$$Me \rightleftharpoons Me^+ + e^\ominus$$

on the basis of the law of mass action by producing a large excess of electrons in the flame or by charge transfer. This can be done very easily in practice by adding a large excess of an easily ionized element (usually potassium or cesium) to sample and reference solutions. The buffer element is largely ionized in the flame, so that ionization of the analyte element is suppressed virtually quantitatively (see figure 96). The ease with which the ionization equilibrium can be shifted allows ionization to become a real interference and is thus a good reason why it should be suppressed. If a reference solution containing the analyte element merely as a pure salt is used, the element will be ionized to a given extent. If the sample solution contains metals other than the analyte, which is to be expected, and these metals are more easily ionized or are present in excess, the ionization equilibrium will be shifted to the left and a higher measure will be obtained. In this way real errors can occur, often of substantial extent.

Since condensed-phase interferences decrease with increasing temperature but ionization increases, it is often necessary to judge which one is the more serious. Frequently a flame of higher temperature is chosen, since ionization is in general much easier to control. The decision also often depends on the matrix; the choice of the best means of atomization in dependence on the matrix will therefore be discussed as appropriate in later chapters.

KORNBLUM and DE GALAN [698] found that elements with an ionization potential of less than 5.5 eV were more or less completely ionized in a nitrous oxide/acetylene flame. An addition of at least 10 g/L of another easily ionized element (cesium or potassium) is required to permit determinations free of ionization interferences. Elements with an ionization potential of up to about 6.5 eV require the addition of 1–2 g/L of an ionization buffer, while elements with an ionization potential over 6.5 eV only require about 0.2 g/L.

In determining the necessary quantity of an ionization buffer, it is usual to try to reach the plateau of the curve and at the same time to optimize the sensitivity. The quantity determined for pure solutions may in fact be unnecessarily high for real sample solutions, since concomitants can also have a buffer action [906]. Use of lower buffer concentrations nevertheless presupposes good knowledge of the matrix. The use of ionization buffers in the nitrous oxide/acetylene flame has been described for many elements [377] [794] [2162] and is generally applied to avoid ionization interferences. Since ionization in the flame is preceded by volatilization and both are equilibrium reactions, the actual processes can have a complex relationship that often evades quantitative evaluation.

In theory, excitation of the analyte element can also lead to an interference similar to ionization. Since an element can only exist in the excited state for a very short time in the flames normally used in AAS, the number of excited atoms is always less than 1%. This interference is therefore virtually unknown in AAS, or it can be neglected.

8.1.6. Spatial-Distribution Interferences

Spatial-distribution interferences are caused by changes in the flowrate or flow pattern of the sample in the flame. This results in a change in concentration of the analyte (in all its forms) in the observation zone, although the total concentration in the flame is constant. The observation zone is that region of the flame through which the optical radiation beam passes on its way to the monochromator exit slit and detector.

It is a prerequirement that the concentration of the analyte element (in all its forms) must change in the observation zone to distinguish between spatial-distribution interferences and solute-volatilization interferences; the latter solely require a change in the portion that is volatilized. A solute-volatilization interference cannot occur if the sample is quantitatively volatilized, but a spatial-distribution interference is quite possible.

Changes in the mass flowrate can be caused through alterations in the quantity of combustion products and thereby through changes in the volume of the flame. The flame may remain unchanged, on the other hand, and only the flowrate or flow pattern of the sample in the flame changes. BOSS and HIEFTJE [2095] and also L'VOV *et al* [2580] examined the flow pattern of samples and sample atoms in flames. The results indicated that the patterns are influenced mainly by the size and rate of volatilization of the particles. The diffusion of molecules and atoms appears to be of less significance. It may thus be concluded that the spatial distribution of atoms in an unchanged flame has a close connection with the volatilization of the condensed phase. Nevertheless, this interference should not be confused with solute-volatilization interferences.

An alteration in the spatial distribution of the atoms can lead to an increase or a decrease of the sensitivity, depending on the observation zone [1018] [2762]. The narrower this zone is, the more likely such interferences can occur; they depend on the specific spectrometer used, its optical system, and on the burner system.

The gas mixture containing the sample aerosol streams vertically from the burner slot at a given velocity. The gas volume changes substantially in the primary reaction zone, mainly due to its thermal expansion. Consequently its velocity and flow pattern also change; as well as the vertical flow pattern a horizontal component becomes noticeable. This sudden change in the flow pattern exerts a horizontal force on the particles that causes a deviation of the linear movement. Because of their mass, however, the lateral spread of the particles will be less than that of the flame gases and becomes smaller with increasing size of the particles. A spatial-distribution interference can take place when an interferent in the sample causes the particles to be bigger after desolvation, so that with a sufficiently long residence they are concentrated more in the middle of the flame than if no matrix was present.

L'vov et al [2581] [2582] examined this effect quantitatively and found that it could be neglected for normal, lower analytical concentrations. In extreme cases with high total salt content of refractory compounds it can attain considerable extent. Interferences of the spatial distribution of atoms in the hot nitrous oxide/acetylene flame are much more frequent, since this flame has a lateral expansion that is five times as large as that of an air/acetylene flame. Spatial-distribution interferences were reported even for the very first work on flames of higher temperature [48]. KOIRTYOHANN and PICKETT [688] explained the signal enhancement of Al, Ba, Ca, Li and Sr through perchloric acid as an interference of this type. They found that the interference disappeared when the slot burner was turned through 90° and verified their supposition by measuring the lateral absorption profile for strontium. L'vov et al [2582] later measured the lateral flame profile for numerous elements using an optical beam with a very narrow cross-section of 0.3 mm.

On closer scrutiny of the problem there is a strong dependence of the observed effects on the experimental conditions and technical details [823] [2808] [3065] [3066], as well as a close relationship to solute-volatilization interferences [1306]. To classify an interference as a spatial-distribution interference, two criteria must be met: The interference must disappear when the burner is turned 90° to the optical axis, and the relative increase should become smaller with increasing observation height.

8.2. The Graphite Furnace Technique

The graphite furnace technique permits the direct analysis of liquid, dissolved, and solid samples. A measured or weighed quantity of sample is dispensed into the graphite tube and through a series of stepwise or ramped temperature increases, as many of the concomitants as possible are removed before the analyte element is atomized by a final, rapid increase in temperature. The first marked difference to atomization in the flame is that a portion of the concomitants (e.g. the solvent) is removed before atomization of the analyte element, so that these cannot cause interferences as in the flame. As will be discussed in this section, the better the separation of the concomitants from the analyte element before atomization, the freer the determination will be from interferences.

A further advantage is the inert gas atmosphere in which atomization in graphite furnaces takes place. This is further enhanced by the strongly reducing properties of glowing carbon. While in the flame the risk always exists that the atoms produced can form compounds with the highly reactive components of the flame (e.g. oxygen and hydoxide radicals) and thereby be lost to the determination, in the graphite tube the inert, reducing atmosphere promotes atomization. Nevertheless, the theoretically better freedom from interferences of the graphite furnace technique often cannot be attained in practice. The main reason is that in the graphite tube the buffer action of the flame gases is absent, so that the environment is determined to a large extent by the sample constituents.

Similar to the flame technique, non-spectral interferences are divided into condensed-phase, volatilization, and vapor-phase interferences. A distinction is not always possible, however, since solid/vapor reactions can also take place.

The special problems and possibilities of the direct analysis of solids will also be treated. The graphite furnace technique is the sole technique in atomic absorption spectrometry with which solid samples can be analyzed without major difficulties.

8.2.1. Atomization in Graphite Furnaces

Certainly the most thorough theoretical and practical investigations on atomization in graphite furnaces have been carried out by L'vov [8], who has been occupied with this subject since 1959 [775] [776].

For the following discussion it is important to consider how the sample is distributed over the graphite surface. A simple calculation shows that the space required for the analyte element is in the range 0.0001 mm^2 to 10 mm^2, according to the nature and quantity of the sample. This is considerably less than the surface area of the graphite tube. If it is assumed that there is an even distribution of the element over the area wetted by the sample solution, then a monoatomic or monomolecular layer should exist. Even when small heaps of atoms or molecules (crystals) are formed during the drying process, which can be assumed as likely, there is still always good contact of the analyte element with the graphite surface. This may not be the case when solid samples are introduced directly into the tube or when highly concentrated salt solutions are pipetted, but this will be fully discussed later.

If the atomization itself is considered, it can emanate from either molecules or atoms, according to the nature of the sample and the behavior of the analyte element. If it emanates from molecules, atomization can be either a simple thermal decomposition (dissociation) of a compound, or the reduction of an oxide on the glowing graphite surface. The difference between these two mechanisms is solely the active participation of the tube material (carbon) on the dissociation of the sample molecules, and frequently a clear distinction is not simple. If atomization emanates from the metal, it can be either desorption or volatilization. Because of the small quantity and wide distribution of the sample, boiling as the atomization mechanism can be excluded by deduction.

If a monoatomic layer exists after· drying of the sample, i. e. every atom is isolated from the next, atomization will be a desorption from the graphite surface and will be determined by the adsorption isotherm. It must be directly stated, however, that this mechanism is very unlikely and that up to now no indication of such an atomization could be found. If, on the other hand, the atoms are present as small heaps (e. g. crystals), which can be assumed as probable, then the atomization is a pure volatilization. The volatilization temperature can lie considerably below the boiling point or even the melting point. Because of the extremely small sample mass and the correspondingly low partial pressure of the atom cloud which is formed, quantitative volatilization can take place even at very low vapor pressures.

According to the kinetic gas theory, the mass G which the sample loses per unit time and area when it is heated to temperature T is given by equation (8.3).

$$G = P \sqrt{\frac{M}{2\pi RT}} \quad [\text{g}/\text{cm}^2 \text{ s}], \tag{8.3}$$

where P is the saturated vapor pressure at temperature T and M is the molecular mass. Nevertheless, it should be taken into account that equation (8.3) is only valid for atomization *in vacuo*, but that in the presence of a gas which does not react with the sample, G can be up to two orders of magnitude lower.

The vapor pressure at which an element can be atomized within 0.1 second (a requirement that will be explained later) may be quickly calculated. If it is assumed that a sample mass of 10 ng with a mean atomic mass of 50 is distributed over an area of 1 mm² at an atomization temperature of 3000 K, from equation (8.3) it can be calculated that the pressure P is about 10 to 15 pascal. For most elements, however, the saturated vapor pressure at 3000 K is greater than this, with the exception of tantalum, tungsten and rhenium, and a few others that form carbides with extremely low vapor pressures, such as hafnium, niobium, thorium and zirconium. The majority of the other elements can be atomized either in less than 0.1 second or at temperatures below 3000 K.

Under given prerequirements, an element can be atomized within about 0.1 second when the saturated vapor pressure is about 10 to 15 pascal. Atomization can already begin at considerably lower vapor pressures (i. e. temperatures) but the atomization time is then correspondingly longer. This consideration is important, because during thermal pretreatment the sample may not be heated above the point at which atomization just begins since losses of the analyte element will otherwise take place. Furthermore, it can be deduced that the time it takes to reach the atomization temperature also plays an important role. With a slow rate of heating, a considerable portion of the sample will have volatilized before the actual atomization temperature is reached (i. e. the atomization takes place slowly). Volatilization will take place within the specified time of 0.1 second only with an infinitely fast increase to the atomization temperature.

The atomization time is of greatest importance since the maximum density of the cloud of atoms can be achieved only when the atomization time is shorter than the residence time of the atoms in the graphite tube. As will be explained, the residence time of the atoms in the graphite tube is only a few tenths of a second, so that an atomization time of 0.1 second is absolutely essential. L'vov showed that it really is possible to achieve the maximum atom cloud density and thereby the maximum peak height sensitivity. He also pointed out that atomization in the graphite tube is around a thousand times slower than in the flame, and that this is the reason for the relative freedom from interferences of determinations in the graphite tube since refractory substances have more time to dissociate.

The actual atomization process will now be considered again; i.e. the discussion on the mechanism of atomization in the graphite tube and the question whether it emanates from metal or from molecules; and how far hot graphite takes an active part in the atomization. For the experimental clarification of this question, a system of measurements can be employed that has been tested in the present author's laboratory [3048] for a large number of elements in the presence of various anions. Double curves are established in which the signal height is plotted versus the temperature. In the first curve the height of the signal at the optimum atomization temperature is plotted versus the pretreatment temperature as the variable and in the second curve the signal height is plotted versus the temperature of atomization, where this is variable (figure 97). The first curve shows the temperature to which a sample can be thermally pretreated without loss of the analyte element. Furthermore, from the first curve it is possible to derive the temperature at which the element is quantitatively volatilized. From the second curve it is possible to read off the temperature at which atomization is first evident, the appearance temperature, and also the optimum atomization temperature at which the maximum atom cloud density is attained.

Figure 97. Thermal pretreatment/atomization curves for the graphite furnace technique.—Curve **A**: the absorbance measured for atomization at the optimum atomizaton temperature **4** is plotted against the thermal pretreatment temperature, where this is the variable. **1** is the highest thermal pretreatment temperature that can be applied without losses of the analyte element taking place. **2** is the lowest temperature at which the analyte element is quantitatively volatilized. Curve **B**: the absorbance is plotted versus the atomization temperature. **3** is the appearance temperature of the analyte element and **4** is the optimum atomization temperature.

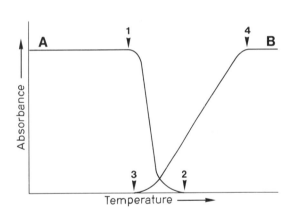

If physical data, such as melting point, boiling point and decomposition point, of the analyte element and its compounds are entered into these experimentally determined curves, it is frequently possible to draw conclusions about the atomization mechanism. This is a purely empirical method, but it has led to very useful results [1297]. In the following, a number of examples will be mentioned which give indications of differing mechanisms.

Figures 98 and 99 show the pretreatment/atomization curves for gold and iron, respectively. In both examples, the first losses during thermal pretreatment take place at 100 to 200 °C below the melting points of the respective elements, and at the same time a small atomiza-

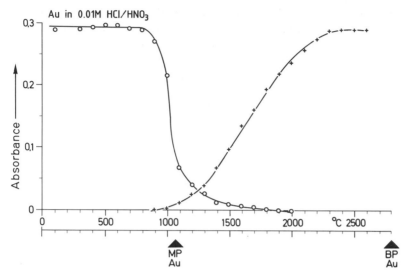

Figure 98. Thermal pretreatment/atomization curves for gold.—Initial preatomization losses and the appearance temperature are below the melting point of gold. Atomization emanates from the metal.

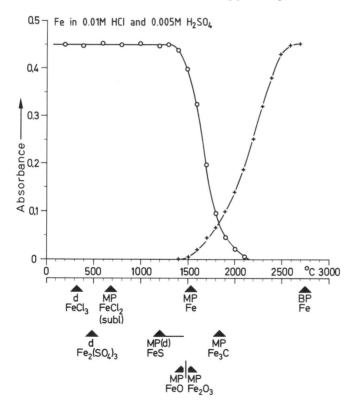

Figure 99. Thermal pretreatment/atomization curves for iron.—Initial preatomization losses and the appearance temperature are below the melting point of iron. Atomization emanates from the metal.

tion signal can be measured. Both curves are independent of the salt used or the acid added. This finding, which incidently is very similar for all the noble metals examined and various other elements, is a clear indication that atomization emanates from the metal. Recent investigations by ROWSTON and OTTAWAY [2804], using thermogravimetric measurements, X-ray diffraction studies, etc., have verified this supposition for the noble metals. This finding is also in agreement with earlier deliberations that with very small masses of metal to be atomized, very low partial pressures are already sufficient for atomization and that this takes place below the melting point.

This however means that the sample must be reduced to metal early in the thermal pretreatment phase, as assumed by FULLER [393] [2277].

Accordingly, the difference between the previously discussed mechanisms is solely the time point at which reduction or dissociation takes place. If the decomposition temperature of the salt (with or without the participation of the hot graphite) is clearly lower than the volatilization temperature, the sample will be reduced to metal during thermal pretreatment, and atomization is then a direct volatilization. If, on the other hand, the decomposition temperature of the compound is higher than the vaporization temperature of the metal, the decomposition temperature is also the appearance temperature.

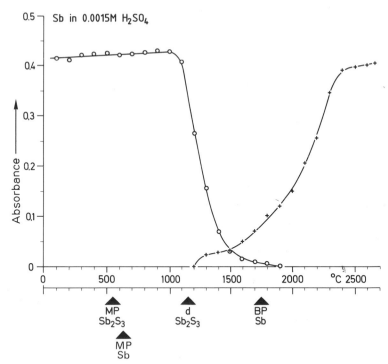

Figure 100. Thermal pretreatment/atomization curves for antimony in sulfuric acid.—Initial preatomization losses and the appearance temperature coincide with the decomposition temperature of antimony(III) sulfide.

An example of the second mechanism is the thermal pretreatment/atomization curve for antimony in sulfuric acid solution shown in figure 100. The point at which the first losses occur and at which the atomization signal can be simultaneously observed (appearance temperature) is virtually identical to the decomposition point of antimony(III) sulfide. This is obviously formed during thermal pretreatment by reduction of the sulfate on the hot graphite.

A particularly interesting curve profile is exhibited by the element beryllium. If this element is determined in acid solution (figure 101), even at relatively low temperatures there is a fall-off in the thermal pretreatment curve, but no atomization signal can be measured. The point at which the first losses occur coincides with the decomposition point of beryllium sulfate or the boiling point of beryllium chloride, respectively. Since no atomization can be measured at the same time, the beryllium must be transported from the tube in a molecular form and hence lost to the subsequent determination.

By increasing the temperature further, the pretreatment curve reaches a plateau, i.e. with further increase in the temperature, no further loss of beryllium takes place. This indicates the formation of a stable compound. The renewed fall-off in the curve and the simultaneous appearance of the first atomization signal coincides with the decomposition point of beryllium carbide. It is then clear that at the first decomposition point, beryllium sulfate forms beryllium carbide and a volatile compound, which has yet to be identified.

The curve profile is completely different when an alkaline beryllium solution is used (i.e. the hydroxide is present), as is shown in figure 102. Beryllium oxide, which is thermally very stable, is formed during thermal pretreatment and can therefore be more strongly heated in the graphite tube without losses. In this case, atomization clearly appears to be a reduction of the oxide on the hot graphite surface. A very similar curve profile is also found for such elements as aluminium and silicon. They can only be satisfactorily determined if the oxide is formed during thermal pretreatment.

CAMPBELL and OTTAWAY [205] have postulated the reduction mechanism

$$MeO + C \rightarrow Me + CO$$

for many metals. They calculated the free energy for the corresponding reaction and compared the values determined thermodynamically with the appearance temperatures and found good agreement. From the above observations, however, this oxide reduction is not the only atomization mechanism coming into question.

A further indication that the active participation of the hot graphite is not in all cases necessary is shown by experiments on the direct determination of easily atomized elements in solid rock samples [768] [1295]. A few milligrams of the powdered rock are placed into the graphite tube and heated to around 1400–1700°C. In this procedure, the analyte element (e.g. cadmium, lead, mercury, silver, thallium, zinc) can hardly come into direct contact with the graphite, nevertheless the first atomization signals can be recorded below the melting points of the elements. The sensitivities here, however, are considerably lower than when working with solutions.

The concept proposed by CAMPBELL and OTTAWAY [205] certainly brought very useful information on the reactions involved in the atomization process, but it is nevertheless not

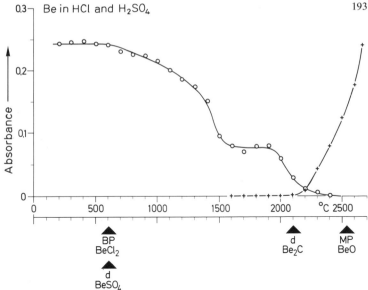

Figure 101. Thermal pretreatment/atomization curves for beryllium in acid solution.—Initial preatomization losses coincide with the boiling point of the chloride or the decomposition point of the sulfate. The appearance temperature coincides with the decomposition point of beryllium carbide.

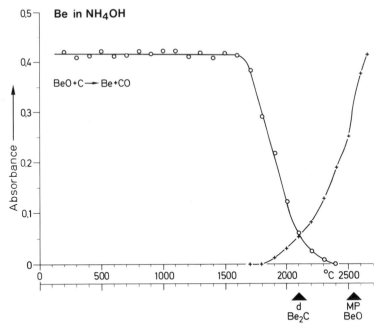

Figure 102. Thermal pretreatment/atomization curves for beryllium in ammonia solution.—Initial preatomization losses and the appearance temperature coincide, and this temperature bears no correlation with the transformation temperature of any compound. Atomization is a reduction of the oxide on the hot graphite.

possible to describe the time-dependent signals of the graphite furnace technique purely on a thermodynamic basis (as shown by L'VOV [8]). The analytical signals generated in a graphite furnace are normally curves having a maximum. Their profile for a given element depends on the physical and chemical properties of the matrix, on the geometry and design of the furnace, the graphite material, the gas flowrate, and the rate of heating.

STURGEON et al [2938] investigated atomization mechanisms on the basis of a thermodynamic/kinetic approach. Under the supposition that an equilibrium exists between the surface of the sample and the vapor phase in the furnace and that the production of atoms is characterized by a unimolecular velocity constant, the logarithm of absorbance plotted versus the reciprocal temperature $1/T$ gives a straight line. The activation energy E_a of the determining step for the atomization can be obtained from this plot. Three possible atomization mechanisms can be deduced from these E_a values: Thermal dissociation of the oxide; thermal dissociation of the halide; reduction of the oxide on hot graphite with subsequent volatilization of the metal.

In later work, STURGEON and CHAKRABARTI [2140] [2936] refined their procedure and proposed four mechanisms:

1) Reduction of the solid oxide on the graphite surface according to:

$$MO_{(s)} \xrightarrow{C} \underset{\underset{\ldots\ldots\ldots\ldots\ldots\ldots\ldots\uparrow}{\mid}}{M_{(l)}} \rightleftharpoons \tfrac{1}{2} M_{2\,(g)} \rightarrow M_{(g)}$$

where M = Co, Cr, Cu, Fe, Mo, Ni, Pb, Sn, V.
(s) = solid, (l) = liquid, (g) = gaseous.

2) Thermal decomposition of solid oxides according to:

$$MO_{(s)} \xrightarrow{T} M_{(g)} + \tfrac{1}{2} O_{2\,(g)}$$

where M = Al, Cd, Zn.

3) Dissociation of oxide molecules in the vapor phase according to:

$$MO_{(s)} \rightleftharpoons MO_{(g)} \rightarrow M_{(g)} + \tfrac{1}{2} O_{2\,(g)}$$

where M = Cd, Mg, Mn, Zn.

4) Dissociation of halide molecules in the vapor phase according to:

$$MX_{2\,(s)} \rightarrow MX_{2\,(l)} \rightarrow MX_{(g)} + X_{(g)}$$
$$\underset{M_{(g)} \quad X_{(g)}}{\swarrow \quad \searrow}$$

where M = Cd, Fe, Zn,
      X = Cl.

Reactions 1 and 2 require intimate contact with the graphite surface and depend strongly on the rate of heating used for atomization. This was verified experimentally by using a capacitor discharge for the atomization step [2140]. Reactions 3 and 4, in contrast, are via dissociation in the vapor phase and should therefore depend strongly on the temperature of the inert gas and whether the graphite surface from which the sample is volatilized is in thermal equilibrium with the gas phase in which atomization takes place. These requirements will be treated in detail later, since they have a significant bearing on the occurrence of interferences.

In 1974, FULLER [2277] proposed a kinetic theory for atomization in graphite furnaces and was thereby able to explain the anomalies of the thermodynamic approach. For the atomization of copper, a two-stage reaction was called for, in which the reduction of CuO to Cu by the hot graphite is a slow first order reaction, followed by rapid volatilization of the metallic copper. The rate of concentration change of copper atoms with time is independent of the rate of loss of copper atoms from the graphite tube and is only dependent on the rate of formation. The much higher sensitivity for copper in a graphite tube lined with tantalum foil can be explained by the rapid reduction of CuO by tantalum. The reduction of CuO is thus the rate determining step. PAVERI-FONTANA *et al* [2734] also attempted to explain the atomization mechanism kinetically; their approach is however limited to open systems such as a graphite rod.

Later, FULLER [2280] put forward a generally valid equation (8.4) which describes the concentration of metal atoms M in a graphite furnace at time t.

$$\text{Absorbance } (M) = p\, M_0\, \frac{k_1}{k_2 - k_1}\, [\exp(-k_1 t) - \exp(-k_2 t)] \tag{8.4}$$

where M_0 is the original quantity of analyte element dispensed into the furnace, k_1 is the first order rate constant for the metal atoms, k_2 is the first order rate constant for their transport from the graphite tube, and p is a proportionality constant, being a function of the oscillator strength (a constant) for the given element and the efficiency of atom production. With this equation it is possible to include all factors influencing the form of the signal profile. The dependence of the signal on the rate of formation of metal atoms in the atomizer is reflected in k_1. It is thus possible to elucidate physical effects such as a change of contact of the matrix with the graphite and the resulting alteration in the rate of reduction of the metal salt, or when the matrix causes an acceleration or retardation of volatilization of the metal. Chemical effects, such as the formation of a more stable or less stable compound of the analyte with the matrix, are also included in this constant.

The dependence of the rate of transport of atoms from the furnace is represented by k_2. The flowrate of the inert gas through the tube, or a changed rate of diffusion of atoms through the wall of the tube due to ageing, or condensation of atoms on cooler parts of the furnace are included in k_2.

The dependence of the signal on the extent with which the analyte is volatilized in comparison to other species (e.g. oxides, chlorides, etc.) is included in p. This dimension reflects any alterations in the efficiency of atomization.

FULLER emphasized that, in contrast to peak height evaluation, peak area integration is independent of factors that influence the rate of formation of atoms (k_1). The integrated signal is nevertheless dependent on the rate of transport of atoms from the furnace and on the efficiency of atomization; it is linearly proportional to p and inversely proportional to k_2. During his investigations on the formation and transport of atoms in graphite furnaces, SMETS [2886] found two groups of elements for each. For the formation of atoms, the rate-determining step is either the reduction of the oxide on carbon, or volatilization of the metal released by reduction of the oxide on carbon. He found that the only exception was aluminium, for which he assumed a vapor phase dissociation of AlO. The energy of activation for the atomization of this element agrees very well with the dissociation energy of $AlO_{(g)}$. SMETS found no general difference in the kinetic behavior of the analyte element, whether determined as the chloride or as an oxygen-containing salt. He concluded that a metal oxide is always formed as an intermediate step.

For the transport of atoms from the furnace, SMETS also proposed two mechanisms. Elements such as Ag, As, Au, Bi, Cd, Hg, Pb, Se and Zn are transported by diffusion from the tube. The integrated absorbance is directly proportional to the integrated rate of formation of atoms; in other words to the portion of the element actually atomized. The gas stream through the tube has a considerable influence on the sensitivity for these elements. In contrast, elements such as Ba, Be, Ca, Cr, Cu, Fe, K, Li, Mn, Mo, Na, Ni, Sr, U and V are not transported from the tube by diffusion. They are repeatedly volatilized and condensed in a type of short-range distillation resulting from the prevailing temperature gradient or specific "sticking probability". In this respect the formation of stable carbides or intercalation compounds with graphite can be considered. An alteration in the gas flowrate has no marked influence on the sensitivity for these elements. Resulting from the longer residence time of the atoms in the radiation beam, the peaks generated are clearly broader and consequently the integrated absorbances higher.

FRECH et al [2266] investigated atomization processes in complex chemical systems on the basis of high temperature equilibria calculations. Such calculations nevertheless require thermal equilibrium to be established at the time of atomization, and this is not given in a non-isothermal furnace. If the sample is atomized during the heating phase, a portion of the vapor is lost. The equations used by FRECH can still be used semiquantitatively under these conditions, however.

The free energies of formation for all possible species are entered into a computer program, i. e. for all reaction products that can be present in concentrations sufficient to influence the mass equilibrium. These are in particular O_2, H_2, and N_2 from traces of water or nitric acid, or Cl_2 from hydrochloric acid, that are retained in the tube. Using this computer program, a number of interesting elements were examined and inferences made on the observed phenomena.

FRECH and CEDERGREN [2260–2262] examined lead with special reference to its determination is steel. They found that even small quantities of chlorine caused rearrangement, so that volatile PbCl and $PbCl_2$ were formed. These species are driven from the furnace at 700 K by argon, so that substantial losses occur. The authors found that in the presence of H_2, greater concentrations of chloride could be tolerated without the formation of noticeable amounts of lead chlorides. The reaction mainly responsible for the removal of chloride is:

$$FeCl_{2\,(g)} + H_{2\,(g)} \rightarrow Fe_{(s)} + 2\,HCl_{(g)}$$

The higher the partial pressure of H_2, the more effectively chloride is removed from the system. In a non-coated graphite tube, H_2 is formed from water retained after thermal pretreatment [2261]. The quantity of H_2 produced is fully sufficient to remove all chlorine from the tube and to prevent the formation of volatile lead chlorides.

PERSSON *et al* [2742–2744] examined the atomization of *aluminium* and found that during thermal pretreatment, CO and H_2 as well as Cl_2 interfere. Although $Al_2O_{3(s)}$ is stable to 1800 K, in the presence of CO and H_2 losses occur at 1500 K. These species are the main reaction products between graphite and the water remaining in the tube. The state of the graphite is therefore very important for the determination of aluminium, since virtually no water remains on an intact layer of dense pyrolytic graphite. Additionally, O_2 and N_2 interfere during atomization, so that the exclusion of these elements is a prerequirement for good results.

During investigations on *silicon,* FRECH *et al* [2263] observed losses at pretreatment temperatures above 1600 K in the form of $SiO_{(g)}$. The first atomization signal was registered from about 2200 K, so that atomization is a vapor phase dissociation of SiO according to:

$$SiO_{2\,(s)} \rightleftharpoons SiO_{(g)} \xrightarrow{T} Si_{(g)} + O_{(g)}$$

At somewhat higher atomization temperatures silicon carbide, that may have formed, is also dissociated. At temperatures above 2900 K $SiC_{2\,(g)}$ is increasingly formed, so that higher atomization temperatures are not necessarily advantageous.

For *iron,* FRECH *et al* verified the atomization mechanism already proposed by the present author [3048]:

$$FeO_{(s)} + C_{(s)} \rightarrow CO_{(g)} + Fe_{(s)} \rightarrow Fe_{(g)}$$

The pretreatment/atomization curves shown in figure 99 (page 190) are in excellent agreement with the theory.

An especially interesting element thoroughly investigated by PERSSON and FRECH [2741] is *phosphorus*. The volatile oxides $PO_{(g)}$ and $PO_{2\,(g)}$ prevail at temperatures above 1800 K, so that substantial losses must be expected at lower pretreatment temperatures. At higher temperatures the formation of phosphorus atoms is determined by the equilibrium between monoatomic and diatomic phosphorus. Larger quantities of phosphorus are first formed at relatively high temperatures.

Non-coated graphite tubes are sufficiently reactive to reduce the partial pressure of oxygen to such a level that losses of phosphorus as gaseous oxides during thermal pretreatment are insignificant. Such tubes are, nevertheless, not an ideal environment for the atomization of phosphorus, since hydrogen is formed from retained water at higher temperatures, aiding the formation of methino phosphide ($HCP_{(g)}$).

It has been shown that the most favorable conditions for phosphorus are obtained by atomizing from a solid pyrolytic graphite platform mounted in a non-coated graphite tube [3061]. The non-coated tube ensures that the partial pressure of oxygen is low enough to prevent the formation of PO and PO_2. The formation of hydrogen on the platform is insignificant, since no water is retained. This arrangement also ensures that the furnace is in thermal equilibrium. PERSSON and FRECH [2741] have emphasized that acceptable and reproducible results for phosphorus can only be obtained when the heating rate and final temperature of the furnace, as well as the gas atmosphere, are well under control during the determination. A stabilized temperature furnace should bring the best results.

In a series of recent investigations, L'VOV et al [2577] [2578] [2586] [2588] [2589] [2592] examined the "thermochemistry of gaseous media" more closely. They came to the conclusion: "For a long period of time we have been living in an 'oxidizing' world and have somewhat got used to the idea that monoxides and hydroxides represent the only obstacles on our way to solving the problem of complete and overall atomization. With the advent of high-temperature reducing flames and graphite furnaces a hope began to dawn that free carbon present in atomizers of these kinds will help to solve this problem.

However, this did not happen. Having eliminated the previous obstacles we stumbled upon others. The reducing medium did not remain inert with respect to free atoms and 'issued' in place of monoxides and hydroxides their carbon analogs, i.e. dicarbides and monocyanides. In this respect, the 'reducing' world turned out to be a symmetric image of the 'oxidizing' one.

Nature continues to place ever new obstacles on the way to our goals, yet it cannot prevent us from revering and admiring the logic and beauty of these obstacles."

L'VOV et al draw attention to the similarity between CN radicals and halogen atoms, which in fact has been known for some time. The energies of dissociation of monocyanides lie between those of chlorides and fluorides. The electronegativity of the OH radical is somewhat lower than that of the CN radical. Thus, the dissociation energies of hydroxides must be lower than those of monocyanides.

L'VOV and PELIEVA [2586] [2589] investigated 42 elements and found that 30 of them form monocyanides. Nevertheless, a substantial difference between argon and nitrogen as inert gas can be observed. The spectra of numerous monocyanides can be clearly identified in nitrogen, while in argon virtually nothing can be seen [2588]. L'VOV also pointed out that the spectra of oxides and hydroxides scanned by various authors in graphite furnaces are in reality incorrectly interpreted spectra of monocyanides and halides [2577].

The significance of isothermal conditions during atomization will now be treated again. In graphite furnaces of the MASSMANN type, on which most commercial furnaces are based, the furnace is not in equilibrium with respect to either volume or time at the moment of atomization. Rather, its temperature changes in a complex manner with time. At a tube wall temperature that is characteristic for the analyte element and matrix, the element is volatilized and atomized while the temperatures of the graphite tube and the inert gas continue to change. If the element is determined in a different matrix, it can volatilize and atomize at another temperature. The peak shaped atomization signal will consequently be shifted on the time axis. The shape and size of the absorption signal depends on a number of variables, e.g. the temperature and residence time in the graphite tube. This means that the absorption signals for both samples will be different, both in peak height and peak area, even when the concentration of the analyte element is the same. This leads to a whole series of interferences, and these will be treated in the following sections.

SLAVIN et al [2882] proposed combining a number of already recommended procedures into one package and thus improving the properties of MASSMANN-type furnaces significantly.

First of all it is important that the graphite tube is heated as quickly as possible to the required final temperature. This can be achieved with an ultrafast rate of heating of about 2000°C/s, as explained in section 3.2.4 [3064]. At the same time, atomization of the analyte element is delayed by use of the platform proposed by L'vov *et al* [2590] (refer to section 3.2.5). The graphite tube can attain the final temperature and the inert gas in the tube has time to reach this temperature before the platform reaches the atomization temperature and the analyte element is volatilized.

It is naturally important that the inert gas stream through the tube is interrupted during, or even better slightly before, atomization. Thereby, premature transport of the atoms from the graphite tube and interferences to the thermal equilibrium caused by flowing, cold gas are avoided. If the inert gas is in thermal equilibrium with the tube wall at the moment of atomization or of volatilization of the sample, no further expansion of the gas takes place, thus avoiding a further loss of atoms. Transport of atoms from the tube is then only via diffusion.

A further important criterion for achieving thermal equilibrium in a MASSMANN furnace at the moment of volatilization is that the temperature step from the thermal pretreatment temperature to the atomization temperature should not be too large. SLAVIN [2882] quotes a temperature of 1000°C, which should not be exceeded if possible. If the temperature step is larger, thermal equilibrium will not be attained and the sample is volatilized into an environment whose temperature is changing. The desired effect will thus not be achieved.

The ultrafast rate of heating permits optimum sensitivity to be attained at substantially lower atomization temperatures [2245] [3064]. The values are often lower for peak area measurements than for peak height. If matrix modification [2217] is used in addition, it is often possible to raise the thermal pretreatment temperature. With these measures combined, it is possible to meet the above requirement of a temperature step of less than 1000°C for most elements.

Another reason for preferring peak area measurements to peak height is that the measures are independent of the rate of formation of the atoms. Varying contact of the sample atoms with the graphite surface or varying rates of volatilization play no further part [2280]. The ability of obtaining undistorted signals of high resolution is very important during the development of ideal conditions for atomization at thermal equilibrium. BARNETT and COOKSEY [2053] developed a computer program that permitted highly resolved signals to be displayed on a video screen. The signals can also be stored on floppy disk, so that they can be recalled and viewed as required.

The major advantage of the stabilized temperature furnace will become evident in the following section. Next to the theoretical considerations on atomization, it has been established that with this concept determinations essentially free of vapor-phase interferences can be performed in graphite furnaces.

8.2.2. Spectral Interferences

Genuine spectral interferences caused by direct overlap of atomic lines are just as rare in the graphite furnace technique as in the flame technique, as explained in chapter 7. Background attenuation due to absorption of the primary radiation by molecular bands of volatilized concomitants and radiation scattering on sample particles in the absorption volume are nevertheless frequent in the graphite furnace technique.

The extent of the background attenuation depends to a very large degree on the design of the furnace, the temperature program, and other operating parameters. Added to this, the same concomitants frequently cause not only spectral interferences but also interferences due to reactions with the analyte element in the condensed and vapor phases. Often it is possible to employ the same measures against both types of interference.

Early in the development of electrothermal atomization, L'vov [8] [778] recognized that molecular absorption by alkali halides and radiation scattering on "mist" at the colder ends of the furnace were major causes of interferences. He established that an interference must occur when the interferent is present in a concentration four or five orders of magnitude higher than the analyte element. Due to the very high sensitivity of the graphite furnace technique and its use in trace and nanotrace analysis, this relationship is very quickly reached.

MASSMANN *et al* [447] [2630] [2631] were the first to make thoroughgoing investigations on the processes responsible for background attenuation. They found that molecular absorption is generally reproducible, but that radiation scattering can vary from sample to sample (figure 103). They also observed that radiation scattering always follows molecular absorption, and this they attributed to the molecules migrating to the cooler tube ends where they condense and cause scattering. This observation depends very strongly on the type of furnace used, however.

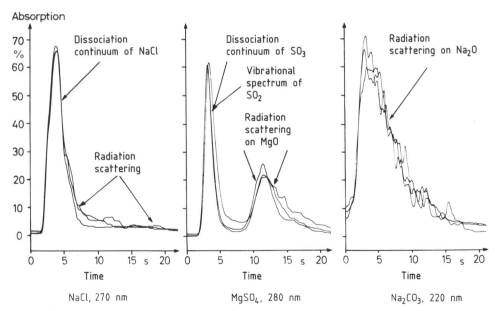

Figure 103. Time profiles of the absorption signals in a graphite furnace with NaCl, MgSO$_4$ and Na$_2$CO$_3$ as samples (according to [447]).

MASSMANN *et al* [2630] divided molecular spectra into dissociation continua and electronic band spectra. Dissociation spectra exhibit characteristic, broad maxima that arise because the molecules dissociate in the vapor phase due to radiation absorption (photodissociation). The long wavelength maximum corresponds to the dissociation of the molecule into neutral atoms, while the short wavelength maxima correspond to dissociation into excited atoms. The energy differences between the individual maxima correspond to the various states of excitation (energy levels) in the atoms.

The absorption spectra of various diatomic alkali halides have also been observed and reported by other authors [2002] [2179] [2764]. GÜÇER and MASSMANN [447] were also able to study similar processes with polyatomic oxygen-containing anion salts. They were able to observe the following photodissociation processes for a large number of metal sulfates:

$$SO_3 + h\nu \rightarrow SO_2 + O \quad \text{(limiting wavelength 330 nm)}$$
$$SO_2 + h\nu \rightarrow SO \;\; + O \quad \text{(limiting wavelength 190 nm)}$$
$$SO \;\; + h\nu \rightarrow S \;\;\; + O \quad \text{(limiting wavelength 245 nm)}$$

The electronic band spectra that can be observed in graphite furnaces exhibit a fine structure (which can only be discerned with high spectral resolution). GÜÇER and MASSMANN [447] and also at a later date L'VOV [2577] observed CN bands in a furnace purged with nitrogen. These bands have a maximum at 390 nm and their intensity increases exponentially with the furnace temperature. Strong C_2 bands with maxima at 430 nm and 470 nm also contribute to background absorption in the furnace. As established by L'VOV and MASSMANN, radiation scattering is most frequently caused by volatilized concomitants condensing on cooler parts of the furnace. The extent of this interference depends largely on the design of the furnace. A further cause can be smoke or "soot" formed during the thermal decomposition of biological or other organic materials in an inert gas atmosphere. Particles sublimed from the graphite at high temperature are another possible cause of interference. Generally, though, the contribution due to radiation scattering in a well designed tube furnace is relatively low; it can reach a significant level with carbon rod furnaces, however.

The design and the selected parameters of the furnace are decisive for the occurrence of background attenuation. Temperature and gas stream are especially important. Virtually all commercial furnaces exhibit a temperature gradient of varying extent from the middle to the ends. Consequently there is a risk that when volatilized sample expands in the tube, it can reach cooler zones and recondense. This risk even exists in tubes heated uniformly over their entire length [2539], since cold gas streams around them.

ISSAQ and ZIELINSKI [2418] tried to get round this problem by making large slots at the ends of the graphite tube and only using the hot center for analytical purposes. They found that this change brought a substantial reduction in the background. As already explained in section 3.2, an exceedingly effective measure for the reduction of background attenuation is a controlled, symmetrical gas steam from each end of the graphite tube and the expulsion of volatilized substances at the hottest part of the tube (refer to figures 31 and 32, pages 58 and 59). In comparison, background attenuation in furnaces without a gas stream, or with only a poorly regulated gas stream, is frequently more than an order of magnitude higher.

A further important factor is temperature. The appearance of molecules means simply that the temperature in the graphite tube is insufficient for their dissociation. Thus, when using capacitor discharge for tube heating, CHAKRABARTI *et al* [2141] observed virtually no background attenuation. Other authors [2199] [2929] have also noted the requirement for better isothermal conditions between the tube wall and the gas to permit better control of background interferences. It could also be shown that background attenuation was significantly lower in a number of cases when the gas temperature was higher than the temperature of the surface from which the sample was volatilized. This can be achieved by an ultrafast rate of heating and atomization off the L'vov platform (figure 104).

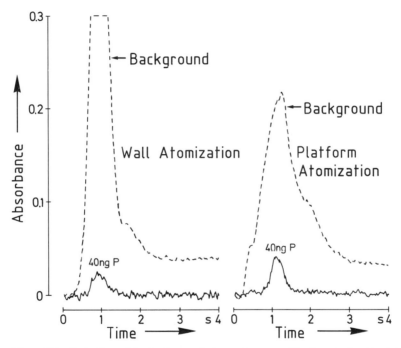

Figure 104. Reduction of the background attenuation caused by iron during the determination of phosphorus (213.6 nm) by atomizing off the L'vov platform in a stabilized temperature furnace.

Careful programming of the rate of temperature increase in the graphite tube is especially important to avoid excessive background attenuation [88]. By selection of suitable temperature, numerous concomitants can be volatilized and thus removed before atomization of the analyte element. Biological and other organic materials can mostly be thermally decomposed (charred) at temperatures between 500 °C and 800 °C. The use of a temperature ramp with a slow, even increase of temperature has proved very effective for the removal or thermal decomposition of complex matrices [1294].

After introduction of the sample into the graphite tube, the solvent is evaporated during the first temperature step. Temperature and time should be determined such that the solvent is evaporated quantitatively, but not too slowly and without spattering. Higher boiling constituents that are frequently present, such as oxyacids, must be removed in a second step in a similar manner. It is important to realize that such acids, or their decomposition products, can be retained in the graphite lattice or at active sites to temperatures well above their boiling points. For their complete removal, temperatures of to near 1000°C are often required. These problems essentially do not arise with pyrolytically coated graphite tubes due to the non-porous and largely inactive surface.

For thermal pretreatment, a temperature just below the appearance temperature of the analyte element is important. Thermal pretreatment can, and should, be taken to this temperature to exploit every possibility to separate or decompose the matrix. It should be noted, however, that the matrix can strongly influence the volatility of the analyte element. The *manner* in which the maximum thermal pretreatment temperature is reached is decided largely by the concomitants. It is important that separation should be as quick as possible, but that spattering of the sample or other violent reactions should be avoided. For multicomponent mixtures, or generally for high matrix concentrations, multistep programs with ramping of the temperature are imperative. A controlled gas stream through the graphite tube during thermal pretreatment is also decisive. The volatilized sample constituents must be expelled from the graphite tube and out of the radiation beam, without their recondensing on cooler zones. If volatilized matrix is not expelled from the optical beam, pretreatment times will be unnecessarily extended and condensed sample constituents will revolatilize during atomization, leading to the well-known interferences.

If the matrix can be successfully removed by thermal pretreatment, the analyte element can generally be atomized without interferences. If the analyte element is so volatile that separation of the matrix is not completely possible, then low temperature atomization of the analyte is a potential remedy. Various authors found that the sensitivity was reduced and thus the signal lower, but that volatilization of the matrix was retarded, so that non-specific signals were also lower and appeared later.

LUNDGREN et al [771] developed a temperature controlled power supply with a very fast rate of temperature increase. They atomized cadmium in the presence of 2% sodium chloride at 750°C. At this temperature virtually no sodium chloride vaporizes, so the interference due to background attenuation is negligible. As mentioned earlier, this technique was refined [3064] and used by various authors for the determination of cadmium in sea-water [2788], sediments and particulate matter [2775], or lead in sea-water [2788], etc.

Frequently, however, thermal pretreatment alone is not sufficient to achieve satisfactory separation of the analyte element and the matrix. If the volatilities of the analyte element and the matrix constituents are very similar, even sophisticated temperature programs will not bring successful separation.

MATOUŠEK [2634] made a detailed report on the possibilities of chemical pretreatment of the sample and the reduction of background interferences. Phosphoric acid [2632] and nitric acid [2108] [2262] are particularly suitable for this purpose. The latter has long been known as an "ashing additive" for biological samples and other organic materials and often reduces the background attenuation in the graphite furnace quite substantially. FRECH and CEDERGREN [2262] nevertheless draw attention to the fact that the use of oxyacids can also lead to a loss in sensitivity if the acid and its decomposition products are not eliminated by the application of a sufficiently high temperature.

EDIGER *et al* [320] were the first to propose adding a reagent, such as ammonium nitrate, in high concentrations to the sample to achieve a matrix modification effect. This reagent can serve either to increase the volatility of the concomitants or to improve the thermal stability of the analyte element, i.e. to reduce its volatility; in especially favorable cases it is possible to achieve both goals simultaneously. Matrix modification serves to make the volatilities of the analyte element and the concomitants sufficiently different to permit easier separation through thermal pretreatment.

In their first experiments, EDIGER *et al* proposed an addition of ammonium nitrate as a means of eliminating interferences due to sodium chloride in the analysis of sea-water. According to the equation

$$NaCl + NH_4NO_3 \rightleftharpoons NaNO_3 + NH_4Cl$$

sodium nitrate and ammonium chloride are formed, and these decompose or sublime at temperatures below 400 °C, so that the equilibrium is shifted fully to the right. Excess ammonium nitrate is also easily removed from the graphite tube during thermal pretreatment. As indicated in figure 105, background attenuation during atomization can be very markedly reduced. Since matrix modification can also aid in controlling chemical interferences in the condensed and vapor phases in addition to reducing background attenuation, it will be treated in detail in the following sections (8.2.3 and 8.2.4).

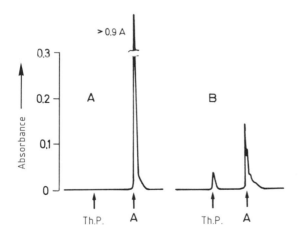

Figure 105. Reduction of the background attenuation caused by 0.15 mg NaCl (A) through matrix modification with 6 mg ammonium nitrate (B); measured at 229 nm (according to [2217]). Th. P. = Thermal pretreatment at 300 °C, A = Atomization at 2200 °C.

The use of other gases, such as hydrogen as the purge gas, to reduce spectral interferences was first proposed by AMOS *et al* [46]. AMOS used a carbon rod and the hydrogen ignited on the hot surface. ROUTH [2803] also reported a reduction of background attenuation for the determination of chromium in urine when the graphite tube was operated in a hydrogen diffusion flame. BEATY and COOKSEY [2066] found that the background signals for sodium chloride and calcium oxide could be substantially reduced if hydrogen was passed through the graphite tube for 10 seconds before atomization.

KUNDU and PREVOT [2514] reported the rapid decomposition of the organic matrix in the determination of copper in oil when they mixed the inert purge gas with oxygen at low temperature. At the same time the marked background attenuation due to the formation of smoke was strongly reduced. The use of air or oxygen as the purge gas during thermal pretreatment has proved very valuable during the analysis of larger quantities of biological materials. Real ashing in the graphite tube then takes place and not the charring of the matrix usual in an inert gas atmosphere. The latter led to considerable difficulties in the determination of nickel in serum, for example, due to the carbon residues [2066]. On the other hand, when oxygen was used during thermal pretreatment, virtually no background attenuation was observed [2065] [2066]. This technique was also successfully applied to the determination of cadmium in blood [2198].

Several authors found the interferences due to concomitants in the graphite furnace technique to be too large, or the means of reducing or eliminating them to be too complicated. They therefore employed separation techniques before introduction into the graphite tube. The best known technique is the extraction of the complexed metal with an organic solvent [2702] [2703] [2705] and occasionally their re-extraction back into aqueous media [2604] [2704]. Since the analyte element is almost completely freed from concomitants, the determination is usually fully free of interferences. Nevertheless, the problems discussed in section 6.4 on trace analysis should on no account be forgotten. In this concentration range, every pretreatment step brings the risk of losses (incomplete extraction) and contamination (labware, reagents and solvents) of the analyte element.

Another possibility for separating the analyte element from concomitants is electrolytic deposition. The proposed procedures nevertheless have a number of practical difficulties; a survey has recently been published by MATOUŠEK [2634]. It is also expedient to mention the technique of partial solution, precipitation and extraction of the matrix that has been very thoroughly investigated by JACKWERTH et al [2424] as a versatile concept for element preconcentration.

All the techniques described thus far, perhaps with the exception of separation techniques, reduce the background attenuation but cannot eliminate it entirely. This means that a background corrector is usually required for the graphite furnace technique.

On the other hand, the extreme concentration ratios between the analyte element and the concomitants mean that the limits of a background corrector are often quickly reached. This clearly indicates that even with a background corrector, the measures for reducing background attenuation mentioned above are indispensable. Generally, background attenuation must be reduced to the working range of the background corrector by the use of suitable temperature programs and/or matrix modification before the remaining interference is eliminated instrumentally.

It must again be emphasized that the sole use of a background corrector cannot guarantee accurate results, on the contrary, it can even introduce errors. The main cause for such errors is a background signal that is too high. The optimum way of recognizing this interference is to record the non-corrected (or the background) and the corrected signals simultaneously (or sequentially). For most continuum source background correctors, the limit for complete correction of background signals lies between 0.5 A and 0.8 A. Since in practice it is often not possible to achieve optimum coincidence of the beams of both radiation sources, it should be attempted to keep background attenuation to below 0.5 A.

Since measurement of the background signal is time offset to measurement of the total absorption, interferences can arise in the graphite furnace technique if the background signal appears too quickly, so that background attenuation changes markedly from measurement to measurement. It is then often possible to observe a sinusoidal crossing of the baseline.

A further substantial error can arise due to the fact that continuum source background correctors are unable to completely correct electronic band spectra. Since these consist of many narrow rotational lines, the actual background attenuation depends on the degree of overlap between the element absorption line and the individual molecular rotational lines [2365] [2630]. With a continuum source background corrector, only the mean value of the molecular rotational lines across the spectral bandpass is taken and subtracted from the total absorption (refer to figure 79, page 139).

Despite the fact that the limitations of continuum source background correction have been known for many years, surprisingly few authors have reported spectral interferences that they were unable to completely eliminate. MANNING [2614] found spectral interferences during the determination of selenium in iron at the 196.0 nm line. He attributed this to the fact that iron has several resonance lines in the region around 196 nm and although these do not absorb selenium-specific radiation, they absorb radiation from the continuum source. VAJDA [2995] reported numerous further element combinations in which high concentrations of matrix elements interferred with background correction when they had a resonance line directly neighboring the analytical line. SAEED and THOMASSEN [2815] found that calcium phosphate caused considerable overcompensation in the determination of antimony, arsenic, selenium and tellurium; they attributed this to a possible spectral interference through P_2 molecular rotational bands.

The only technique that clearly brings a substantial improvement for all these interferences is Zeeman-effect background correction. And conversely, the graphite furnace technique is the only technique in atomic absorption spectrometry in which spectral interferences occur to such an extent that the application of the Zeeman effect is worthwhile.

As already explained in detail in section 7.1.2, Zeeman-effect background correction is suitable for the correction of high background attenuation up to about 2.0 A and for accurate measurements when the background has fine structure (provided the inverse Zeeman effect with the magnetic field at the furnace is applied).

This is indicated in figure 106 on the example of cadmium in the presence of high aluminium concentrations. For continuum source background correction, the high background signal of greater than 1 A causes marked interferences which make evaluation impossible. With Zeeman-effect background correction such effects cannot be observed; the analytical measure can be clearly evaluated. The background signal generated by a 2% iron solution at the 213.6 nm phosphorus line is depicted in figure 107. Normally, a continuum source background corrector should cope with a background signal of about 0.3 A, but in fact overcompensation occurs here and small phosphorus signals disappear completely. A perfect baseline is obtained with Zeeman-effect background correction and the extremely low content of phosphorus in pure iron (about 0.001%) can be recognized as a signal.

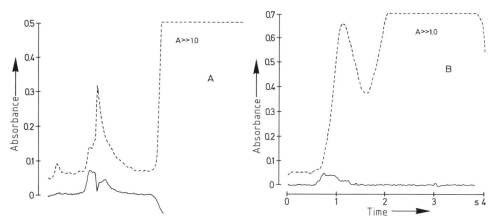

Figure 106. Highly resolved signals for cadmium with and without background correction in the presence of 0.5% aluminium. Atomization off the L'voᵥ platform in a stabilized temperature furnace. **A**: Continuum source background correction (D₂ lamp); **B**: Zeeman-effect background correction.

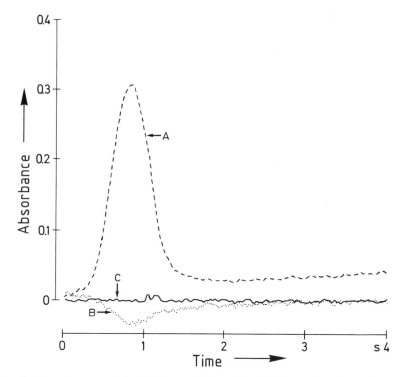

Figure 107. Background attenuation caused by 2% iron solution at the 213.6 nm phosphorus line. **A**: Without background correction; **B**: With continuum source background correction (D₂ lamp); **C**: Zeeman-effect background correction (according to [3061]).

8.2.3. Volatilization Interferences

Volatilization interferences, or interferences in the condensed phase, include all influences on the analyte element to the moment when it volatilizes and leaves the graphite surface. The processes are in part very complex and have not been clarified by a long way. Included under volatilization interferences are losses of the analyte element during thermal pretreatment, and formation of carbides, intercalation compounds and similar processes that lead to incomplete atomization. Kinetic effects are also included if they influence volatilization of the element into the vapor phase.

SEGAR and GONZALEZ [1109] were the first to surmise that the considerable losses in sensitivity observed in the determination of trace elements in sea-water could be due to co-volatilization with the sodium chloride during thermal pretreatment. CZOBIK and MATOUŠEK [2182] refer to the carrier effect long known in emission spectrometry that leads to simultaneous volatilization of, for example, sodium chloride and lead; sodium chloride can volatilize in microgram quantities at temperatures little above 700°C and carry lead with it. These authors also established that in the presence of sodium chloride the lead signal virtually coincides with the volatilization of sodium chloride from the tube wall.

YASUDA and KAKIYAMA [3090] [3091] assumed that elements such as cadmium, lead and zinc form gaseous metal chlorides in the presence of higher chloride concentrations and these are partially transported from the tube at the start of atomization. They also observed the spectra of numerous metal halides at relatively low temperatures, which however did not appear at higher temperatures. KARWOWSKA et al [2457] found that the signal for iron was strongly depressed in the presence of halogenated organic solvents and attributed this to the formation of volatile iron(II) chloride.

SMEYERS-VERBEKE et al [2888] also found that alkaline-earth chlorides can lead to losses of the analyte element during thermal pretreatment. They reported that through the careful choice of a thermal pretreatment program, the interferent could be removed and the interference thus reduced or even eliminated.

Several authors [2072] [2441] have reported the strong interferences caused by perchloric acid. This acid strongly attacks the surface of graphite, especially when uncoated tubes are used, and also leads to very marked signal depression. FULLER [2279] ascribed the interference of perchloric acid on the determination of thallium to the formation of the volatile chloride. The addition of excess sulfuric acid eliminated the interference due to perchloric acid, and also the interferences due to hydrochloric acid and sodium chloride.

KOIRTYOHANN et al [2481] observed that the signals for aluminium, gallium and thallium were depressed by more than 95% in the presence of 0.5 M perchloric acid. These authors found that the effect remained even when the tube was heated to well above the boiling point of perchloric acid; it is necessary to heat to above 1700°C to eliminate the effect. Perchloric acid, or a decomposition product, apparently reacts with graphite to form a thermally stable compound that later volatilizes and interferes with the atomization. ALDER and HICKMAN [2022], however, have challenged the theory that chlorides are volatilized during thermal pretreatment. These authors consider the formation of a compound in the graphite furnace that leads through hydrolysis of chlorides to the oxide as more likely.

FULLER [2282] found substantial differences in the interference of magnesium chloride on the determination of lead in dependence on the graphite surface. While strong interference could be observed in new, uncoated tubes and in pyrocoated tubes, no interference was observed in "aged" tubes. The interference due to magnesium chloride could also be eliminated by the addition of oxalic acid. Both oxalic acid and finely dispersed carbon in aged tubes reduce lead oxide to metallic lead. FULLER thus concluded that the interference due to magnesium chloride is a vapor-phase interference. To eliminate the influence of nitric acid in the determination of lead in biological materials, BEHNE et al [2072] added formic acid. They found that neither formic acid nor oxalic acid resulting from oxidation interfered with the determination.

Losses during thermal pretreatment become a problem, or a source of error, when the volatility of the analyte element is increased relative to the reference solution by concomitants in the sample. For this reason the requirement is frequently stated that the temperature curve for thermal pretreatment for every element in every matrix should be recorded. This would naturally make the graphite furnace technique exceedingly complicated; for samples with continuously changing compositions it is simply not feasible.

The matrix modification technique proposed by EDIGER [2217] represents a major step for making the conditions plain and analytically controllable. As mentioned earlier, a reagent, usually an inorganic salt, is added to sample and reference solutions in large excess. The reagent causes either the interfering concomitant to become more volatile and thus more easily separated, or the analyte element to be converted to a less volatile form. In especially favorable cases it is possible to achieve both effects with a *single* reagent.

The separation of sodium chloride as sodium nitrate and ammonium chloride through the addition of ammonium nitrate has already been cited as the first example of matrix modification (refer to section 8.2.1). For control of interferences in the condensed phase, the second possibility, namely conversion of the analyte element into a defined and thermally stable compound, is the most important. Matrix modification permits all samples containing the analyte element to be treated essentially in the same way, regardless of the original bonding form and concomitants.

Nickel is among those reagents already proposed by EDIGER [2217] for matrix modification; it stabilizes arsenic up to about 1400°C and selenium up to about 1200°C. This is presumably due to the formation of the thermally very stable nickel arsenide or nickel selenide. CHAKRABARTI et al [2144] found that nickel sulfate was the most suitable salt. GLADNEY [2302] also used nickel to stabilize bismuth up to 1200°C. Next to nickel, copper, silver and other noble metal salts have been proposed as stabilizers for arsenic and selenium, and found as equivalent. WEIBUST et al [3041] found palladium and platinum to be the most suitable to stabilize both inorganically and organically bound tellurium. Pretreatment temperatures up to about 1050°C can be used. HENN [2366] proposed molybdenum as a stabilizer for selenium and was able to apply pretreatment temperatures up to 1300°C. He also successfully used the same reagent to reduce matrix interferences on the determination of lead and cadmium [2367].

To stabilize cadmium, EDIGER [2217] originally investigated ammonium fluoride, ammonium sulfate and ammonium hydrogen phosphate, and found all three reagents to be largely equivalent. They all permitted pretreatment temperatures up to 900°C. HINDERBERGER *et* al [2372] made a thoroughgoing examination of primary ammonium phosphate and found that as well as for cadmium it could also be used effectively for chromium, lead and nickel. SPERLING [2900] found the addition of ammonium peroxodisulfate and sulfuric acid greatly improved the direct determination of cadmium.

In 1973, MACHATA [783] [784] proposed the addition of lanthanum for the determination of lead in blood. ANDERSSON [2037] also found that lanthanum eliminated the interference of sulfate on the determination of lead and increased the sensitivity.

BRODIE and MATOUŠEK [2108] [2632] recommended the addition of phosphoric acid to avoid losses of cadmium during thermal pretreatment and to permit the use of higher temperatures. CZOBIK and MATOUŠEK [2181] later investigated the use of phosphoric acid for numerous other elements. For elements with an atomization temperature lower than tin they found that phosphoric acid increased the atomization temperature, while for other elements it had no influence. A number of authors used phosphoric acid for matrix modification in the determination of lead [2122] [2379] [3040]. They were able to increase the pretreatment temperature to 900°C to 1000°C and eliminate numerous interferences. The reagents used most frequently and with the greatest success for lead are primary and secondary ammonium phosphate.

BELLING and JONES [2073] observed that sodium nitrate depressed the signal for manganese more strongly than potassium nitrate. EBDON et al [2215] reported that although potassium nitrate depresses the manganese signal, it was enhanced by calcium nitrate. SLAVIN et al [2882] recommended magnesium nitrate as an especially effective reagent for matrix modification. Thereby it is possible to stabilize manganese up to 1400°C, aluminium to 1700°C, chromium to 1650°C, cobalt to 1450°C and nickel to 1400°C, and to eliminate numerous interferences.

Mercury is without doubt the most volatile element and it must be questioned whether it is at all logical to determine it by the graphite furnace technique. Nevertheless, EDIGER [2217] proposed matrix modification with ammonium sulfide, with which it can be stabilized to 300°C. ISSAQ and ZIELINSKI [2417] were able to use thermal pretreatment temperatures up to 200°C when they stabilized mercury with hydrogen peroxide. Nevertheless, ALDER and HICKMANN [2021] doubted whether hydrogen peroxide had any effect at all. They found that only hydrochloric acid and hydrogen peroxide together stabilized mercury, and hydrochloric acid had the main effect. Hydrogen chloride is present in excess in the vapor phase and possibly stabilizes mercury as $HgCl_2$.

KIRKBRIGHT et al [2470] found that potassium dichromate in nitric acid solution was the most suitable matrix modifier for mercury. It permits thermal pretreatment up to about 250°C. Using the same reagent, selenium can be stabilized up to 1200°C. A mixture of potassium permanganate and silver nitrate in nitric acid solution also had a similar effect for mercury, but is not suitable for the stabilization of selenium.

For a long period, work on the elimination of interferences in the graphite furnace technique was largely empirical. The possible reason for interferences, such as losses during thermal pretreatment, were not investigated on a systematic basis. MATSUSAKI et al [2637] found that the interference caused by sodium chloride or potassium chloride on the determination of aluminium disappeared if thermal pretreatment temperatures of over 1000°C were applied. HOCQUELLET and LABEYRIE [2378] reported that for tin, lower pretreatment temperatures lead to lower sensitivity, but that losses of SnO_2 can occur at higher temperatures. MONTASER and CROUCH [2674] pointed out that in those cases where atomization is via the oxide, and this is volatile at a temperature much lower than the atomization temperature, substantial losses can occur. In such cases it is important to select an atomization temperature that is not too high and to use a fast rate of heating. The marked reduction in matrix interferences observed by CHAKRABARTI *et al* [2141] when they used capacitor discharge for atomization is probably due to this.

Systematic investigations on the chemical reactions taking place in graphite furnaces were first undertaken by FRECH *et al* [2266]. They studied the atomization processes for numerous elements in complex chemical systems on the basis of high temperature equilibria calculations. Such calculations are of special significance for vapor-phase interferences, but provide information on reactions in the condensed phase and are very informative about losses during thermal pretreatment.

FRECH *et al* found, for example, that in an uncoated graphite tube, water is retained in sufficient quantity, even after 15 minutes at 1200°C *in vacuo*. At higher temperatures the water gas equilibrium is rapidly established

$$CO + H_2O \rightleftharpoons CO_2 + H_2$$

so that in an uncoated tube there is always a relatively high partial pressure of hydrogen. The conditions in a pyrocoated graphite tube are different, however.

FRECH and CEDERGREN [2260] were able to show for the determination of lead in steel that in the absence of hydrogen, volatile lead chlorides are formed at 400°C, leading to losses during thermal pretreatment. In the presence of a sufficiently large quantity of hydrogen, chlorine is expelled from the graphite tube at 600°C as hydrogen chloride. If, on the other hand, the sample is present in nitric acid solution and a relatively high thermal pretreatment temperature is not applied, the partial pressure of oxygen during atomization can be quite large. Gaseous lead oxides can then be formed, leading to a loss in sensitivity [2262].

Although aluminium oxide is stable up to 1500°C, in the presence of CO and H_2 it is only stable up to 1200°C [2743] [2744]. This explains the influence of the graphite surface and also the marked differences in thermal pretreatment temperatures reported by various authors. The action of an oxidizing additive for matrix modification, such as magnesium nitrate, is thus also clear. Additionally, the interference due to chloride on the determination of aluminium becomes smaller the higher the thermal pretreatment temperature.

The determination of phosphorus by the graphite furnace technique presents a number of problems. EDIGER *et al* [2219] found that calcium phosphate could be determined quite well, but that phosphoric acid gave virtually no signal. These authors suspected that phosphorus was lost in molecular form and suggested the addition of lanthanum for matrix modification [2218].

On the basis of high temperature equilibria calculations, PERSSON and FRECH [2741] showed that at lower temperatures the prevailing species are the volatile PO, PO_2 and P_2. Larger quantities of phosphorus atoms are only formed at high temperatures. In the presence of calcium, a thermodynamically stable compound is formed ($Ca_3(PO_4)_2$), which then reacts with carbon $Ca_3(PO_4)_{2(s)} \xrightarrow{C_{(s)}} CaO_{(s)} + P_{(g)} + (PO, PO_2, P_2)_{(g)}$

Lanthanum reacts in the same way and forms an even more stable phosphate than calcium.

The graphite material also plays a role in the atomization of phosphorus. PERSSON and FRECH [2741] state the requirement that atomization should take place under reducing conditions to keep the formation of PO as low as possible. Uncoated graphite tubes guarantee a sufficiently low partial pressure of oxygen, but offer nevertheless an unsuitable environment for the atomization of phosphorus. The hydrogen produced in relatively large quantities in these tubes leads to the formation of HCP, a further gaseous phosphorus compound. The most suitable conditions are obtained through atomization off a pyrolytic graphite platform in an uncoated graphite tube [3061], since there is a low partial pressure of oxygen and only a small quantity of hydrogen is formed, because water can hardly penetrate the platform of pyrolytic graphite.

KOREČKOVÁ et al [2494] made a very thorough examination of the various factors influencing the determination of arsenic in the graphite furnace. They compared a normal, uncoated graphite tube with a tube made of vitreous carbon and found some substantial differences. In both tube types the appearance temperature of 1100°C was the same and the peak maxima coincided, but the concentration of free atoms in the vitreous carbon tube was much lower. This could be seen from the lower peak height and area measured with the vitreous carbon tube, and in substantial tailing of the signal measured with the normal graphite tube, presumably due to retention effects.

KOREČKOVÁ et al indicate the requirement that arsenic should form a compound with graphite. As a number of authors have established, at lower temperatures oxygen undergoes chemisorption on graphite and forms carbon-oxygen complexes which build active sites and attract arsenic. Hydrogen must be present to stabilize the graphite-arsenate compounds. The building of active sites is catalyzed by water, i.e. the surface of uncoated graphite tubes offers good conditions for the formation of intercalation compounds when aqueous samples are analyzed.

Water and oxygen are removed as the tube temperature increases, so that the stability of the interlamellar arsenic compounds decreases; the optimum is at about 500°C. Nevertheless, some arsenic is retained on normal graphite at higher temperatures, even at 2500°C, as indicated by tailing. Lamellar compounds can be formed at flaws in the crystal lattice and these can bind arsenic strongly. Such compounds can also be formed on the graphite surface with volatilized arsenic after atomization; the characteristics of the graphite surface are naturally of significant importance.

The surface characteristics can change markedly through reactions of concomitants with graphite. Nitric acid is known to increase the distance between graphite layers, thus increasing the number of active sites, leading to an increase in the arsenic signals. In contrast, phosphoric acid depresses the signal since arsenic and phosphorus compete in the formation of interlamellar compounds. Hydrogen peroxide stabilizes arsenic very markedly; thermal pretreatment up to 1400°C is possible and the appearance temperature is increased to 1800°C. Other oxidants, such as nitric acid or potassium permanganate, have a similar effect. As to be expected, however, these reagents are only effective for uncoated graphite tubes, but not for tubes made of vitreous carbon. This means that they do not react with arsenic, but create further active sites on the graphite surface.

Nickel stabilizes arsenic regardless of the graphite material and other concomitants, and permits thermal pretreatment temperatures up to at least 1100°C. KOREČKOVÁ *et al* propose the formation of $Ni(AsO_3)_2 \cdot NiO$. SMETS [2886] also stated the requirement for a similar reaction of volatilized atoms with graphite, as already described for arsenic, for numerous other elements. The formation of intercalation compounds and also a type of short range distillation explain the longer residence times and broadened peaks for these elements. VEILLON *et al* [3005] established through investigations with ^{51}Cr that a considerable portion of this element is irreversibly bound to graphite; marked differences nevertheless occur, depending on the sample matrix.

The active participation of graphite, or of active sites occupied by oxygen, is only of advantage for atomization in particular cases. In general, it should be attempted to exclude the influence of the tube material, especially if stable carbides can be formed. Carbide formation is not an interference in the strictest sense, since the interference does not result from the sample. Nevertheless, it should be eliminated, since the extent of carbide formation in uncoated tubes can be influenced by sample constituents if they also react with graphite.

ORTNER *et al* [2272] [2712] [2713] found, for example, that better sensitivity and repeatability were obtained for a number of elements when molybdenum or tungsten was present in the matrix. Consequently they developed a technique for impregnating graphite tubes with sodium tungstate. Various other research groups [2353] [2559] [2813] [2970] [3097] also proposed coatings with different carbide-forming metals, such as lanthanum, tantalum, zirconium or molybdenum (refer to section 3.2).

L'VOV and PELIEVA [2584] lined a graphite tube with tantalum foil and found, as was expected, that the sensitivity for elements that form carbides or graphite intercalation compounds was improved. Often it was possible to lower the atomization temperature as well. These authors also reported that a number of elements, such as Co, Ir, Mo, Ni, Pd, Pt, Rh and Ru, form intermetallic compounds with tantalum and consequently volatilize more slowly and incompletely, even at the highest temperatures.

Without doubt the best and most effective coating for graphite tubes is pyrolytic graphite (refer to section 3.2.2 for details). Initially there were considerable difficulties due to the varying quality of the pyrocoating [3020], so that various authors [2617] [2618] [2881] proposed treating the tubes with molybdenum to seal flaws and cracks. With improvements in pyrocoating techniques, however, tubes can now be produced that are coated with a dense, even layer of pyrolytic graphite.

FULLER [2280] referred to the kinetic aspect of the pyrocoating, since this clearly increased the rate of atom release per unit time. The integrated signal can also become larger through more effective production of atoms. Nevertheless, the efficiency of atom production can be falsified or completely blanketed due to the "adhering effect" of numerous atoms on uncoated graphite described by SMETS [2886]. As a result of multiple condensation and volatilization due to short range distillation or the formation of intercalation compounds with graphite, an atom can pass many times into the radiation beam and cause a deceptively high integrated atom concentration. This effect is not observed with pyrocoated tubes.

L'VOV et al [2579] developed a macro kinetic theory of sample volatilization for the graphite furnace technique. According to these authors, the sample is not distributed as an even monomolecular layer on the surface, but in the form of individual microparticles (crystals, droplets, etc.) that are separated from each other. With uncoated graphite, the sample solution penetrates the graphite due to capillary action. After drying, the sample is thus distributed in the width of the tube wall. Atomization "off the wall" really means atomization out of the graphite lattice. In contrast, with pyrocoated graphite the sample is only distributed on the surface. This explains the effect of the pyrocoating and other surface coatings for elements that are relatively inactive with carbon, such as copper, indium or tin [2272] [3012].

According to VAN DEN BROEK and DE GALAN [2997] the volatilization of atoms from the graphite surface is described in good approximation by an Arrhenius equation. The rate at which the element passes into the gas volume of the graphite furnace depends on the temperature of the graphite tube wall, the frequency factor and the energy of activation. The latter two factors can be drastically altered by concomitants.

For the determination of lead in blood, GARNYS and SMYTHE [2295] established that after thermal pretreatment most of the lead was present as oxide or metal, embedded in a film of carbon of open structure that also contained the oxides and carbides of iron, calcium and silicon in large quantities. Examination under the electron microscope showed the ash lattice to be 0.2 to 2 µm thick, permitting the lead to volatilize quantitatively at 1700 °C. After each successive sample, a portion of the sample is absorbed into the carbon layer and lost to the determination, until a fairly continuous and dense surface layer has been formed. Brief introduction of oxygen into the tube during thermal pretreatment destroys the carbon lattice and permits an interference-free determination [2198].

SLOVÁK and DOČEKAL [2885] reported a reciprocal effect. For the determination of copper and iron in matrices of higher concentration, the matrix can act as a "miniplatform", delaying atomization and thus increasing the sensitivity.

L'VOV [8] [2576] has repeatedly emphasized that for quantitative volatilization of the sample under isothermal conditions, peak area integration can eliminate kinetic influences on the signal form. This fact was also established by SMETS [2886], and he found that the integrated absorbance signal is independent of the graphite structure. Nevertheless, the emphasis must be placed clearly on isothermal conditions. SLAVIN et al [2882] have also laid stress on this condition, which is a major constituent of the packet of conditions stipulated for a stabilized temperature furnace.

In conclusion, a phenomenon observed for a number of volatile elements will be mentioned, namely double peaks or a time offset of the signal. MCLAREN and WHEELER [2657] found that for the determination of lead in various sample solutions, the addition of reagents such as ascorbic acid, hydrofluoric acid, or hydrogen peroxide caused double peaks of varying appearance time. REAGEN and WARREN [2784] found an influence of the tube ageing process on double peaks; ascorbic acid prevented this but caused an early peak. FULLER [2282] also reported that the condition of the graphite tube had an influence on how strongly double peaks occurred.

CLARK et al [235] attributed the occurrence of double peaks to a temperature gradient between the middle and the ends of the graphite tube. SLAVIN and MANNING [2879] as well as

L'VOV [2576] proposed the same mechanism of condensation and revolatilization. As shown by SMETS [2886], however, this process is continuous and can only lead to peak broadening. Double peaks have only been observed for a few elements such as lead and zinc, while condensation and revolatilization can be expected for more than half of all elements (according to SMETS [2886] lead and zinc excluded). SALMON *et al* [2817] also mention that double peaks are obtained for the atomization of lead and zinc from carbon rods. With this type of atomizer, condensation and revolatilization are extremely unlikely. SALMON *et al* thus proposed that oxygen chemisorbed onto active sites on the graphite is responsible for double peaks and a shift in the appearance temperature. Volatile metals exhibit this effect because their volatilization temperatures lie between the temperature for optimum oxygen adsorption at 500°C and the total desorption temperature at 950°C. It has been proposed that atomization of lead is preceded by reduction on the graphite surface. This takes place in the temperature range around 500°C, the optimum temperature for the formation of surface oxides. When the active sites are occupied by oxygen (very stable surface oxides are formed), however, a different atomization mechanism with another activation energy can be the result. It may be a requirement that the few remaining active sites participate, but that they are only effective at higher temperatures, or it may be that the later peak is due to a direct thermal dissociation of the metal oxide, since no active sites are available.

Various nitrates and phosphates have been recommended for the matrix modification of lead and provide stabilization up to about 950°C. It is easily possible that these reagents directly influence the quantity of chemisorbed oxygen by decomposing with the release of oxygen and occupying the active sites. On the other hand, ascorbic acid, for example, could release active sites and thus promote the reduction of lead oxide at a lower temperature.

Similarly, the finely divided carbon abundantly available in "aged" graphite tubes can offer a large number of active sites; this would explain the low appearance temperature. On the other hand, metal carbides or pyrocoating can block active sites that would normally react with lead oxide.

SALMON *et al* [2817] point out that double peaks do not only cause errors for peak height evaluation. Time-shifted signals caused by volatilization at varying temperatures are responsible for noticeable errors even with peak area integration, since the requirement for isothermal conditions is not met. Such double peaks and time-shifted signals can only be contered effectively through matrix modification. In addition, use of the optimum amount of oxygen during thermal pretreatment could lead to effective ashing of the sample with minimum carbon deposition in the tube.

8.2.4. Vapor-Phase Interferences

Already in the preceding section on volatilization interferences, mention was made of vapor phase reactions and a distinction is often not easy. Further, the condensed phase is continuously in interaction with the vapor phase, and the latter can react with the graphite surface. Since all these processes take place rather rapidly and at increased temperature, it is anything other than easy to follow them and to describe them unambiguously.

As early as 1971, AGGETT and WEST [2014] showed that vapor-phase interferences exist. They used a double carbon rod to atomize the analyte element and the interferent independently. The interferences were identical, regardless whether both species were present in the condensed phase or first came into contact in the vapor phase. These authors also found that the extent of the interferences increased with increasing height above the carbon rod, i.e. with decreasing temperature. Nowadays it is known that in the vapor phase in a graphite furnace, similar to in a flame, processes such as dissociation and ionization take place, and these processes can be described by the law of mass action and the SAHA equation. Nevertheless, there are substantial and decisive differences with respect to the nature and extent of the reactions and interferences. For example, ionization in the graphite furnace is very low, among other things, due to the high concentration of free electrons produced in the furnace [2930]. The usual method in the flame of adding an easily ionized element in excess to suppress ionization interferences is ineffective in the graphite furnace. LUECKE *et al* [768] even found a marked loss in sensitivity when cesium salts were added to easily ionized elements.

In a graphite furnace, in contrast to the flame, the species introduced with the concomitants are the main reaction partners and not the flame gases. The only notable exception is when nitrogen is used as the purge gas instead of the more usual argon. HUTTON *et al* [2402] observed strong CN and C_2 emission bands in the furnace when nitrogen was used.

As mentioned earlier, L'VOV *et al* [2577] [2589] have pointed out that cyanides and dicarbides are nothing other than carbon analogs of the monoxides and monohydrides. It is necessary to change one's way of thinking from the oxidizing world of the flame to the reducing world of the graphite furnace. L'VOV and PELIEVA [2586] found that of 42 elements investigated, 30 formed monocyanides in a nitrogen atmosphere. These authors mention the dependence on the purge gas [2588]; for example, the spectra of alkaline earth cyanides are also clearly identifiable in nitrogen but not in argon. L'VOV and RIBZYK [2592] reported the absorption spectrum of AlCN for the determination of aluminium with nitrogen as purge gas, while they attributed the molecular spectrum observed in argon to Al_2C_2. If the above exceptions are excluded, then using a truly inert purge gas, the atmosphere in a graphite furnace is largely determined by the materials introduced with the sample. In his early work, L'VOV [8] used thermodynamic equilibria calculations to make quantitative statements about possible reactions and interferences. These calculations are based on the prerequirement that the furnace must be in thermal equilibrium.

FRECH *et al* [2266] point out that under non-isothermal conditions the sample volatilizes while the furnace is still heating, which means that a portion of the vapor phase is lost. Even though no genuine state of equilibrium is reached, the equations can nevertheless still be used to obtain a semi-quantitative statement. These authors applied high temperature equilibria calculations to the study of atomization processes in complex chemical systems. The free energies of formation were entered for all species that could be present in concentrations high enough to influence the mass equilibrium. Oxygen, hydrogen and nitrogen, and also possible sulfur from traces of water and sulfuric acid, etc. are particularly important. Some of the results have already been discussed in section 8.2.3 on volatilization interferences. One of the main causes for vapor-phase interferences in graphite furnaces is the formation of monohalides. A shift of the dissociation equilibrium through decomposition products of the solvent, which are always present in large excess, is a further important factor. Most elements form stable molecules with even the lowest concentrations of interfering elements.

Lead is probably the element that has been investigated most thoroughly. Even small quantities of chloride cause a redistribution of lead so that it is present as the volatile monochloride PbCl and dichloride $PbCl_2$ [2260]. These forms are mostly expelled with the purge gas, so that substantial losses of lead occur. In the presence of hydrogen, on the other hand, higher chloride concentrations can be tolerated before noticeable quantities of lead chlorides are formed.

For the determination of lead in steel, FRECH and CEDERGREN [2261] showed that chloride is expelled from the system according to:

$$FeCl_{2(g)} + H_{2(g)} \rightarrow Fe_{(s)} + 2\,HCl_{(g)}$$

As explained earlier, in uncoated tubes hydrogen is formed from water retained in the tube [2261]. The thermal pretreatment temperature is consequently very critical for the occurrence or non-occurrence of interferences [2265].

According to the findings of FRECH and CEDERGREN [2262], interferences should hardly occur for the determination of lead in the presence of higher concentrations of sodium chloride. Sodium chloride is a relatively stable molecule and the partial pressure of chlorine is low. $Pb_{(g)}$ is the preferred species above 1200 °C; $PbCl_{2(g)}$ only occurs in larger quantities below 900 °C. As frequently emphasized, these conditions are only valid for an isothermal furnace. It is therefore hardly surprising that in a WOODRIFF-type furnace sodium chloride hardly interferes in the determination of lead at 1600 °C, while strong signal depression is observed in a conventional furnace [2338]. As to be expected, atomization off the L'VOV platform [2590] or from a tungsten loop [2621] brings a significant improvement, since a good approximation to isothermal conditions for atomization is achieved.

In the presence of sodium sulfate a competitive reaction takes place with the formation of $PbS_{(g)}$ that hinders quantitative atomization of lead. In a WOODRIFF-type furnace the rate of recovery at 1600 °C is 96%, while in a non-isothermal furnace it is 40% [2338]. The rate of recovery for atomization off the L'VOV platform in a stabilized temperature furnace is likewise over 90%.

HOLCOMBE et al [2223] [2386] [2818] draw especial attention to the influence of oxygen on vapor-phase interferences. By measuring the absorbance at various levels in the graphite tube, these authors found that the signal for lead decreased markedly at a small distance above the tube wall when oxygen was mixed with the purge gas. Simultaneously they observed a higher appearance temperature and a smaller integrated signal, i.e. fewer atoms were produced.

Excess nitrate or sulfate in a sample can lead to errors when comparison is made to aqueous reference solutions. In many cases errors can be attributed to predictable vapor phase oxidation of the analyte element by decomposition products of nitrates or sulfates [2223]. The extent of signal depression is proportional to the degree of decomposition of the nitrate or sulfate; the greatest signal depression is caused by salts that are easiest to decompose. Many interferences reported in the literature can be explained by this effect. BELLING and JONES [2073] reported, for example, that sodium nitrate depresses the signal for manganese more than potassium nitrate, even when the latter is present in higher concentrations. EBDON *et al* [2215] showed that while potassium nitrate depresses the manganese signal, calcium nitrate causes signal enhancement. FRECH and CEDERGREN [2262] demonstrated for the determination of lead that thermal pretreatment of the sample to just below the volatilization temperature of lead effectively eliminated the interference due to nitrate, since nitrates decompose at this temperature. Even for very simple matrices, nitrates can be formed from sodium or potassium salts after addition of nitric acid. This can then lead to the interferences mentioned above and cause signal depression. Careful choice of conditions for thermal pretreatment can help to volatilize or decompose interfering species before atomization and thus effectively eliminate interferences.

EKLUND and HOLCOMBE [2222] also demonstrated a number of instances in which preferential bonding of a concomitant element with oxygen in the vapor phase leads to signal enhancement for the analyte element. In this type of interference the authors included the signal enhancement of lead and chromium by 30–60% through lanthanum or of silver by manganese. HOLCOMBE *et al* [2386] point out that vapor-phase interferences in furnaces cannot be adequately described by equilibrium or thermodynamic principles. At the partial pressures coming into consideration in the furnace, reaction kinetics play a significant role in a prediction on the extent to which the analyte element and the interferent react.

FRECH [2259] observed for the determination of antimony in steel that in the presence of traces of oxygen, both hydrochloric acid and nitric acid caused strong signal depression. These acids do not interfere when oxygen is excluded.

PERSSON *et al* [2743] [2744] established that even the smallest quantities of oxygen, hydrogen, chlorine, sulfur or nitrogen interfere in the determination of aluminium during the atomization phase. The application of the highest possible pretreatment temperature and the exclusion of oxygen and nitrogen (from the atmosphere, for example) are essential prerequirements for good results.

The determination of phosphorus [2741] is also critically dependent on the partial pressures of oxygen, hydrogen and nitrogen in the graphite tube. Added to this, reproducible signals can only be obtained when the heating rate, final temperature and atmosphere in the furnace are exactly controlled. Meaningful signals for phosphorus cannot be expected from a furnace that is not in thermal equilibrium.

KOREČKOVÁ *et al* [2494] investigated possible vapor-phase interferences on the determination of arsenic. In the presence of chloride, $AsCl_{3(g)}$ is one of the most stable compounds up to 1700 °C. In the presence of excess hydrogen, in contrast, the equilibrium partial pressure of chlorine is insufficient to form large quantities of $AsCl_{3(g)}$. The authors thus found the use of uncoated graphite tubes to be more of advantage.

$AsS_{(g)}$ is formed in the presence of sulfate. The formation of this species is best suppressed through higher temperatures and a low partial pressure of oxygen. A further possibility is the addition of lanthanum, since lanthanum sulfide is formed and thus simultaneous volatilization of arsenic and sulfur prevented. Interferences in the determination of arsenic can be minimized by raising the temperature, since most molecular species are less stable and the formation of $As_{2(g)}$ decreases significantly above 1700 °C. Optimum conditions are obtained through use of the platform technique and the addition of suitable buffers for stabilization, such as nickel.

In general, vapor-phase interferences can be reduced or eliminated by the same means used for the elimination of many spectral or volatilization interferences. These measures include careful temperature programming for thermal pretreatment and selective volatilization of concomitants, chemical pretreatment or matrix modification, and the highest possible thermal pretreatment temperature. Another very important factor for the elimination of vapor-phase interferences is thermal equilibrium in the furnace; the L'VOV platform plays a significant role in this respect.

Nevertheless, it must be very strongly emphasized that the use of the L'VOV platform alone does not create isothermal conditions [2880]. Further decisive factors are: The graphite furnace must be heated very rapidly to the atomization temperature (with a rate of heating of $2000\,^\circ C/s$ or more); the L'VOV platform should have minimum mechanical contact with the tube so that atomization is delayed until the tube wall and the gas phase have reached their final temperature and are in equilibrium; the purge gas stream through the graphite tube should be interrupted during atomization; and the temperature step from thermal pretreatment to atomization should be not more than $1000\,^\circ C$. The latter requirement can generally be met by matrix modification to stabilize the analyte element. Once the system is in thermal equilibrium, peak area integration brings further advantages and eliminates kinetic influences.

Frequently, peak area integration permits even lower optimum atomization temperatures to be selected, so that the requirement for a small temperature step can be met more easily. Nevertheless, this should not be taken too far, since from thermodynamic considerations the extent of vapor-phase interferences is lower at higher temperatures [2138]. In an ideal system, the sample would be volatilized into an environment of high and constant temperature, and it would also be possible to monitor the composition of the vapor phase.

A number of recent publications show how successfully interferences can be eliminated in a furnace operating on the above principle at thermal equilibrium. HINDERBERGER *et al* [2372] found for the determination of cadmium, chromium, lead and nickel in biological materials that only a combination of matrix modification and atomization off the L'VOV platform eliminated the interferences almost completely. For the determination of lead in twelve different urine samples under these conditions (secondary ammonium phosphate for matrix modification), a mean gradient of 0.97 ± 0.03 was obtained, i.e. a virtually identical gradient for all samples, permitting direct calibration against aqueous reference solutions. FERNANDEZ *et al* [2243] observed that for the determination of lead and cadmium in various biological reference materials with atomization off the L'VOV platform, correct results were only obtained with peak area integration. Direct calibration against aqueous reference solutions was not possible when evaluation was via peak height.

MANNING and SLAVIN [2619] were not able to observe the signal enhancement for manganese in the presence of magnesium chloride reported by SMEYERS-VERBEKE *et al* [2888]; the recovery was the same as that reported by HAGEMAN *et al* [2337] for a WOODRIFF-type furnace. A similar effect is found for the interference of calcium chloride on manganese, where signal depression is normally observed. In a WOODRIFF-type furnace, HAGEMAN *et al* [2337] found less than 10% interference at a 10^4-fold excess of calcium chloride, while in a Model CRA-63 graphite furnace the interference was 90% at a 10^3-fold excess and the manganese signal was suppressed completely with a 10^4-fold excess of calcium chloride.

SMEYERS-VERBEKE *et al* [2888] also reported strong signal depression in a Model HGA-72 graphite furnace. In contrast, MANNING and SLAVIN [2619] found that in a stabilized temperature furnace the results were virtually identical to those obtained by HAGEMAN *et al* in an isothermal furnace.

SLAVIN and MANNING [2880] succeeded in reducing the interference of chlorides on the determination of cadmium, lead and thallium using atomization off the L'VOV platform into a stabilized temperature furnace to a level that could be handled by a background corrector. This is in agreement with the findings of L'VOV [2576] that the vapor phase equilibria of metal chlorides at higher temperature are shifted toward stronger dissociation. MANNING *et al* [2621] also reported that thallium at 2700 °C could tolerate ten times more chloride than at 1800 °C. SLAVIN *et al* [2882] also made a similar finding for the interference of calcium chloride on the determination of aluminium; a 10^4-fold excess caused no interference when a stabilized temperature furnace was used.

In conclusion, it would appear that at the present state of the art using careful thermal pretreatment, the addition of suitable reagents for matrix modification, and atomization off the L'VOV platform in a stabilized temperature furnace, condensed-phase and vapor-phase interferences can be largely eliminated. Nevertheless, there is still a great deal of work to be done before all problems have been clarified. Suitable matrix modification and thermal pretreatment can also reduce spectral interferences (background attenuation). But it is still a fact that most measurements using the graphite furnace technique require background correction. Not infrequently, the interferences are so large, either because of high background attenuation or because molecular absorption with rotational fine structure is present, that complete correction can only be achieved with the Zeeman effect.

8.2.5. Analysis of Solid Samples

When compared with other methods of instrumental analysis, the disadvantage of atomic absorption spectrometry, namely that it is only possible to work with diluted liquid or dissolved samples, is frequently mentioned. This is certainly largely true for the flame technique where the nebulizer/burner system is hardly suitable for the analysis of solid samples. Similarly, the hydride technique and the cold vapor technique are also largely limited to solutions.

While colloidal solutions can be analyzed in the flame without too much difficulty [469], suspensions can cause considerable difficulties, as shown by the example of the determination of iron particles in used lubricating oils [83] [709]. Care must always be taken to ensure the thorough mixing of suspensions if reliable results are to be obtained [626]. The direct introduction of solid samples into the flame, using a screw feed for example [240] [435] [436], brings a variety of problems and is only practicable for a few elements and samples.

L'VOV [2575] undertook a number of basic investigations and calculations for the direct analysis of solids in flames and concluded that these were unsuitable. To obtain a relative detection limit of 10^{-6}% in a flame (at a mean absolute detection limit of 10^{-8}g/s for an element), it is necessary to introduce the sample into the flame at a rate of 1 g/s, and this is hardly practicable. The detection limits attainable for direct solid sampling into a flame will not therefore be very favorable. Further, if the particles introduced into the flame are of average volatility, they must have a grain size of 1 µm or less to be volatilized with sufficient completeness. With the usual grinding procedures used for geological samples, grain sizes of more than one to two orders of magnitude larger are obtained. This means that for atomization in the flame, substantially more effort is required to pulverize the sample, which is just not acceptable due to the high risk of contamination as well as the time required.

In contrast, the graphite furnace technique offers relatively favorable conditions for the direct analysis of solids. Liquid or dissolved samples are not nebulized continuously, but rather an aliquot is dispensed into the tube and dried, so that the sample is present in solid form. Consequently, solid samples can also be dispensed directly into the tube; further thermal pretreatment steps and atomization will be analogous for both sample types. The advantages of direct solid sampling are obvious. The omission of a digestion step can, depending on the type of sample, save a substantial amount of time and simplify the entire analytical procedure. Additionally, the sensitivity of AAS can only be fully exploited with solid samples, since the sample, and thus the concentration of the analyte element, is not diluted. Thereby further separation and enrichment steps are superfluous, leading to a further saving of time. In trace analysis, every sample handling step, every reagent and all laboratory ware introduce the risk of contamination or loss of the analyte element. Thus, direct solids analysis offers minimum systematic errors. In addition, the risk exists that the analyte element can be unfavorably influenced by digestion and other reagents, or that the tube lifetime is impaired or background attenuation introduced.

Nevertheless, against these advantages a number of disadvantages must be taken into account, such as the difficulty of accurately weighing and introducing a few milligrams of sample into the tube. With these small weighings, homogeneity of the sample very rapidly becomes a problem, leading to poor precision of the determination. Difficulties can also be expected for calibration since solid samples cannot always simply be compared with aqueous reference solutions. When solutions are dried in the furnace, a relatively thin layer having good contact with the graphite surface is produced, but solid samples are mostly in the form of loose heaps in the tube, so that the heat transfer is poorer. In the case of biological or other organic samples, higher background attenuation than for dissolved samples can be expected.

L'vov [8] investigated the problem of direct solids analysis very thoroughly and examined the causes of poor reproducibility. He found that the typical error for dispensing liquids was around 0.025 µL, while it was about 0.1 mg for solids. Apart from this, homogeneity of solid samples plays a crucial role.

For example, with a grain size of 3 µm a grain has a volume of 1.4×10^{-11} cm^3, i.e. a weight of about 4×10^{-11} g (with a specific weight of 3 g/cm^3). With a concentration of 1×10^{-4}% of the analyte element, 1 mg of sample then contains 25 grains with element. The relative standard deviation in the number of grains n in a single sample corresponds to

$$\frac{\Delta n}{n} = \frac{1}{\sqrt{n}},$$ (8.5)

i.e. when $n = 25$, the relative standard deviation for 1 mg of sample is 20% solely due to the statistical distribution. L'VOV also gives a quantitative relationship between the relative standard deviation s_r and the mean grain size P of the sample in which

$$s_r \sim \frac{1}{\sqrt{P}}. \tag{8.6}$$

The obvious requirement for the finest possible division of the sample is evident. Nevertheless, this is not always without problems since every grinding procedure is a potential source of contamination. LANGMYHR [2520] made a comprehensive list on the magnitude of dispensing errors for the most unfavorable condition, namely that the trace element is not evenly distributed but is in the form of discrete particles. Table 16 presents the calculated relative standard deviation for mercury (as mercury(II) sulfide in a matrix of iron(II) sulfide), based on 5 mg sample quantity with, 1, 3, 10 and 30 µg/g Hg and various grain sizes.

Table 16. Relative standard deviation for mercury(II) sulfide in a matrix of iron(II) sulfide in dependence on the grain size and mercury content [2520]

Particle size µm	ASTM mesh	Approx. No. of particles per 5 mg	Relative Standard Deviation of mercury content (%)			
			1 µg/g	3 µg/g	10 µg/g	30 µg/g
105	140	1.6×10^3	232	134	73	42
53	270	1.3×10^4	81	47	26	15
37	400	3.8×10^4	48	27	15	9
20	—	2.4×10^5	19	11	6	3.5
10	—	1.9×10^6	6.7	3.9	2.1	1.2

From the table it is clear that under a combination of unfavorable conditions substantial errors or scatter can occur. On the other hand, it should be noted that under favorable conditions satisfactory precision can be obtained for solid sampling. LANGMYHR also pointed out that for most analyses, the sample quantities used correspond to those volatilized in spark emission spectrographs. The scatter of the analytical measure due to sample inhomogeneity is then about the same for both techniques [2520].

Naturally, not all samples have such an unfavorable distribution of the trace element in the form of discrete particles. Trace elements are frequently distributed relatively evenly in a sample, so that it is not necessary to grind the sample too finely. Homogeneous distribution is frequently found in polymers [634] [637], so that they can often be analyzed directly with success. MARKS *et al* [2626] and also LUNDBERG and FRECH [2570] [2571] introduced whole turnings (1 to 5 mg) of steel or nickel alloy into the graphite furnace and determined elements such as antimony, bismuth, lead, selenium, tellurium and thallium with a relative standard deviation of less than 10%. Good precision was similarly found for the determination of beryllium in coal dust [2301] and heavy metals in sewage sludge [2133]. GONG and SUHR [2307] found for the determination of cadmium in rocks that the samples had to be

ground finely (~40 μm), since cadmium is incorporated in the structure. For river sediments, in contrast, the cadmium is more adsorbed on the surface, so that coarser grinding
(>100 μm) is fully adequate.
The relative standard deviations quoted in the literature for direct solid sampling of the
most varying materials in the graphite furnace lie between 5% and 20%. It should nevertheless be borne in mind that inhomogeneity of the sample is not the only factor influencing
the precision of determinations. L'VOV [2575] drew attention to the fact that under some circumstances the signal can assume an arbitrary form when solid samples are atomized. Reliable results can only be obtained when the peak area is integrated over the entire period
during which the analyte element is atomized. The integral is proportional to the total number of atoms of this element and their residence time in the atomizer.
Although peak area integration brings a number of distinct advantages, it can also cause errors, as recent work has shown [2320]. For the direct analysis of solid samples, larger quantities of concomitants are also frequently volatilized during atomization. Thus although the
same atom cloud density and consequently the same signal height as the reference solution
without matrix is achieved briefly, the residence time of the atoms in the radiation beam is
significantly shorter. As a result of this displacement effect, peaks are generated that have a
substantially narrower width at half height than for pure salts of the same element (see figure 108).

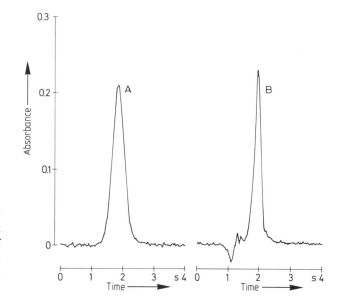

Figure 108. Influence of the matrix on signal form during direct
solids analysis. Determination of
cadmium, **A**: in a reference solution free of matrix, and **B**: in
beef liver.

Both CHAKRABARTI et al [2142] and PRICE et al [2761] found a marked improvement in the
precision of determinations when solid samples —biological, geological and metallurgical— were dispensed onto the L'VOV platform rather than onto the tube wall. Apparently
the influence of sample distribution in the tube and contact with the tube wall is reduced,
since on the platform the sample is heated by radiation. GROBENSKI et al [2320] did not find

this difference in the precision when all parameters had been optimized and the sample wetted in the graphite tube with nitric acid.

After the above deliberations on sample homogeneity, it is worth mentioning that the microanalysis permitted by the graphite furnace technique can provide very useful information on the distribution of trace elements in various areas of a sample. The smallest particles can be removed from the surface or from inside of a metallurgical, geological or biological sample and examined. The determination of numerous trace elements in a single segment of hair is a well-known example [785] [2024]. Such investigations can provide information on the diet, metabolism and environmental influences on people in long past ages.

Much more important than the above considerations on the precision of direct solids analysis is the question of accuracy. This is in close correlation with calibration. The most reliable technique, as in other analytical methods of instrumental trace analysis (X-ray fluorescence, optical emisson with arcs and sparks), is the comparison with preanalyzed samples. The closer the reference material is to the sample, with respect to both the composition of the main constituents and the concentration of the analyte element, the higher will be the probability of a correct measure. This technique has been successfully employed with the graphite furnace, especially for the analysis of metallurgical samples such as complex nickel alloys [2053] [2571] [2626].

Standard reference materials are available in relatively large numbers for geological and metallurgical samples, but the values quoted for trace elements are frequently not very reliable. Such reference materials are hardly available for environmental samples, however, so that this method of calibration can seldom be used. Naturally it is also possible for a laboratory to prepare its own reference materials. Well ground and homogenized samples can be analyzed by as many independent analytical methods as possible until good agreement is obtained for the values of the analyte trace element. These samples can then be analyzed by the new technique and, as shown in figure 109 on the example of mercury, the absorbance measured in the graphite furnace can be entered against the concentration determined by neutron activation [1295]. If the points are on a straight line, this plot can be used as the analytical curve for the calibration of other similar samples.

A further possibility of calibration is the preparation of synthetic reference materials by mixing the main constituents in the form of pure compounds with the analyte element. LANGMYHR *et al* [2533] determined lead in teeth against pure hydoxyapatite (the main constituent of teeth) to which lead had been added in known quantity. Normally the preparation of such solid reference materials is difficult and lengthy. Various authors have examined the possibility of adding aqueous reference solutions to solid samples as a means of calibration. As far as possible, similar quantities of the solid sample should be dispensed into the graphite tube and spiked with varying concentrations of the analyte element. The solution is allowed to soak well into the pulverized sample before the drying step is started. Thereafter the sample is thermally pretreated and atomized in the normal manner. It has proved to be of advantage to wet the non-spiked sample portion with a diluted acid to achieve a similar and even distribution in the graphite tube [2320]. This technique has been applied with success to various biological, geological and industrial samples [1295] [2320] [2521] [2522] [2526] [2748] and also gave good agreement with certificate values for standard reference materials.

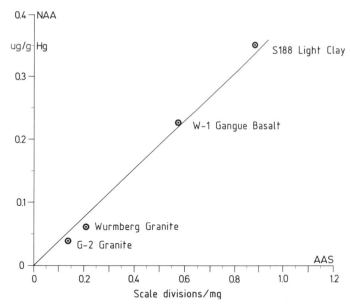

Figure 109. Determination of mercury in rocks by graphite furnace AAS using direct solids analysis compared to neutron activation analysis (NAA).

GROBENSKI *et al* [2320] found for investigations on biological samples that by very carefully optimizing the graphite furnace temperature program, the analyte addition curve was parallel to the analytical curve for slightly acidified solutions. This finding was verified by other authors, at least for various elements in biological [2562], geological [2307] [2859] and metallurgical [2041] samples, and also in graphite [2505] and sewage sludge [2133].

This finding cannot, of course, be generalized, since the chemical form of the analyte element is often different in the sample and reference solutions. The atomization mechanisms are also frequently different, because after drying of the reference solution the element is present in a thin salt layer, while in the solid sample it is surrounded by an enormous excess of concomitants. Exact control of the peak form and especially of the appearance time and time point of peak maximum provide additional valuable information that guards against false measurements [2320].

In some cases, matrix modification appears to be of advantage for accuracy of the determination. LANGMYHR and KJUUS [2525] found for the determination of lead and manganese in bones that every trace of chloride must be removed from the sample. They therefore heated the sample in the graphite furnace with an excess of nitric acid. For the determination of cadmium in bovine liver, CHAKRABARTI *et al* [2142] added ammonium sulfate to stabilize this volatile element.

The question of background attenuation in the direct analysis of solid samples is hardly mentioned in the literature. Most authors merely note that background correction was applied without making any comment on the magnitude of the background signal. MARKS *et al* [2626] presumed that lower background should occur for the direct analysis of nickel alloys since no salts are formed. They in fact found for the determination of bismuth that at a relatively low atomization temperature (2200 °C) the signal/background ratio for the solid sample was more than an order of magnitude better than for the digested sample. At a higher atomization temperature (2500 °C) the difference was no longer significant.

In contrast, GROBENSKI et al [2320] found a distinctly higher background signal for the direct analysis of solid biological samples than for the digested samples. The background attenuation was at times so large that it could no longer be corrected by a continuum source background corrector (refer to figure 110). They solved this problem by ashing the solid sample in situ in the graphite tube by passing oxygen for ten seconds through the tube during thermal pretreatment.

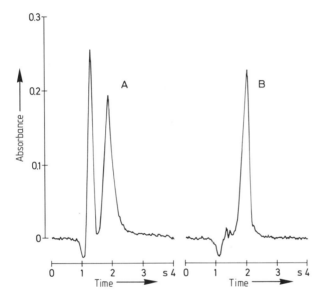

Figure 110. Reduction of background attenuation (**A**) during the determination of cadmium in liver tissue by ashing with oxygen in situ (**B**).

In conclusion it can be said that the direct analysis of solid samples in graphite furnaces offers a number of advantages, which have not been exhausted by a long way. With suitable calibration, the same accuracy can be achieved as with other methods of trace analysis. For the determination of one or a few elements in samples that are difficult to digest, direct solids analysis is certainly quicker. For the determination of many elements, however, digestion is preferable.

A number of authors have reported lower chemical interferences and signal depression for the analysis of solid samples compared to digested samples. Certainly the risk of contamination from reagents and laboratory ware is lower for the direct analysis of solids, so that this appears better for the extreme trace range. Further, for nanotrace analysis, solid samples are not "diluted" so that the sensitivity of the graphite furnace technique can be fully utilized.

On the other hand it must be said that solutions can be pipetted and measured with far greater precision and that dissolution of a sample offers the best means of homogenization.

Calibration of solutions is also generally easier than for solid samples. Also, the dispensing of solutions can be readily automated, but the dispensing of solid samples into the graphite tube will always be somewhat more difficult.

8.3. The Hydride Technique

The greater majority of the work done on the hydride technique is orientated toward practical applications and the elimination of interferences is largely treated empirically. The mechanisms of interferences and atomization are thus far well less understood than in the two techniques treated previously. Further, many of the investigations have been performed in homemade apparatus which differ in detail, often significantly. Frequently, these differences in the apparatus have a substantial influence on the atomization mechanism, and thus the sensitivity, and also on the observed interferences. Added to this, the experimental conditions, such as sample volume, acid volume, concentration and composition, and many other parameters, often differ substantially. In the following section an evaluation of interferences will be presented rather than a complete summary of all interferences reported in the literature. It is difficult to classify non-spectral interferences in the hydride technique since there is still a good deal of uncertainty as to their mechanisms. Kinetic interferences will be treated; these interferences occur in direct, dynamic systems, but not when the hydride is collected. The oxidation states of the analyte element also play a certain role and may, in part, be kinetic interferences, but the effect of oxidation state will be treated separately. Depending on where they occur, other interferences will be divided into chemical interferences in the reaction vessel and gas-phase interferences in the atomizer.

8.3.1. Atomization Mechanisms

Initially in the hydride technique, atomization was performed mainly in argon/hydrogen diffusion flames [362] [524]. Later these were replaced by graphite furnaces [678] [2651] or, more usually, by heated quartz tubes [1271] [2158] [2312] [2967]. Commercial hydride systems use almost exclusively electrically heated or flame heated quartz tubes. Very good results have also been obtained for atomization in fuel gas rich oxygen/hydrogen flames burning in an unheated quartz tube [2188] [2854] [2856] [2857].
A variety of circumstances, especially the relatively easy atomization of the gaseous hydrides in heated quartz tubes and the increase in sensitivity with increasing tube temperature, led to the conclusion that atomization of the hydrides was a relatively simple process. Various observations further led to the assumption that atomization is a simple thermal dissociation; the analyte element reaches the atomizer as the gaseous hydride and then decomposes with the release of free atoms.
A series of recent observations, however, lead to serious doubt that thermal dissociation is the only atomization mechanism. DĚDINA and RUBEŠKA [2188] investigated the atomization of selenium in a cool oxygen/hydrogen flame burning in an unheated quartz tube. They found for all hydrogen flowrates that with increasing oxygen flowrate there was an in-

itial steep rise in the sensitivity, followed by a slow decrease. The position of the maximum was only dependent on the diameter of the quartz tube. The slow decline in the sensitivity corresponded to the increase in temperature and to the expansion of the gas streaming through the tube. These results clearly demonstrate that an increase in the oxygen flowrate above the sensitivity maximum does not enhance the efficiency of atomization. Even heating the side entrance tube of the T-tube to temperatures between 300°C and 900°C at an oxygen flowrate below optimum did not increase the peak area for selenium. The atomization thus cannot be the result of thermal dissociation.

DĚDINA and RUBEŠKA concluded that atomization is brought about by free radicals that are produced in the primary reaction zone of the diffusion flame according to the following equations:

$$H + O_2 \rightleftharpoons OH + O \tag{1}$$

$$O + H_2 \rightleftharpoons OH + H \tag{2}$$

$$OH + H_2 \rightleftharpoons H_2O + H \tag{3}$$

In the presence of excess hydrogen it can be assumed that only OH and H radicals are formed and in quantities corresponding to the total amount of oxygen, i. e. two radicals per oxygen molecule. Since the recombination of radicals is much slower than their formation, there is a quantity far above the equilibrium concentration in the secondary reaction zone. Since reaction (3) is very rapid, equilibrium between H and OH radicals is established very quickly. As the equilibrium constant of reaction (3) is very high and also the hydrogen concentration in the atomizer is much higher than the water concentration, the concentration of H radicals is several orders of magnitude higher than that of OH radicals. In all probability, atomization is via a two-step mechanism with the predominating H radicals:

$$SeH_2 + H \rightarrow SeH + H_2 \quad (\Delta H = -189 \text{ kJ/mol}) \tag{4}$$

$$SeH + H \rightarrow Se + H_2 \quad (\Delta H = -131 \text{ kJ/mol}) \tag{5}$$

Corresponding reactions with OH radicals are also conceivable, but owing to their low concentration, any role these reactions might play is negligible. A further reaction that should be taken into account is recombination:

$$Se + H \rightarrow SeH \quad (\Delta H = -305 \text{ kJ/mol}) \tag{6}$$

If reactions (4), (5) and (6) are now taken into consideration, it can be shown that after a sufficiently large number of collisions with H radicals, equilibrium is attained. SeH_2 is not included in the equilibrium and the ratio between SeH and Se corresponds to the ratio of the constants of formation of reactions (5) and (6). Since recombination reaction (6) is strongly exothermic and needs a third partner to take up the energy, it can reasonably be assumed that this reaction is very much slower than the formation of Se atoms. The probability for the formation of free selenium atoms from SeH_2 is thus proportional to the number of collisions with free radicals and the efficiency of atomization should increase with increasing number of radicals. When the oxygen flowrate is optimum, a state of equilibrium will be reached in which all the selenium is present as free atoms. These observations and conclusions made by DĚDINA and RUBEŠKA permit analogous conclusions to be drawn for the atomization of selenium and the other hydride-forming elements in hydrogen diffusion flames. The fact that the sensitivities for most elements in these flames of lower temperature are as good, or even better, than in the hotter air or nitrous oxide/acetylene flames gives rise to considerable doubt about the thermal dissociation of the hydrides. It has already been shown that there is a large number of H radicals present in an argon/hydrogen diffusion flame, so the same reaction mechanism can be expected.

The question now open is atomization in quartz tubes. Sporadic publications report on the influence of the quartz tube surface on which a "catalytic film" must first be formed before maximum sensitivity can be attained [2232]. If this film is destroyed, considerable signal depression occurs and if the tube becomes poisoned it cannot be reconditioned. Other authors also attributed sensitivity losses to devitrification of the quartz through sodium hydroxide traces in the carrier gas and, especially, burnt-in metal traces [2666].

Such effects have also been observed in the present author's laboratory. New, untreated quartz tubes only reach full sensitivity for an element after several hours or even several days use. This effect can be eliminated by bathing new quartz tubes in hydrofluoric acid for a few minutes; they then exhibit maximum sensitivity from the first determination. This refutes the theory of a catalytic film that must first develop. In all probability, there are active sites, or a "catalytic film", on the surfaces of *untreated* quartz tubes that must first be removed. Even if the sensitivity of a determination deteriorates after prolonged use, rebathing in hydrofluoric acid restores the quartz tube to pristine condition.

During detailed investigations on the temperature dependence of atomization in quartz tubes in the author's laboratory [3056], it was observed that the sensitivity exhibited a dependence on the time the apparatus was preflushed with inert gas. Longer preflush times, especially in the temperature range 700°C to 800°C, brought substantially lower sensitivity than shorter preflush times. The cause is apparently due to the decrease in the oxygen content in the sample solution during flushing. If measurements are performed largely with the exclusion of air, all elements investigated, with the exception of bismuth, exhibited no measurable signal or distinctly reduced sensitivity at 700–800°C (see figure 111). With increasing temperature the sensitivity also increases and for a number of elements does not appear to have reached a maximum even at a tube temperature of 1000°C. In the presence of oxygen, in contrast, maximum sensitivity is obtained spontaneously at 700°C and a further increase in temperature brings no further increase in the signal (as shown in figure 111). These findings show quite clearly that oxygen plays an active role in the atomization of gaseous hydrides in heated quartz tubes. Here, as in the experiments of DĚDINA and RUBEŠKA, an increase in the oxygen content above an optimum brings no further improvements. As well as oxygen, there is also adequate hydrogen in the quartz tube from the reaction of the borohydride with the acidified sample solution, so it is conceivable that similar reactions to those in a fuel gas rich hydrogen flame take place.

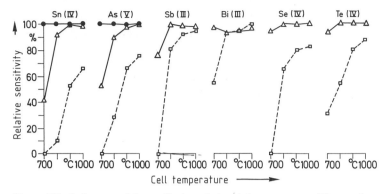

Figure 111. Influence of the preflush period and the gas composition on the sensitivity of the hydride technique at various cell temperatures (according to [3056]). △ 25 s preflush; □ 60 s preflush; • argon with 1% oxygen as purge gas.

The significance of hydrogen for atomization could be demonstrated impressively for arsenic [3056]. In a pure argon atmosphere, arsenic hydride (arsine) could not be measurably atomized, even in a quartz tube heated to 1000 °C. The hydride is thermally dissociated, but its dissociation does not lead to free arsenic atoms. In all probability at these temperatures, the more stable tetramer As_4 or dimer As_2 is formed. Atomization takes place as soon as hydrogen is mixed with the argon. In the absence of oxygen, the same dependence of the sensitivity on the temperature of the quartz tube was observed as in earlier experiments, while when oxygen was introduced (argon with 1% oxygen), maximum sensitivity was attained at temperatures from about 600 °C.

From these observations it is clear that hydrogen is essential for the atomization of arsine (and other gaseous hydrides) in a heated quartz tube and that oxygen plays a supporting role, at least at lower temperatures. Measurements of the gas phase temperature under various experimental conditions showed that the addition of oxygen did not lead to an increase in the temperature [3056].

It can reasonably be assumed that in a heated quartz tube the atomization of gaseous hydrides is via collisions with hydrogen radicals, just as proposed by DĚDINA and RUBEŠKA [2188] for atomization in fuel gas rich oxygen/hydrogen flames. Hydrogen radicals are also produced in the quartz tube through reaction with oxygen. The reaction occurs relatively spontaneously between 500 °C and 600 °C, as shown by the complete atomization signals obtained at these temperatures when sufficient oxygen is present.

The influence of temperature on the sensitivity in the absence of oxygen can be explained by the fact that under these conditions at lower temperatures fewer H radicals are formed, or their rate of formation is slower. At higher temperatures the very small residual amount of oxygen, still remaining in the solution even after long purging or present in the argon employed, appears to be sufficient to produce an adequate number of H radicals.

It is a prerequisite for the free radical mechanism of atomization of gaseous hydrides that the concentration of H radicals in the quartz tube is far higher than the equilibrium concentration, or that the lifetime of the radicals is relatively long. This can be assumed in the pure argon atomsphere, provided there are no active sites on the quartz surface in the tube that could catalyse recombination of the radicals. This would plausibly explain the influence of the surface properties and of the "catalytic film".

Within a main group of the periodic table, decreasing amounts of oxygen appear to be necessary for atomization, with increasing atomic weight (e. g. in the order As-Sb-Bi), as shown in figure 111. For bismuth, even under almost complete exclusion of air, 50% of the maximum sensitivity is attained at 700°C. This effect can no doubt be attributed to the higher efficiency of the collision of the larger and less stable hydride molecules, so that fewer free radicals are required for atomization.

8.3.2. Spectral Interferences

In the hydride technique, the analyte element passes into the atomizer as the gaseous hydride, while concomitants normally remain in the reaction vessel. Consequently, due to the relatively small number of components in the gas phase in the atomizer, spectral interferences can be virtually excluded. For atomization in flames, variations in the transparency occasionally occur when the hydride, together with the hydrogen generated, enters the flame; the use of a background corrector is recommended in such cases [362]. For atomization in heated quartz tubes, no interferences due to background attenuation occur, providing that the hydrogen is not allowed to ignite at the open tube ends [2967]. Solely SINEMUS [2869] reported very minor background attenuation during the determination of selenium traces in 5 M hydrochloric acid. Since the signal was very small and also very constant, and caused by the hydrochloric acid, it could be subtracted like a blank value.

8.3.3. Kinetic Interferences

Kinetic interferences are caused by varying rates of development or liberation of the hydride from solution. These interferences only occur in direct systems where the measurement is performed "on line" with the generation of the hydride; they do not occur when the hydride is collected, e. g. by trapping out in a cold trap. SIEMER and KOTEEL [2856] reported that peak area integration instead of the more usual peak height evaluation eliminates most kinetic interferences as well.

A typical kinetic interference is the retardation of hydride liberation through excessive foam in the sample solution [3051]. If biological materials that have not been fully decomposed are analyzed by the hydride technique, dense foam is produced when alkaline borohydride solution is added and this foam retains a portion of the hydride. The signal is thereby distinctly lower and broader than for acidified reference solutions. The best remedy is the use of a reliable antifoaming agent, which also makes the analyte addition technique unnecessary. For some elements, however, antifoaming agents cause a slight signal depression.

A further kinetic interference appears to be the influence of the volume of measurement on the sensitivity (see figure 46, page 74). Under constant conditions, the hydride is not liber-

ated so easily from larger measurement volumes as from smaller ones. It is sometimes difficult to eliminate this effect via peak area integration. However, in practice this effect does
not play a significant role since constant volumes of measurement are generally employed
and the differences in sensitivity between large and small measurement volumes are not
very notable.

8.3.4. Oxidation State Influences

The hydride technique is the sole technique in AAS in which the oxidation state of the analyte element has a significant influence on the sensitivity. For the representative elements of
Group IVA in the periodic table—germanium, tin and lead—nothing is reported in the literature on the influence of oxidation state on sensitivity. For the elements of Group VA, the
sensitivity difference in peak height between the +3 and +5 oxidation states is less than a
factor of two. This difference can partly be eliminated through peak area integration. For
the elements in Group VIA, selenium and tellurium, only the tetravalent oxidation state
gives rise to a measurable signal, so that it is hardly possible to speak here of a kinetic interference. The sensitivity of the pentavalent oxidation state of arsenic is about 70 to 80% of
the trivalent oxidation state; for antimony the higher oxidation state has about 50% sensitivity of the lower state (figure 112). For convenience, a number of authors chose to work
with the higher oxidation state of the analyte element, if this arose during digestion of the
sample, and calibrate with corresponding reference solutions, rather than to perform a prereduction [2232] [3052] [3054]. However, this does not appear to be completely without
problems, since under certain circumstances the pentavalent state of arsenic is more subject
to chemical interferences than the trivalent state [3060].

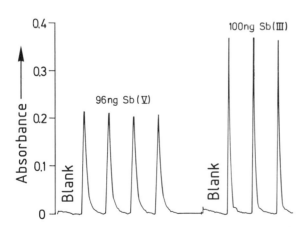

Figure 112. Signals obtained for trivalent and pentavalent antimony by
the hydride technique (according to
[3052]).

Most authors perform a prereduction to the trivalent state to obtain the full sensitivity of
the method. As reductant, either a mixture of potassium iodide and ascorbic acid [2735]
[2858] or potassium iodide in strongly acidified solution [2643] [2869] is used. The reduction
of antimony is almost spontaneous, but arsenic is somewhat slower, so warming or a longer

reaction time is necessary [2869]. Bismuth exists virtually only in the trivalent form, so that there is no problem.

Various authors found that both oxidation states of antimony [3052] and arsenic [2455] [2857] gave the same sensitivity in peak area, i. e. a genuine kinetic interference could be the cause. Nevertheless, SIEMER et al [2857] maintain that this is only the case under very specific analytical conditions. When using larger quantities of acid, these authors also observed lower sensitivity for the higher oxidation state.

Apart from the concentration of acid, the difference in signal height between the two oxidation states also appears to depend on the borohydride concentration. THOMPSON and THOMERSON [2967] found that the difference for arsenic was only 10% when they went from 1% borohydride to 4%. HINNERS [2374] found that with a sufficiently large quantity of borohydride there was no difference between the oxidation states of arsenic in 2 M to 10 M hydrochloric acid.

The difference between the +4 and +6 oxidation states of the Group VI elements is very pronounced. For both selenium and tellurium, the hexavalent state does not appear to give a measurable signal. At least, nothing has been reported on whether the higher oxidation state is reduced to the hydride after a sufficiently long period, in which case it would be possible to speak theoretically of a retarded reaction.

In practice, prereduction of selenium and tellurium is always required when they exist in the hexavalent state. For selenium, a prereduction in hot, 4 M to 6 M hydrochloric acid has been proposed [2180] [2785] [2786]. A dependence of the reduction efficiency on both the acid concentration and the boiling period were reported, however [2180] [2869]. Tellurium can be reduced more easily than selenium to the tetravalent state; short boiling in semiconcentrated hydrochloric acid is sufficient for complete reduction [2869].

8.3.5. Chemical Interferences

Acids generally have little influence on the hydride-forming elements; only tin and lead exhibit a relatively strong pH dependence, so that they are normally determined in buffered solutions. Of the various acids normally encountered in practice, only hydrofluoric acid causes interferences at relatively low concentrations (figure 113), so that it should be fumed off if possible before the determination. Hydrochloric acid (figure 114), nitric acid (figure 115) and sulfuric acid (figure 116) only depress the signal at relatively high concentrations. These acids are thus excellently suited for making up the sample solution to the final measurement volume in the reaction vessel of the hydride system. Residual acids remaining from digestion procedures (e. g. hydrofluoric acid or perchloric acid) should be so diluted hereby that they cannot interfere in the determination. In practice, the lowest acid concentration permitting a rapid, homogeneous and quantitative reaction with sodium borohydride should be chosen. For most hydride-forming elements this is 0.5 M (1.5%) hydrochloric acid or similar. Higher acid concentrations are correspondingly more expensive, especially in the degree of purity required for trace analysis, and since they are normally present in large excess also bring the risk of contamination. Further, working with high acid concentrations is not particularly pleasant and affords increased safety measures for the user and the apparatus.

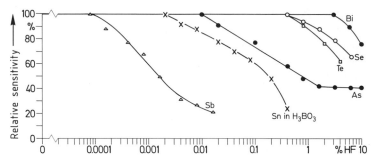

Figure 113. Influence of increasing concentrations of hydrofluoric acid on the sensitivity of hydride-forming elements.

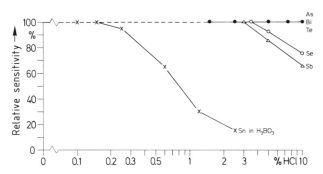

Figure 114. Influence of increasing concentrations of hydrochloric acid on the sensitivity of hydride-forming elements.

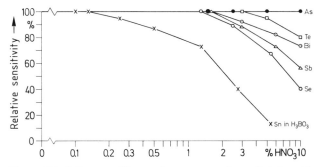

Figure 115. Influence of increasing concentrations of nitric acid on the sensitivity of hydride-forming elements.

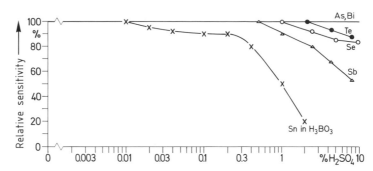

Figure 116. Influence of increasing concentrations of sulfuric acid on the sensitivity of hydride-forming elements.

Nevertheless, it is a major advantage of the hydride technique that higher acid concentrations can be tolerated without the occurrence of greater losses in sensitivity or of other interferences. Various authors have established that cation interferences (these will be treated later) are strongly dependent on the acid concentration [2084] [2666] [3054] [3058]. In the presence of higher concentrations of hydrochloric acid or nitric acid, the interference-free range can often be extended by several orders of magnitude. In pratice, therefore, it is often preferable to dilute the sample solution with a more highly concentrated acid and thereby accept any minor disadvantages or sensitivity losses that this brings.

KANG and VALENTINE [2455] found that more acid interferences occur for measurements in peak height than in peak area; via the latter virtually no more interferences were observed. This indicates that the minor interferences observed at higher acid concentrations are probably kinetic interferences. The strong pH dependence of tin and lead, however, is more likely due to other causes.

Metallic elements frequently present, such as the alkali and alkaline earth metals, aluminium, titanium, vanadium, chromium(III), manganese(II) and zinc, do not interfere in the determination of hydride-forming elements, or at least only in extremely high concentrations. Thereby a very important prerequirement is fulfilled, namely that for the examination of geological and biological samples, and also for the analysis of sea-water, surface waters, etc. by the hydride technique, no major interferences are to be expected. These types of sample contain elements in high concentration that do not interfere in the determination. The major interferences come from the elements of Groups VIII and IB of the periodic table. During their experiments on automated arsenic and selenium determinations, PIERCE and BROWN [2749] found that cation interferences were more pronounced when sodium borohydride was added to the samples first, followed by hydrochloric acid, than *vice versa*. They observed that a precipitate was immediately formed when they added borohydride to neutralized sample solutions. They connected this precipitation with the interferences postulated by SMITH [2890] in an earlier publication.

KIRKBRIGHT and TADDIA [2471] drew attention to the fact that nickel and elements in the platinum group are hydrogenation catalysts and can absorb a large amount of hydrogen.

They are also capable of capturing and decomposing the hydride, especially in the finely divided form in which they are produced during reduction. The addition of 500 mg nickel powder to the sample solution before reduction to the hydride caused total signal suppression.

It is thus clear that there is a direct relationship between the precipitation of finely divided metal and the observed interferences on the generation and release of hydride from the sample solution. Various authors have also established that depression of the signal is independent of the analyte-to-interferent quantity ratio [2658] [2666] [2778], but only depends on the concentration of the interferent in the final analysis solution [3054] [3058], and on the acid concentration and composition.

The fact that the signal depression is not dependent on the analyte-to-interferent quantity ratio nor on the contact time of the reaction partners before reduction excludes the formation of non-reduceable, or difficult to reduce, compounds before the reduction. TÖLG et al [2666] thus came to the assumption that the hydride after its formation reacted at the gas/liquid phase boundary with the free interfering ions present in the acidified sample solution to give insoluble arsenides, selenides, tellurides, etc. Such a reaction depends only on the diffusion speed of the covalent hydride from inside the gas bubbles to the phase boundary. According to the laws of diffusion, the quantity of hydride-forming element transported would be proportional to the initial concentration of the hydride in the bubble and largely independent of the content of the element in the sample solution. An estimation of the rate of diffusion of the hydride in a nitrogen/hydrogen carrier gas mixture and the length of time a gas bubble is in solution indicate the high probability for such a reaction mechanism.

RAPTIS et al [2778] also made observations in this direction. They found that correct results were obtained for low sample weighings, but for high sample weighings, even under identical conditions, low results were obtained. This indicates quite clearly that signal depression does not depend on the analyte-to-interferent ratio, but only on the concentration of the interferent in the analysis solution. MELCHER et al [2658] verified this finding; they nevertheless found that correct results could be obtained for higher sample weighings when the analyte addition technique was used.

The question is now open whether the observed interferences are caused by reaction of the covalent hydride with ions in solution or with the finely divided, precipitated metal. The experiments of KIRKBRIGHT and TADDIA [2471] mentioned previously, in which nickel was added directly, indicate that the reaction gas/solid at least plays a part. Also, the qualitative connection between signal depression and precipitation of metal indicates such a reaction.

Detailed experiments in which the hydride, after generation in hydrochloric acid solution, was conducted through a gas washbottle containing a solution of the interfering element in ionic form clearly showed that both reaction mechanisms are possible [3057]. Bivalent transition metal ions such as cobalt, copper, iron or nickel, interfere largely in the reduced, metallic form, but not in solution. In this case the interference can be dramatically reduced by preventing precipitation of the metal after borohydride addition. It is then clear, at least for these elements, that the reduction to metal is the first, decisive reaction. The finely divided metal then captures and decomposes the hydride in a catalytically supported reaction to give an insoluble compound. The observations of PIERCE and BROWN [2749] that many more interferences occur when borohydride is added to the sample before hydrochloric acid are also a clear indication in this direction.

Transition elements, however, also interfere in ionic form in solution, at least in high concentrations [3057]. In this case the mechanism proposed by TÖLG et al [2666] that the hydride reacts at the gas/solution boundary appears to be valid. Decisive for all these observations, however, is the fact that interferences caused by transition metals on the hydride technique can be relatively easily eliminated, or at least reduced by several orders of magnitude. The following measures are available: Increase of the acid concentration (figure 117); use of an acid mixture (figure 118); dilution of the sample solution, either by using a smaller sample aliquot under otherwise identical experimental conditions, or by using the original aliquot and diluting to a larger final volume (figure 119). Utilizing these possibilities, interference-free determinations of hydride-forming elements have been performed in foodstuffs [2857], biological materials [2735] [2893], geological and environmental samples [2893] and also in high purity metals and alloys [2254] [3054] [3058].

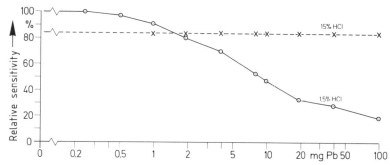

Figure 117. Extension of the interference-free range for the hydride technique through the use of higher acid concentrations. Determination of arsenic in the presence of lead in 1.5% and 15% hydrochloric acid, respectively.

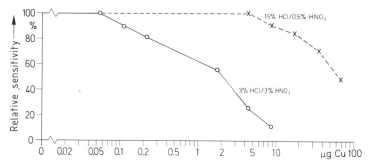

Figure 118. Extension of the interference-free range for the hydride technique through the use of acid mixtures. Determination of tellurium in the presence of copper(II) in 3% HCl/3% HNO_3 and 15% HCl/0.5% HNO_3, respectively.

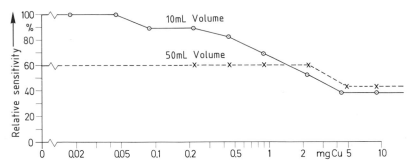

Figure 119. Extension of the interference-free range for the hydride technique through the use of larger analysis volumes. Determination of antimony in the presence of copper(II).

BERNDT *et al* [2084] found for the determination of arsenic in lead that the range of interference-free determination could be extended by almost four powers of ten when 6 M hydrochloric acid was used instead of the more usual 0.5 M. Similar findings were made in the present author's laboratory for other hydride-forming elements in the presence of iron and nickel [3054], and also in copper and several other transition metals [3058].

Frequently it is possible to reduce the influences of interfering elements to a common denominator, so that the range for an interference-free determination can be calculated in advance and the acid concentration and composition, as well as sample dilution, can be correspondingly selected [3054] [3058]. Nevertheless, this is not possible in all cases since varying interactions come into play, especially for samples with complex compositions. RAPTIS *et al* [2778] found for the determination of selenium in vegetable and animal materials, for example, that interferences occurred even though the absolute quantity of an interfering element could not be made responsible for the low results. They therefore assumed that synergetic influences of various elements play a part. Investigations performed by FLEMING and IDE [2254] have shown that the interaction of several elements does not just bring a higher susceptibility to interferences, but can also have a positive influence. While nickel even in relatively low concentrations interferes in the determinations of hydride-forming elements, the influence in the presence of high concentrations of iron is substantially lower. These authors were able to perform interference-free determinations of antimony, arsenic, bismuth, selenium, tellurium and tin in steel even in the presence of higher quantities of nickel. For the analysis of high alloy steels they added iron to eliminate the influences of nickel and other interfering elements. These findings were verified in the present author's laboratory and other alloys were systematically examined [3054] [3058]. The effect of nickel on the determination of arsenic is shown in figure 120. The addition of iron also eliminates the interference due to nickel on the determination of arsenic in waste water [3060].

A further interesting possibility for reducing interferences is based on the varying stability of the compounds that hydride-forming elements form with the elements of Groups VIII and IB. TÖLG *et al* [2666] found that silver influences the determination of selenium in the presence of tellurium in a different way. In the absence of tellurium, silver interferes in the

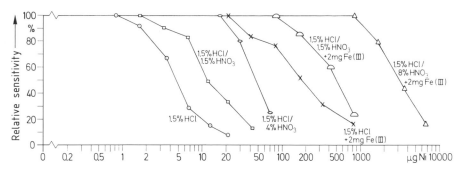

Figure 120. Extension of the interference-free range for the determination of arsenic in the presence of nickel through the addition of iron(III) in various acid mixtures (according to [3049]).

determination of selenium in concentrations from 25 µg/L (in 0.3 M hydrochloric acid), while in the presence of 200 µg tellurium, the influence of silver is first observed at 500 µg/L. This is an extension of the interference-free range by a factor of 200.

Based on this relationship, KIRKBRIGHT and TADDIA [2472] developed a procedure for reducing the interferences of a variety of metal ions on the determination of selenium. The Te^{2-} ion produced by reduction with sodium borohydride can form very stable tellurides with numerous interfering ions. The constants of stability of the tellurides of copper, nickel, palladium, platinum and other elements are lower than those of the corresponding selenides. The addition of excess tellurium(IV) extends the interference-free range for the determination of selenium in the presence of these elements quite considerably.

This is the sole reference in the literature to the addition of another hydride-forming element to the analyte element to reduce interferences. With careful scrutiny of the constants of stability, it is almost certainly possible to find other element combinations where similar effects occur.

Various authors have attempted to prevent the reduction and precipitation of the interfering metal, and thereby eliminate the interference, by adding various reagents, in particular complexing agents. Potassium iodide eliminates the influence of cadmium, iron, copper, cobalt and silver on the determination of arsenic at 1000-fold excess and reduces the influence of nickel [3086]. Similarly, an addition of potassium iodide reduces the interference of iron on the determination of antimony in rocks [2146] and the influence of copper on the determination of bismuth in biological materials [2797].

The sodium salt of ethylenediaminetetraacetic acid (EDTA) forms stable complexes with numerous transition elements. Using this reagent, DRINKWATER [2206] was able to determine bismuth in nickel alloys and LINDSJÖ [2550] bismuth in cobalt. The EDTA-cobalt(III) complex is stable in 5 M hydrochloric acid, but the corresponding bismuth complex is not. MAUSBACH [2642] was able to completely eliminate the interferences of cobalt and nickel on the determination of arsenic and selenium using EDTA. The influence of copper on the determination of arsenic could also be eliminated, but not on the determination of selenium. MAUSBACH explained this as being due to varying pH dependence of the complexes of the metals examined.

GUIMONT et al [2328] applied EDTA, cyanides and thiocyanates successfully to mask nickel during the determination of arsenic in rocks and sediments. Nevertheless, RUBEŠKA and HLAVINKOVÁ [2809] have pointed out that these complexing agents can only be utilized when the final pH value of the solution after addition of borohydride is alkaline, or when borohydride pellets are added as the alkalinity is sufficient in their direct neighborhood. In 1 M hydrochloric acid solution, neither EDTA nor thiocyanates have an effect since the complexes are not stable enough.

LINDSJÖ [2550] found that thiourea was very suitable for complexing copper during the determination of antimony and VIJAN and CHAN [3016] utilized sodium oxalate to eliminate the interferences of copper and nickel on the determination of tin. KIRKBRIGHT and TADDIA [2471] added thiosemicarbazide and 1,10-phenanthroline to reduce the influences of copper, nickel, platinum and palladium on the determination of arsenic. DORNEMANN and KLEIST [2205] found that pyridine-2-aldoxime had the greatest effect for eliminating interferences on the determination of arsenic; 300 mg of this complexing agent added to the analysis solution eliminated the interferences caused by 40 mg of copper, cobalt or nickel (in each case as the oxide).

A last possibility of avoiding interferences in the hydride technique is separation of the hydride-forming element through co-precipitation with lanthanum hydroxide [2068] [2069] [2550] [2971], iron hydroxide [2068], aluminium hydroxide [2550], or hydrated manganese dioxide [3016]. The greatest problem with this technique is the purity of the metal salts, which are added in fairly large excess, so that high blank values must frequently be expected.

8.3.6. Gas-Phase Interferences

Virtually all hydride-forming elements interfere mutually, and in some cases even very low concentrations can depress the signal. The extent of the interference is only dependent on the concentration of the interfering element and not on the interferent-to-analyte ratio. Thus, for many samples these interferences do not appear after sufficient dilution, so that various authors saw no need to eliminate them. VERLINDEN and DEELSTRA [3009] found that for the determination of selenium, the following elements interfered at the given quantities: Bismuth and tin in twofold to fivefold excess; antimony from about twentyfold excess; tellurium and arsenic from a hundredfold or fivehundredfold excess, respectively. No attempt was made to eliminate the interferences since the freedom from interferences was sufficient for the analysis of biological materials.

YAMAMOTO and KUMAMARU [3086] eliminated the influences of lead, selenium(IV), tellurium(IV) and bismuth on the determination of arsenic by adding potassium iodide. An excess of these elements up to 1000-fold did not interfere. ROMBACH and KOCK [2797] used the same reagent to eliminate the influences of arsenic and selenium on the determination of bismuth. VIJAN and CHAN [3016] added sodium oxalate to the sample solution to control the interferences of antimony and arsenic on the determination of tin.

Up to the present, very few authors appear to have considered the mechanisms of mutual interferences of the hydride-forming elements. SMITH [2890] presumed that the formation of compounds in the cool argon/hydrogen diffusion flame that he used could be an explanation for the mutual interferences, but he made no comment on the nature of such compounds.

A type of "competitive reaction" is also conceivable, i.e. the interferent and the analyte compete for the sodium borohydride. The more easily reduced element would react more rapidly and use the borohydride preferentially, so that less borohydride would be available for the slower reacting element, for which a lower signal would then be obtained. This supposition appears to obtain verification in the fact that arsenic(III) interferes in the determination of selenium five to ten time more strongly than arsenic(V), which is known to be reduced more slowly [2666].

A detailed investigation of arsenic and selenium [3055] verified that arsenic(III) interfered in the determination of selenium at a tenfold lower concentration (> 0.01 mg/L) than arsenic(V), which only showed an influence above 0.1 mg/L (figure 121). On the other hand, selenium interfered much more strongly in the determination of arsenic at concentrations above 0.001 mg Se/L (figure 122). The interference is independent of the oxidation state of arsenic. Both interferences are independent of the concentration of the analyte and only depend on the concentration of the interferent. To be able to judge the reaction mechanisms, the signals for Se(IV), As(III) and As(V) were rapidly recorded and superimposed, as depicted in figure 123. Selenium(IV) reacts the quickest, followed by arsenic(III), with arsenic(V) as the slowest.

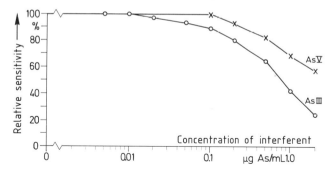

Figure 121. Influence of increasing concentrations of arsenic on the determination of selenium (according to [3055]).

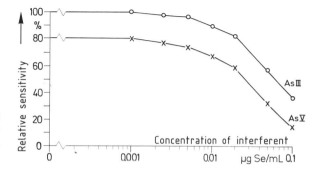

Figure 122. Influence of increasing concentrations of selenium on the determination of arsenic(III) and arsenic(V).

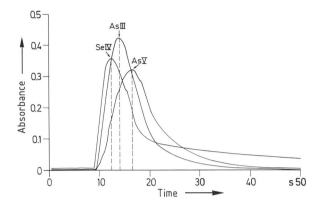

Figure 123. Atomization signals for arsenic(III), arsenic(V), and selenium(IV) in the hydride technique.

If a competitive reaction for the borohydride was in fact the explanation for the interference, then the more quickly reacting arsenic(III) should have stronger influence on selenium than the slower arsenic(V), and this corresponds with the observations. Further, selenium, which reacts the fastest, should have a stronger influence on arsenic, and this is again in agreement with the observations. Finally, there should also be a distinct difference between the interference caused by selenium on each of the oxidation states of arsenic, since Se(IV) and As(III) appear almost simultaneously, while As(V) appears much later and more slowly. As the experiment shows, however, there is no difference in the influence of selenium on either oxidation state of arsenic.

Consequently the mechanism of a competitive reaction for the sodium borohydride is rather improbable. Additionally, a large excess of sodium borohydride is usually employed, so that no shortage of reductant should occur, considering the very low concentrations of the hydride-forming elements normally present in the samples. Also, other reactions in solution, which could lead to signal depression, are very unlikely since no insoluble compounds between arsenic and selenium are known.

It can thus be assumed that both hydrides are formed and transported to the atomizer. As explained in section 8.3.1, atomization of the hydrides is brought about by collisions with H radicals. From the flowrates of the gases and the lifetimes of radicals, DĚDINA and RUBEŠKA [2188] deduced that all these are concentrated in a relatively thin layer. The number of collisions between radicals and hydride molecules, and thus the efficiency of atomization, depends only on the cross-section of the radicals in this layer.

The concentration of radicals in a heated quartz tube purged by an inert gas is certainly not large and it is possible to imagine that a shortage of radicals can often occur, especially when oxygen is largely excluded [3056]. If the mutual interference of arsenic and selenium is now taken into consideration, it is clear that selenium hydride, which volatilizes earlier than arsine, increases this shortage of radicals. Consequently for arsenic, which reaches the quartz tube later, not enough radicals are available to cause the same degree of atomization that is achieved in the absence of selenium. This effect is largely independent of the appearance time of arsine, which explains why the influence of selenium is virtually the same for arsenic(III) and arsenic(V). The fact that arsenic interferes less in the determination of selenium than *vice versa* and that arsenic(V) interferes less than arsenic(III) supports this theory. The later arsine reaches the heated quartz tube, the less its influence will be on the atomization of the faster selenium hydride.

If the mutual interference between arsenic and selenium is in fact a gas-phase interference in the atomizer and not a reaction in the reaction vessel, it should then be possible to eliminate the interference by preventing the interfering hydride from reaching the atomizer. It should be possible to achieve this with one of the known interfering ions of Groups VIII or IB. Naturally the "buffer" should only influence the formation or liberation of the interfering hydride, but not the analyte.

Copper(II) is an ion that interferes much more strongly with selenium than with arsenic (see figure 124). The addition of, say, 50 mg/L copper should suppress development of selenium hydride fully, but have no influence on arsenic. Figure 125 demonstrates that this effect does in fact take place and that the addition of a sufficiently high concentration of copper extends the interference-free range for the determination of arsenic by nearly three orders of magnitude [3055].

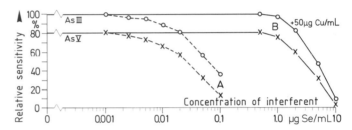

Figure 124. Influence of copper(II) on the determination of arsenic and selenium in 1.5% hydrochloric acid.

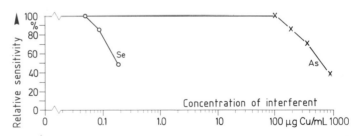

Figure 125. Extension of the interference-free range for the determination of arsenic in the presence of selenium through the addition of 50 mg/L copper.

8.4. The Cold Vapor Technique

The cold vapor technique can naturally only be used for the determination of mercury, since it is based on the unique properties of this element. Since mercury can be easily reduced to metal from its compounds and exhibits a significant vapor pressure of 0.0016 mbar at 20 °C, it can be determined without a special atomizer. After reduction it must merely be transported into the vapor phase by a stream of gas. A discussion on the atomization mechanism is thus superfluous for this technique.

Virtually no spectral interferences occur with the cold vapor technique. It was occasionally reported in earlier work that water vapor caused background attenuation; a detailed examination however revealed that water does not absorb at the resonance line of mercury. The observed interferences were due to droplets of solution carried along by the gas stream or by condensation of water vapor in the absorption cell [2926]. This can be prevented by a suitable drying agent or more simply by warming the absorption cell.

A certain dependence of the signal height on the volume of solution has been observed when mercury is not determined via amalgamation. This is a kinetic influence which does not interfere when samples and reference solutions are made up to the same analysis volume. Thus only chemical interferences through concomitants in solution, which can prevent reduction of the mercury or which react to give difficultly reduced compounds, remain for the cold vapor technique. Nevertheless, it is also important to discuss systematic errors, since these can be very serious owing to the particular characteristics of mercury. It must be pointed out, however, that these systematic errors are not characteristic for the method of determination. They are quite independent of AAS as an analytical method and of the cold vapor technique.

8.4.1. Systematic Errors

The causes for the systematic errors that occur to such a large extent in the determination of mercury all depend to a lesser or greater degree on the mobility of this element and its compounds. These errors include blank values and contamination due to reagents, laboratory ware, or from the atmosphere, and losses due to volatilization, adsorption, or chemical reaction. In the extreme trace range, in which mercury must often be determined, these phenomena can lead to results that are wrong by several orders of magnitude. Even in the "normal" range, substantial errors can occur if extreme care is not taken. In non-contaminated regions in the atmosphere the concentration of mercury is very low and rarely exceeds a few nanograms per cubic meter. But in laboratory atmospheres it is frequently enriched and values of 100 ng/m^3 are not uncommon [2291]. Moreover, unlike most other contaminating elements, mercury is not bound to dust particles but is present as a vapor, so that it cannot be held back even by the high efficiency filters of an ultra clean room. This can lead to uncontrollable blank values and to contamination of samples, the surfaces of laboratory apparatus, and reagents [2975].

KAISER et al [2448] reported substantial problems that occur during the storage of samples. Soil samples that had been stored in air for 30 days at 20 °C took up mercury from air in quantities many times over the original content. The values increased to a hundredfold, but without any regularity. This uptake of mercury can also take place through the diffusion of mercury vapor through plastic foil. If samples are stored in plastic vessels at locations where there is a concentration gradient between the environment and the sample, exchange of mercury through the walls of the vessel can occur. This process depends on the type and thickness of plastic, as well as the temperature and other factors. Substantial errors occur due to contamination of apparatus from the laboratory atmosphere. Thorough cleaning by fuming out with nitric acid and water, for example, should be a matter of course, but even then not all mercury traces can be removed. It should thus be attempted to use as little laboratory apparatus as possible and the apparatus should consist of inert materials with the minimum possible surface area. The use of compound techniques can bring success in this respect.

The best-known source of element blank values is the reagents. In the first place, there are very few reagents that can be prepared very pure or be purified without a great deal of effort. Among these are water and some acids; all other reagents hardly meet the demands of extreme trace analysis [2975].

Acids can be very effectively purified by distillation at a temperature below their boiling point (subboiling distillation), but it should be borne in mind that such solutions do not remain pure over prolonged periods. Mercury obeys an effective distribution function and an equilibrium between a gas and an aqueous phase is rapidly established [2482]. This means that in an open vessel a high concentration of mercury is very quickly obtained when mercury is present in the laboratory atmosphere. It has even been found that opening a bottle of ultrapure acid briefly can markedly increase the blank value [3053].

The second source for systematic errors in the determination of mercury is losses due to volatilization, adsorption, or chemical conversion. These losses can occur during sampling, during sample preparation or digestion, during storage, or during the actual measurement. Due to the mobility of mercury, most losses are in principle the result of exchange reactions. Thus for example, when a sample has a higher concentration of mercury, some of this metal is deposited on the walls of the vessel. If this vessel is used again without prior cleaning, and is now filled with a sample of low mercury content, a reversed exchange takes place and mercury from the walls of the vessel passes into solution. The loss of mercury from one sample can be the cause of contamination in the next.

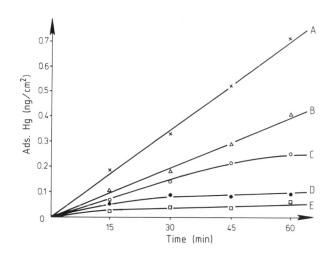

Figure 126. Adsorption of mercury vapor on various materials. **A**: Graphite—PTFE; **B**: PTFE; **C**: Borosilicate glass; **D**: Quartz; **E**: Vitreous carbon. Experimental conditions: Temperature 20°C; nitrogen carrier gas, 5 L/h; mean mercury concentration in the carrier gas 1 ng/mL±10%; material surfaces flushed with double-distilled water (according to [2448]).

A number of authors have dealt with the question of the most suitable materials for the collection and storage of samples (refer to figure 126). Normal laboratory glass definitely has the worst characteristics; PTFE is better, but by no means satisfactory unless it has been previously fumed out with 65% nitric acid [2447]. In general, quartz and vitreous carbon are the most suitable materials in which to store samples containing mercury. Fuming out with 65% nitric acid reduces the readiness of these materials to adsorb mercury [2448]. Nevertheless, it is only possible to store solutions for longer periods without losses if potassium iodide, cyanide or bromide is added to complex the mercury [2447]. LITMAN et al [2551] reported high adsorption rates of mercury on glass, PTFE and especially polyethylene in the concentration range below 1 µg/L. They presumed that the losses could be attributed to a reduction to the metal. KOIRTYOHANN and KHALIL [2482] found using polypropylene vessels that low results were frequently found for reference solutions, but not for samples containing an excess of an oxidizing agent from digestion procedures. The cause was found to be di-t-butyl-methylphenol, which is added to polypropylene as an antioxidant and was found to act as a reducing agent. The problem could be eliminated by adding an oxidizing agent, such as potassium permanganate, to the reference solutions.

TÖLG [2975] reported precipitation exchange reactions for mercury, for example amalgamation on non-noble metal surfaces or chemical exchange reactions on the surface of tubing. Red rubber tubes, which are vulcanized with antimony sulfide, bind mercury ions to the surface as HgS, and PVC tubes bind mercury to non-saturated chlorine sites.

With regard to sample collection and storage, STOEPPLER and MATTHES [2921] reported that non-acidified sea-water samples lost their mercury content very rapidly. At pH 2.5, no change in the total mercury content occurred over more than two months, although there was a decrease in the methylmercury content with a proportionate increase in the inorganic mercury content. MATSUNAGA et al [2635] found that at a concentration of 0.5 µg/L the mercury rapidly decreased even in acidified solution. The presence of sodium chloride prevented adsorption losses on the walls of the vessel.

KAISER et al [2448] draw urgent attention to the fact that during sample preparation, particulate matter should not be removed by filtration, since the larger portion of the mercury will be adsorbed owing to the large surface area of the filter. The portion of mercury remaining adsorbed on the precipitate after centrifugation can also be substantial for determinations in the lower ng/g range. Systematic errors occur for solid samples during grinding, sieving or homogenizing. Care should be taken that the apparatus is so designed that losses due to amalgamation cannot occur. Apparatus made of inert plastic and cooled in liquid nitrogen has proved successful for homogenization [2448]. Reports on losses during freeze drying are very conflicting [2551]. RAMELOW and HORNUNG [2771] found no significant differences before and after lyophilization for 49 fish samples and 8 mussel samples.

To bring solid biological or geological samples into solution, a wet digestion in perchloric acid, sulfuric acid, nitric acid, or hydrofluoric acid is usually employed, sometimes under the addition of potassium permanganate or vanadium pentoxide. (Details are provided in the chapter on specific applications.) The biggest problems with such digestion procedures are contamination of the reagents used and, especially, possible losses of mercury due to volatilization. For determining the lowest mercury contents, a minimum quantity of reagents that can be easily purified should only be used and the surface area of the apparatus should be small. The use of autoclaves has become popular for digestions since low quantities of acid are sufficient, the ratio of surface area to sample volume is favorable, and losses largely excluded because the system is closed. For the digestion of biological materials semi-concentrated nitric acid, either alone [3050] or with vanadium pentoxide as catalyst [2502], is generally sufficient. TöLG [2975] has pointed out, however, that PTFE is by no means an inert plastic and its surface can change after repeated digestions under pressure. Further, the contamination and surface quality of PTFE change from batch to batch, so that precon-ditioning and frequent blank value measurements are essential. The use of vitreous carbon as a possible alternative to PTFE has been proposed [2448].

Next to digestions in closed systems, two procedures for digestion in open vessels, apparently without losses, have been recently reported. STUART [2925] digested fish tissue in sulfuric acid/nitric acid in a round bottom flask with a high efficiency condensor under strong heating and reported 100% recovery of radioactively marked mercury. At the same time he observed that for varying partial ashing procedures between 50°C and 75°C, only small quantities of mercury were released. This led to substantial retardation effects for release of mercury from solution.

KNAPP *et al* [2476] [2477] reported a mechanized system for digesting organic materials in chloric acid/nitric acid. The very high oxidation potential of this mixture appears to be the reason why mercury remains in solution at digestion temperatures of over 100°C. Nevertheless, the shape of the digestion vessel plays a significant role [2448].

The dependence of mercury losses on the heating block temperatures and the shape of the digestion vessel are shown in figure 127. While mercury losses are significant at 120°C with a low vessel, the temperature can be increased to 200°C when a long neck vessel is used. A portion of the digestion mixture vaporizes at these temperatures, but there is no loss of mercury, presumably due to oxidation of volatilized mercury in the vapor phase by ClO_2 liberated by the chloric acid [2448].

Based on the volatility of mercury and the thermal instability of its compounds, a number of authors have reported successful ashing of biological, petrochemical or geological materials in a stream of oxygen. These are mostly modifications of the Schöninger or Wickbold procedures. The volatilized mercury is collected either by freezing in a cold trap [2305] [2798] or by dissolving in acidified permanganate solution [2478] [2986]. To carry out such procedures free of errors nevertheless requires a high degree of skill and experience [2478]; in the hands of an experienced analyst, however, good results can be obtained [2798].

TöLG [2975] described a further digestion technique that is free of losses. The sample is ashed in a stream of oxygen in a quartz apparatus and the mercury is collected on a cold finger cooled with liquid nitrogen. Thereafter the mercury is dissolved by refluxing with a little nitric acid in the same apparatus.

All the points mentioned earlier for storage of samples are equally valid for the actual apparatus in which the determination is performed. Materials that bind mercury, such as rubber and PVC, should be avoided [2975]. STUART [2926] has also drawn attention to the fact that every component coming into contact with mercury after its reduction is a potential source of difficulty. To prevent losses, the surfaces coming into contact with mercury should be as small as possible. Even with the greatest of care, however, some mercury will remain in the system and can be released during the next determination. Frequent blank measurements

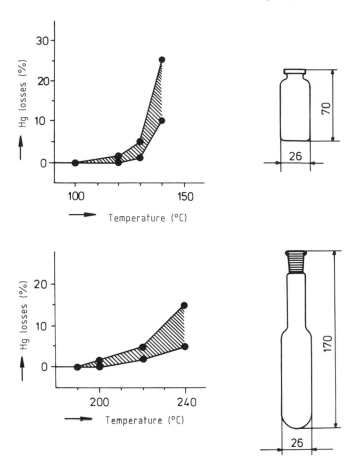

Figure 127. Losses of mercury during digestion in chloric acid and nitric acid in dependence on the temperature and shape of the digestion vessel (according to [2448]).

are thus essential, especially when the mercury concentrations of the samples vary from each other more markedly.

The drying agent can be a particular problem. Calcium chloride is particularly unsuitable since when moist it adsorbs most of the mercury [2926]. Magnesium perchlorate is substantially better since it only binds a few percent of the mercury. Concentrated sulfuric acid appears to be best, although a large dead volume in a wash bottle can cause problems [2503].

8.4.2. Chemical Interferences

Chemical interferences are seldom for the determination of mercury by the cold vapor technique. High concentrations of most acids and very many cations do not interfere. The few known interferences are summarized briefly in table 17 [2448]. It can be noted that the reductants, tin(II) chloride and sodium borohydride, complement each other in an ideal manner. Only silver interferes equally for both reductants, while copper, arsenic and bismuth only cause a certain, minor interference with sodium borohydride. Selenium and iodide interfere markedly with tin(II) chloride, but only interfere with sodium borohydride at very high concentrations.

Table 17. Limiting concentrations of interfering elements above which significant interferences occur in the determination of mercury by the cold vapor technique [2448]

Element	Limiting concentration (weight %) for reduction with:	
	$SnCl_2$ solution	$NaBH_4$ solution
Ag	0.005	0.005
As	10	0.25
Bi	10	0.25
Cu	10	0.25
I	0.003	1
Sb	1	2.5
Se	0.0005	0.5

ROONEY [2799] reported complete suppression of the mercury signal in the presence of 1 g/L gold, palladium, platinum, rhodium and ruthenium when using sodium borohydride as reductant. He attributed these interferences to the reduction of these metals and subsequent amalgamation of the mercury; the same can be expected for copper and silver. WHITE and MURPHY [3067] eliminated the interference due to silver in the determination of mercury in silver and silver nitrate by adding bromide. Silver is precipitated as silver bromide, while mercury remains in solution as the $HgBr_4^{2-}$ complex and can be determined free of interferences in the filtrate.

KULDVERE and ANDREASSEN [2512] found when determining mercury in seaweed that iodine, even in large amounts, did not interfere with the samples, but interfered with the reference solutions. The reason is that in the absence of organic materials the nitric acid is not used up and iodide is oxidized via iodine to iodate. In the presence of organic materials, in contrast, so much nitric acid is used that iodide is only oxidized to iodine and this escapes from the flask in one form or another with the nitrose gases.

STUART [2927] warns of several sources of error that can arise from the use of common reagents. Hydroxylamine hydrochloride, which is added to solutions to reduce excess permanganate, can have a substantial influence on mercury in higher concentrations. 25 mg cause a signal depression of 15% and 100 mg a depression of 65%. Since this interference affects peak height and peak area measurements to equal degrees, it is a clear indication that mercury is not liberated at all. A further interference, which is only observed in peak height however, is caused by sulfhydryl groups. An example is cysteine, which is often added to complex mercury to obtain better reproducibility, but which is also often present in numerous biological matrices. It only interferes, though, when the sample is not completely ashed. Reagents containing sulfhydryl groups retard the release of mercury from the sample; an interference that can be eliminated by peak area integration or by amalgamation.

AGEMIAN et al [2007] [2009] reported that permanganate and persulfate, frequently used to digest organo-mercury compounds, cannot be used for samples with a high chloride content. The chloride is oxidized to chlorine gas, which causes substantial interferences. They used sulfuric acid/dichromate for the digestion of sludges and sediments and photo-oxidation with UV radiation for water with a high salt content.

TOFFALETTI and SAVORY [2974] established that organo-mercury compounds are volatilized and partially reduced by sodium borohydride, but that their sensitivity is different from that of ionic, inorganic mercury. Using a heated quartz tube, these authors found that the sensitivities for the species Hg^{2+}, CH_3Hg^+ and $C_6H_5Hg^+$ decreased because of the short residence time. Above 500°C, however, the sensitivity for methylmercury chloride increased markedly and reached the same value as for ionic mercury at about 700°C. This indicates a thermal decomposition of a volatile compound, which is only complete at 700°C. It was further found that in the absence of copper the phenylmercury chloride reacted to slowly that practically no measurable signal was obtained. In the presence of 10 mg/L copper(II) the peak areas for all three species were virtually identical.

If mercury is collected by amalgamation on a noble metal before determination, the kinetic interferences mentioned earlier are eliminated, provided the collection time is long enough, but not the chemical interferences. YAMAMOTO et al [3087] found using sodium borohydride as reductant that silver, copper and nickel are initially precipitated, but that they redissolve in the acid medium. The signal depression caused by these elements could be eliminated, at least partially, by a sufficiently long reduction time (15 minutes).

A further interference in this technique is a partial or complete coating or poisoning of the gold surface by other gaseous reaction products, so that amalgamation is incomplete. The risk is much higher for sodium borohydride than for tin(II) chloride, since more substances are released using the stronger reductant. Due to the more violent reaction of sodium borohydride, the probability that solution droplets will be carried to the amalgamation surface is also greater [3053]. Frequent conditioning of the gold adsorber in nitric acid is recommended to obtain optimum amalgamation. Many of the materials proposed as adsorbers, such as silver or gold wool, or quartz wool coated with gold, do not withstand frequent washing. A gold/platinum gauze has proved to be very effective due to its mechanical stability [2448] [3053].

9. Related Analytical Methods

In this chapter two other methods will be mentioned briefly because they are also based on the absorption and emission of radiation, namely atomic emission spectrometry (AES) and atomic fluorescence spectrometry (AFS).

Before the individual methods are more fully described, a preliminary remark would appear to be necessary to remove any false preconceptions. All three methods, AAS, AES and AFS, employ comparable atomizers, for all three methods the sensitivity (among other things) is dependent upon the total number of atoms N formed in the ground state, and for all three methods non-spectral interferences are nothing more than an influence on the total number N of atoms formed. This influence on the free atoms takes place in the atomizer, and the free atoms are unaware whether we wish to determine them by means of atomic absorption spectrometry, atomic fluorescence spectrometry or atomic emission spectrometry. Therefore, the same transport, solute-volatilization and vapor-phase interferences occur to the same degree in all three methods, provided the same type of atomizer is employed. The frequently expressed view that flame emission spectrometry is much more strongly subject to non-spectral interferences than atomic absorption spectrometry comes from the use of unsuitable flames and burners, which would incidentally also cause the same interferences in atomic absorption.

9.1. Atomic Emission Spectrometry

It is quite outside the scope of this book to treat AES with its various excitation sources (sparks, arcs, flames, plasma, glow discharge, etc.) in detail. Likewise, it is not possible to discuss the various systems, from classical high-resolution spectrographs with photographic plates as detectors, to polychromators and rapid sequential multielement instruments, nor to handle the various monochromators. Instead, three techniques will be picked out because they have similarities to AAS techniques, so that a direct comparison can be made. These techniques are flame emission, which can be performed with virtually all atomic absorption spectrometers, sequential inductively coupled plasma atomic emission spectrometry (ICP-AES), and graphite furnace atomic emission spectrometry.

9.1.1. Flame Atomic Emission Spectrometry

Much has been written against flame AES compared to flame AAS that has later proved untenable. Thus equation (1.7) (page 7), which represents the dependence of the ratio of excited atoms to non-excited atoms upon the energy of excitation and the absolute temperature, has been used to demonstrate the lower sensitivity of flame AES. It is correct that the energy of excitation is inversely proportional to the wavelength (equation 1.4, page 2), and the proportion of excited atoms, i. e. the ratio N_j/N_0 (equation 1.7), decreases exponentially with increasing energy of excitation. Thus the ratio of excited to non-excited atoms is more unfavorable in the short wavelength range of the spectrum than in the long wavelength range (compare figure 4, page 8). However, it is not correct to state that AAS is always more sensitive than AES because the number of excited atoms is always smaller than the number of atoms in the ground state (see table 2, page 8). Atomic absorption spectrometry is only dependent upon the number of atoms N_0 in the ground state; for flame AES, as well as the number of atoms N_j in an excited state, the lifetime of these atoms is important. Experience has shown that AAS is only more sensitive than flame AES when the excitation potential is greater than 3.5 eV; with lower potentials flame AES is usually more sensitive.

The detection limits that can be achieved with various techniques of atomic emission spectrometry are compiled in table 18 for comparison. It can be seen that flame AAS and flame AES are, if anything, complementary rather than competitive. Both methods should in fact be considered in this way, especially since the majority of atomic absorption spectrometers nowadays permit both methods to be performed. Flame AES is no longer an independent instrumental technique, except perhaps for the determination of sodium and potassium, and occasionally calcium or lithium, in biological fluids using flame photometers.

PICKETT and KOIRTYOHANN [987] found that AAS and AES together are much better and more universally applicable than the majority of other analytical techniques. Like other authors [351], they found that with the use of hot, premixed flames, such as the nitrous oxide/acetylene flame, emission is just as free of interferences as absorption.

A further censur derived from equation (1.7) is the high temperature dependence of flame AES compared to flame AAS; small temperature changes in the flame cause a considerable alteration in the number of excited atoms, while the number of atoms in the ground state remains relatively unaltered. It is true that there is an exponential dependence of the number of excited atoms on the temperature, but false conclusions should not be drawn from this. Temperature changes do not occur in the flame unless someone alters the stoichiometry. If the stoichiometry is altered, the chemical properties are also changed and these can have a strong influence on the production of free atoms; flame AAS and flame AFS are affected to the same degree as flame AES by such alterations.

It is occasionally considered an advantage of flame AES that no additional sources (hollow cathode lamp, etc.) are required. For elements that must only be very occasionally determined, this is certainly a financial factor that can be of importance from case to case. Generally however, the specificity of atomic absorption spectrometry is worth the price of a hollow cathode lamp. The most important difference between the two techniques is still the frequent occurrence of spectral interferences due to overlapping of spectral lines in flame AES which are practically unknown in AAS [176].

Such spectral interferences occur in emission when the monochromator is unable to separate the emission lines of two different elements and the determination is thus not element specific. This is essentially a monochromator problem when genuine overlapping of two lines does not take place [176], and many interferences of this type disappear when monochromators having a spectral bandwidth of 0.03 nm, or better, are employed. Naturally such a monochromator must have a correspondingly good reciprocal linear dispersion of at least 2 nm/mm to enable such narrow bandpasses to be used satisfactorily. With such slit widths it is often more meaningful to scan a section of the spectrum with a motorized wavelength drive rather than to carry out the determination at a fixed wavelength, since problems with drift, etc. can otherwise occur. BARNETT et al [78] reported the excellent results they obtained for emission measurements using a commercial atomic absorption spectrometer.

Table 18. Typical detection limits (in µg/L) attainable with various techniques of atomic emission spectrometry

Element	Flame AES	Graphite Furnace AES	ICP-AES
Ag	20	0.45	
Al	10	1	20
As	50 000		50
Au	500	160	
B	30 000	200	4
Ba	1	4	0.5
Be	40 000	460	
Bi	40 000	30	
Ca	0.1		
Cd	2000	50	4
Co	50	10	6
Cr	5	1	5
Cs	8	18	
Cu	10	2	3
Fe	50	7	3
In	5	0.65	
Ir	100 000	860	
K	3	0.0015	
Li	0.03	0.07	
Mg	5	1	
Mn	5	1.5	1
Mo	100	16	8
Na	0.1	0.0025	
Ni	30	15	10
P			50
Pb	200	27	
Pd	50	60	
Rb	0.3	0.1	
Si	5000	90	
Sn	300	15	30
Sr	0.1	1	
Ta	18 000		20
Ti	200	17	2
Tl	20	1	
U	10 000	2500	50
V	10	9	5
W	500		40
Zn	50 000	1500	2
Zr	3000		4

A second disadvantage of flame AES, apart from the possible occurrence of line overlap, is the occurrence of background emission from the flame or from the sample matrix. This effect can influence the detection limit in emission considerably more in the presence of a complex matrix than in AAS. The detection limits quoted in table 18 are valid for pure aqueous solutions and can be noticeably poorer in the presence of complex or concentrated matrices.

9.1.2. ICP Atomic Emission Spectrometry

At the end of the 1970s, the interest in AES concentrated increasingly toward inductively coupled plasma (ICP) as the excitation source. Regrettably, premature comparisons were made with AAS and the rapid decline of this technique was prophesied. These comparisons were to a large extent based on simplifications and only took flame AAS into consideration. When merely detection limits are compared, flame AAS has virtually no advantages over ICP-AES, as scrutiny of table 18 quickly reveals. In the following section, it will be attempted to make a realistic comparison between sequential ICP-AES and the various techniques of AAS.

Quite generally, a plasma is a "luminous gas mixture". The luminescence is caused by the practical ionization of gas molecules or atoms, which upon recombination release the energy taken up in the form of radiation. A plasma is thus a gas whose atoms or molecules are dissociated to a certain degree into positive (ions) and negative (electrons) charge carriers, so that as well as neutral particles, mobile electrically charged particles are also present.

In the case of an ICP, ionization of the gas takes place in a quartz tube inside the induction coil of an RF generator. Argon is generally used as the plasma gas since it is relatively easy to ionize. The plasma torch comprises three concentric quartz tubes (figure 128). The nebulized sample is transported by an argon stream through the central tube to the plasma. Auxiliary argon streams at a relatively low flowrate through the middle tube, while the plasma argon streams through the outer tube. The special "tulip" shape of the middle tube causes a dynamic pressure in the plasma argon so that it streams with high velocity along the inside wall of the outer tube. This configuration causes the argon atoms in the central turbulent zone to be ionized in the induction field, while the fast gas stream along the outer tube wall acts as coolant. However, this principle only functions up to an RF power level of about 2 kW. At higher power levels the fast gas stream is also ionized so that the cooling effect is lost and the quartz tube begins to melt.

A particular characteristic of the ICP is that the plasma is circular. The sample aerosol transported by the carrier gas through the central quartz tube thus enters the plasma axially. Resulting from the relatively high acceleration of the carrier gas a tunnel is formed in the center of the plasma in which the temperature is 6000 K to 8000 K. This high temperature, together with the relatively long residence time of the sample inside the plasma, is decisive for the efficiency of energy transfer from the ionized gas to the sample and thus for its atomization and excitation. The ICP differs quite distinctly from other plasmas, such as the microwave induced plasma (MIP), where the sample does not enter the center of the plasma but only resides at the surface.

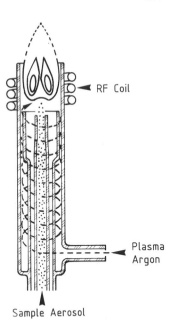

Figure 128. Cross-section of a plasma torch for ICP-AES.

The high temperature and the long residence time of the sample in the plasma, together with the virtually inert chemical environment, guarantee a high degree of atomization and freedom from solute-volatilization and vapor-phase interferences. Even very stable compounds are broken down by the high temperature and dissociated into atoms. It should be noted that this is due to the atomizer (ICP torch compared to a conventional flame) and has nothing to do with the difference between absorption and emission.

As well as temperature and residence time, the chemical environment is also of significance. In all flames there is a relatively high partial pressure of oxygen or oxygen compounds. This means that for elements having a high affinity for oxygen the degree of atomization is not as high as would be expected at the prevailing temperature. These elements (alkaline-earths, rare earths, boron, silicon, etc.) form oxide or hydroxide radicals with the flame gases which cannot be further dissociated. A plasma, in contrast, consists initially only of the inert gas argon; oxygen is only produced in small quantity from the dissociation of water in the sample. There is thus only a very low partial pressure of oxygen and the formation of oxides is very unlikely. Consequently, elements with a high affinity for oxygen are effectively atomized in an ICP.

A further advantage of the ICP is the very high density of electrons prevailing in the plasma. In flames and other atomizers at high temperature the sample atoms are ionized to varying degrees; the higher the temperature, the greater the degree of ionization. Since ionized atoms absorb and emit at wavelengths other than neutral atoms, ionization is a genuine interference which must be eliminated by the addition of suitable ionization buffers. As a result of the high electron density in the plasma, ionization is virtually completely suppressed even at the high prevailing temperature. Solely the alkali elements are an exception and relatively poor detection limits are attained with ICP-AES.

In absorption the sample must only be atomized, but in emission additional energy is required to raise the atoms to an excited state. A temperature of 3000 K or less is usually adequate to obtain a good degree of atomization. In emission, such temperatures are only sufficient to excite elements whose wavelengths lie in the visible and near UV ranges. Elements in the far UV require much higher excitation temperatures. With a temperature of around 8000 K, the ICP offers good conditions for intense emission over the entire spectral range.

With increasing temperature, however, the number of emission lines increases, since through the higher energy more of the excited states will be populated. An absorption spectrum, in contrast, has only few lines because under the conditions normally prevailing the spectral lines originate from the ground state. An absorption spectrum is therefore comparitively clear. The selection of the analytical line (i.e. the spectral line at which the measurement is performed) in AAS is via the element-specific radiation source and the modulation principle. The analyte radiation is thus "marked" while foreign radiation is not modulated and consequently ignored by the selective amplifier. The monochromator in AAS has the sole task of separating several absorption lines of the analyte element from each other; emission lines from other elements do not interfere. A spectral bandpass of 0.2 nm to 2 nm is normally sufficient.

In emission, the monochromator has the task of separating the analytical line from *all* foreign radiation. Otherwise, not only the radiation from the analyte element, but the sum of all radiation passing the exit slit and falling on the detector will be measured. Thus, in general, the quality of an emission spectrometer is directly proportional to the resolution of the monochromator.

The narrower slits required for higher resolution mean that less radiant energy reaches the detector and that the risk of drift is greater. For these reasons there is a certain limiting value for routine instruments; nowadays a resolution of 0.02–0.03 nm is considered as a practicable value.

The advantage of AAS is its high specificity, which makes it a reliable, easy-to-use analytical method. On the other hand, AAS is a single-element method. It is necessary to first determine one element in a series of samples, then change and optimize the parameters for the next element and repeat the series, and so on. Atomic emission spectrometry offers the advantage of simultaneous or sequential multielement determinations. Since all elements in the sample emit their characteristic radiation simultaneously, it is only necessary to make the requisite optical and electronic provisions to allow determination of the analyte elements simultaneously or sequentially. The principle disadvantage of AES is that the respective emission lines must be separated using very narrow slits and that the risk of false measurements due to background radiation and spectral interferences is present.

Simultaneous multielement atomic emission spectrometers have been successfully used for many years, particularly in metallurgical laboratories. In these instruments the emission source and dispersive element (grating) are permanently fixed. A large number of exit slits are also fixed in a circle in positions for selected emission lines. A photomultiplier with electrical measuring system, data handling equipment, etc. is mounted behind each slit.

The distinct advantage of these largely fully automatic spectrometers is the speed with which a large number of elements can be determined simultaneously in a single sample. The time required for a full analysis from sampling to readout is about two minutes. The disadvantage of such a system is its fixed configuration, which makes it inflexible. There is a limit to the proximity with which the secondary slits can be mounted, so that not all spectral lines lying close together can be selected (the minimum spectral separation between two slits is about 1–2 nm). Subsequent alteration of the slit positions is hardly possible, or only at great expense and effort. The analytical laboratory must have prior knowledge of the required task, including the number of elements to be determined, the matrix and concomitant elements, possible line overlapping, the required and attainable determination limits in the matrix, etc. In other words, a simultaneous multielement spectrometer is usually a dedicated instrument designed for a specific analytical task and can only be used for this task.

The situation is quite different for a sequential multielement spectrometer where the monochromator is driven by a stepping motor, for example. The elements are determined one after the other, the grating being slewed to the next wavelength at each change of element. Such a system offers complete flexibility in the choice of lines chosen for element and background emission measurements. Sequential ICP emission measurements can be used for quantitative and qualitative trace analyses in a similar manner to flame AAS. However, ICP-AES generally offers lower detection limits and a greater linear working range for most elements. Especially for elements having a high affinity for oxygen or other components in the flame, ICP-AES provides results that are several orders of magnitude better.

Flame AAS is clearly faster than ICP-AES when only one element per sample solution is to be determined, and the achievable precision is also better. As soon as more than two or three elements are to be determined, ICP-AES is definitely faster. Sequential spectrometers are offered that can determine 15 elements per minute with a precision of a few percent.

If the graphite furnace, hydride, or cold vapor techniques are considered, then ICP-AES is about as fast for the determination of one element per sample solution, but the advantage of better detection limits for most elements is no longer valid. These AAS techniques are often superior to ICP-AES by more than two powers of ten, so that their significance for trace analysis remains unchallenged.

The widespread occurrence of spectral interferences in emission due to line overlap is a decisive difference between AAS and AES. Since a sequential multielement spectrometer using a stepping motor driven monochromator offers good prerequisites for correction of these influences, the associated problems will be treated in some detail. An exact knowledge of the radiation background and of emission lines of concomitants in the neighborhood of the analytical line is essential to achieve satisfactory correction and thus accurate results.

Spectral interference is a general term used to cover various forms of overlapping of lines from the matrix and the concomitant elements with the analytical line of the analyte element. This overlapping can take the form:
– Direct coincidence of lines;
– Overlapping of very close-lying lines;
– Overlapping of broadened lines;
– Continuous radiation.

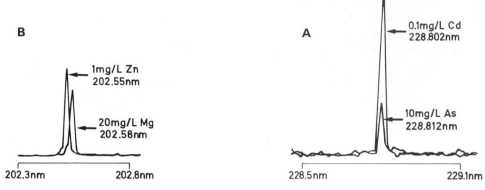

Figure 129. Spectral interferences in AES due to direct overlapping of two close-lying spectral lines of different elements.

The first two cases are examples of the classical form of spectral interferences; examples are presented in figure 129. Both forms only differ in principle by the separation of lines from each other, since there are virtually no cases known where two lines coincide exactly. With a very high resolution monochromator (0.01 nm), the lines of Mg and Zn (example B) and also the even more closely lying lines of As and Cd (example A) can be separated. The extent of these spectral interferences is thus a function of the monochromator.

The second form of spectral interference is caused by strong line broadening or continuous radiation. Intensely emitting elements such as calcium, magnesium and aluminium exhibit substantial line broadening when they are present in the sample in higher concentration. This leads to overlapping, even though the lines are so far separated that they should be resolved by the monochromator (figure 130). Occasionally, these elements exhibit a spectral

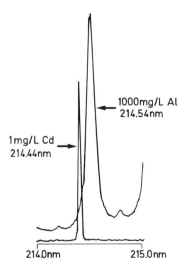

Figure 130. Spectral interferences in AES caused by overlapping of a strongly broadened line of a concomitant element present in high concentration.

continuum in the far UV range which naturally cannot be separated by even the highest resolution monochromator (figure 131).

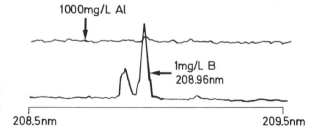

Figure 131. Spectral interference in AES due to superposition of the spectral continuum of a concomitant element.

Overlapping of non-broadened, closely lying lines can be prevented by the use of a high resolution monochromator. Such instruments, however, place very high requirements on the working location and show a strong tendency toward drift, etc. For practical reasons, monochromators with a resolution of 0.02–0.03 nm are chosen, and these have proved worthwhile. Several possibilities are available for the measurement and for the correction of spectral interferences:
– Measurement of a matrix solution;
– Selection of an alternate line;
– Oscillating quartz refraction plate;
– Computer program for the stepwise measurement and subtraction of the background.

The ideal and most reliable method for recognizing and eliminating spectral interferences is to measure a matrix solution (zero member compensation solution). This is a solution containing the entire matrix and concomitant elements in the same concentrations as in the sample, but not the analyte element itself. Since such matrix solutions are not available in the majority of cases, this method of correction is rarely used.

Once the presence of spectral interferences has been recognized, an attempt should always be made to see whether the analyte element can be determined free of interferences at another emission line. Every spectral interference, even when measures for correction have been taken, can adversely influence both the precision and the accuracy of the determination. It can thus be of advantage to change to a less sensitive line when no spectral interferences occur at this line. Every determination should therefore be optimized with respect to the choice of analytical line.

The use of a matrix solution permits spectral interferences to be measured exactly at the analytical line, thus allowing their reliable elimination. When this is not possible, it should be attempted to measure the background as close to the analytical line as possible, to achieve good correction. One method of accomplishing this is to use an oscillating or rotatable quartz refraction plate. A quartz plate is mounted vertically in the optical beam in front of the monochromator exit slit. When the plate is turned at an angle to the optical beam, there is a slight shift of wavelength. It is thus possible to measure the background very closely to either side of the analytical line without having to alter the monochromator setting. This technique can be used successfully when the background to either side of the analytical line is clearly defined. A further, more universal method is a computer program for the stepwise measurement and subtraction of the background. A computer-controlled stepping motor drive for the grating offers the possibility of freely selecting the wavelength to either side of the analytical line for the measurement of the background.

A knowledge of the spectral environment of the analytical line is also necessary for this correction technique. But this knowledge is in any case necessary if correct results are to be obtained by ICP-AES for changing matrices. If the spectral environment is known, then a programmable stepping motor drive offers the most reliable and elegant possibility for the correction of various spectral interferences.

From the deliberations made thus far it is clear that the answer to the question "ICP-AES or AAS?" must be answered that both techniques have their respective fields of application. ICP-AES extends atomic spectrometry as an analytical method quite considerably. But ICP-AES should be seen as a complement to AAS and not as an alternative to replace or supersede the latter.

ICP-AES exhibits excellent detection limits for a number of refractory elements, such as B, Ta, Ti, U, W etc., that can only be determined in the flame with lower sensitivity. Penetration into the start of the vacuum UV range with ICP-AES is also relatively easy, so that the determination of the non-metals phosphorus and sulfur with detection limits of better than 1 mg/L is possible.

As with other emission techniques, a distinct advantage of ICP-AES is the possibility of performing multielement determinations. The rapid, simultaneous multielement instruments cannot be included in this comparison because their design restricts them to dedicated applications. Flexible, sequential multielement instruments, on the other hand, are faster than AAS when more than 2–3 elements per sample are to be measured.

Because of the high temperature and the inert environment, ICP-AES exhibits hardly any matrix dependence. Both solute-volatilization and vapor-phase interferences are extremely seldom, so that complex samples can be analyzed without difficulty. The only interferences that can cause considerable difficulties in ICP-AES are spectral interferences. With markedly varying sample compositions, spectral interferences can cause problems and cost the analyst a good deal of time in their elimination. Nevertheless, most spectral interferences either can be avoided by changing to another spectral line, or can be corrected with a suitable program. An astute analyst will recognize spectral interferences and be able to obtain accurate results.

As already mentioned on a number of occasions, AAS is characterized by its unparalleled selectivity and specificity, and these make it a facile spectrometric method. Moreover, it has a very favorable price-to-performance ratio, so that the purchase of an atomic absorption spectrometer is worthwhile even for small laboratories. Flame AAS is still unsurpassed in its speed and precision for the determination of one, or a few, elements per sample and a high sample throughput. This is largely due to the excellence of the nebulizer/burner system in use. A stable signal is obtained one to two seconds after aspiration is started and a measurement time of only a few seconds is sufficient to attain a precision of 0.2% or better.

It is true that the temperatures of the flames generally used are not always adequate to atomize the nebulized sample completely but this seldom leads to serious interferences. One difficulty is the low sensitivity with which elements such as boron, tantalum, tungsten or uranium can be determined. A flame temperature that is too low is also responsible for the occurrence of solute-volatilization interferences, such as the interference of phosphate on the determination of calcium. Nevertheless, such interferences can generally be eliminated fairly easily; in the above case by using the hotter nitrous oxide/acetylene flame, or chemically by the addition of another cation (barium or lanthanum as the chloride) that binds more strongly to the interfering anion (phosphate) than calcium. In flame AAS the sample matrix is only of subordinate significance and a change between various sample types is mostly without problems. In fact, this is one of the major advantages of flame AAS, since even for very complex samples reliable and accurate results can be achieved without too much effort. There is virtually no other analytical technique that can be adapted so rapidly to changing sample types.

The graphite furnace technique offers the best detection limits for the majority of elements. It is superior to ICP-AES mostly by two orders of magnitude and to flame AAS by more than three. Graphite furnace AAS is thus the technique of choice when trace to nanotrace analysis is required.

Graphite furnace AAS is not as free of interferences as flame AAS. Some of the problems arise from the fact that work is performed in a concentration range in which false measurements due to adsorption on the walls of vessels or contamination of the reagents by the analyte can reach substantial proportions. However, the majority of volatilization and vapor-phase interferences mentioned in the literature appear to be eliminated when a stabilized temperature furnace with L'vov platform and suitable matrix modification additives are used. The background attenuation frequently observed with this technique can be well corrected by utilizing Zeeman-effect background correction. This technique is definitely more reliable and universal than all methods for the elimination of spectral interferences in AES.

Independent of these possibilities of interference, the graphite furnace technique is frequently the only technique with which trace or nanotrace determinations can be made without prior preconcentration of the analyte element. It should also be borne in mind that the graphite furnace technique benefits from the selectivity and specificity of AAS, which can frequently be of considerable significance in the nanotrace range.

Further, graphite furnace AAS only requires extremely small sample volumes, thus permitting true microanalysis, even for trace elements. Also, solid samples can be analyzed directly in the graphite furnace.

Finally the hydride-generation and cold vapor techniques should be mentioned since they offer better detection limits than graphite furnace AAS with lower susceptibility to interferences. These techniques are optimum when low concentrations of antimony, arsenic, bismuth, selenium, tellurium or mercury are to be determined.

Since each of the techniques described has its own specific advantages and disadvantages, it is certainly possible to select the technique that is optimum for a given, limited field of application. In many laboratories the elements to be determined and the types of sample are constantly changing, so that a commitment to one technique would be a serious limitation, frequently demanding compromises. The advanced laboratory would then endeavor to have all spectrometric techniques of elemental analysis available, provided that this is financially justifiable. Only then is it possible to select the optimum technique for a given analytical problem, and in cases of doubt a further technique would be available to independently check the accuracy of the results.

9.1.3. Graphite Furnace AES

In 1975, OTTAWAY and SHAW [2718] examined the possibility of using a graphite furnace for emission measurements. They expected that problems due to background radiation of species such as C_2, CH, CN and OH would be far fewer than in a flame, even though a graphite furnace emits an intense continuum. In the first experiments, these authors found detection limits for sodium and potassium that were one and two orders of magnitude, respectively, better than in absorption. EPSTEIN et al [2229] used wavelength modulation with an oscillating quartz refraction plate for background correction in graphite furnace AES. They were thereby able to improve the detection limits for several elements by one to two powers of ten. They also found that for emission measurements barium was ionized by up to 35%, but that ionization could be supressed by addition of potassium.

ALDER et al [2023] observed the emission lines of 20 elements and concluded that a thermal excitation mechanism was the most probable. They obtained detection limits of 5 pg for aluminium and 140 pg for calcium and molybdenum. The authors found that the emission from the tube walls caused considerable interferences, especially in the visible range. LITTLEJOHN and OTTAWAY [2553] showed that with an optimized optical configuration and careful alignment of the graphite furnace, these interferences could be kept to a minimum. At all costs, radiation from the tube wall must be prevented from entering the monochromator. Later OTTAWAY and SHAW [2719] modified the shape of the tube and were thereby able to attain a temperature 370°C higher than before; they were then able to obtain detection limits for refractory elements that were up to a power of ten better. These authors were able to show that there was an exponential dependence between the emission signal and the temperature. They assumed that the low excitation temperature and the pure thermal character of the excitation were principal advantages compared to other excitation sources in AES. They reported very low radiation background, only weak ionization, and substantially fewer spectral interferences.

In later investigations, LITTLEJOHN and OTTAWAY [2555] [2558] modified the tube further and found that the tube form has a major influence on the emission intensity and thereby on the attainable detection limits. If the wall temperature decreases toward the tube ends, the mean temperature of the atomic vapor is lower than the wall temperature. This leads to relatively poor detection limits for volatile elements. If the temperature gradient is inverted such that the tube ends heat up more rapidly and reach a higher temperature than the center of the tube, the gas temperature will be higher than the wall temperature. The emission intensity for many elements can thereby be improved. For elements that are difficult to atomize and excite, such as gadolinium, molybdenum or titanium, the best detection limits are attained in a graphite tube that is hotter in the center, however. An ultrafast heating rate for atomization brings a further improvement for refractory elements [2554].

In subsequent work, LITTLEJOHN and OTTAWAY [2556] found that for instruments without automatic correction of the background emission, optimization of the furnace temperature is necessary to obtain the best detection limits. For most elements, the optimum temperature is lower than the maximum attainable temperature, but only when the rate of heating is independent of the selected final temperature. If the heating rate is determined by the final temperature, the relationships become complex and an optimization of the temperature hardly brings a measurable effect.

In studies on the mechanism of excitation in a graphite furnace, LITTLEJOHN and OTTAWAY [2557] came to the conclusion that under practical analytical conditions and an interrupted gas stream through the graphite tube, a local thermal equilibrium was established. The graphite furnace is in consequence a unique emission source. Since the furnace is relatively cool compared to other excitation sources, favorable signal-to-noise ratios, good signal stability and a long residence time of the atoms result. Consequently, very good detection limits can be obtained for numerous elements (table 18). It is certain that further optimization of the graphite furnace and the spectrometer for emission measurements will bring an even greater improvement.

9.2. Atomic Fluorescence Spectrometry

The principles of atomic fluorescence spectrometry (AFS), like those of AAS, have been known for a long time, but the significance of AFS as an analytical method was first recognized by WINEFORDNER and VICKERS [1347] in 1964.

Atomic fluorescence is the sequel to and in principle the reverse of atomic absorption. Atomic fluorescence spectrometry is based on the absorption of optical radiation of suitable frequency (wavelength) by gaseous atoms and the resultant deactivation of the excited atoms with the release of radiation. The frequencies (wavelengths) emitted are characteristic of the atomic species.

A distinction is made between various types of atomic fluorescence, depending on whether the excitation line and the fluorescence line arise from the same or different energy levels, or terminate at the same or different energy levels [867] [1343] [2952]. It is further of importance when the energy levels differ whether the fluorescence line has more energy or less energy than the excitation line. The various types of fluorescence are depicted schematically in figure 132. In addition, OMENETTO and WINEFORDNER [2709] enumerate a number of seldomly met types of fluorescence.

The most common form is *resonance fluorescence* in which the excitation and fluorescence lines have the same upper and lower energy levels (A and B in figure 132). The excitation and fluorescence lines thus have the same frequency. The lines can originate from the ground state or from an excited state.

If the excitation line and the fluorescence line have the same upper energy level but different lower energy levels, we speak of *direct line fluorescence*. If the energy of excitation is greater than the fluorescence energy, the effect is termed STOKES *direct line fluorescence* (C and D in figure 132). The reverse case, when the fluorescence energy is greater, is termed *antistokes direct line fluorescence* (E and F in figure 132). All these lines can originate from the ground state or from an excited state.

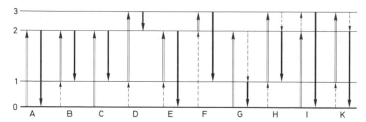

Figure 132. Schematic of the various possible fluorescence transitions between four energy levels. **A** and **B**: Resonance fluorescence; **C** and **D**: STOKES direct line fluorescence; **E** and **F**: Antistokes direct line fluorescence; **G** and **H**: STOKES stepwise line fluorescence; **I** and **K**: Antistokes stepwise line fluorescence.

If the excitation line and the fluorescence line have different upper energy levels, the effect is termed *stepwise line fluorescence,* and depending on the energy relationships, this is divided into *Stokes* (G and H) and *antistokes* (I and K) stepwise line fluorescence. If, after radiation excitation, further collisional excitation occurs (I in figure 132), a thermally assisted process takes place [258] [939] [940].

The intensity of atomic fluorescence, and thereby the achievable detection limit, is dependent upon the number of atoms in the ground state and the intensity of the incident radiation source. For this reason, research into atomic fluorescence is closely connected with the examination of newer and more intense radiation sources [252] [304] [305] [763] [818] [1264] [1342] [1345] [1346], especially electrodeless discharge lamps [171] [225] [258] [377], and also lasers [937] [938].

The fluorescence signal, as in atomic absorption, is usually differentiated from an emission signal (thermally produced) by employing alternating light. The radiation source is modulated at a given frequency and the amplifier is tuned to the same frequency. The observation of the fluorescence generally takes place at right angles to the incident beam. One difficulty is to distinguish between genuine fluorescence and stray radiation; if radiation from the radiation source is scattered on particles in the flame, this is also modulated and leads to a false signal just like background attenuation in atomic absorption. RAINS *et al* [1013] therefore suggested a correction system for atomic fluorescence similar to that used in atomic absorption. By alternate irradiation with the continuum of a xenon arc lamp and a spectral source, the interferences due to stray radiation are eliminated. For the reasons given above, a great deal of work has been carried out on burners and flames in connection with fluorescence [32] [159] [160] [161] [252] [579] [838] [1156]. The considerable influence of the flame on fluorescence analyses has been emphasized by various authors [578] [837]; attention must be paid to more than just the highest possible degree of atomization and the lowest possible formation of compounds through the chemical composition of the flame. The fluorescence intensity can be strongly reduced in a flame if atoms excited by radiation energy lose their energy through non-radiative collisions (quenching). These non-radiative collisions are strongly dependent upon the composition of the flame. For example, nitrogen exhibits a much stronger quenching effect than argon, and cooler flames a lesser effect than hotter ones.

An advantage of atomic fluorescence spectrometry over atomic absorption spectrometry is that the lowest detection limits can be more easily achieved (at least in theory) by increasing the radiant intensity of the radiation source. Also a slight instability of the radiation source should not affect the detection limit to the same degree as in AAS. In atomic fluorescence spectrometry the detection limit is the technically simple measurement of a very small radiation signal (similar to AES), while in AAS it is the detection of very small differences between two intense radiation signals. With AFS a fluctuation in the radiation source is conveyed proportionally to the signal, while in AAS it appears in its full magnitude in the detection limit and is therefore frequently the limiting factor for a determination.

While the width of the line of excitation from the radiation source can enter directly into the sensitivity of a determination in atomic absorption spectrometry (see page 19 *et seq*) and the presence of non-absorbable radiation from the radiation source within the observed spectral bandpass can lead to considerable curvature of the analytical curve (see page 83 *et seq*), this is not possible in atomic fluorescence spectrometry. In AAS, the radiation from the radiation source (after spectral dispersion) falls directly onto the detector; in AFS on the other hand, measurement is at right angles to the incident radiation. Thus it is possible to work with continuum sources and without extremely small slit widths [243] [1238] since the non-absorbable radiation does not fall on the detector.

In principle, it is even possible to work without a monochromator, as suggested initially by JENKINS [577]. By using an element specific radiation source and a selective amplifier, the fluorescence of the atoms in a flame was directly measured by a photomultiplier. Nevertheless, it was necessary to use a filter to reduce the strong direct radiation from the flame which caused higher noise in the detector signal. VICKERS and VAUGHT [1268] used a "solar blind" detector which was only sensitive between 160 nm and 320 nm. LARKINS and WILLIS [733] [734] used a similar non-dispersive system, with good sucess, for their measurements and obtained detection limits that were up to an order of magnitude better than with a monochromator. Other authors [2047] [2687] followed this example and also obtained good results.

Since the radiation from the source does not fall directly on the detector in AFS, analytical curves that are linear over three to five orders of magnitude are obtained, while in AAS they are frequently only linear over two orders of magnitude. Nevertheless, the profiles of analytical curves in AFS are not always as simple as would appear here. Self-absorption and other effects can occasionally have a considerable influence on the profile of the analytical curves [76] [935] [1375].

Although atomic fluorescence spectrometry appears to have a number of advantages over AAS and AES, it has nevertheless not found recognition as a general analytical method. AFS has similar selectivity and specificity to AAS, and with comparable atomizers permits substantially better detection limits to be attained. It would appear that this analytical technique cannot tolerate such complex matrices as AAS. Furthermore, hot flames, which eliminate many interferences and make the determination of many elements possible, bring certain problems to AFS, such as background emission and quenching [76].

On the other hand, the ability to use continuum sources and to work without a dispersive element are attractive arguments. And AFS is also suitable for multielement determinations [2991]. Nevertheless, OMENETTO and WINEFORDNER [2709] are of the opinion that AFS will not assert itself over the established spectrometric techniques (AAS and AES). For special applications, however, it can be a very useful technique.

For the interested reader the following publications are highly recommended: WINEFORDNER and MANSFIELD [1343], WEST [1309], BROWNER [2113], OMENETTO and WINEFORDNER [2709], and an excellent monograph by SYCHRA, SVOBODA and RUBEŠKA [2952].

10. The Individual Elements

The most important characteristics of the elements determinable by AAS will be presented. The various atomization techniques, possible interferences and their avoidance or elimination, and characteristic concentrations and attainable detection limits will be discussed. A series of applications manuals published by the PERKIN-ELMER CORPORATION [1] [2035] [2223] serves as the basis for the information. This basic information has been substantially extended.

10.1. Aluminium

Prior to 1966, a number of authors attempted to determine aluminium in oxygen/acetylene flames of varying stoichiometry [223] [310] [791] [1146]; organic solvents were frequently used to increase the sensitivity. A completely satisfactory determination of aluminium was first possible after the introduction of the nitrous oxide/acetylene flame by WILLIS [1333]. Practically no interferences are present in this flame; RAMAKRISHNA [1016] reported that acetic acid increased the absorption of aluminium by about 10% and the presence of titanium is said to increase the absorption by around 25%. According to WEST et al [1306] this is due to a reduction of the lateral diffusion and thereby to an increased atom concentration in the middle of the flame. Silicon on the other hand has a slightly depressive influence on the absorption of aluminium [238] [365].

The characteristic concentration for aluminium in a nitrous oxide/acetylene flame is approximately 1 mg/L 1% at the 309.3 nm line, and the detection limit is about 0.03 mg/L for an aqueous solution. Comparitive values for other resonance lines are compiled in table 19.

Since aluminium is ionized by about 10% in a nitrous oxide/acetylene flame, about 0.1% potassium (as chloride) or another easily ionized metal should be added to sample and reference solutions.

Aluminium can be complexed with oxine at pH 8 and extracted with MIBK. The extraction must be executed within three minutes, otherwise the risk of co-precipitation exists [373].

For the determination of aluminium in the graphite furnace, argon should be used as purge gas [802], since aluminium forms a stable monocyanide in the presence of nitrogen [2576]

Table 19. Al resonance lines

Wavelength nm	Energy level (K)	Characteristic concentration mg/L 1%
309.3	112–32437	1.2
396.1	112–25348	2.0
308.2	0–32435	2.5
394.4	0–25348	4.0
257.5	112–38934	16

which leads to a substantial loss in sensitivity. PERSSON *et al* [2743] [2744] reported that Al_2O_3 volatilizes and is reduced in the vapor phase. They found that even small amounts of O_2, H_2, Cl_2, N_2 and S cause interferences during atomization; H_2 and Cl_2 also cause interferences during thermal pretreatment. The interferences were smallest when the highest possible pretreatment temperatures were used. The presence of H_2 lowers the pretreatment temperature markedly.

MANNING *et al* [2620] investigated the atomization of aluminium from the L'VOV platform in a stabilized temperature furnace. They used 50 µg magnesium nitrate as a matrix modifier and were thereby able to use a thermal pretreatment temperature of 1700°C [2877].

JULSHAMN [2441] reported the total suppression of the aluminium signal by 1 M perchloric acid in an uncoated graphite tube. KOIRTYOHANN *et al* [2481] studied the mechanism of this interference and found that the signal is depressed by 95% by 0.5 M perchloric acid. They established that the volatilization of aluminium is not influenced, but that aluminium is transported from the tube as a compound in the presence of perchloric acid. MANNING *et al* [2620] also found strong interferences through perchloric acid when graphite tubes with a poor-quality pyrolytic layer were used; with good quality pyrotubes, no interferences were observed up to 0.5 M perchloric acid. These authors also found that many of the interferences reported in the literature, such as by $CaCl_2$ or $CuCl_2$ [2637], could be drastically reduced or eliminated by using good quality pyrotubes, magnesium nitrate as matrix modifier, and atomization off the L'VOV platform in a stabilized temperature furnace.

The optimum atomization temperature for aluminium off the L'VOV platform is 2500°C when an ultrafast heating rate is used.

10.2. Antimony

Antimony can be determined in an air/acetylene flame virtually free of interferences. However, little has been published about this element, so that possibly not everything is known. MOSTYN and CUNNINGHAM [888] found it of advantage to match the matrix to avoid minor interferences. For the determination of antimony, three resonance lines of similar sensitivity are available as shown in table 20. For the majority of determinations the 217.6 nm line is employed. High iron [3083], copper [888] and lead [1337] concentrations can cause weak spectral interferences at this line. In the presence of these elements, and also with higher antimony concentrations, it is recommended to use the 231.1 nm line.

Table 20. Sb resonance lines

Wavelength nm	Energy level (K)	Characteristic concentration mg/L 1%
217.6	0–45 945	0.35
206.8	0–48 332	0.5
231.1	0–43 249	0.8

CHAMBERS and MCCLELLAN [2145] described the extraction of antimony with APDC in MIBK, followed by back extraction into the aqueous phase to increase the sensitivity for subsequent determination in the flame. SUBRAMANIAN and MERANGER [2941] used the same procedure and found that only Sb(III) could be extracted without problems; Sb(V) could only be extracted in 0.3–1.0 M HCl.

Very little is reported in the literature on the determination of antimony in the graphite furnace. KUNSELMAN and HUFF [2515] reported signal enhancement by hydrochloric acid, nitric acid and sulfuric acid. In contrast, FRECH [2259] observed that in the presence of oxygen (diffused into the tube), hydrochloric acid and nitric acid caused signal depression. With the exclusion of oxygen, antimony could be determined without interferences. FRECH also found that chromium, nickel and iron influence the determination, but the influences disappeared when a fast heating rate (> 900 °C/s) is used.

Using ammonium dichromate as matrix modifier, antimony can be thermally pretreated up to about 1000 °C without losses. The optimum atomization temperature off the L'VOV platform is 1800 °C.

Antimony can be determined with advantage by the hydride technique; the determination limit is 1.0 ng absolute or 0.02 µg/L. CHAN and VIJAN [2146] reported an interference due to iron which could be eliminated through the addition of potassium iodide. In the author's laboratory, iron in concentrations up to 0.2 g/L was found to interfere in the presence of 3% hydrochloric acid [3054]. ALDUAN *et al* [2028] eliminated an interference due to copper by adding KSCN.

The atomization signal for antimony in the hydride technique depends on the oxidation state. The signal for Sb(III) is nearly twice as high as that for Sb(V) [3052] (refer to figure 112, page 232); there is virtually no difference for peak area integration. A number of authors nevertheless determined antimony in the pentavalent state if this was present after digestion procedures [2232] [3054]. If the oxidation state is ambiguous, a prereduction should definitely be performed. Potassium iodide in hydrochloric acid solution is the most suitable reductant [2869].

YAMAMOTO *et al* [3084] reported that at pH 8 antimony(III) is selectively reduced to the hydride and that even a 100-fold excess of antimony(V) does not interfere.

10.3. Arsenic

The determination of this element is considerably influenced by the fact that its resonance lines are at 193.7 nm and 197.2 nm at the start of the vacuum UV range and that the available hollow cathode lamps are of rather low performance. Furthermore, since many photomultipliers only exhibit low sensitivity below 200 nm, and radiation losses owing to increased absorption of lenses and low reflectance of mirrors can occur, arsenic can only be determined satisfactorily in the best atomic absorption spectrometers. Correspondingly, little was initially published about this element. Only after the development of high performance electrodeless discharge lamps has the determination of arsenic become largely free of problems (figure 133). The characteristic concentration with these lamps in an air/acetylene flame is 0.5 mg/L 1% and the detection limit is 0.15 mg/L. With normal hollow cathode lamps, the corresponding values are 1.5 mg/L 1% and 2.3 mg/L, respectively.

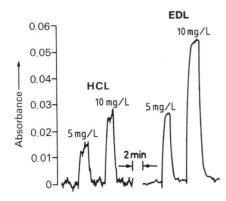

Figure 133. Determination of low arsenic concentrations employing a hollow cathode lamp (HCL) and an electrodeless discharge lamp (EDL) under identical analytical conditions and similar instrumentation in an air/acetylene flame.—A considerably better signal-to-noise ratio is obtained with the electrodeless discharge lamp [74].

Very few interferences can be observed for arsenic in an air/acetylene flame [1154]. The disadvantage of this flame is that it absorbs about 60% of the radiant energy emitted from the primary source; the air/hydrogen flame exhibits virtually the same absorption (see figure 15, page 32).

The use of an argon/hydrogen diffusion flame brings a considerable improvement in the sensitivity and detection limit for arsenic (table 21). The probability of interferences is higher in this flame, although nothing has been reported on this up to the present. Background effects are to be expected to a considerable degree for the determination of arsenic, especially with the argon/hydrogen diffusion flame, so that great caution is recommended for the analysis of complex materials.

Table 21. Comparison: Air/acetylene and argon/hydrogen flames for the As resonance line at 193.7 nm

	Air/Acetylene	Argon/Hydrogen
Absorption of the flame	60%	15%
Characteristic concentration (mg/L 1%)	0.5	0.15
Detection limit	0.15	0.02

KASZERMAN and THEURER [2458] reported that in an argon/hydrogen diffusion flame the sensitivity for arsenic(III) is two to three times better than for arsenic(V) and that the presence of nitric acid amplifies this effect. Significant differences between the two oxidation states are also said to exist in the nitrous oxide/acetylene flame.

CHAMBERS and McCLELLAN [2145] extracted arsenic with APDC in chloroform, followed by back extraction into the aqueous phase to increase the sensitivity for subsequent determination in the flame. SUBRAMANIAN and MERANGER [2941] found that with APDC as the complexing agent only arsenic(III) could be extracted into MIBK.

For the determination of arsenic in the graphite furnace, EDIGER [2217] proposed nickel as matrix modifier. Thermally stable nickel arsenide is formed, permitting thermal pretreatment up to about 1400°C. Various interferences, for example from different acids [2515], can be virtually completely eliminated.

KOREČKOVÁ *et al* [2494] investigated the various factors influencing the atomization of arsenic. They observed that the signal was much broader when using uncoated graphite tubes and concluded that this was due to the formation of a compound between graphite and arsenic. The signal enhancement caused by nitric acid and the signal depression brought about by phosphoric acid can both be explained via the active participation of graphite. Nickel binds arsenic independently of the graphite material, which is why the interferences disappear. Optimum conditions are obtained by atomization off the L'vov platform in a stabilized temperature furnace, with an atomization temperature of 2300°C.

POLDOSKI [2752] reported a spectral interference by silicate in the determination of arsenic in river and lake waters. The interference is dependent upon the selected bandpass. It can be controlled by adding hydrofluoric acid and using the narrowest possible bandpass.

The hydride technique is excellently suited for the determination of arsenic; the determination limit is 1 ng absolute or 0.02 µg/L. Various reports have been published on interferences caused by elements of Groups VIII and IB. These influences depend strongly on the composition and concentration of the acid employed and on the dilution of the sample. For example, iron(II) only interferes in the determination of arsenic(III) at concentrations above 2 g/L when a mixture of 1.5% hydrochloric acid and 1.5% nitric acid is used [3054]. BERNDT *et al* [2084] found that the interference-free range for the determination of arsenic in the presence of lead could be extended by four orders of magnitude when 6 M hydrochloric acid instead of 0.5 M was used.

FLEMING and IDE [2254] observed that even relatively low concentrations of nickel interfere, but that this interference was not apparent in the analysis of steels. During extensive investigations in the present author's laboratory [3049] [3058] it was found that in the presence of sufficient iron and a suitable acid mixture (1.5% HCl/8% HNO_3) the interference-free range could be extended by three powers of ten, so that up to 0.1 g Ni/L had no influence. Similar measures could also be taken to eliminate the influences due to iron, copper and nickel [3060].

BÉDARD and KERBYSON [2069] eliminated the influence of copper through co-precipitation with lanthanum hydroxide; DORNEMANN and KLEIST [2205] eliminated the interferences due to cobalt, copper and nickel through the addition of pyridine-2-aldoxime; and GUIMONT *et al* [2328] added potassium thiocyanate to eliminate interferences due to nickel. Many other authors [2751] [2809] [2857] [2893] found that dilution of the sample was sufficient to eliminate any real problems caused by the influence of these interfering elements.

Low concentrations of selenium are sufficient to interfere in the determination of arsenic [3055]. This is a gas-phase interference in the heated quartz tube. The interference can be eliminated by preventing selenium hydride from entering the atomizer. This can be achieved, for example, by adding a suitable quantity of copper to the analysis solution to prevent the formation of selenium hydride.

The two oxidation states of arsenic exhibit slightly different sensitivities. Potassium iodide [2168] is generally used for reduction of the higher oxidation state, occasionally under the addition of ascorbic acid [2858]. MAY and GREENLAND [2643] found that interferences could thereby be eliminated. HINNERS [2374] established that in the presence of sufficient borohydride there was no difference between the oxidation states.

Various authors attempted to determine the oxidation states of arsenic separately. Thereby, the fact that arsenic(III) can be more easily reduced at pH 4 to 5, while arsenic(V) requires a stronger acid medium, was utilized [2011] [2480] [2688] [2844]. ANDREAE [2038] and also BRAMAN et al [2101] determined various methylarsenic compounds by selective reduction and volatilization.

10.4. Barium

Barium exhibits numerous chemical interferences in an air/acetylene flame [635] which are strongly reduced or completely eliminated in a nitrous oxide/acetylene flame [2603]. Also in a hot flame, the strong background absorption caused by CaOH bands in the presence of a calcium matrix, which BILLINGS [123] observed in an air/acetylene flame, disappears (see figure 93, page 171). The strong emission of the nitrous oxide/acetylene flame near the Ba line, which brings a considerable contribution to the noise, can occasionally have an interfering influence (see figure 134). Increasing the lamp current and reducing the monochromator slit can be a remedy.

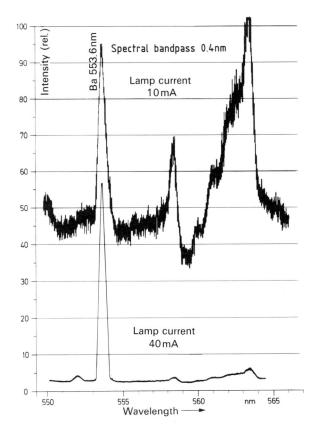

Figure 134. Influence of the lamp current on the determination of barium.—The upper spectrum shows the high portion of emission radiation emitted by a nitrous oxide/acetylene flame when a low intensity hollow cathode lamp is employed. The lower spectrum was scanned under identical conditions with an Intensitron® high intensity hollow cathode lamp. For both spectra, zero balance was carried out before the flame was ignited.

RUBEŠKA [1066] found that barium could be determined without difficulties in a nitrous oxide/acetylene flame shielded by an inert gas, even in the presence of a large excess of calcium (CaOH emission).

Barium is ionized to 80–90% in a nitrous oxide/acetylene flame (see figure 96, page 183) so that it is necessary to add 0.2–0.5% potassium (as chloride) or another easily ionized metal to all sample and reference solutions. Barium has only one usable resonance line at 553.6 nm. The characteristic concentration in an air/acetylene flame is about 10 mg/L 1% and in the nitrous oxide/acetylene flame about 0.4 mg/L 1%. The detection limits in these flames are 0.7 and 0.01 mg/L, respectively. A ten times better detection limit can be obtained with flame emission spectrometry [635].

For the determination of barium by the graphite furnace technique, the direct emission from the graphite tube can interfere. An atomization temperature that is unnecessarily high should thus not be selected. A further problem can arise due to the fact that a deuterium lamp has very little energy at the wavelength of barium and consequently cannot be used for background correction. This problem can be effectively eliminated when the spectrometer uses either a halogen lamp or the Zeeman effect for background correction.

Barium forms a rather stable carbide and cannot therefore be satisfactorily determined in an uncoated graphite tube. BENSHAW [1031] found that the sensitivity for barium could be improved by a factor of 20 by lining the graphite tube with tantalum foil. The same result can be obtained by using pyrocoated graphite tubes.

In 0.2% nitric acid solution, a thermal pretreatment temperature of 1500°C can be applied. The optimum atomization temperature using an ultrafast heating rate is 2700°C.

10.5. Beryllium

Beryllium can only be very poorly determined in an air/acetylene flame, but exhibits excellent sensitivity with a characteristic concentration of approximately 0.03 mg/L 1% and a detection limit of about 0.002 mg/L in a nitrous oxide/acetylene flame. The most important resonance line is at 238.4 nm; up to the present, nothing has been reported on the use of other lines.

The determination of beryllium appears to be largely free of interferences in the nitrous oxide/acetylene flame. BOKOWSKI [137] reported a minor interference in the presence of high silicon and aluminium contents, but found no influences from phosphate or sulfate. RAMAKRISHNA [1016] eliminated the depressive effect of aluminium by adding about 1.5 g/L fluoride. He found a further increase in the signal of about 20% by adding acetic acid. Beryllium is not noticeably ionized in a nitrous oxide/acetylene flame, so the addition of alkali is not required.

NAKAHARA *et al* [903] also found the determination of beryllium in the nitrous oxide/acetylene flame to be quite free of interferences. For the analysis of alloys, palladium and silicon occasionally caused interferences. Phosphoric acid enhanced the signal somewhat, while hydrochloric and nitric acids depressed it.

A pyrocoated graphite tube is the most suitable for the determination of beryllium by the graphite furnace technique. THOMPSON *et al* [2965] found that the beryllium signal is enhanced by the addition of calcium. MAESSEN *et al* [2600] established that in the presence of aluminium nitrate, beryllium is not volatile until a temperature of 1600 °C, so that this reagent is a suitable matrix modifier. HURLBUT [2401] found few interferences in the determination of beryllium when lanthanum is added to the sample solution acidified with nitric acid and the graphite tubes were treated with lanthanum.

Using magnesium nitrate as matrix modifier, a thermal pretreatment temperature of 1500 °C can be applied. The optimum temperature for atomization off the L'VOV platform is 2400 °C.

10.6. Bismuth

Bismuth can be determined without interferences in an air/acetylene flame. The characteristic concentration at the 222.8 nm line is 0.2 mg/L 1% and the detection limit is about 0.02 mg/L. Various other resonance lines are compiled in table 22. An improved signal-to-noise ratio can be obtained in an air/hydrogen flame with a detection limit of about 0.015 mg/L.

Table 22. Bi resonance lines

Wavelength nm	Energy level (K)	Characteristic concentration (mg/L 1%)	Spectral bandpass nm
222.8	0–44865	0.2	0.07–0.2
306.8	0–32588	0.6	0.7
206.2	0–48489	1.6	0.7
227.7	0–43912	3	0.7

Bismuth can be determined with good sensitivity by the graphite furnace technique. Interferences are hardly to be expected for atomization off the L'VOV platform is a stabilized temperature furnace.

KANE [2454] determined bismuth in rocks after digestion in hydrofluoric acid/perchloric acid and extraction as the iodide in MIBK and back extraction into the aqueous phase with EDTA. Bismuth must be determined relatively frequently in metallurgical samples. BARNETT and MCLAUGHLIN [2060] analyzed iron, copper and zinc alloys after solution in nitric acid using the analyte addition technique. MARKS and WELCHER [2625] analyzed alloys used in the manufacture of gas turbines after solution in hydrofluoric acid/nitric acid. FORRESTER et al [2257] analyzed ultrapure nickel by dissolving it in ultrapure nitric acid and dispensing directly into the graphite tube. The analytical results were in excellent agreement with the certificate values for bismuth.

MARKS et al [2626] analyzed complex nickel alloys directly in the solid form as turnings without pretreatment and found little background attenuation since no metal salts are formed. HEADRIDGE and THOMPSON [2358] also determined bismuth in nickel alloys and LUNDBERG and FRECH [2572] in steel directly by solid sampling.

HAMNER *et al* [2344] determined bismuth in metallic chromium after a fusion. They found that nickel had a stabilizing effect and increased the sensitivity. GLADNEY [2302] investigated matrix modifiers and found that bismuth could be thermally stabilized up to 1200 °C through the addition of nickel. Without nickel, thermal pretreatment temperatures only up to 600 °C can be used. With ammonium dichromate as matrix modifier bismuth can be thermally treated up to 900 °C without losses; the optimum temperature for atomization off the L'vov platform is 1800 °C.

Bismuth can be determined with excellent sensitivity by the hydride technique. The determination limit is 1.0 ng absolute or 0.02 µg/L.

ROMBACH and KOCK [2797] determined bismuth in organic materials. They found that the addition of urea to the digestion solution eliminated the interferences due to nitrogen oxides. Potential interferences due to copper, arsenic or selenium were eliminated by adding potassium iodide.

BÉDARD and KERBYSON [2068] determined bismuth in metallic copper and found that the signal was depressed when more than 1 mg copper was present in the measurement solution. They therefore separated bismuth through co-precipitation with iron hydroxide or lanthanum hydroxide. FLEMING and IDE [2254] determined bismuth in steel in the presence of 2 M sulfuric acid and found that under these conditions the interferences due to copper, nickel or molybdenum were reduced through the iron that was present. They therefore kept the amount of iron constant at 20 mg and added iron for the analysis of high-alloy steels.

Investigations in the present author's laboratory showed that in the presence of an acid mixture of 1.5 % hydrochloric acid and 2.5 % nitric acid, up to 1 g Fe/L does not interfere in the determination of bismuth [3054]. The other metals also present in low-alloy steels did not influence the bismuth signals under these conditions, so that determinations could be performed directly against aqueous reference solutions in which the acid content was matched.

DRINKWATER [2206] determined bismuth in complex nickel alloys and found that a fine black precipitate formed when sodium borohydride was added, causing substantial interferences. He eliminated these interferences by adding EDTA and observed that the bismuth signal increased in proportion to the volume of EDTA solution added.

SINEMUS *et al* [2869] determined bismuth in surface waters in hydrochloric acid solution. They found no influence due to oxidation state, since the pentavalent state is highly improbable.

10.7. Boron

Boron can only be determined in a nitrous oxide/acetylene flame; with a characteristic concentration of about 15 mg/L 1 % and a detection limit of 1 mg/L it is one of the least sensitive elements in atomic absorption spectrometry. The resonance lines at 249.7/249.8 nm are used almost exclusively.

Several authors [65] [795] reported the determination of boron in aqueous solution, but various refractory substances interfere in the analysis. For this reason, and because of the low sensitivity, various authors proposed extraction procedures for enrichment and for a separation of the interfering matrix. AGAZZI [21], HOSSNER [536] and WEGNER *et al* [1288] extracted a 2-ethyl-1,3-hexandiol complex with chloroform. However, this solvent is not very suitable for combustion in a nitrous oxide/acetylene flame. MELTON *et al* [858] and HOLAK [526] therefore used MIBK as solvent since it proved to be very suitable for this extraction. ELTON-BOTT [2227] increased the sensitivity for boron by converting it to the volatile boric acid methyl ester and conducting this in gaseous form into a nitrous oxide/acetylene flame. CHAPMAN and DALE [2149] volatilized boron as the fluoride and were thereby able to increase the sensitivity.

MANNING and SLAVIN [808] found that an isotope analysis by means of atomic absorption spectrometry is only possible for very light and very heavy atoms. MROZOWSKI [891] attempted to separate the ^{10}B and ^{11}B lines by cooling the hollow cathode lamp in liquid air, but he was unsuccessful.

The isotope shift is about 0.001 nm while the line width in a nitrous oxide/acetylene flame is about 0.006 nm, so that separation is impossible. GOLEB [426] was also unable to obtain different absorption values for the two isotopes by using isotope lamps and an absorption tube.

HANNAFORD and LOWE [2346] finally succeeded in determining the isotope ratio by using a neon-filled discharge lamp and a water cooled cathodic nebulization chamber as the absorber at the 208.89/208.96 nm doublet, since here there is a higher isotope shift.

Boron can only be determined by the graphite furnace technique in tubes having a perfect pyrocoated layer. A fast heating rate for atomization increases the sensitivity and reduces tailing. SZYDLOWSKI [2954] determined boron in water and found that the addition of 1000 mg/L barium as the hydroxide gave the best sensitivity.

A thermal pretreatment temperature of up to 1000°C can be applied in the presence of 500 mg/L calcium; the optimum atomization temperature is higher than 2700°C.

10.8. Cadmium

Cadmium can be easily determined without any interferences, worth mentioning, in an air/acetylene flame. At the 228.8 nm line the characteristic concentration is 0.02 mg/L 1% and a detection limit of 0.0005 mg/L can be easily achieved. In an argon/hydrogen diffusion flame the sensitivity can be improved by a factor of two and a detection limit of 0.0003 mg/L can be achieved. However, the occurrence of spectral and non-spectral interferences is more probable in this flame.

For the determination of higher cadmium concentrations, the resonance line at 326.1 nm is suitable because the characteristic concentration is about 20 mg/L 1%; so that excessive dilutions can be avoided.

Since cadmium is very easily atomized, it can also be determined very well by the boat technique and the DELVES system. The detection limits are 1×10^{-10} g and 2×10^{-11} g, respectively [218].

For a number of years the high volatility of cadmium presented difficulties for a determination by the graphite furnace technique. LUNDGREN *et al* [771], however, utilized this high volatility to atomize cadmium out of a sodium chloride matrix at 800 °C using a fast rate of heating; at this temperature, sodium chloride is not noticeably volatile.

Interferences due to chloride are not so marked for cadmium as with lead or thallium, since the energy of dissociation of CdCl is rather low at 49 kcal/mol. Nevertheless, numerous authors have reported interferences due to chloride [2367] [2632] or to perchloric acid [2441], which they attempted to eliminate by the addition of phosphoric acid [2108] [2632] or molybdenum [2367].

EDIGER [2217] investigated various ammonium salts as matrix modifiers and proposed secondary ammonium phosphate as this stabilizes cadmium up to about 900 °C. SLAVIN *et al* [2880] [2882] and also HINDERBERGER *et al* [2572] established that an interference-free determination of cadmium is possible for atomization off the L'vov platform in a stabilized temperature furnace with primary ammonium phosphate as matrix modifier. A thermal pretreatment temperature up to about 800 °C can be applied; the addition of magnesium nitrate permits a somewhat higher temperature. The optimum temperature for atomization off the L'vov platform is 1700 °C.

One of the greatest problems for the determination of cadmium by the graphite furnace technique is the risk of contamination. SALMELA and VUORI [2816] found that proprietary yellow pipet tips were heavily contaminated by cadmium. High cadmium blanks are also found for various reagents, so that purification is frequently necessary (refer to section 6.4).

10.9. Calcium

Calcium is one of the elements most frequently determined by atomic absorption spectrometry, and the number of publications is correspondingly large. The first work by WILLIS [1324] [1328] and DAVID [268] [269] between 1959 and 1961 was concerned with the determination of calcium in serum and urine, and in plants and soil samples, respectively.

Calcium can be determined virtually free of interferences in a nitrous oxide/acetylene flame [3297] with a characteristic concentration of 0.09 mg/L 1% and a detection limit of about 0.001 mg/L. Interferences can only be observed in this flame in the presence of high silicon and aluminium concentrations, and slight ionization can easily be removed by adding small quantities of alkali [800].

In spite of this fact, calcium is frequently determined with an air/acetylene flame and then serves as a prime example for elements disturbed by numerous interferences. RAMA-KRISHNA [1014] published a systematic study in which the effect of more than 50 ions on the absorption of calcium was investigated. Only two anions (from 20 investigated) and 15 cations (from 32 investigated) were without any great influence on the absorption of calcium. Nevertheless, it should be emphasized that this work was carried out with a direct injection burner; with a premix burner these interferences are much less marked [301]. In a premix burner the influences of up to 500 mg/L silicon and 1000 mg/L aluminium or phosphorus can be well controlled by the addition of 1% lanthanum (as hydrochloric acid solution of the oxide) or 1% EDTA (as di-sodium salt). Calcium can then be determined in an air/acetylene flame at the 422.7 nm resonance line with a characteristic concentration of 0.1 mg/L

1% and a detection limit of 0.001 mg/L. Although large spectral bandpasses (about 2.0 nm) are used for the determination of calcium in an air/acetylene flame, this is not recommended for the nitrous oxide/acetylene flame. As shown in figure 135, there is an intense CN emission band between 422 nm and about 410 nm in this flame. If this is not separated in the monochromator, the signal-to-noise ratio is worsened noticeably.

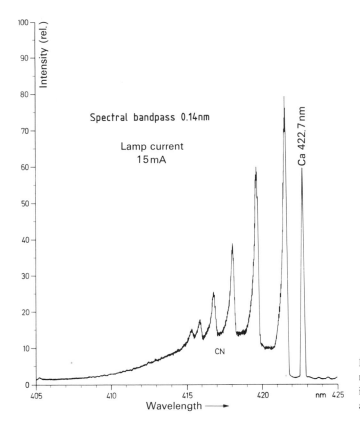

Figure 135. Emission of the nitrous oxide/acetylene flame in the region of the Ca resonance line at 422.7 nm.

Higher calcium concentrations can be determined at the 239.9 nm resonance line with a characteristic concentration of about 10 mg/L 1%. The signal-to-noise ratio is not so good at this line, however, and it is not recommended for precision analyses.

Calcium is a very common element and the sensitivity of the graphite furnace technique is rarely required. CARRONDO *et al* [2133] determined calcium in sewage sludge by dispensing a suspension in diluted nitric acid directly into the graphite furnace. The results were in good agreement with those for the flame technique. SMITH and COCHRAN [2891] determined calcium in brines using uncoated graphite tubes and a high argon flowrate during atomization to reduce sensitivity.

The best sensitivity for calcium is obtained in pyrocoated graphite tubes. Nevertheless, the determination is very prone to contamination and should therefore be carried out in a dust free area. The maximum thermal pretreatment temperature for calcium is 1200 °C and the optimum atomization temperature 2400 °C.

10.10. Cesium

The resonance line for cesium is at 852.1 nm, at the start of the infrared range, and is therefore outside the normal working range of a number of atomic absorption spectrometers. Up to the present no usable hollow cathode lamps have been developed for cesium, but excellent electrodeless discharge lamps have been on the market for some years [2061]. These have largely replaced the vapor discharge lamps used previously. Cesium can be determined in an air/acetylene flame with a characteristic concentration of 0.1 mg/L 1% and a detection limit of about 0.02 mg/L without noticeable interferences. Cesium is, however, strongly ionized in this flame, and large quantities of other alkali salts must be added to suppress the ionization [569]. The ionization is considerably lower in an air/hydrogen flame, so that working with this flame is easier. A number of authors prefer the air/propane flame [296], in which other interferences can occasionally occur. For the determination of cesium in rocks and minerals, LUECKE [2566] found interferences due to higher aluminium and phosphate contents, as well as spectral interferences.

For higher cesium concentrations the resonance line at 455.6 nm is suitable since the characteristic concentration is about 20 mg/L 1% [407].

A detection limit of 0.008 mg/L can be achieved in emission using an air/acetylene or an air/hydrogen flame when 100 mg/L sodium salt is added [363].

The determination of cesium by the graphite furnace technique was first described by BARNETT et al [2061] in connection with investigations on electrodeless discharge lamps. They obtained a detection limit of 0.3 µg/L. FRIGERI et al [2271] determined cesium in river and lake waters and found interferences due to iron and cobalt. The maximum thermal pretreatment temperature is about 900 °C and the optimum atomization temperature is 1900 °C.

10.11. Chromium

Chromium can be determined both in a fuel gas rich air/acetylene flame and in a nitrous oxide/acetylene flame. A comprehensive study on the optimum atom concentration and the absorption profile in fuel gas rich and weak air/acetylene flames, respectively, has been made by RANN and HAMBLY [1018]; the greatest sensitivity was found with a three-slot burner head [1172] and a faintly luminous flame. Chromium exhibits a large number of resonance lines of similar sensitivity. For the majority of work the 357.9 nm line is used. At this line the characteristic concentration is 0.04 mg/L 1% and the detection limit is about 0.002 mg/L. According to GATEHOUSE and WILLIS [407], the resonance line at 425.4 nm with a characteristic concentration of 0.1 mg/L 1% can be used for higher chromium concentrations.

AGGETT and O'BRIEN [2012] [2013] have published very comprehensive papers on the formation of chromium atoms in air/acetylene flames.

While chromium can be determined virtually interference-free in numerous samples [989], in the presence of iron and nickel the absorption is considerably reduced. ROOS [1045] also found that this interference was strongly dependent on the gas mixture; furthermore, he found that chromium(III) exhibited a higher sensitivity than chromium(VI), so that the oxidation state also plays a part. These effects can be largely eliminated by adding 2% ammonium chloride [410], or 2% potassium persulfate [1272], or even 1% ammonium hydrogen fluoride, either alone or together with 0.2% sodium sulfate [1008], to the sample and reference solutions. ROOS [1049] attributed the elimination of these interferences to a carrier distillation, i. e. iron and other interfering ions distil off much more rapidly in the flame with the added salts. KRAFT *et al* [2506] found that the intensity of the chromium signal bore a complex relationship to the oxidation state and the flame temperature. To eliminate interferences, other authors extracted chromium in MIBK [825] or used ion exchangers to remove the interfering ions [581]. 8-Hydroxyquinoline has also been proposed as a releasing agent to suppress interferences [945].

WOLF [3076] used a combination of chelation, extraction and chromatographic separation prior to determination in a flame and obtained a detection limit of 1 ng chromium. RAO and SASTRI [2777] have published a good review on extraction procedures for chromium.

YANAGISAWA *et al* [1361] separated chromium(III) and chromium(VI) by extraction before determining the two oxidation states by atomic absorption. They used hydroxyquinoline at pH 6 as chelating agent for Cr(III) and diethylthiocarbamate at pH 4 for Cr(VI).

CRANSTON and MURRAY [2172] and also VAN LOON *et al* [3003] reported the separate determination of the two oxidation states of chromium in water. DE JONG and BRINKMANN [2189] extracted both oxidation states at different pH values with Aliquat-336, a mixture of methyl-tri-n-alkylammonium chlorides.

These interferences, which are often difficult to control, can be most easily eliminated by using a nitrous oxide/acetylene flame [547] [1053]. Hardly any interferences are known in this flame and both oxidation states exhibit the same sensitivity [2506]. The addition of potassium is recommended for determinations in complex matrices. For precision analyses, even in this flame it is still necessary to match the sample matrix, since acceptable results cannot otherwise be obtained [1299]. The characteristic concentration in the nitrous oxide/ acetylene flame at the 357.9 nm line is 0.5 mg/L 1%.

SLAVIN [2876] found that the graphite furnace technique was the most suitable for the determination of chromium in environmental samples, because it brings the best detection limits, and other techniques are more expensive and not so simple to use. For a long period, chromium was regarded as a difficult element for the graphite furnace technique. One reason is certainly that a deuterium lamp has insufficient radiant intensity at the 357.8 nm chromium resonance line to provide satisfactory background correction. This problem is eliminated when the spectrometer uses either a halogen lamp or the Zeeman effect for background correction.

In addition, reports were made on the high volatility of a chromium compound that led to substantial losses during thermal pretreatment in the furnace [2071]. Nevertheless, other authors were unable to find volatile chromium compounds [2873]. There are also reports in the literature on relatively strong interferences due to chloride [2441] [2638] which could not be verified by other authors [2513]. CARRONDO *et al* [2133] determined chromium in sewage sludge by dispensing a suspension in nitric acid solution directly into the graphite tube. Calibration was against acidified reference solutions. The agreement with samples determined in the flame after prior digestion was good.

HINDERBERGER *et al* [2372] were able to determine chromium in body fluids and tissues free of interferences when they used primary ammonium phosphate as matrix modifier and atomized off the L'VOV platform. SLAVIN *et al* [2877] [2882] subsequently found that magnesium nitrate is the best matrix modifier. Chromium can be thermally pretreated up to 1650 °C without losses, so that most interfering concomitants can be removed successfully. The determination is virtually free of interferences when atomization is off the L'VOV platform in a stabilized temperature furnace. The optimum atomization temperature is 2500 °C.

10.12. Cobalt

Cobalt can be determined without difficulties by atomic absorption spectrometry in an air/ acetylene flame. Although it exhibits a large number of resonance lines (the most important are compiled in table 23), only a few are suitable for analytical purposes. Clear spectral interferences due to overlapping can occur with a number of the lines given in table 23, especially when using multielement lamps which contain nickel and when the spectral bandpass is too large (greater than 0.2 nm).

Table 23. Co resonance lines

Wavelengtht nm	Energy level (K)	Characteristic concentration (mg/L 1%)
240.7	0–41 529	0.1
242.5	0–41 226	0.12
252.1	0–39 649	0.15
241.1	816–42 269	0.15
352.7	0–28 346	1.8
345.4	3483–32 431	2

The most suitable line for lower cobalt concentrations is the 240.7 nm line with a characteristic concentration of 0.1 mg/L 1% and a detection limit of about 0.06 mg/L. Because of weak self-absorption, which could be observed much more intensely with older lamps [471], and a number of lesser ionization lines that cannot be separated from the resonance line, there is a distinct curvature of the analytical curve at higher absorbance values. For higher cobalt concentrations the 352.7 nm line is more suitable; it exhibits good linearity and a very favorable signal-to-noise ratio, thus making good precision possible.

MCPHERSON [851] showed that the following elements interfere with the determination of cobalt in an air/acetylene flame: 2000 mg/L Cr, Ni, W; 1000 mg/L Cu, Mo; 500 mg/L Si;

200 mg/L Mn, V; 100 mg/L Ti and 50 mg/L P and S. With high matrix concentrations, matching the matrix can become necessary. With a refractory matrix it can be of advantage to determine cobalt in a nitrous oxide/acetylene flame. The characteristic concentration in this flame is about 0.7 mg/L 1% at the 240.7 nm line.

SIMMONS [1127] complexed cobalt with 2-nitroso-1-naphthol and extracted it with chloroform from plant material digestions. Because of the poor combustion properties of chloroform, he evaporated the solution and dissolved the residue in MIBK.

Various authors reported the determination of cobalt in soil and plant samples and in lake water after extraction [2419] [2737] [2866] or after separation of the matrix on an exchange column [2468]. BATLEY and MATOUŠEK [2062] separated cobalt by direct electrolytic deposition into a graphite tube. SLAVIN *et al* [2874] found that the direct determination of cobalt in meat with the graphite furnace technique was better and simpler than via extraction. PETROV *et al* [2746] also preferred the simpler direct procedure for the determination of cobalt in soil extracts. MAIER *et al* [2608] did not observe any interferences for the determination of cobalt in surface waters.

LIDUMS [2547] observed distinct matrix effects for the direct determination of cobalt in blood and urine. He therefore separated the element in an ion exchanger. JULSHAMN [2441] found a slight signal depression due to perchloric acid. SLAVIN *et al* [2877] [2882] investigated the effect of magnesium nitrate as a matrix modifier and found that it permitted cobalt to be thermally pretreated up to 1450°C. The optimum atomization temperature off the L'vov platform is 2500°C.

10.13. Copper

Copper is one of the elements most frequently and easily determined by atomic absorption. It exhibits no interferences in an air/acetylene flame [646] [989], is virtually independent of the stoichiometry of the flame and the lamp current, and is therefore frequently used as a standard to test an instrument or a procedure.

Copper exhibits a number of resonance lines, all of which are analytically suitable (table 24). By selecting the correct resonance line, even higher copper concentrations can be precisely determined without excessive dilution. For lower copper concentrations the 324.7 nm

Table 24. Cu resonance lines

Wavelength nm	Energy level (K)	Characteristic concentration (mg/L 1%)
324.7	0–30784	0.03
327.4	0–30535	0.07
222.6	0–44916	0.5
249.2	0–40114	2.5
244.2	0–40944	10

line with a characteristic concentration of 0.03 mg/L 1% and a detection limit of about 0.001 mg/L is the most suitable.

FUJIWARA *et al* [2276] made detailed investigations on the atom distribution of copper in an air/acetylene flame in the presence of various acids.

A somewhat better sensitivity can be achieved by using an air/propane flame, but certain interferences cannot be completely excluded in this flame.

Copper can be complexed with a large number of chelating agents over a wide pH range and extracted with virtually all organic solvents. It is frequently employed as a reference element for extractions to check the completeness of the extraction procedure [39]. Copper can be freed from its matrix—even under extreme conditions—by multiple extraction [1243].

Copper is frequently determined by the graphite furnace technique and only very minor interferences have been reported. Copper can be determined in serum [2235] directly with excellent accuracy and in urine without interferences after prior low temperature ashing [2956]. Good results are also obtained for the analysis of meat after digestion in nitric acid [2874], and also for the direct solids analysis of liver and fish meal [2522]. CARRONDO *et al* [2133] determined copper in sewage sludge by dispensing a suspension directly into the graphite tube. PETROV *et al* [2746] also used a direct procedure for the determination of copper in soil extracts. SUZUKI *et al* [2951] observed an interference due to chloride in a molybdenum tube furnace which they eliminated by adding thiourea. JULSHAMN [2441] found a signal depression due to perchloric acid; this same interference caused SIMMONS and LONERAGAN [2867] to extract copper before the determination during the analysis of plant samples. Various other authors also reported the extraction of copper during the analysis of surface waters [2100] [2432] [2468] [2894] and soil extracts [2419] [2737].

Interferences due to chloride should not occur for atomization off the L'VOV platform in a stabilized temperature furnace and the determination should be largely free of interferences. In 0.2% nitric acid copper can be thermally pretreated up to about 1200°C without losses. The optimum temperature for atomization off the L'VOV platform is 2300°C.

10.14. Gallium

Gallium can be determined simply in a stoichiometric air/acetylene flame; the greatest problem is a good hollow cathode lamp. Since gallium melts at only 30°C, it is difficult to manufacture a hollow cathode lamp of usual design. MULFORD [892] described a hollow cathode lamp in which the metal was in the molten state, and he also examined various resonance lines for their sensitivities (table 25). At the 287.4 nm line the characteristic concentration is 1.3 mg/L 1% and the detection limit is about 0.1 mg/L. With a nitrous oxide/acetylene flame the best sensitivity is observed at the 294.4 nm line with a characteristic concentration of 1.0 mg/L 1%.

Hitherto, little has been reported on the determination of gallium. GUPTA *et al* [449] investigated the influence of 45 different ions on the absorption of gallium in an air/acetylene flame. They found that 12 ions depressed the signal slightly when their concentrations exceeded determined values. In general, though, gallium appears to be largely determinable without interferences.

Table 25. Ga resonance lines

Wavelength nm	Energy level (K)	Characteristic concentration (mg/L 1%)	Spectral bandpass nm
287.4	0–34782	1.3	0.2
294.4	826–34782	1.1	0.2
417.2	826–24788	1.5	0.2
403.3	0–24788	2.8	0.7
250.0	826–40811	10	0.7
245.0	0–40803	12	2.0

ALLAN [2030] reported a spectral interference due to manganese at the 403.298 nm resonance line.

Several authors describe interferences due to chloride on the determination of gallium by the graphite furnace technique. In the author's laboratory it was found that an excess of ammonia solution added to the sample in the graphite tube caused numerous interferences. EDIGER [2217] reported that a mixture of nitric acid and hydrogen peroxide doubled the gallium signal and improved the reproducibility. In 0.2% nitric acid a thermal pretreatment temperature of up to 800°C may be applied; the optimum atomization temperature is 2700°C when an ultrafast heating rate is used.

10.15. Germanium

Germanium can be determined at 265.1 nm in a nitrous oxide/acetylene flame with a characteristic concentration of 2.5 mg/L 1% and a detection limit of approximately 0.2 mg/L [795]. POPHAM and SCHRENK [997] have published a comprehensive study on the absorption of germanium in a nitrous oxide/acetylene flame and in a fuel gas rich oxygen/acetylene flame. They found numerous interferences, especially in the nitrous oxide/acetylene flame, but these were probably due to the fact that they made the measurements only 2 mm above the burner slot, which is certainly insufficient for complete atomization.

Germanium can be determined with greater sensitivity using the hydride technique with $NaBH_4$ as the reducing agent. The detection limits are 2×10^{-7} g or 0.01 mg/L, respectively [358].

A flame must be used for this element, however, since germanium hydride is insufficiently atomized in a heated quartz tube.

Up to the present, very little has been published on the determination of germanium by the graphite furnace technique. EDIGER [2217] found that oxidizing acids improve the sensitivity for germanium and he proposed perchloric acid. JOHNSON *et al* [2438] reported that germanium cannot be determined using a carbon rod, but that a graphite furnace is very suitable. ROZEMBLUM [2806] determined germanium in ultrapure water indirectly via the germanomolybdic acid.

In 0.2% nitric acid germanium can be thermally pretreated up to about 800°C; the optimum atomization temperature is 2600°C using an ultrafast heating rate.

10.16. Gold

Gold is very frequently determined in soil samples and galvanic baths by atomic absorption spectrometry. The most sensitive resonance line is at 242.8 nm with a characteristic concentration of 0.25 mg/L 1% and a detection limit of about 0.006 mg/L; frequently the slightly less sensitive 267.6 nm line (0.4 mg/L 1%) enables better and more accurate analyses to be carried out because of its higher intensity.

Normally, gold is determined in a sharp, fuel gas weak air/acetylene flame in which no interferences worth mentioning occur. Certain differences arise only when cyanide gold solutions are compared with hydrochloric acid solutions; in such cases the sample and reference solutions should be matched. Nevertheless, according to experiences gained in the present author's laboratory, minor interferences occur during the determination of gold in complex matrices—even though no clear chemical interferences can be confirmed—which are strongly dependent upon alterations in the composition of the flame. Thus it is frequently impossible to determine gold in complex matrices with an accuracy better than ±5%, even with the analyte addition technique. Although a nitrous oxide/acetylene flame reduces the sensitivity of a gold determination, it partially eliminates these difficult-to-control interferences.

The determination of gold in the presence of other noble metals is often relatively difficult. ADRIAENSSENS and VERBEEK [17] found that the majority of interelemental interferences observed in acid solution do not occur in a 2% potassium cyanide solution. SEN GUPTA [1114] eliminated the interelemental interferences by adding a copper/cadmium buffer solution. He has also published a good review on the various methods in noble metal analysis [1116].

Since gold must frequently be determined with even greater accuracy and especially in smaller concentrations, numerous extraction methods are in use [439] [445] [992] [1182] [1231] [1232] which will be described, in part, later.

MOJSKI [2671] used di-n-octylsulfide to extract gold into cyclohexane and RUBEŠKA *et al* [2810] used dibutylsulfide to extract gold and palladium from geological samples into toluene. They used both a flame and a graphite furnace for the determination.

ROYAL [2805] attempted to determine gold by the graphite furnace technique by nebulizing the sample solution as an aerosol into the graphite tube; he found severe interferences due to sodium, potassium and calcium. SCHATTENKIRCHNER and GROBENSKI [2824] determined gold in blood and urine, and KAMEL *et al* [2452] examined protein fractions and found no interferences. ROWSTON and OTTAWAY [2804] established optimum conditions for the determination of gold and other noble metals and found a characteristic concentration of 0.9 µg/L 1%.

In 0.2% nitric acid gold can be thermally pretreated up to about 650°C; the optimum atomization temperature is 1600°C using an ultrafast heating rate.

10.17. Hafnium

Virtually nothing is to be found in the literature on the determination of hafnium; it can be determined in a nitrous oxide/acetylene flame at the 286.6 nm line with a characteristic concentration of 15 mg/L 1% and a detection limit of 2 mg/L. A number of further resonance lines of lower sensitivity are compiled in table 26.

The signal for hafnium is clearly enhanced in the presence of hydrofluoric acid, so that sample and reference solutions should contain at least 0.1% hydrofluoric acid to control this effect and to obtain the best possible sensitivity. BOND [141] favored the addition of 0.1 M ammonium fluoride solution to increase the signal.

Table 26. Hf resonance lines

Wavelength nm	Energy level (K)	Characteristic concentration (mg/L 1%)
286.6	0–34877	15
307.3	0–32533	25
289.8	2357–36850	70
296.5	2357–36075	50
295.1	2357–36237	100
294.1	0–33995	150
290.5	2357–36773	150
377.8	0–26464	150

10.18. Indium

Indium may be determined in an oxidizing air/acetylene flame at the 304.0 nm line with a characteristic concentration of 0.5 mg/L 1% and a detection limit of about 0.02 mg/L. Further resonance lines are compiled in table 27. The signal-to-noise ratio, and thereby the achievable detection limit, for indium, like most other easily melted elements, is dependent on whether the construction of the hollow cathode allows the molten metal to be used or

Table 27. In resonance lines

Wavelength nm	Energy level (K)	Characteristic concentration (mg/L 1%)	Spectral bandpass nm
304.0	0–32892	0.5	0.7
325.6	2213–32916	0.6	0.2
410.5	0–24373	1.3	0.7
451.1	2213–24373	1.5	1.4
256.0	0–39048	5	2.0
275.4	0–36302	11	2.0

not [892]. The introduction of an electrodeless discharge lamp for indium has brought a distinct improvement.

There is little literature on the determination of indium, but no serious interferences seem to exist. While SATTUR [1079] was unable to detect any chemical interferences, MULFORD [892] found that various elements, especially in higher concentrations, increasingly depressed the indium absorption.

FUJIWARA *et al* [2276] made a thorough investigation of the influence of various acids on the atomization of indium in an air/acetylene flame. NAKAHARA and MUSHA [2685] found that in an argon/hydrogen diffusion flame the determination of indium with a characteristic concentration of 0.08 mg/L 1% was much more sensitive than in an air/acetylene flame. With the exception of silicon and vanadium, the numerous interferences could be eliminated by the addition of magnesium chloride.

DITTRICH *et al* [2200] found severe vapor-phase interferences due to halides for the determination of indium by the graphite furnace technique. L'VOV *et al* [2579] explained the enhancing effect of a pyrolytic coating (indium hardly reacts with graphite) by a macrokinetic theory of sample volatilization. In uncoated tubes the sample penetrates the graphite and is distributed over the thickness of the wall by capillary action. Atomization off the wall is in this case atomization out of the graphite matrix. With pyrocoated tubes, on the other hand, the sample is only distributed on the surface.

In 0.2% nitric acid indium can be thermally pretreated up to about 800°C; the optimum atomization temperature is 1500°C using an ultrafast heating rate.

10.19. Iodine

At 183.0 nm, the resonance line for iodine is already so far in the vacuum UV that it is virtually unreachable with commercial instruments. Nevertheless KIRKBRIGHT *et al* [654] [669] [2473] succeeded in determining iodine by making a number of minor alterations. They flushed the monochromator of a commercial instrument with nitrogen, used a nitrous oxide/acetylene flame shielded with nitrogen and an electrodeless discharge lamp for iodine, and reduced the passage of the beam through air to a minimum by using flushed tubes. In this way they achieved a characteristic concentration of 12 mg/L 1% and a detection limit of about 6 mg/L. The determination was free of interferences and no differences between the various types of bonding of iodine could be established. Later, the same authors described an improvement of 38 times in the sensitivity by employing the LEIPERT amplification [670]. The iodine is oxidized to iodate which is then treated with an excess of iodide to release six equivalents of iodine. This is extracted in MIBK and nebulized directly into the flame. The characteristic concentration is then 0.32 mg/L 1%.

L'VOV and KHARTSYZOV determined iodine in a graphite furnace [780]. They used the line at 206.2 nm, which is an excited line whose lower energy level is 0.94 eV above the ground state. At an atomization temperature of 2400°C the detection limit for iodine was about 2×10^{-9} g.

Using the spectral line at 183.0 nm, KIRKBRIGHT and WILSON [2474] determined iodine in a commercial double-beam spectrometer and obtained a detection limit of 1×10^{-9} g. Numerous ions enhanced the signal by up to 100%; the authors presumed that this was due to background attenuation.

NOMURA and KARASAWA [2697] observed that when solutions containing mercury(II) nitrate and iodine were heated in a graphite furnace, two peaks were obtained for mercury. One of these peaks could be attributed to the more stable HgI_2. In this way, the authors were able to perform good, reproducible determinations for iodine. Interferences were caused by cyanides, sulfides and thiosulfates, however.

10.20. Iridium

For a long period, iridium was considered indeterminable by atomic absorption spectrometry. The first success was obtained by MULFORD [893] who, on WILLIS's suggestion, examined this element more exactly. Later MANNING and FERNANDEZ [801] made a thorough study and found a characteristic concentration of 4 mg/L 1% in an air/acetylene flame somewhat rich in fuel gas at the 264.0 nm resonance line. The 208.9 nm line was found to be twice as sensitive, but the signal-to-noise ratio was considerably worse so that the better detection limit of about 1 mg/L and better precision were obtained from the first line. Several further resonance lines are compiled in table 28.

Table 28. Ir resonance lines

Wavelength nm	Energy level (K)	Characteristic concentration (mg/L 1%) [801]	+La+K [390]
264.0	0–37872	4	6.5
208.9	0 47858	1.5	3
266.5	0–37515	5	
285.0	0–35081	6	8
237.3	0–42132	6	
250.3	0–39940	7	
351.4	0–28452	35	

FUHRMAN [390] found that the sensitivity of the iridium determination could be doubled by adding 100 mg/L potassium plus 1000 mg/L lanthanum; the addition of sodium or potassium alone, on the other hand, reduced the signal considerably. FUHRMAN further reported that in the presence of potassium and lanthanum, no influences due to platinum or titanium could be observed on the absorption of iridium.

JANSSEN and UMLAND [571] favored the addition of 1% sodium and 1% copper for the determination of iridium in the presence of other platinum metals. The sensitivity was said to have been increased by ten times and the interferences caused by other platinum metals eliminated.

GRIMALDI and SCHNEPFE [443] used the same additions for the determination of iridium in rocks. To eliminate the same interferences, TOFFOLI and PANNETIER [1234] preferred the addition of 0.5% lithium with, under certain circumstances, some copper, and SEN GUPTA [1114] used 0.5% copper and 0.5% cadmium. This author has also published a good review on the determination of noble metals by atomic absorption spectrometry [1116].

LUECKE and ZIELKE [769] found that in a nitrous oxide/acetylene flame the numerous interferences present during the determination of iridium disappeared completely. The sensitivity, which was reduced by about 50%, could be brought back practically to the original value by chlorination. In the $(IrCl_6)^{2-}$ complex, iridium exhibits the same sensitivity in the nitrous oxide/acetylene flame as in the air/acetylene flame, and the determination is completely interference free.

In later work, however, these authors found that various iridium complexes can exhibit markedly differing sensitivities.

For the determination of iridium by the graphite furnace technique, ADRIAENSSENS and KNOOP [16] found that only minor interelemental effects occurred in the presence of other noble metals and that hydrochloric acid had no influence. ROWSTON and OTTAWAY [2804] optimized the determination of iridium by the graphite furnace technique and obtained a characteristic concentration of 38 μg/L 1%. In the presence of 0.2% nitric acid, iridium can be thermally pretreated up to about 1000°C; the optimum atomization temperature is 2500°C using an ultrafast heating rate.

10.21. Iron

Iron is one of the elements most frequently determined by atomic absorption spectrometry; it is determined in a large number of different samples and mostly in lesser concentrations. The determination of iron appears to be largely free of interferences in a stoichiometric air/acetylene flame. ALLAN [37] only found an interference caused by silicon, which can be easily eliminated by adding 200 mg/L calcium [989]. TERASHIMA [1223] found that besides silicon, strontium, aluminium, manganese, citric acid and tartaric acid also depressed the iron signal and that the effect decreased with increasing height above the burner slot. ROOS and PRICE [1051] also found a strong influence on the iron signal by citric acid and eliminated the interference by adding phosphoric acid or sodium chloride. OTTAWAY *et al* [944] observed a strong reduction of the iron signal by cobalt, copper and nickel. The interference, however, is strongly dependent upon the flame conditions, such as fuel gas/oxidant ratio, height of observation in the flame, and the anion. Addition of 8-hydroxyquinoline was recommended to eliminate this interference. To eliminate the same interference, MARTIN [826] added lanthanum. For the examination of biological materials, ZETTNER [1380] found no influences from a large number of cations and anions, from the oxidation states of iron and from various chelating agents. FERRIS *et al* [365] observed a slight depression of the iron signal by aluminium and silicon.

In strong nitric acid solutions, especially in a slightly reducing flame, a distinct reduction of the iron signal can be observed that is only weakly seen in strongly oxidizing flames. With a nitrous oxide/acetylene flame the effect disappears completely, although the sensitivity is somewhat lower, which is acceptable in many cases.

BALL and GOTTSCHALL [2054] reported numerous interferences, for example from nitric acid, oxalic acid and citric acid, when using a three-slot burner and an air/acetylene flame. It should be noted that this burner is operated with a fuel gas rich flame. In the course of a careful study, THOMPSON and WAGSTAFF [2968] found that in a fuel gas rich air/acetylene flame higher sensitivity is attained, but that there is a distinct difference between the oxidation states of iron and that interferences due to silicon and calcium also occur. They therefore recommended a fuel gas weak flame in which these interferences could not be observed.

MALONEY et al [2611] reported the extraction of iron from aqueous thiocyanate solutions by sorption onto polyurethane foam. The thiocyanate complex can be easily desorbed with 0.1 M nitric acid.

For the determination of iron, a large number of resonance lines of different sensitivities are available. The most important are compiled in table 29. The most frequently used line at 248.3 nm gives a characteristic concentration of 0.04 mg/L 1% and a detection limit of about 0.005 mg/L at a spectral bandpass of 0.2 nm.

Table 29. Fe resonance lines

Wavelength nm	Energy level (K)	Characteristic concentration (mg/L 1%)
248.3	0–40 257	0.04
248.8	416–40 594	0.07
252.3	0–39 626	0.07
271.9	0–36 767	0.12
302.1	0–33 096	0.14
	416–33 507	
250.1	0–39 970	0.4
216.7	0–46 137	0.5
372.0	0–26 875	0.27
296.7	0–33 695	0.32
246.3	0–40 594	0.48
386.0	0–27 167	0.49
344.1	0–29 056	0.65
293.7	0–34 040	0.80
382.4	0–26 140	5.0

Iron is normally contained in relatively high concentration in dust particles; the determination of iron by the graphite furnace technique is thus very prone to contamination. SOMMERFELD et al [2897] found, for example, high iron concentrations in pipet tips which could not be completely eliminated even by cleaning. Thorough cleansing of laboratory ware by fuming out (refer to section 6.4) and the use of an autosampler (refer to section 3.2.6 and figure 41, page 66) are strongly recommended.

An interference due to chloride during the determination of iron has also been mentioned. VOLLAND et al [3021] found a serious signal depression due to halogenated hydrocarbons.

These are presumably intercalated in the graphite matrix and only released at relatively high temperatures. KOIRTYOHANN *et al* [2481] found a substantial influence due to perchloric acid and this finding was verified by JULSHAMN [2441].

Magnesium nitrate is a very suitable matrix modifier for the determination of iron; it permits thermal pretreatment temperatures up to 1450°C to be applied. The optimum temperature for atomization off the L'vov platform is 2400°C.

10.22. Lanthanum, Lanthanides

The lanthanides are very similar in their behavior in atomic absorption spectrometry, even though they are determinable with widely varying sensitivities. All the rare earths require a nitrous oxide/acetylene flame for their determination. They are more or less strongly ionized in this flame so that an addition of approximately 0.1 to 0.3% potassium is always required to suppress this effect. The most important resonance lines of *lanthanum* and the rare earths are compiled in table 30 together with their attainable characteristic concentrations and detection limits.

Because the individual rare earth metals do not mutually influence one another in atomic absorption spectrometry and no interferences of note occur in the hot flame, the ideal means for an absolute specific determination is given for these elements, which are normally so difficult to separate. Emission measurements are more sensitive for a number of rare earths, but because of the copius number of lines an assignment is often made difficult.

Despite various attempts [48] [885], it was not possible to determine *cerium* by AAS for a long period [574]. According to a brief reference, cerium is said to exhibit a minimal absorption at 522.4 nm or 569.7 nm; the achievable detection limit is around 100 mg/L. L'vov *et al* [2577] [2587] determined cerium in a graphite furnace with good sensitivity. The difficulties in the determination of cerium had previously been ascribed to ionization, the formation of volatile oxides, or to the complexity of the spectrum. L'vov found that the problem was due to the formation of a monocyanide in the vapor phase. At 2700°C, cerium is atomized by only 5% in an argon atmosphere. Cerium is completely atomized from a tungsten wire in a furnace lined with tantalum. Under these conditions the detection limit is 4 ng at the 567 nm line. JOHNSON *et al* [582] recommended an indirect determination via the hetero-poly-molybdo-cerium-phosphoric acid which contains six atoms of molybdenum for each atom of cerium. After extraction with isobutyl acetate, the molybdenum is determined, and as a result of the enriching effect a characteristic concentration of 0.1 mg/L cerium is obtained.

Because it does not occur naturally, *promethium* has not yet been examined by atomic absorption spectrometry.

A comprehensive study on the determination of *rare earth metals* in the nitrous oxide/acetylene flame has been published by AMOS and WILLIS [48]. MOSSOTTI and FASSEL [885] examined the use of continuum radiation sources and a turbulent oxygen/acetylene flame for the determination of lanthanides. A great deal of information, especially on the various resonance lines, is given by MANNING [793] [794] and FERNANDEZ [360].

Table 30. Resonance lines of La and the lanthanides

Element	Wavelength nm	Characteristic concentration (mg/L 1%)	Detection limit (mg/L)
La	550.1	48	2
	418.7	63	
	495.0	72	
	403.7	170	
Dy	421.2	0.7	0.05
	404.6	1.0	
	418.7	1.0	
	419.5	1.3	
	416.8	7.3	
Er	400.8	0.7	0.04
	415.1	1.2	
	389.3	2.3	
	408.8	3.4	
	393.7	3.6	
Eu	459.4	0.6	0.02
	462.7	0.8	
	466.2	0.9	
	321.1	7	
Gd	368.4	19	1.2
	407.9	19	
	405.8	22	
	405.4	25	
	371.4	28	
Ho	410.4	0.9	0.04
	405.4	1.1	
	416.3	1.4	
	417.3	5	
	404.1	7	
Lu	336.0	6	0.7
	331.2	11	
	337.7	12	
	356.8	13	
	298.9	55	
Nd	492.4	7	1
	463.4	11	
	471.9	19	
	489.7	35	

Pr	495.1	39	5
	513.3	60	
	492.5	80	
	505.3	100	
	504.6	110	
	502.7	110	
Sm	429.7	7	2
	476.0	12	
	511.7	14	
	472.8	16	
	520.1	17	
Tb	432.7	8	0.6
	431.9	10	
	390.1	13	
	406.2	15	
	433.8	16	
Tm	371.8	0.5	0.01
	410.6	0.7	
	374.4	0.7	
	409.4	0.8	
	420.4	1.5	
Yb	398.8	0.1	0.005
	346.4	0.5	
	246.5	1	
	267.2	5	

Many of the lanthanides can be determined with good sensitivity by the graphite furnace technique. GROBENSKI [2319] investigated all of the rare earths and the influence of pyrocoated graphite tubes on their determination. He found a direct relationship between the energy of dissociation of the gaseous monoxides and their sensitivity in the graphite furnace. He observed a signal depression for samarium in the presence of nitric acid and sulfuric acid.

L'VOV and PELIEVA [2584] and SEN GUPTA [2841] determined the characteristic concentrations of numerous lanthanides. SEN GUPTA used a pyrocoated graphite tube while L'VOV and PELIEVA lined the graphite tube with tantalum foil. SEN GUPTA avoided interferences due to concomitants in rock samples by co-precipitating the lanthanides with calcium and iron.

MAZZUCOTELLI and FRACHE [2648] determined *europium* in silicate matrices and observed signal enhancement due to sodium, potassium, calcium, aluminium and iron. HORSKY [2393] determined *praseodymium* in a furnace with an ultrafast heating rate and obtained a characteristic mass of 5 ng 1%. This is three orders of magnitude better than in a flame.

10.23. Lead

Lead can be determined in the most widely varying flames without noticeable interferences occurring. Usually an air/acetylene flame is used, although a number of authors prefer the air/propane flame. DAGNALL and WEST [262] found minor interferences from aluminium, beryllium, zirconium, phosphate and sulfate in this flame, but these could be largely eliminated by adding EDTA.

The most important resonance lines of lead are compiled in table 31. Although the 217.0 nm line is noticeably more sensitive than the 283.3 nm line, it does not give a better detection limit owing to a poorer signal-to-noise ratio. Furthermore, more background attenuation effects occur at the 217.0 nm line, so that the 283.3 nm line is used with greater success.

Table 31. Pb resonance lines

Wavelength nm	Energy level (K)	Characteristic concentration (mg/L 1%)
217.0	0–46069	0.08
283.3	0–35287	0.2
261.4	7819–46061	5
368.4	7819–34960	17
364.0	7819–35287	40

ISSAQ and ZIELINSKI [2416] found that when dilute, aqueous solutions containing lead were stored in glass or polyethylene vessels, substantial losses occurred within a relatively short time. Acidification with nitric acid prevents these losses.

Since lead must often be determined in very low concentrations, the detection limit of about 0.01 mg/L in the flame is often not adequate. Early on, the boat technique [616] or the DELVES system [292] were applied to increase the sensitivity. The real breakthrough came with the graphite furnace technique, and this is used principally nowadays for the determination of lead.

Early publications on the determination of lead in various samples reported numerous interferences, both spectral and non-spectral, which in part were dependent on the volatility of lead.

OTTAWAY [2717], for example, reported considerable interferences due to chlorides. Simultaneously with the occurrence of signal depression he observed the spectrum of lead chloride and thus concluded that lead is volatilized in molecular form. He also observed an interference when lead and magnesium chloride were placed at different locations in the graphite tube, thus indicating a vapor-phase interference. HENN [2367] found substantial interferences through 5% hydrochloric acid; he was able to reduce these effectively by adding molybdenum to the sample and reference solutions. VICKREY *et al* [3011] found a similar improvement when using zirconium coated graphite tubes.

BEHNE *et al* [2072] reported strong signal depression due to perchloric acid; this acid also attacks the graphite tube and thus cannot be used. Nitric acid also interfered in the determination of lead, so these authors removed it from the sample solution with formic acid. JULS-HAMN [2441] also found an interference due to perchloric acid. WEGSCHEIDER *et al* [3040] reported that all acids depressed the signal for lead except phosphoric acid, which caused signal enhancement. HODGES and SKELDING [2380] found that matrix interferences on the determination of lead in urine were lowest when orthophosphoric acid was added to the sample solution and the graphite tube was conditioned with molybdenum.

FRECH and CEDERGREN [2260-2262] employed high temperature equilibria equations to examine the interferences and found that chloride interferences were distinctly lower in the presence of hydrogen. These investigations were applied to steel analysis, and it would appear that iron plays an important role. In the presence of sodium chloride, hydrogen did not influence the interference adequately. Other authors verified the positive influence of hydrogen during the analysis of steels [2600] and geological samples [2363].

A solution to the problem was first found when the influence of the graphite surface and matrix modification were thoroughly investigated [2617] [2618]. SLAVIN and MANNING [2880] then atomized off the L'VOV platform and found that the majority of the interferences reported in the literature no longer occurred. HINDERBERGER *et al* [2372] verified that lead could be determined free of interferences in numerous biological materials, including urine, by using primary ammonium phosphate as matrix modifier and atomizing off the L'VOV platform.

Under these conditions, thermal pretreatment temperatures up to 950°C can be applied; magnesium nitrate as matrix modifier permits somewhat higher temperatures. The optimum temperature for atomization off the L'VOV platform is 1900°C.

Lead can also be determined by the hydride technique, but the sensitivity is not especially good. For atomization in a heated quartz tube, THOMPSON and THOMERSON [2967] obtained a detection limit of 0.1 mg/L. FLEMING and IDE [2254] were able to increase the sensitivity significantly by adding tartaric acid and potassium dichromate, but they reported that copper and nickel caused severe interferences. VIJAN and SADANA [3018] used the hydride technique to determine lead in drinking water after co-precipitation with manganese dioxide.

Many authors have described the combination of chromatographic techniques and AAS to determine organolead compounds in petroleum (gasoline) [1108] [2096] [2165] [2191] [2796] and in air [2153] [2796]. SIROTA and UTHE [2873] determined tetraalkyl lead in tissue after extraction.

BRIMHALL [163] and KIRCHHOF [647] reported a possible method for the determination of the three lead isotopes 206, 207 and 208 by atomic absorption spectrometry, and the construction of isotope hollow cathode lamps.

10.24. Lithium

Lithium can be effectively determined in an air/acetylene flame practically without interferences. At the 670.8 nm line the characteristic concentration is 0.03 mg/L 1% and the detection limit about 0.0005 mg/L. High lithium concentrations are more conveniently determined at the 323.3 nm resonance line with a characteristic concentration of 10 mg/L 1%.

Although lithium is in the first main group in the periodic table, due to its chemical behavior it belongs to the alkaline-earth elements [689] rather than to the alkali elements. In contrast to the alkali metals, in cooler flames (e. g. air/propane) lithium can only be determined with considerably reduced sensitivity rather than increased. Lithium also exhibits no ionization in the air/acetylene flame.

PATASSY [961] found no interferences for the determination of lithium in an air/acetylene flame. This was verified by FISHMAN and DOWNS [374] for a number of elements, but they found a minor interference from strontium. This finding was verified by SLAVIN [12], who found an increase of 5% in the lithium absorption in the presence of 100 mg/L strontium.

For the determination of lithium in rocks and minerals, LUECKE [2566] observed slight signal depression due to aluminium, phosphate, fluoride and perchlorate in concentrations up to 10% in an air/acetylene flame. All these interferences disappeared in a nitrous oxide/acetylene flame, so he recommended the hotter flame for the determination of lithium in geological samples.

For the determination of lithium by the graphite furnace technique, KATZ and TAITEL [2460] observed a severe signal depression due to 0.05 M calcium chloride. This interference could be eliminated by adding sulfuric acid. Phosphoric acid also eliminates interferences, but leads to poorer precision. STAFFORD and SAHAROVICI [2906] determined lithium in serum without interferences. In the present author's laboratory, distinctly lower sensitivity was found when lithium was determined directly in serum compared to aqueous solutions. At the same time, however, the thermal stability increased, so that a thermal pretreatment temperature of 1100 °C could be used. In 0.2% nitric acid a thermal pretreatment temperature of up to 1000 °C can be applied; using an ultrafast heating rate, the optimum atomization temperature is 2200 °C.

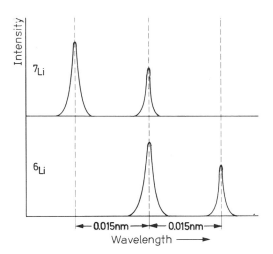

Figure 136. Doublet separation of the lithium isotopes ^6Li and ^7Li. The weaker doublet line of the more abundant ^7Li isotope coincides with the more intense doublet line of the rarer ^6Li isotope.

It is possible to determine the isotope ratio ^6Li to ^7Li by atomic absorption spectrometry. WALSH [1282] suggested this initially and it was later carried out in practice by ZAIDEL and

KORENNOI [1369], MANNING and SLAVIN [808], and GOLEB and YOKOYAMA [427]. ZAIDEL determined the total concentration of lithium by flame emission and then the ^7Li concentration using a hollow cathode lamp enriched with this isotope. MANNING used a flame as radiation source and sprayed the pure isotopes into it to determine the natural isotope ratio. GOLEB employed cooled hollow cathode lamps and a flameless atomization cell. RÄDE [2770] determined the isotope ratio using a commercial double-beam spectrometer with lithium isotope lamps and an air/acetylene flame; the precision was 0.5 atom%. CHAPMAN and DALE [2148] [2150] used a double-beam instrument to measure the absorbance ratios directly.

The basis of the isotope determination by atomic absorption spectrometry is the isotope shift in the absorption spectrum, which for lithium is 0.015 nm, i.e. considerably greater than the natural width of an absorption line. However, a disadvantage is that the 670.8 nm resonance line is a doublet with a separation of 0.015 nm, so that the weaker doublet line of the more abundant ^7Li isotope coincides with the more intense doublet line of the rarer ^6Li isotope. This fact, which is shown schematically in figure 136, prevents an independent determination of both isotopes. However, investigations by WHEAT [1313] [1315] and BUTLER and SCHROEDER [196] show that it is easily possible to determine the isotope ratio with a relative accuracy of better than 2% by using a dual-channel atomic absorption spectrometer with Li isotope hollow cathode lamps. Therefore, atomic absorption spectrometry offers a very attractive alternative to mass spectrometry for these determinations.

10.25. Magnesium

Magnesium is one of the elements most frequently determined by atomic absorption spectrometry. This is certainly due in part to the very high sensitivity that can be obtained for this element. At the 285.2 nm resonance line the characteristic concentration in an air/acetylene flame is 0.003 mg/L 1%, and because of the ideal signal-to-noise ratio, detection limits of 0.0001 mg/L can be achieved.

This high sensitivity is naturally not always of advantage. Higher magnesium concentrations are more advantageously determined at the 202.5 nm resonance line which gives a characteristic concentration of about 0.1 mg/L 1% with a three slot burner head and about 1–2 mg/L 1% [973] with a single slot burner transverse to the beam path. In a laminar air/acetylene flame, magnesium appears to be largely free of interferences. ALLAN [36] found no interferences due to sodium, potassium, calcium or phosphate. WILLIS [1325] [1332] verified this finding, although he found that protein enhanced the magnesium signal by about 8%. This effect can be eliminated by adding EDTA. DAWSON and HEATON [279] came to the same results. BELCHER and BRAY [98] found that phosphate and silicon up to 100 mg/L had no influence on magnesium, but that aluminium interfered somewhat. For the analysis of fertilizers McBRIDE [846] found that the absorption of 0.8 mg/L magnesium is not influenced by even 2000 mg/L phosphorus (as phosphate). Even with an excess of phosphorus of 10^5 times, SLAVIN [1145] could only find a slight influence due to viscosity, but no interferences.

Work by ZETTNER [1379] showed how dependent these interferences are on the burner design. He modified the burner employed by SLAVIN so that larger droplets reached the flame, and found that with an addition of 10 mg/L phosphorus the magnesium signal was depressed by 10%. DAVID [269] and STEWART *et al* [1175] similarly found that interferences due to phosphate and sulfate occurred with other burner designs and that calcium enhanced the absorption of magnesium. HUMPHREY [541] observed no interferences from numerous cations in considerable excess; only 500 mg/L silicon reduced the absorption of magnesium by 4%.

These manifold investigations make it clear that through the use of an air/acetylene flame and a well designed burner, only the highest concentrations of silicon and aluminium can interfere with the determination of magnesium somewhat, and that no phosphate interferences, even in the highest concentrations, are to be expected. A very comprehensive study on magnesium spinel ($MgO \cdot Al_2O_3$) interferences was made by HARRISON and WADLIN [475]. RUBEŠKA and MOLDAN [1069] demonstrated that this mixed oxide really is formed by collecting the particles in an air/acetylene flame and identifying them by direct X-ray analysis. HARRISON and WADLIN found this interference is very strongly dependent on the height of observation above the burner slot and on the stoichiometry of the flame. While 100 mg/L aluminium depressed the magnesium absorption by 65% directly above the burner slot, the effect was less than 5% at a height of 20–30 mm. Although the sensitivity for magnesium increases with an increasing excess of fuel gas, the interference due to aluminium is more marked in a reducing flame, so that it is preferable to work with somewhat reduced sensitivity in an oxidizing flame in the presence of this element. Titanium and zirconium give similar interferences with magnesium and exhibit similar behavior to aluminium.

AMOS and WILLIS [48] found that the sensitivity for magnesium was reduced by about 50% in a nitrous oxide/acetylene flame. Furthermore, magnesium is ionized by about 6% in this flame so that the addition of potassium (1000 mg/L) is recommended. Although this flame is not necessary for the determination of magnesium, it can be of use to eliminate the above interferences [150]. NESBITT [906] verified this for the aluminium interference occurring during the determination of magnesium in silicate rocks. FLEMING [378] observed that spinel formation was dependent on the stoichiometry of the nitrous oxide/acetylene flame and this was verified by HARRISON and WADLIN [475]. According to the fuel gas/oxidant ratio, an increase as well as a decrease in the magnesium signal can take place in the presence of aluminium. HARRISON and WADLIN found that an increase in the signal was dependent upon the height of observation above the burner slot and that at a certain height the influence was practically zero.

While the sensitivity for magnesium is the same in an air/acetylene as in an air/propane flame, a large number of interferences occur in the latter. HORN and LATNER [533] were unable to find interferences due to Na, K, Ca and phosphate in this flame, but JONES and THOMAS [586] reported both enhancement and depression due to these elements in the same flame. HALLS and TOWNSHEND [459] found a very marked interference due to phosphate and various other ions in the air/propane flame. WALLACE [1281] found a reduction of 90% in the magnesium signal due to 200 mg/L aluminium and considerable interferences from Cu, Mn, Ni and Pb. ANDREW and NICHOLS [54] found that even 0.5 mg/L Al and 0.2 mg/L Si caused a reduction of more than 50%. FIRMAN [370] and also ELWELL and GIDLEY [7] observed influences from a large number of acids and cations. The work of ANDREW and

NICHOLS [55] shows how unsuitable cool flames are for the determination of magnesium. These authors examined the influence of two "identical" gases on various interferences and found considerable variations with different gas mixtures. SUNDERMAN and CAROL [1202] examined an air/hydrogen flame for the determination of magnesium and found—in contrast to an air/acetylene flame—a depression of the signal due to protein. FLEMING and STEWART [379] found numerous interferences in a turbulent oxygen/acetylene flame, in part with non-uniform courses. In a turbulent oxygen/acetylene flame, RAMAKRISHNA [1014] discovered all interferences that also occur during the determination of calcium, but to a somewhat reduced extent.

YOZA and OHASHI [1364] used atomic absorption as a detector for magnesium in gel chromatography.

Magnesium is also one of the most sensitive elements for determination by the graphite furnace technique. Since it is present in most samples in concentrations that can easily be determined by the flame technique, there is little requirement for the increased sensitivity of the graphite furnace technique. The greatest problem with this technique is the risk of contamination; this problem has been treated by TSCHÖPEL *et al* [2987]. SMITH and COCHRAN [2891] determined magnesium in saturated salt solutions using uncoated graphite tubes and an increased argon flowrate during atomization to reduce the sensitivity.

10.26. Manganese

Manganese can be determined in an air/acetylene flame without major interferences. The best characteristic concentration of 0.03 mg/L 1% can be achieved at the 279.5 nm resonance line; in practice, the triplet 279.5/279.8/280.1 nm is mostly employed since the use of greater slit widths brings a marked improvement to the signal-to-noise ratio. When using the triplet (approx. 0.7 nm spectral bandpass), the characteristic concentration is 0.1 mg/L 1% and the detection limit is about 0.001 mg/L. Manganese is an exception in that several lines can be used simultaneously without the usual disadvantages. The three triplet lines have almost equal sensitivity so that a greater loss in sensitivity and strong curvature of the analytical curve are avoided. Each individual line naturally gives better linearity. For the determination of higher manganese concentrations, the 403.1 nm resonance line with a characteristic concentration of about 0.3 mg/L 1% is more suitable. In this case, a spectral bandpass of 0.2 nm is necessary to separate the 403.3 nm resonance line, otherwise there is considerable curvature of the analytical curve.

ALLAN [37] found no interferences from Na, K, Ca, Mg, and phosphate for the determination of manganese in an air/acetylene flame. PLATTE and MARCY [989] found a 20% reduction of the manganese signals when small quantities of silicon were present; by adding 200 mg/L calcium (as chloride), no further interferences due to 1000 mg/L silicon and numerous other cations could be observed. BELCHER and KINSON [100] found no interference from silicate when only the inner flame cone without the edges was observed. Silicate interference can be eliminated by using a nitrous oxide/acetylene flame in which the characteristic concentration is about 0.3 mg/L 1%. HUSLER [547] determined manganese in steel in this flame without interferences and BARNETT [73] found virtually no interferences from acids in this flame or in a fuel gas weak air/acetylene flame. Interferences first became noticeable from acid concentrations between 5% and 10% (v/v).

MANSELL [814] described the extraction of manganese in detail and FELDMANN, BOSSHART and CHRISTIAN [354] examined the sensitivity of manganese in four different solvents. MANSELL and EMMEL extracted manganese and other trace elements from a concentrated brine solution [815]. CALKINS [202] separated manganese from a large excess of aluminium by complexing it with 8-hydroxyquinoline and extracting with chloroform.

OLSEN and SOMMERFELD [928] have drawn attention to the instability of many metal chelates, especially those of manganese. Immediately after extraction, precipitation begins, leading to a loss of sensitivity and blockage of the nebulizer. Directly after the extraction of the diethyl-dithiocarbamate complex in MIBK, these authors evaporated the solution to dryness and then took the residue up in 0.1 M HCl/acetone (1:1). This solution was stable for at least two weeks.

GENC *et al* [2298] found that atomization of manganese in a graphite furnace is via the solid oxide which volatilizes and is atomized in the vapor phase. They observed that the appearance temperature is 1205 °C, regardless of whether the metal was present as chloride, sulfate or nitrate.

In the earlier literature, numerous references are made to severe interferences due to chlorides [72] [1109] [2073]. More recent investigations by SMEYERS-VERBEKE *et al* [2887] [2888] and by HAGEMANN *et al* [2337] also refer to severe signal depressions due to magnesium chloride and calcium chloride. DOKIYA *et al* [2202] compared the interferences due to calcium chloride in various graphite furnaces and found that they were markedly lower in larger tubes than in smaller ones. MCARTHUR [2649] found no interferences for the determination of manganese in sea-water with 3.5% salt content when ammonium nitrate was added and a slow rate of heating was used. KLINKHAMMER [2475], on the other hand, preferred extraction with 8-hydroxyquinoline in chloroform and back extraction of the manganese with nitric acid for the determination in sea-water. BONILLA [2091] found few interferences for the determination of manganese in tissue in nitric acid solution.

MANNING and SLAVIN [2619] found that for atomization off the L'VOV platform in a stabilized temperature furnace, virtually all of the interferences reported in the literature disappeared. Using this system, practically the same freedom from interferences was attained as HAGEMANN *et al* [2337] reported for a WOODRIFF-type furnace.

In later work, SLAVIN *et al* [2877] [2882] investigated magnesium nitrate as a matrix modifier for manganese and found that the thermal pretreatment temperature could be increased to 1400 °C. Using matrix modification and atomization off the L'VOV platform in a stabilized temperature furnace, manganese can be determined free of interferences in virtually all matrices. FERNANDEZ *et al* [2244] found that for certain samples, such as sea-water, Zeeman effect background correction is necessary. The optimum temperature for atomization off the L'VOV platform is 2300 °C.

10.27. Mercury

Because of the unique position that mercury occupies, both with respect to its ecological significance and its analytical determination, it will be treated in somewhat greater depth.

The natural abundance of mercury is very low at $8 \times 10^{-6}\%$ and bears no relationship with the significance it has gained in recent years. The most outstanding property of this element is without doubt its volatility. All compounds are volatile at temperatures below 500°C and decompose easily to the free metal, especially in the presence of reducing agents. Mercury has a high vapor pressure at ambient temperature; 0.0016 millibar at 20°C, corresponding to a concentration of about 14 mg mercury per cubic meter air. The main sources of mercury are weathering and volcanic activity. Resulting from these natural processes, about 40 000 tons of mercury are released worldwide every year and taken up by water and the atmosphere. In addition to this, industrial pollution increasingly plays a significant role; approximately 15 000 tons annually are additionally released into the environment [2978]. Since emissions from industrial plants (alkaline chloride and electrical industry, paints, pesticides, chemicals, etc.) are frequently concentrated locally, they can be particularly dangerous. The most spectacular and without doubt the most catastrophic case in this respect was the accident at Minamata Bay in Japan where at least 46 people lost their lives. A number of chemical works conducted mercury-containing effluents into the bay. The toxic element, accumulated in fish, thus reached human foodstuffs.

Mercury from fossil fuels also plays a given role. Although the mercury content in coal and oil is quite low, nevertheless considerable quantities of the metal are emitted into the atmosphere. Volatilized mercury normally returns relatively quickly from the atmosphere back to the surface of the earth and is taken up by water and by the top layers of soil. The greater part of the mercury reacts to give insoluble compounds such as HgS and $HgSe$ and is thus remineralized. Under influence of microorganisms, a certain portion nevertheless reaches biological cycles and thus takes part in one way or another in ecological processes. It is of great significance in this respect that the various chemical bonding forms of mercury can have substantially different biological effects. The mobile phases are of particular importance; the metallic vapor, which is quickly absorbed into the body via the mucous membranes when inhaled, and various organometallic compounds. In other words, in forms in which all anthropobiological mercury is enriched in our biosphere.

Since mercury must be generally determined in the smallest traces and atomic absorption spectrometry only gives a characteristic concentration of 5 mg/L 1% and a detection limit of 0.2 mg/L at the 253.7 nm line, it is little suited for the determination of mercury. This low sensitivity is due to the resonance line, corresponding to the transition from the ground state to the first excited state, being at 184.9 nm in the vacuum UV range and thus making the determination impossible with normal instruments. KIRKBRIGHT [654] successfully carried out a determination of mercury at this line by using a modified atomic absorption spectrometer with a flushed monochromator and a nitrous oxide/acetylene flame shielded with nitrogen. He achieved a characteristic concentration of 0.05 mg/L 1% and a detection limit of 0.02 mg/L.

Because of the great significance of mercury and because of the simplicity and specificity of atomic absorption spectrometry, various procedures have been developed which improve the sensitivity of the determination. Already before the rediscovery of atomic absorption spectrometry, WOODSON [1354] had described a measurement system for mercury which was later employed by many authors [70] [71] [755] [847] [874] [948] [949] [1388]. BRANDEN-BERGER and BADER [153] [154] also developed a flameless method with which 0.2 ng of mercury could be determined. The boat technique described by KAHN and co-workers [616] also improved the detection limit and 20 ng mercury can be determined in a commercial atomic absorption spectrometer. WHEAT [1314] employed the method described by BRANDENBERGER for the determination of mercury in radioactive samples.

The most successful procedure for the determination of mercury traces, and which is used most frequently nowadays, was proposed by POLUEKTOV *et al* [996] in 1964 and was later thoroughly investigated by HATCH and OTT [481]. The mercury is reduced to free metal in an acid solution by tin(II) chloride or sodium borohydride [2659] and then conducted to an absorption cell mounted in the optical beam of an atomic absorption spectrometer. The detection limit, with this procedure, is 0.02 µg/L mercury and is thus more than three orders of magnitude better than in the flame. Through accumulation on silver or gold, a further increase in sensitivity is possible, with a detection limit of less then 0.1 ng absolute or about 1 ng/L [2448] [2640].

A detailed description of the cold vapor technique for the determination of mercury has already been presented in section 3.4, so that further details in this section are unnecessary. Likewise, interferences and systematic errors have been treated in section 8.4. Next to the numerous investigations in the fields of medicine, toxicology and environmental control [57] [61] [576] [753] [786] [895] [1024] [1251], further areas include the analysis of air [2291] [2434] [2755] [2826], coal [2292] [2679], petrochemicals [2478] and ores [2104].

The various bonding forms of mercury have quite different effects. Mercury sulfide or mercury selenide are virtually insoluble and thus non-poisonous, so that they have little significance for the environment. In contrast, the mobile phases, mercury vapor and certain organometallic compounds, are of great significance. Mercury vapor is absorbed rapidly through the mucous membranes, enters the blood system and is distributed as compounds, leading to chronic poisoning. Soluble, inorganic mercury compounds are equally poisonous.

Mercury is especially biologically active in the form of its organometallic compounds. Virtually only methylmercury chloride and dimethylmercury chloride are produced naturally through the action of microorganisms; other alkyl or aryl mercury compounds only occur in higher concentrations at locations where they are released into the environment in industrial effluents.

It is thus clear that the determination of the total mercury content in a sample often is insufficiently expressive. Unlike virtually any other element, it is necessary to determine not only the total mercury content but also the individual species, particularly for the ecological examinations of biological and geochemical samples. It must also be borne in mind that ionic mercury and organically bound mercury are in equilibrium with each other and that this equilibrium can change rapidly under given circumstances once the sample has been collected. This has been reported for sea-water by STOEPPLER and MATTHES [2921] and also by MATSUNAGA *et al* [2635]; both groups observed that the methylmercury content decreased with a concomitant increase in the ionic mercury content.

Methylmercury chloride is the prevailing species in sea-water. It is formed through biological transformation from ionic mercury on the lower plains in tropical regions and then accumulated in fish, etc. It is in this form that mercury most frequently enters human foodstuffs, although there are other sources. Methylmercury chloride is converted in part to ionic mercury in the human body.

For the determination of the total mercury content, it is important that organomercury compounds are converted quantitatively to the ionic form during digestion of the sample, otherwise they will not be measured using the cold vapor technique. The commonest procedure is an oxidative digestion with potassium permanganate or potassium dichromate in sulfuric acid solution. Some authors first dissolve the sample in nitric acid/sulfuric acid [2512], with the addition of vanadium pentoxide as catalyst if necessary [2220], or in nitric acid under the addition of sodium molybdate [2213], before oxidizing with potassium permanganate, with the addition of sulfuric acid if need be.

VELGHE *et al* [3007] developed a very rapid, semiautomatic procedure for fish samples in which the sample is heated briefly with the double quantity of solid potassium permanganate and concentrated sulfuric acid. The sample dissolves in less than a minute and organomercury compounds are quantitatively broken down.

Other authors proposed digestions in autoclaves using nitric acid/sulfuric acid under the addition of vanadium pentoxide as catalyst and an oxidation with hydrogen peroxide [2137] or with perchloric acid and potassium permanganate [2836]. In both cases the recovery was between 80% and 90%. AGEMIAN *et al* [2007] have pointed out that neither permanganate nor persulfate can be used as the oxidizing agent when relatively high concentrations of chloride are present in the sample. Under these conditions, chloride is oxidized to chlorine gas, so that this procedure cannot be used for sea-water, for example. These authors thus proposed photo-oxidation with UV radiation for water samples with high chloride contents. They investigated seven organomercury compounds and found rates of recovery between 91% and 102%. For sludges and turbid waters with high solids content they proposed an initial digestion in sulfuric acid/dichromate, which extracts mercury very efficiently from particulate matter, before photo-oxidation [2009].

Two procedures are given in the literature for determining the species methyl-, ethyl-, and ionic mercury; one is via the extraction of the organomercury compounds, while the other uses a sequential determination of the individual compounds.

Methylmercury can be extracted from hydrochloric acid solution in the presence of sodium chloride with benzene or toluene and then back extracted with aqueous cysteine acetate solution or a solution of glutathione in ammonia [2636]. The methylmercury chloride in this solution is then oxidized with permanganate/sulfuric acid [2186] or permanganate/peroxosulfate [2087]. The mercury now in inorganic form can be determined in the usual manner. The extracted organic mercury can be reduced directly with tin(II) chloride in the presence of sodium hydroxide and copper(II) [2636].

The sequential procedure is based on the fact that in alkaline medium only inorganic mercury is reduced by tin(II) chloride. The best results are obtained in the presence of 15% or stronger sodium hydroxide solution. Better reproducibility is obtained by adding cysteine to the sample solution before reduction to complex the mercury [2129] [2214] [2445] [2552].

If the sample solution is shaken briefly with 10 M sulfuric acid, the temperature rises to 85 °C. At this temperature mercury is released from phenylmercury without losses, while methylmercury and ethylmercury are unaffected. It is thus possible to determine the inorganic mercury and phenylmercury [2129]. The content of phenylmercury is then obtained by difference between this value and the value for inorganic mercury alone. If the total mercury content is then determined after oxidation with permanganate/sulfuric acid, the difference gives the methylmercury content [2129].

A further variant of the sequential procedure is possible in that alkylmercury in alkaline medium in the presence of cadmium(II) is selectively reduced to mercury by tin(II) chloride [2605]. Cadmium salts can react with organomercury compounds and release inorganic mercury ions; cadmium replaces mercury in the organic compound. For a sequential determination, the procedure is as follows: Inorganic mercury in alkaline medium is first determined by adding tin(II) chloride; a mixture of tin(II) chloride/cadmium(II) chloride is then added to the same solution and the methylmercury determined [2214]. A fresh sample solution can also be used for the second step, so that both inorganic mercury and methylmercury are released by the tin(II) chloride/cadmium(II) chloride mixture at high pH; the methylmercury content can be obtained by difference [2552].

A final possibility is the addition of 18 M sulfuric acid to the sample (so that phenylmercury is released), followed by reduction with tin(II) chloride/cadmium(II) chloride. Since the measurement is performed in acid solution, inorganic mercury and phenylmercury can be determined. If sodium hydroxide is then added in excess, methylmercury can be determined in alkaline solution in the presence of cadmium [3008].

It is important to note that these determinations can only be performed with tin(II) chloride as reductant. If sodium borohydride is used as reductant, inorganic mercury, methylmercury and phenylmercury are all volatilized. The sensitivities of the three species are not identical, however [2974].

The cold vapor technique is the most successful and without doubt the most widely employed technique for the determination of mercury. Nevertheless, attempts have been made to determine mercury by the graphite furnace technique. Except for direct solids analysis [1295], it is essential to stabilize mercury by matrix modification because of its high volatility. ISSAQ and ZIELINSKI [2417] found that 1% hydrogen peroxide stabilized mercury and permitted thermal pretreatment up to 200°C. ALDER and HICKMAN [2021], on the other hand, maintain that hydrogen peroxide alone has no influence, but only in the presence of hydrochloric acid. Hydrogen chloride is present in excess in the vapor phase and stabilizes mercury as $HgCl_2$; hydrogen peroxide has an additional influence, possibly through the formation of an addition compound.

EDIGER [2217] proposed 5% ammonium sulfide solution as stabilizer and was able to thermally pretreat mercury up to 250°C. KIRKBRIGHT *et al* [2470] found that the following modifiers were equally as effective with respect to sensitivity: 1% HNO_3/5% Na_2S; 1% HNO_3/ 0.1% $KMnO_4$/0.5% $AgNO_3$; 1% HNO_3/0.05% $K_2Cr_2O_7$. All three mixtures permit thermal pretreatment temperatures up to 250°C to be applied. Despite the available possibilities, it is unlikely that the graphite furnace technique will gain any great significance for the determination of mercury.

Besides the above procedures, many extraction methods to increase the sensitivity of mercury have been described. Frequently ammonium pyrrolidine dithiocarbamate is used as complexing agent and MIBK as extracting solvent [111] [1323].

OSBORN and GUNNING [942] determined the mercury isotope ratio by atomic absorption spectrometry.

10.28. Molybdenum

Molybdenum can be determined both in a fuel gas rich air/acetylene flame and in a nitrous oxide/acetylene flame. At the 313.3 nm resonance line the characteristic concentration in an air/acetylene flame is about 1 mg/L 1% and in a nitrous oxide/acetylene flame about 0.1 mg/L 1%; the detection limits are 0.1 mg/L and 0.03 mg/L, respectively. Besides its high sensitivity, the nitrous oxide/acetylene flame is also largely free of interferences, so that it is used almost exclusively nowadays.

The earlier studies of DAVID [271] and MOSTYN and CUNNINGHAM [887] were all carried out with a very fuel gas rich air/acetylene flame. However, as the investigations of RANN and HAMBLY [1018] have shown, even under fuel gas rich conditions the flame zone with the highest atom concentration is very limited, so that a large portion of the hollow cathode radiation passes through areas in which the formation of compounds is possible. ROOS [1045] also found that iron caused considerable interferences in the determination of molybdenum; also DAVID [271] and MOSTYN and CUNNINGHAM [887] found interferences in the analysis of steel which they could control by adding 2% ammonium chloride.

In a detailed study on the determination of molybdenum in various flames, STURGEON and CHAKRABARTI [2932] found that the highest free atom concentration was always located within a very narrow area in the flame. An adequate atom density was only found in strongly reducing flames, either a very fuel gas rich air/acetylene flame or a nitrous oxide/ acetylene flame. The authors concluded that atomization is principally via MoO.

The possible interferences in the determination of molybdenum in fuel gas rich air/acety- lene flames have been little investigated and matching of the matrix is frequently necessary. A number of interferences occur in a nitrous oxide/acetylene flame which can lead either to depression or to enhancement of the signal. VAN LOON [1254] found that the addition of 0.1 to 0.3% aluminium eliminated all interferences. WEST et al [1306] accounted this effect of aluminium to an inhibition of the lateral diffusion of molybdenum, leading to an increased atom concentration in the middle of the flame. Especially with organic solvents, which give considerable difficulties in the "fat" air/acetylene flame, the hot nitrous oxide/acetylene flame shows its advantages.

KIRKBRIGHT [655] complexed molybdenum with 8-hydroxyquinoline in the presence of fluoride and EDTA as masking reagents and extracted it into butanol, since this burns well in the nitrous oxide/acetylene flame. URE [1250] treated soil extracts with tri-N-octylamine and 2% n-octanol in petrol and then back extracted the molybdenum with ammonia gas into the aqueous phase from which he determined it in a nitrous oxide/acetylene flame. DELAUGHTER [289] used toluene-3,4-dithiol as the complexing agent for extraction into MIBK, and BUTLER and MATHEWS [195] complexed molybdenum with ammonium pyrrol- idine dithiocarbamate and also extracted into MIBK. However, it was found necessary to mask iron with citric acid before the extraction.

Considering the importance of molybdenum, it is surprising that so little is reported in the literature on its determination by the graphite furnace technique. SNEDDON et al [2895] in- vestigated the atomization mechanism and used X-ray analysis to identify the compounds formed from ammonium molybdate at various temperatures in the graphite furnace. They found that the appearance temperature was 1900°C. The atomization is from molybdate via MoO_3, MoC, Mo_2C, and possibly solid Mo, to gaseous Mo.

RUNNELS *et al* [2813] showed that molybdenum carbide is formed when molybdenum solutions are dispensed into an uncoated graphite tube. They treated the surface of the tube with elements such as lanthanum or zirconium to prevent contact of the sample with graphite. DAGNALL [249] reported an increase of 70% in the sensitivity when the graphite tube was saturated with tungsten and KUZOVLEV *et al* [719] found an increase in the sensitivity by a factor of 2.5 when the tubes were coated with a mixed carbide of tantalum and niobium. As can be seen from figure 137, pyrocoating of the graphite tubes and a fast rate of heating bring a substantial improvement in the sensitivity for molybdenum.

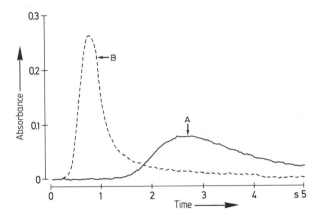

Figure 137. Highly resolved signals for molybdenum with the graphite furnace technique. **A**: "Normal" heating with voltage control; **B**: Ultrafast heating with optical temperature control (Perkin-Elmer HGA-500 Graphite Furnace).

BODROV and NIKOLAEV [2089] determined molybdenum in steel using a L'VOV furnace and found no interferences other than from high concentrations of titanium and niobium. NAKAHARA and CHAKRABARTI [2684] determined molybdenum in synthetic sea-water after selective volatilization of the salt matrix. Initially, marked signal depression was caused by NaCl, KCl and Na_2SO_4, and signal enhancement by $MgCl_2$ and $CaCl_2$. According to the authors, the addition of $MgCl_2$ eliminated the depressing influences.

Using magnesium nitrate as matrix modifier, thermal pretreatment temperatures up to about 1700°C can be applied; for an ultrafast heating rate the optimum atomization temperature is 2700°C.

10.29. Nickel

Nickel is one of the metals frequently determined by atomic absorption spectrometry. The best characteristic concentration of 0.04 mg/L 1% is obtained in an air/acetylene flame at the 232.0 nm resonance line; the detection limit is 0.004 mg/L. To be able to use the 232.003 nm line meaningfully, a spectral bandpass of 0.2 nm is necessary since the two strong emission lines at 231.716 nm and 232.138 nm otherwise cause a considerable curvature in the analytical curve and a strong reduction in the sensitivity. Since the 232.0 nm line, even with a spectral bandpass of 0.2 nm, gives a markedly non-linear analytical curve, it is not very suitable for the determination of higher nickel concentrations. For these, the 341.5 nm resonance line with a characteristic concentration of 0.2 mg/L 1% is more suitable. This line even permits the use of wider spectral bandpasses (up to 0.7 nm) and then gives a good signal-to-noise ratio. Further resonance lines are compiled in table 32. However, slight spectral interferences can occur at several of these lines, especially with multielement lamps which also contain cobalt; a critical check of the lines and the signals is recommended.

Table 32. Ni resonance lines

Wavelength nm	Energy level (K)	Characteristic concentration (mg/L)
232.0	0–43 090	0.04
231.1	0–43 259	0.07
341.5	205–29 481	0.2
305.1	205–32 973	0.25
346.2	205–29 084	0.35

The determination of nickel in an oxidizing (fuel gas weak) air/acetylene flame appears to be largely free of interferences. KINSON and BELCHER [645] found no interferences due to high concentrations of Co, Cr, Cu, Mn, Mo, Al, V and W. Equally, HCl, HNO_3, H_2SO_4 and H_3PO_4 had no influence. These findings were verified by PLATTE and MARCY [989], although great care must be taken in an exact adjustment of the burner and flame gases [73] [1199]. SUNDBERG [1199] related the effects (enhancing and depressing) of other metals (Co, Cr, Cu, Fe, Mn, Zn) found with non-exact gas settings and height of observation to the dissociation energies of the corresponding monoxides. In a less sharp flame, interferences from iron and chromium become noticeable. For high precision analyses, WELZ and SEBESTYEN [1299] found it necessary to match the matrix to maintain a relative accuracy of 0.3%. The potential interferences of iron and chromium can be eliminated by the use of a nitrous oxide/acetylene flame; the characteristic concentration at the 232.0 nm resonance line is about 2 mg/L 1%.

Nickel can be complexed with ammonium pyrrolidine dithiocarbamate and extracted with MIBK [1170]. BURREL [183] used multiple enrichment steps such as co-precipitation with iron(III) hydroxide, which is then later removed [307], to determine the smallest traces of nickel.

For the determination of nickel by the graphite furnace technique, CRUZ and VAN LOON [245] found substantial signal enhancement through potassium nitrate, iron nitrate, and several other nitrates. Calcium chloride suppressed the signal completely. FULLER [393] found nickel losses during thermal pretreatment and JULSHAMN [2441] observed signal depression due to perchloric acid.

For the analysis of sea-water, KINGSTON *et al* [2468] separated nickel on an ion exchange column, while other authors preferred extraction with an organic solvent and back extraction with nitric acid [2137] [2432]. BOYLE and EDMOND [2100] used co-precipitation with cobalt pyrrolidine dithiocarbamate for separation.

MAIER *et al* [2608] determined nickel directly in surface waters. CARRONDO *et al* [2133] dispensed suspensions of sewage sludge in nitric acid directly into the graphite tube. The direct determination against reference solutions acidified with nitric acid brought good results.

IU *et al* [2419] determined nickel in soil extracts after extraction with chloroform, while PEDERSEN *et al* [2737] extracted with xylene. PETROV *et al* [2746] determined soil extracts directly and found the method both quick and reliable.

VÖLLKOPF *et al* [3022] developed a direct procedure for the determination of nickel in serum; they found that a good quality pyrocoating was essential. BROWN *et al* [2112] developed the IUPAC reference method for the determination of nickel in serum and urine via extraction with APDC in MIBK. ADAMS *et al* [2000] also found that extraction procedures for nickel in urine were superior to the direct procedure. DUDAS [2207] established, however, that extraction solutions should not be dried immediately after dispensing into the graphite tube, but after a pause of about two minutes.

HINDERBERGER *et al* [2372] established that nickel can be determined directly in blood, liver tissue and urine without interferences when primary ammonium phosphate is used as matrix modifier and atomization is off the L'VOV platform. SLAVIN *et al* [2877] [2882] investigated magnesium nitrate as matrix modifier and found that nickel could be thermally pretreated up to 1400 °C. The determination is virtually free of interferences when atomization is off the L'VOV platform in a stabilized temperature furnace. Under these conditions the optimum atomization temperature for nickel is about 2500 °C.

VIJAN [3015] attempted to volatilize nickel as the carbonyl by reducing it to metal from its compounds with sodium borohydride and then applying carbon monoxide under pressure. He conducted the resulting gas into a heated quartz tube and obtained a characteristic mass of 20 pg 1 %. However, he found that the reaction was not quantitative since an equilibrium concentration of nickel carbonyl is obtained.

10.30. Niobium

Niobium can be determined in a nitrous oxide/acetylene flame with a characteristic concentration of about 40 mg/L 1 % and a detection limit of about 2 mg/L at the 334.3 nm resonance line [795] with a spectral bandpass of 0.2 nm. Since niobium is noticeably ionized in the hot flame, the addition of about 0.1 % potassium (as chloride) to sample and reference solutions is recommended. WALLACE *et al* [3029] found an improvement in the linearity, sensitivity, precision and detection limit for niobium by adding aluminium to hydrofluoric acid solutions. In the presence of 1 % HF and 0.2 % Al the detection limit is 0.75 mg/L. It is possible that aluminium reduces the tendency to forming refractory oxides and also facilitates atomization. A reduction of the lateral diffusion is a possible explanation for the enhanced sensitivity. Up to the present, nothing has been reported on interferences.

10.31. Non-Metals

Iodine, phosphorus and sulfur are treated individually since there are direct atomic absorption procedures for these non-metals. The remaining non-metals cannot be determined directly by atomic absorption spectrometry since their resonance lines lie in the vacuum UV range.

BECKER-ROSS and FALK [2067] nevertheless reported the successful determination of *bromine* in a special graphite furnace with a vacuum monochromator at 148.86 nm. They obtained a detection limit of 1.5 ng Br.

DAGNALL, THOMPSON and WEST used molecular emission for the determination of *sulfur* at an S_2 band [254] and *phosphorus* at an H—P—O band [259]; in the presence of indium they determined the *halogens* by means of InCl, InBr and InI emission bands [261]. Similar work with special application to biological samples has been published by GUTSCHE and HERRMANN [450–453]. GUTSCHE *et al* [2330] used a graphite furnace as a bromine-specific detector for gas chromatography using the indium method.

For the indirect determination of non-metals by atomic absorption spectrometry, two different ways have been followed: The utilization of chemical interferences, or the use of known precipitation reactions with metals. CHRISTIAN and FELDMAN [252] used the first method to determine *orthophosphates, sulfates, sulfides, iodates, iodides, glucose, protein, 8-oxiquinoline* and other complexing agents by examining their influence on the determination of calcium, iron and chromium and placing the extent of this influence in a relationship to the concentration of the interfering ions. KUNISHI and OHNO [717] investigated the indirect determination of sulfate very thoroughly and employed the suppressive effect on the determination of iron and its elimination by lanthanum for the determination of sulfate. With an ion ratio of $La^{3+} : SO_4^{2-} = 2:3$ the original sensitivity of iron is again obtained. The authors speak of a sharp "change" which makes very precise determinations possible. BOND and O'DONNEL [142] determined *fluoride* ions in a similar manner by quantitatively evaluating their influence on the determination of magnesium in an air/coal gas flame and on zirconium in a nitrous oxide/acetylene flame.

SAND *et al* [2819] determined silicate, phosphate, and sulfate via their influence on the atomization of calcium.

The second procedure, formation of an insoluble precipitate with a metal cation with subsequent determination of this metal, was employed by EZELL [336] and WEST and ERLUND-HELMERSON [1312] for the determination of *chloride* in various materials. The chloride ion is precipitated with silver nitrate and either the precipitate is dissolved in ammonia solution and analyzed for its metal content, or the excess metal is measured in the solution after an exactly measured addition of silver nitrate [82].

GAMBRELL [404] determined chloride in water through end point determination with silver nitrate by atomic absorption. Two different quantities of silver nitrate are added to the sample solution, the absorbance for silver measured and extrapolated against zero.

MANAHAN and KUNKEL [790] determined *cyanide* by utilizing the solubility of Cu(II) from basic copper carbonate in alkaline medium. The complexed copper, as $[Cu(CN)_3]^-$, was determined by atomic absorption; the characteristic concentration was 2×10^{-15} mol CN^-. JUNGREIS and AIN [2444] also determined cyanide by filtering the solution through silver

wool and then determining the complexed silver using the graphite furnace technique. KOV-
ATSIS [2504] determined *carbon disulfide* via reaction with zinc acetate and N,N-dibenzyl-
amine. The zinc dibenzyldithiocarbamate formed is extracted in toluene and the zinc determined. MANAHAN and JONES [789] also described a specific detector system for *chelating
agents* for liquid chromatography. A solution containing a chelating agent is conducted
through a short column containing a chelate ion exchanger in the Cu form and then directly
to an atomic absorption spectrometer set for copper. The quantity of copper is directly proportional to the quantity of chelating agent; the detection limit is 5×10^{-7} mmol.

KUMAMARU *et al* determined *nitrate* [716] and organic molecules through co-extraction
with metal complexes. Nitrate and *phthalic acid* [713] [714] were both extracted as ion pairs
from bis-(neocuproin)-copper(I) with MIBK. *Pentachlorophenol* [1359] was also extracted as
an ion pair with tris-(1,10-phenanthroline)-iron(II) in nitrobenzene; the concentrations of
copper and iron respectively in the extracts are proportional to the concentrations of the
ions or molecules under study and can be determined by atomic absorption spectrometry.

WOODIS *et al* [3078] determined *biuret* in fertilizer by treating an alcoholic solution of biuret
and copper with a strong base. A copper-biuret complex is formed, the excess copper precipitates, and the copper in the complex is determined. OLES and SIGGIA [2707] determined *aldehydes* by oxidizing with a silver-ammonia complex (Tollen's reagent), separating the reduced silver and dissolving in nitric acid, and then determining by AAS. They determined
1,2-diols [2708] by oxidizing with periodic acid; the iodate formed is precipitated as silver
iodate, dissolved in ammonium hydroxide and the silver determined by AAS.

10.32. Osmium

SLAVIN [12] first reported in 1968 that WILLIS had found strong absorption for osmium at
various wavelengths. Thereupon FERNANDEZ [356] examined ten resonance lines for their
sensitivity and found a characteristic concentration of 1 mg/L 1% at the 290.9 nm reson-

Table 33. Os resonance lines

Wavelength nm	Energy level (K)	Characteristic concentration (mg/L 1%)		
		N$_2$O/Acetylene [356]	Air/Acetylene [356]	Air/Propane [788]
290.9	0–34365	1.0	5.0	17
305.9	0–32685	1.6	6.4	20
263.7	0–37909	1.8		
301.8	0–33124	3.2		
330.2	0–30280	3.6		
271.5	0–36826	4.2		
280.7	0–35616	4.6		
264.4	0–37809	4.8		
442.0	0–22616	20		
426.1	0–23463	30		

ance line employing a nitrous oxide/acetylene flame. The detection limit appears to be little better than 0.1 mg/L. Further resonance lines, with their sensitivities, are compiled in table 33.

By using a fuel gas rich air/acetylene flame a characteristic concentration of about 5 mg/L 1% can be achieved [356]. MAKAROV *et al* [788] found a characteristic concentration of 17 mg/L 1% in an air/propane/butane flame. OSOLINSKI and KNIGHT [943] determined osmium in thiourea complexes and in chloroform extracts. They draw particular attention to the toxicity of OsO_4 and the necessary safety precautions.

GLADNEY and APT [2303] found that neutral solutions of osmium are very instable and must be prepared fresh daily. Solutions prepared in 1 M HCl are stable for two months when stored in glass, quartz or polyethylene bottles.

10.33. Palladium

Palladium can be determined in a very sharp (fuel gas weak) air/acetylene flame with a characteristic concentration of 0.15 mg/L 1% and a detection limit of about 0.02 mg/L. Two resonance lines at 247.6 nm and 244.8 nm give approximately equal results. However, the latter, even with a spectral bandpass of 0.2 nm, gives a very curved analytical curve; for practical work the line at 247.6 nm is to be preferred. Two further resonance lines at 276.3 nm and 340.4 nm each give a characteristic concentration of about 0.4 mg/L 1%. Instead of an air/acetylene flame, LOCKYER and HAMES [759] and STRASHEIM and WESSELS [1181] used an air/propane flame, which is also preferred by a number of authors for the determination of palladium [296]. Although these authors observed no interferences, the determination of palladium, even in the air/acetylene flame, does not appear to be completely free of interferences; especially high concentrations of other noble metals can exert certain influences. The acid concentration should be matched.

JANSSEN and UMLAND [571] eliminated the interferences that occur during the analysis of noble metals by adding 1% sodium and 1% copper. TOFFOLI and PANNETIER [1234] preferred 0.5% lithium with the possible addition of copper. SEN GUPTA [1114], who has published a good review on noble metal analysis [1116], added 0.5% copper and 0.5% cadmium to eliminate interelemental interferences. ADRIAENSSENS and VERBEEK [17] found that in 2% potassium cyanide solution the majority of interferences which are observed in acid solutions did not occur. Only gold and platinum influenced palladium somewhat; these effects disappeared, however, after addition of silver. HEINEMANN [2361] found that the determination of palladium is little influenced by rhodium and not influenced by platinum. The best conditions are obtained when lanthanum is added as a buffer.

HARRINGTON [469] analyzed colloidal palladium solutions and found that these gave the same absorbance values as true palladium(II) solutions when the burner height was optimized.

ERINC and MANGEE [330] described an extraction of palladium from acid solution with hexone after complexing with pyridine thiocyanate. MOJSKI [2671] extracted palladium from chloride, bromide and iodide solutions with di-n-octylsulfide in cyclohexane. RUBEŠKA *et al* [2810] extracted palladium from geological samples with dibutylsulfide in tol-

uene and performed the determination both in the flame and in the graphite furnace. Rowston and Ottaway [2804] established the optimum conditions for the determination of
palladium by the graphite furnace technique and found a characteristic concentration of
4.5 μg/L 1%.

10.34. Phosphorus

The resonance lines for phosphorus are at around 178 nm in the vacuum UV range, so that
it cannot be determined with normal atomic absorption spectrometers. However, Kirkbright [654] successfully determined phosphorus in a modified atomic absorption instrument with a flushed monochromator and a nitrous oxide/acetylene flame shielded with nitrogen. Of the three resonance lines 177.5 nm, 178.3 nm and 178.8 nm, the 178.3 nm line
gave the best characteristic concentration of 5.4 mg/L 1%. Kirkbright determined phosphorus in meat extracts and milk powder by this procedure.
Walsh [1283] determined phosphorus in an atomization cell using a vacuum spectrometer.
This procedure has found no practical applications up to the present. L'vov and Khartsyzov [3103] reported the successful determination of phosphorus at the 213.55 nm/
213.62 nm and 214.91 nm lines which emanate from metastable states. In a nitrous oxide/
acetylene flame and with a phosphorus hollow cathode lamp, Manning and Slavin [809]
found a characteristic concentration of 290 mg/L 1% at the 214 nm line and 540 mg/L 1%
at the 215 nm line. Thus phosphorus concentrations of 0.2–5% can be well determined. The
authors also point out that phosphorus interferes with numerous cations in atomic absorption spectrometry and it is thus equally possible that numerous cations can interfere in the
determination of phosphorus. Kerber *et al* [636] compared the determination of phosphorus by AAS and FES at the HPO band at 526 nm. They found only slight interferences with
AAS. Syty [1213] found that this HPO emission is considerably increased when a cooled
plate is mounted behind the flame and he used this procedure to determine phosphorus in
phosphate rocks [1214].
The determination of phosphorus at the 213.5/213.6 nm doublet was first reported by
L'vov and Khartsyzov [3103]. In their graphite furnace they obtained a detection limit of
0.2 ng P. Wider interest in this direct determination was not forthcoming, however, until a
satisfactory electrodeless discharge lamp was available for phosphorus [2061]. Ediger
[2218] established the optimum conditions for the determination of this element and found
that the addition of lanthanum as matrix modifier increased the sensitivity by a factor of
about six. He obtained a detection limit of about 0.1 mg/L.
In later work, Ediger *et al* [2219] found that phosphoric acid gave virtually no signal, but
that calcium phosphate could be determined very well. They concluded that phosphoric
acid is lost in molecular form, but that the addition of calcium, or even better lanthanum,
prevents this.
Prévôt and Gente-Jauniaux [2760] determined phosphorus in edible oils and found that
an electrodeless discharge lamp, a spectrometer with high performance in the far UV, and a
graphite furnace with good temperature program selection were essential. They diluted the
oil with MIBK and determined phosphorus directly with a characteristic mass of 0.5 ng 1%.

They found that the addition of lanthanum was only necessary for aqueous solutions; the oil had the same stabilizing effect as lanthanum. Nevertheless, SLIKKERVEER *et al* [2884] found that the addition of lanthanum to a variety of oil samples improved the results.

L'VOV and PELIEVA [2585] determined phosphorus by introducing the sample on a tungsten wire loop into a preheated graphite furnace. Sample pretreatment was not required for this procedure, which brought an improvement in the sensitivity by a factor of 20–30 compared to atomization off the tube wall.

PERSSON and FRECH [2741] investigated the theoretical and practical factors influencing the determination of phosphorus in a graphite furnace. They found that the risk of losses during thermal pretreatment and during heat-up is very severe. Depending on the partial pressure of oxygen, the formation of gaseous PO at temperatures below 1300°C is very probable. A reducing environment is essential, so that uncoated graphite tubes are especially advantageous. On the other hand, hydrogen can be formed in uncoated tubes when the sample is dispensed onto the wall. This can lead to the formation of gaseous HCP and thus to losses. Losses of phosphorus as the dimer P_2 have been observed at temperatures above 1350°C, so that a fast rate of heating for atomization is particularly important. The above authors found that reproducible results can only be obtained when the heating rate and the final temperature in the furnace can be optimized independently. Under non-isothermal conditions, as reported by EDIGER *et al* [2219], no signal is obtained for phosphoric acid, while under isothermal conditions the same sensitivity is obtained for phosphoric acid and calcium phosphate.

Investigations in the present author's laboratory have shown that the best results are obtained for atomization off a L'VOV platform made of pyrolytical graphite using an uncoated tube in a stabilized temperature furnace [3061]. Under these conditions the partial pressure of oxygen is very low, so that the formation of PO is highly improbable. The formation of hydrogen is also prevented since the sample solution does not come into contact with uncoated graphite. Resulting from atomization off the platform, heating rate and final temperature in the furnace are independent of each other. With the addition of 0.1% lanthanum as matrix modifier, it was possible to determine phosphorus in steel directly against reference solutions that only contained the same amount of lanthanum. Since the steel causes intense, structured background attenuation, it is necessary to use Zeeman-effect background correction, especially for low phosphorus concentrations.

Using lanthanum as matrix modifier, phosphorus can be thermally pretreated up to about 1350°C without losses; the optimum temperature for atomization off the L'VOV platform is 2600°C.

Various authors determined phosphorus via the emission of monoxide or hydroxide bands [2350]. CAMPBELL and SEITZ [2124] volatilized the sample in a graphite furnace and conducted the vapor into an air/hydrogen flame.

Various indirect procedures for the determination of phosphorus have been proposed [544]. They are based on the formation of ammonium phosphorusmolybdate or the free heteropolyacid with the subsequent determination of molybdenum by atomic absorption spectrometry. WILSON [1338] dissolved the ammonium phosphorusmolybdate precipitate in ammonia solution and determined the molybdenum directly from this solution. ZAUGG and KNOX [1372] shook acid ammonium molybdate solution and 2-octanol with the phosphorus-containing sample and determined molybdenum directly from the organic phase. In a later report [1373] these authors described the application of this extraction procedure to various biological problems. A similar extraction with butyl acetate as solvent has been described by KUMAMARU et al [715].

MANNING and FERNANDEZ [804] observed an absorption for phosphorus in the graphite furnace at a lead ion line at 220.3 nm, which they attributed to a PO_x radical.

10.35. Platinum

Platinum has a multitude of resonance lines of which the line at 265.9 nm with a characteristic concentration of about 1 mg/L 1% is the most sensitive. The detection limit is 0.04 mg/L in a sharp, fuel gas weak air/acetylene flame. Several further important resonance lines are compiled in table 34.

Table 34. Pt resonance lines

Wavelength nm	Energy level (K)	Characteristic concentration (mg/L 1%)
265.9	0–37 591	1
306.5	0–32 620	2
299.8	776–34 122	5
271.9	824–37 591	12
304.3	824–33 681	18

LOCKYER and HAMES [759] found no interferences for the determination of platinum in an air/propane flame. On the other hand, STRASHEIM and WESSELS [1181] reported interferences from numerous cations, which they could largely eliminate by adding 2% copper (as sulfate). According to a number of investigations, both strong acids and other noble metals interfere in the determination of platinum; matching the matrix and the use of a nitrous oxide/acetylene flame largely eliminates these interferences.

JANSSEN and UMLAND [571] found numerous interelemental effects for the determination of platinum in the presence of other noble metals; they eliminated these effects by adding 1% sodium and 1% copper. At the same time they found that these additions increased the sensitivity for platinum by around 50%.

TOFFOLI and PANNETIER [1234] preferred the adition of 0.5% lithium with possibly some copper, while SEN GUPTA [1114] employed 0.5% copper and 0.5% cadmium as buffer.

ADRIAENSSENS and VERBEEK [17] found that in 2% potassium cyanide solution the majority of interelemental interferences observed during the analysis of noble metals do not occur. They used a fuel gas weak, sharp air/acetylene flame and achieved an accuracy of 2%, even in the presence of considerable excesses of other noble metals.

PITTS *et al* [986] [987] found numerous interferences in an air/acetylene flame which they were able to eliminate by the addition of lanthanum. As an alternative, these authors recommend a nitrous oxide/acetylene flame since no more interferences occur, even though the sensitivity is reduced to a fifth. Even high concentrations of other noble metals no longer interfere.

MOJSKI [2671] extracted platinum from chloride, bromide, and iodide solutions with di-n-octylsulfide in cyclohexane. MACQUET and THEOPHANIDES [2597-2599] investigated the effects of cis-trans isomeric complexes of platinum, and also chelated and non-chelated DNA-platinum complexes, on the sensitivity of AAS. They found that there was a direct relationship between the stereochemistry and the sensitivity, the most stable complexes exhibiting the highest absorbance.

ADRIAENSSENS and KNOOP [16] found only slight interelemental effects for the determination of platinum in the presence of iridium and rhodium. PERA and HARDER [2739] determined platinum in biological materials in good agreement with other procedures. JANOUŠ-KOVÁ *et al* [570] determined platinum in aluminium oxide catalysts and found no interferences due to numerous other ions, even in 100-fold excess. Merely large quantities of strontium or 0.5 M nitric acid depressed the signal somewhat. TELLO and SEPULVEDA [2957] determined platinum in sand. Platinum is separated from interferents and enriched by cupellation. The fusion beads are dissolved in *aqua regia* and analyzed directly.

ROWSTON and OTTAWAY [2804] established the optimum conditions for determination of platinum by the graphite furnace technique. They obtained a characteristic concentration of 23 μg/L 1%. Platinum can be thermally pretreated up to about 1400°C in 0.2% nitric acid; using an ultrafast rate of heating, the optimum atomization temperature is 2500°C.

10.36. Potassium

Potassium is frequently determined with sufficient sensitivity and precision in simple flame photometers. However, atomic absorption spectrometry, with the choice of various resonance lines and flames, offers greater variability and hence optimization for each analytical problem. Furthermore, in the presence of complex matrices, emission is subjected more to interferences than absorption.

The detection limits in atomic absorption and flame emission spectrometry are very similar; while the best values in emission are 0.003 mg/L, for atomic absorption the characteristic concentration at the 766.5 nm resonance line is 0.03 mg/L 1% and the detection limit about 0.001 mg/L. Nevertheless, this detection limit is very dependent upon the instrument, since many photomultipliers only have low sensitivity at the potassium wavelength and thus exhibit an unfavorable signal-to-noise ratio.

Beside the most sensitive line at 766.5 nm, the second doublet line at 769.9 nm is frequently of advantage; it gives approximately half the sensitivity (0.1 mg/L 1%) and an analytical curve with better linearity. For higher potassium concentrations the doublet at 404.4 nm and 404.7 nm with a characteristic concentration of about 5 mg/L 1% is available.

Potassium is frequently determined in an air/acetylene flame. There is distinct ionization that should be eliminated by adding 0.1% cesium. The determination of potassium is sometimes simpler in a cooler flame in which no ionization can be observed. The air/hydrogen flame has proved very advantageous. It gives a greatly improved signal-to-noise ratio, especially at the secondary doublet at 404.4/404.7 nm, and thereby better precision. For the determination of potassium in rocks and minerals, LUECKE [2566] found no noticeable interferences due to aluminium, phosphate, fluoride or perchlorate.

A frequent cause for a noisy display during the determination of potassium (and especially sodium) is contamination of the laboratory air through dust or tobacco smoke. If these pass into the flame, considerable interferences can result, especially with trace analyses. This interference can be easily established by measuring the noise signal with and without a flame; a greater difference indicates contamination as the cause. A simple method for the elimination of this interference—in case the cause cannot be eliminated—is the use of a multi-slot burner head which gives a broader flame. The edges of the flame, in which the interference can be principally observed, are not irradiated by the optical beam and thus do not affect the analysis.

MANNING et al [810] made a detailed examination of the difference between vapor discharge lamps and hollow cathode lamps for the determination of potassium and found that the discharge lamps were superior, especially for the secondary line doublet. Later developments led to greatly improved hollow cathode lamps, however. Nowadays, electrodeless discharge lamps are used almost exclusively for potassium, since they bring the best results.

JOSEPH et al [595] reported a procedure for determining ^{40}K by atomic absorption spectrometry.

YOZA and OHASHI [1364] used AAS as a detector for potassium in gel chromatography.

The determination of potassium by the graphite furnace technique is very subject to contamination, since potassium is present in dust particles in relatively high concentration. It has been reported that the potassium signal is depressed in the presence of excess sodium as well as various acids. OTTAWAY and SHAW [2719] determined potassium by atomic emission spectrometry using a graphite furnace. Potassium can be thermally pretreated up to about 950°C without losses. The optimum atomization temperature is 1500°C, using an ultrafast heating rate.

10.37. Rhenium

Rhenium can be determined in a nitrous oxide/acetylene flame at the 346.0 nm resonance line with a characteristic concentration of 15 mg/L 1% and a detection limit of about 1 mg/L. At the 346.5 nm and 345.2 nm resonance lines the characteristic concentrations are 25 and 35 mg/L 1%, respectively. Up to the present, no chemical interferences have been reported in the nitrous oxide/acetylene flame. SHRENK et al [1097] found that in a turbulent oxygen/acetylene flame the rhenium signal was markedly depressed by various cations in concentrations over 200–300 mg/L as follows: Calcium by about 75%, manganese by 50%, aluminium by 35%, iron by 15% and molybdenum, lead and potassium by about 10%. This can almost certainly be traced to the burner and flame.

10.38. Rhodium

Rhodium can be determined in a nitrous oxide/acetylene flame at the 343.5 nm resonance line with a characteristic concentration of 0.8 mg/L 1% and a detection limit of about 0.005 mg/L. A sharp, fuel gas weak air/acetylene flame gives a better sensitivity and detection limit, but considerable interferences are to be expected. A number of other resonance lines, with the characteristic concentrations attainable in both flame types, are compiled in table 35.

Table 35. Rh resonance lines

Wavelength nm	Energy level (K)	Characteristic concentration (mg/L 1%)	
		Air/Acetylene [500]	N₂O/Acetylene [63]
343.5	0–29 105	0.1	0.8
369.2	0–27 075	0.1	1.4
339.7	0–29 431	0.2	1.7
350.3	0–28 543	0.3	2.5
365.8	1530–28 860	0.5	2.5
370.1	1530–28 543	1	5.5
350.7	2598–31 102	2.5	2.5

The published details on rhodium are, in part, quite contradictory. While LOCKYER and HAMES [759] claim to have found no interferences in an air/coal gas flame, STRASHEIM and WESSELS [1181] reported a multitude of interferences in an air/propane flame which they could not eliminate simply, as with platinum. DEILY [288] found no difficulties for the determination of rhodium in organic solvents in a sharp air/acetylene flame. ZEEMAN and BRINK [1376] reported numerous interferences in an air/acetylene flame, and GINZBURG *et al* [417] obtained similar results.

KALLMANN and HOBART [622] found no interferences from phosphoric acid in a sharp, fuel gas weak flame, but found strong enhancement from alkali metal, aluminium and zinc sulfates, which they traced to the formation of a rhodium alum. JANSSEN and UMLAND [571] found that the addition of 1% sodium and 1% copper increased the sensitivity for rhodium by 50% and at the same time interelemental effects from other noble metals were eliminated. TOFFOLI and PANNETIER [1234] preferred 0.5% lithium and some copper to eliminate the same interferences and SEN GUPTA [1114] used a buffer consisting of 0.5% copper and 0.5% cadmium.

ATWELL and HERBERT [63] carried out systematic investigations in which they examined the influence of 14 cations in concentrations up to 300 mg/L and four acids up to a concentration of 10% on a rhodium solution containing 30 mg/L in an air/acetylene flame and in a nitrous oxide/acetylene flame. In the air/acetylene flame only 300 mg/L lead did not cause any interferences while all remaining elements exerted influences to varying degrees. In the nitrous oxide/acetylene flame, on the other hand, only interferences from iridium and ruthenium could be established, while all the other elements and acids examined had no influence. The authors found that the interferences from iridium and ruthenium could be eliminated by adding 0.5% zinc to the sample solutions. These authors further found that the stoichiometry of the flame, the observation height above the burner head and similar parameters were not critical and that the analytical curve is completely linear up to 100 mg/L. GARSKA [2296] found that the sensitivity in the air/acetylene flame was four times higher than in the nitrous oxide/acetylene flame. Careful adjustment of the acetylene flowrate and alignment of the burner head are critical with respect to sensitivity and linearity of the analytical curve. Most interferences can be eliminated by adding lanthanum in hydrochloric acid as spectrochemical buffer. HEINEMANN [2362] found few influences due to palladium and platinum on the determination of rhodium. He also used lanthanum as a buffer, but found uranium more effective.

ADRIAENSSENS and KNOOP [16] determined rhodium in the graphite furnace and found only slight interelemental effects and a slight signal depression due to nitric acid. ROWSTON and OTTAWAY [2804] established the optimum parameters for atomization and obtained a characteristic concentration of 5.8 µg/L 1%.

10.39. Rubidium

Very little has been written on the determination of rubidium. At the 780 nm resonance line, a characteristic concentration of 0.04 mg/L 1% and a detection limit of about 0.002 mg/L can be achieved in an air/acetylene flame. The sensitivity can be increased by a factor of about 2 in cooler flames. The second doublet line at 794.8 nm ought to exhibit about half the sensitivity, but like the doublet at 420.2/421.6 nm, which should be suitable for higher concentrations, nothing has been reported about this.

In one of the few coherent papers published on rubidium, SLAVIN et al [1151] also discuss the problem of suitable radiation sources. Since it had not been possible to develop a satisfactory hollow cathode lamp for rubidium, vapor discharge lamps were used almost exclusively for earlier work. Nowadays, excellent electrodeless discharge lamps are available for rubidium, permitting a very good detection limit to be attained [2061].

LUECKE [2566] determined rubidium in rocks and minerals and observed signal depression due to large excesses of aluminium, phosphate, fluoride and perchlorate. These interferences could only be partially eliminated by adding lanthanum. The interferences could be eliminated by changing to the nitrous oxide/acetylene flame and adding EDTA. Like other alkali metals, rubidium is noticeably ionized in an air/acetylene flame, so that another easily ionized metal—preferably cesium—must be added to suppress this effect. In an air/hydrogen flame rubidium is hardly ionized, so that this flame is possibly to be preferred.

A detection limit of about 0.0003 mg/L can be achieved by flame emission spectrometry [363].

JOSEPH et al [595] used atomic absorption spectrometry for the determination of ^{87}Rb in various natural samples.

Very little has been published on the determination of rubidium by the graphite furnace technique. Using an electrodeless discharge lamp as primary source, BARNETT et al [2061] obtained a detection limit of 0.02 mg/L. Zeeman-effect background correction is likely to facilitate the analysis of complex samples.

10.40. Ruthenium

Ruthenium can be determined in a nitrous oxide/acetylene flame at the 349.9 nm resonance line with a characteristic concentration of 2.5 mg/L 1% and a detection limit of about 0.1 mg/L. SCHWAB and HEMBREE [1100] found that the sensitivity for ruthenium was improved by 100% by the addition of 0.13 mol lanthanum nitrate and 0.8 mol hydrochloric acid. All interferences were eliminated by this addition. ROWSTON and OTTAWAY [1063] determined ruthenium in an air/acetylene flame and used a mixture of 0.5% copper sulfate and 0.5% cadmium sulfate to eliminate the numerous interferences. EL-DEFRAWY *et al* [2224] found that virtually all interferences were eliminated in the presence of potassium cyanide and hydrochloric acid and sulfuric acid. They attributed this to the strong complexing effect of the cyanide. GLADNEY and APT [2303] found that neutral ruthenium solutions were not stable for longer than one day. In 1 M hydrochloric acid the solutions are stable for up to four months when stored in glass, quartz, or polyethylene bottles.

ROWSTON and OTTAWAY [2804] established the optimum conditions for the determination of ruthenium by the graphite furnace technique and obtained a characteristic concentration of 19 μg/L 1%.

10.41. Scandium

Scandium can be determined in a nitrous oxide/acetylene flame at the 391.2 nm resonance line with a characteristic concentration of 0.3 mg/L 1% and a detection limit of about 0.02 mg/L. AMOS and WILLIS [48] and MANNING [794] examined the sensitivities of numerous other scandium resonance lines, some of which are compiled in table 36. Scandium is noticeably ionized in the nitrous oxide/acetylene flame, so that this effect must be eliminated by adding 0.1 to 0.2% potassium (as chloride) to sample and reference solutions.

Table 36. Sc resonance lines

Wavelength nm	Energy level (K)	Characteristic concentration (mg/L 1%)
390.8	0–25 585	0.4
402.4	168–25 014	0.4
402.0	0–24 866	0.6
327.0	0–30 573	1.0
327.4	168–30 707	1.5

KRIEGE and WELCHER [705] investigated the influence of numerous cations and anions on the absorption of scandium in a nitrous oxide/acetylene flame and found considerable interferences. To eliminate these interferences the authors only recommended matching the matrix.

10.42. Selenium

The most sensitive selenium resonance line is at 196.1 nm at the start of the vacuum UV [221] and thus out of the range of normal spectrometers. With good atomic absorption instruments, using an air/acetylene flame and an electrodeless discharge lamp, a characteristic concentration of 0.4 mg/L 1% and a detection limit of 0.1 mg/L can be achieved at this line. With the simultaneous use of a background corrector, concentrations under 0.1 mg/L can be determined. This effect is explained by the fact that the air/acetylene flame absorbs about 55% of the radiation from the primary source at the selenium resonance line and small fluctuations in the flame absorption enter directly into the measure. With simultaneous background correction, the flame absorption and the related noise are eliminated.

ALLAN [42] employed the resonance line at 204.0 nm, which however only gives a characteristic concentration of 2 mg/L 1%. Determinations are easier in an argon/hydrogen diffusion flame because only about 10% of the radiation energy is absorbed at the 196.1 nm line. A characteristic concentration of 0.25 mg/L 1% and a detection limit better than 0.1 mg/L can be achieved in this flame. While no chemical interferences have been established in an air/acetylene flame up to the present [1017], the cooler diffusion flame must be used with caution. In particular, background effects owing to radiation scattering can occur markedly in this flame. Furthermore, organic solvents cannot be burnt.

MULFORD [894] extracted selenium with ammonium pyrrolidine dithiocarbamate in MIBK and obtained a detection limit of 0.1 mg/L. SEVERNE and BROOKS [1118] determined selenium in biological materials by co-precipitation with arsenic and hypophosphoric acid. The precipitate is dissolved in nitric acid and analyzed in the flame.

CHAMBERS and MCCLELLEN [2145] extracted selenium with APDC in MIBK and then back extracted into the aqueous phase to obtain the highest sensitivity and freedom from interferences. SUBRAMANIAN and MERANGER [2941] found that only selenium(V) is extracted with APDC/MIBK; selenium(VI) is not extracted over the entire pH range.

For the determination of selenium by the graphite furnace technique, EDIGER [2217] proposed nickel as matrix modifier. Selenium is noticeably volatile at 300°C, but with the addition of nickel can be thermally pretreated up to 1200°C. The stabilizing effect is probably due to the formation of nickel selenide; copper has a similar effect, but such high pretreatment temperatures cannot be applied.

In the meantime, many authors have reported the successful application of nickel as matrix modifier to the determination of selenium. MARTIN and KOPP [2628] determined selenium in various environmental samples. They found that sulfate and various cations depressed the signal, but this could be largely eliminated by adding larger quantities of nickel. STEIN et al [2907] determined selenium in fresh water and sea-water under the addition of nickel; calcium was additionally added to the fresh water samples. The selenium recovery using this procedure was good for sulfate contents up to 70 mg/L and chloride contents up to 50 mg/L. Without the addition of nickel, KUNSELMAN and HUFF [2515] found that most acids depressed the selenium signal. They employed the analyte addition technique for the determination of selenium in water samples. In contrast, for the determination of selenium in river waters, KOPP [699] found that various concomitants enhanced the signal.

SHUM et al [2851] determined selenium in fish and foodstuffs under the addition of nickel directly against similarly prepared references. THOMPSON and ALLEN [2964] also added nickel for the determination of selenium in tablets and capsules. They also found that a good quality pyrolytic coating of the tube was essential. SZYDLOWSKI [2953] used copper as matrix modifier for the determination of selenium in high carbohydrate foodstuffs and reported that the graphite furnace technique was less prone to interferences than fluorimetry.

HENN [2366] [2367] preferred the addition of molybdenum to increase the sensitivity and eliminate matrix interferences in the determination of selenium in industrial effluents. SEFZIK [2838] found that potassium iodide is the most suitable modifier to stabilize selenium for the analysis of drinking water. This matrix modifier also permits a thermal pretreatment temperature up to 1200 °C; the procedure is very sensitive and precise.

For the analysis of high temperature alloys, WELCHER *et al* [1291] found that there were no losses of selenium during thermal pretreatment when nickel was present. MARKS *et al* [2626] determined selenium in complex nickel alloys by introducing them directly into the graphite tube without pretreatment and then atomizing. Since no metal salts can be formed, the authors found little background attenuation; a distinctly better signal-to-noise ratio was obtained at lower atomization temperatures.

NEVE and HANOCQ [2690] determined selenium in biological materials after extraction with 4-chloro-1,2-diaminobenzene in toluene. ISHIZAKI [2410] extracted selenium from ashed biological samples with dithizone in carbon tetrachloride and added nickel to the extract to stabilize selenium. For the determination of selenium in biological materials and feedstuffs, IHNAT [553] found severe interferences due to acids. He thus precipitated selenium with ascorbic acid and also added nickel to improve the precision.

KIRKBRIGHT *et al* [2470] investigated various matrix modifiers and found that nickel, copper, and potassium dichromate in nitric acid were equally effective. All three stabilize selenium up to about 1200 °C. Silver and potassium permanganate in nitric acid are only effective up to about 1000 °C. If mercury and selenium are both to be stabilized, potassium dichromate in nitric acid solution is the most suitable.

SLAVIN *et al* [2882] found that the determination of selenium is virtually free of interferences when nickel is added as matrix modifier and atomization is off the L'VOV platform in a stabilized temperature furnace. The optimum atomization temperature is 2000 °C.

For the determination of selenium in an iron matrix, MANNING [2614] observed a spectral interference that leads to severe overcompensation. Iron exhibits a number of resonance lines in the neighborhood of the 196.1 nm spectral line and although they are not absorbed by selenium, they absorb radiation from the continuum source of the background corrector. FERNANDEZ *et al* [2244] have shown that this interference can be completely eliminated by using Zeeman-effect background correction.

The hydride technique is even more sensitive than the graphite furnace technique for the determination of selenium. The determination limit is about 1 ng absolute or 0.02 μg/L. MEYER *et al* [2666] have nevertheless drawn to attention that numerous cations interfere in the determination of selenium and that the interferences depend on the acid concentration. These authors further mention that the interferences are not dependent on the selenium-to-interferent ratio, but merely on the concentration of the interferent in the analysis solution. They made no attempts to eliminate the observed interferences.

On the other hand, numerous other authors found for the analysis of real samples that either no interferences occurred or that interferences could be eliminated by suitable dilution. CLINTON [2160] determined selenium in blood and plant materials; he found that only copper from pesticide residues had an interfering influence on the analysis of plant materials. IHNAT [2404] compared the graphite furnace and hydride techniques for the determination of selenium in foodstuffs and found the latter had the better performance. VIJAN and WOOD [3019] determined selenium in plant materials directly against reference solutions and found no interferences for non-contaminated samples. They employed the analyte addition technique to eliminate minor interferences.

For the determination of selenium in foodstuffs, FIORINO *et al* [2249] found that after the usual dilution, neither acids nor cations were present in interfering quantities. PIERCE *et al* [2751] eliminated a possible interference due to copper on the analysis of surface waters by dilution. WALKER *et al* [3028] determined selenium in petroleum products in good agreement with neutron activation analysis. They found that the concomitant elements did not cause any interferences. The determination of selenium in glass [2370] is also free of interferences.

High concentrations of elements principally from Groups VIII and IB interfere in the determination of selenium. The analysis of metallurgical samples is consequently subject to problems. FLEMING and IDE [2254] found that iron reduces the interferences due to nickel, copper, molybdenum, and other elements. For the determination of selenium in steel, they thus maintained the iron content constant at 20 mg. BÉDARD and KERBYSON [2069], in contrast, eliminated the interferences by co-precipitating selenium with lanthanum oxide.

KIRKBRIGHT and TADDIA [2472] eliminated numerous interferences by adding tellurium. After reduction with sodium borohydride, the resulting Te^{2-} forms a stable telluride with numerous interfering ions. The stability constants of the tellurides of copper, nickel, palladium, platinum, and others are all lower than those of the corresponding selenides.

For the analysis of low alloy steels, it was found in the author's laboratory that iron in concentrations up to 0.3 g/L did not interfere when a mixture of 1.5% hydrochloric acid and 1% nitric acid was used [3054]. The high iron content in the measurement solutions also eliminates possible interferences from transition elements, as already remarked by FLEMING and IDE [2254], so that it is possible to work directly against aqueous reference solutions having the same acid concentrations.

VERLINDEN and DEELSTRA [3009] found that all other hydride-forming elements interfere to varying degrees with the determination of selenium. Since the interference is only dependent on the concentration of the interferent in the measurement solution, but not on the analyte-to-interferent ratio, it can mostly be eliminated by dilution [3055]. LANSFORD *et al* [2535] observed an interference due to arsenic on the determination of selenium in water; they used tin(II) chloride in 6 M hydrochloric acid rather than borohydride as reductant and found that the interference was dependent on the quantity of tin(II) chloride.

Since selenium(VI) gives no measurable signal when sodium borohydride is added and is thus presumably not reduced to the hydride, it must be reduced to selenium(IV) prior to the actual determination. CORBIN and BARNARD [2168] used sodium iodide as reducing agent, but found that very little must be added since otherwise the reduction proceeds to elemental selenium. SINEMUS *et al* [2869] preferred reduction with 37% hydrochloric acid at 80°C in an autoclave. For the analysis of surface waters, these authors first determined selenium(IV) selectively in the slightly acidified sample and then total selenium after reduction with hydrochloric acid.

CHAU *et al* [2152] determined the volatile dimethylselenide and dimethyldiselenide produced by biological activity in the atmospheres of sea-water sedimentary systems by a combination of GC and AAS. Atomization occurred in a heated quartz tube into which hydrogen was conducted. HOLEN *et al* [2387] deposited selenium electrolytically onto a platinum wire and volatilized the selenium in an argon/hydrogen diffusion flame under additional heating of the platinum wire. The detection limit was 0.1 µg/L for a 30 minute electrolysis period.

10.43. Silicon

Silicon is determined frequently nowadays by AAS because the determination can be performed easily and with adequate sensitivity in the nitrous oxide/acetylene flame. At the 251.6 nm line the characteristic concentration is 2 mg/L 1% and the detection limit is about 0.05 mg/L. Various other resonance lines are compiled in table 37. To achieve the quoted characteristic concentration and an analytical curve of satisfactory linearity, a spectral bandpass of 0.2 nm is absolutely essential because silicon has a large number of resonance lines lying close together (see figure 51, page 84).

Table 37. Si resonance lines

Wavelength nm	Energy level (K)	Characteristic concentration (mg/L 1%)
251.6	223–39955	2
251.9	77–39760	3
250.7	77–39955	6
252.9	223–39760	6
251.4	0–39760	6
252.4	77–39683	7
221.7	223–45322	8
221.1	77–45294	14
220.8	0–45276	25

The determination of silicon in a nitrous oxide/acetylene flame is virtually free of interferences. However, the fact that silicon is rapidly precipitated from acid solutions, and thus lost to the determination, quite frequently causes difficulties. DE VINE and SUHR [2195] established that in aqueous solutions in a pH range 1–8 and with a silicon concentration up to 120 mg/L, the monomeric molecule is the most stable form of silicon. Polymerization takes place at higher concentrations; this colloidal form cannot be determined colorimetrically, but by AAS. Nevertheless, there is a risk that colloidal SiO_2 adheres to the walls of the ves-

sel, leading to low results. This problem does not occur at higher pH values. MEDLIN *et al*
[855] recommended a lithium metaborate solubilizer, which keeps both silicon and other cations in solution, for the analysis of silicates. PARALUSZ [950] examined various organic silicon compounds for their suitability as references in organic solvents. MUSIL and NEHASI-
LOVA [2681] investigated the influence of numerous substances on the absorption of silicon
and found that sulfuric acid depresses the signal, while alkali metals, aluminium and a
number of other elements in large excess cause signal enhancement. PARKER [2731] determined methyl siloxane compounds in water after extraction with pentan-1-ol in MIBK. The
organic solution was nebulized directly into a nitrous oxide/acetylene flame.

CHAPMAN and DALE [2149] improved the sensitivity for silicon by heating with copper hy-
droxyfluoride to obtain the volatile silicon tetrafluoride, which they conducted directly into
the flame.

FRECH and CEDERGREN [2263] made a very thorough examination of the determination of
silicon by the graphite furnace technique. Theoretical and practical investigations show that
between 1300 °C and 1800 °C silicon is lost as gaseous SiO. They recommended that silicon
should be atomized at 2600 °C under isothermal conditions. Both chloride and sulfate interfered; the latter interference decreased when a prolonged thermal pretreatment period was
used. Even small traces of water depress the silicon signal markedly. The authors verified
the finding made by MANNING and FERNANDEZ [802] that the silicon signal is higher in the
presence of nitrogen than in argon; they attributed this to a hindrance of oxide formation
by nitride.

MÜLLER-VOGT and WENDEL [2677] used scanning electron microscopy to study the reactions of silicon with uncoated graphite tubes between 1000 °C and 2000 °C. They found that
at a pretreatment temperature up to 1650 °C, silicate is reduced to silicon. At higher temperatures it reacts with graphite to give silicon carbide. From the dependence of the absorbance on the pretreatment time and temperature, the authors concluded that the formation
of SiO or SiC below 1650 °C is improbable. Coating the graphite tube with niobium increased the signal distinctly and the authors attributed this to an increase in the rate of reduction and a decrease in the rate of carbide formation.

Lo and CHRISTIAN [2559] reported severe signal depression due to potassium dichromate,
antimony in hydrochloric acid, and selenium in *aqua regia,* and weaker interferences from a
number of further metals in dilute hydrochloric acid. Coating the tube with the carbides of
lanthanum, zirconium, or molybdenum reduced the interferences and enhanced the signal
for silicon.

In an uncoated graphite tube, silicon can be thermally pretreated up to about 1400 °C without losses; the optimum atomization temperature is 2700 °C using an ultrafast heating
rate.

10.44. Silver

Silver can be determined exceedingly well by atomic absorption spectrometry. In an air/
acetylene flame the characteristic concentration at the 328.1 nm resonance line is about
0.02 mg/L 1% and the detection limit is about 0.001 mg/L. In flames of lower temperature
the sensitivity can be improved by a factor of about two.

According to existing experience, the determination of silver is free of interferences. WILSON [1339] and BELCHER *et al* [101] investigated a large number of anions and cations for their possible effect on the absorption of silver and found no noticeable interferences, even in flames of lower temperature. Especially for the determination of silver traces, the absence of chloride ions must be carefully observed since precipitation can otherwise easily take place. RAWLING *et al* [1026] found that in a 6 M hydrochloric acid solution at least 25 mg/L silver remain in solution. Naturally such acid concentrations require a chemically resistant nebulizer and burner. GREAVES [439] added diethylene-triamine to silver solutions to prevent precipitation.

Since silver must frequently be determined in the smallest traces, the detection limit achievable with AAS is occasionally insufficient. WEST *et al* [1308] extracted silver as the dithizone complex with ethyl propanate and were thus able to greatly improve the sensitivity. EMMERMANN and LUECKE [329] [766] employed the boat technique for the determination of silver traces in soil samples. With this technique 10^{-10} g silver can be determined.

ROWSTON and OTTAWAY [2804] established the optimum conditions for the determination of silver by the graphite furnace technique. They obtained a characteristic concentration of 0.2 µg/L 1%. LANGMYHR *et al* [2532] determined silver in silicate rocks directly in the solid samples. Using peak area integration, they obtained good agreement with other techniques and a precision of 10–20%.

In the presence of 0.2% nitric acid, silver can be thermally pretreated only up to 600°C without losses; the optimum temperature for atomization off the L'VOV platform is 1900°C.

10.45. Sodium

Sodium is routinely determined in simple flame photometers, and the detection limit with flame emission is about ten times better than with atomic absorption spectrometry. Nevertheless, AAS offers certain advantages even for this element, since sodium must rarely be determined in such small traces as to make the detection limit of this method insufficient; further, AAS appears to be free of interferences in the trace range in the presence of complex matrices, which can be of great importance in practice; finally, AAS offers a number of less sensitive lines which make the determination of higher sodium concentrations possible without excessive dilutions.

Sodium is usually determined at the 589.0/589.6 nm doublet. The somewhat differing sensitivities of the two lines results in a slightly curved analytical curve and reduced sensitivity, but the signal-to-noise ratio is very favorable. The more sensitive doublet line at 589.0 nm can be resolved with spectral bandpasses of 0.2–0.4 nm when better linearity of the analytical curve is necessary.

Using wider bandpasses and both doublet lines, the characteristic concentration for sodium in an air/acetylene flame is 0.01 mg/L 1% and the detection limit is about 0.0002 mg/L. Higher sodium concentrations can be advantageously determined at the second doublet at 330.2/330.3 nm. The characteristic concentration is about 2 mg/L 1%.

MANNING [796] found that it is possible to determine sodium with a good zinc hollow cathode lamp and thus reduce the sensitivity even further. Zinc has a doublet at 330.259/

330.294 nm whose second line is very close to the second line of the sodium doublet at 330.232/330.299 nm. The distance between both lines of only 0.005 nm causes slight overlapping of the wings so that the sodium atoms are able to absorb a portion of the zinc radiation. The characteristic concentration is about 140 mg/L 1%. MANNING further pointed out that there is no risk of spectral interferences since the observed zinc line is not a resonance line and only occurs in emission.

Sodium is somewhat ionized in the air/acetylene flame so that another easily ionized element should be added to sample and reference solutions to suppress this effect. The addition of a cesium salt is the most effective. Nevertheless, various authors prefer the use of a lower temperature flame, such as an air/propane or an air/hydrogen flame [296] [810], to suppress the ionization. The latter flame has the advantage, especially at the secondary doublet at 330.2/330.3 nm, of giving a considerably improved signal-to-noise ratio, so that the precision is improved and the analyses are easier.

For the determination of sodium in rocks and minerals, LUECKE [2566] reported a slight interference through aluminium and silicon, which must be attributed to the formation of stable alkali aluminosilicates. A hydrofluoric acid digestion for silicates largely removed this problem. Minor signal depressions of a few percent were also found for phosphate, fluoride and perchlorate.

Sodium is one of the "present everywhere" elements in a laboratory so that the greatest care must always be taken, especially with trace analyses. Sodium is easily dissolved from normal laboratory glassware so that "distilled" water, sample and reference solutions should only come briefly, if at all, into contact with glass. Water should only be distilled in quartz apparatus, or a mixed bed ion exchanger should be employed. Particular care is recommended with the usual washing and cleaning agents since these frequently contain large amounts of sodium. For storing sample and reference solutions, only plastic vessels are suitable. Particular care must be taken with all chemicals that either contain sodium or, frequently the case with acids, are withdrawn from glass bottles (blank values!). The effects of dust and cigarette smoke on flame noise have already been fully discussed under potassium. Everything said there is even more valid for sodium (see page 316); the use of a 3-slot burner head is recommended.

Everything said above about contamination with respect to the flame technique is naturally even more valid for the graphite furnace technique. A meaningful determination of sodium is practically only possible in a dust-free area. Contamination problems are somewhat less at the 330 nm doublet, so that this can be utilized for trace determinations of sodium in ultrapure water [2940] or in heating oil.

10.46. Strontium

In its behavior in a flame, strontium lies between calcium and barium, i. e. it can be determined in both air/acetylene and nitrous oxide/acetylene flames. As expected, there are a large number of interferences in the cooler flame which have been investigated by various authors. In the hot flame, strontium is considerably ionized. The most sensitive resonance line is at 460.7 nm; in the air/acetylene flame the characteristic concentration is 0.15 mg/L

1% and the detection limit about 0.002 mg/L; in the nitrous oxide/acetylene flame with suppression of the ionization, the characteristic concentration is 0.1 mg/L 1% and the detection limit about 0.002 mg/L. FULTON and BUTLER [397] found the best characteristic concentration of 0.075 mg/L 1% in a mixed air/nitrous oxide/acetylene flame, but this is hardly suitable for routine applications.

LOKEN *et al* [761] and PARKER [951] found that strontium is ionized by about 10% in an air/acetylene flame. During a detailed investigation, INTONTI and STACCHINI [557] established that an addition of 50–100 mg/L rubidium was sufficient to reliably eliminate this effect. All remaining alkali metals exhibited irregular effects with increasing concentration. AMOS and WILLIS [48] found about 80% ionization in the nitrous oxide/acetylene flame, but this could be eliminated by the addition of 0.1–0.2% potassium (as chloride).

TRENT and SLAVIN [1242] thoroughly examined possible effects in an air/acetylene flame and found numerous interferences. The majority could be eliminated by adding 1% lanthanum to the sample and reference solutions. This was verified by BELCHER and BROOKS [99], who however, added only 0.2% lanthanum. The following are some of the most important interfering substances: Si, Al, PO_4, and also Ca, Fe, Mg and HNO_3. Higher concentrations of HCl, $CuCl_2$, $CoCl_2$, NH_4NO_3, $CdNO_3$ also have a marked depressive influence [1242].

Very little has been reported in the literature on the determination of strontium by the graphite furnace technique. BEK *et al* [96] determined strontium in blood, and HELSBY [2364] strontium in tooth enamel. Calcium and phosphate in tooth enamel caused interferences, but these could be eliminated by using the analyte addition technique. Hydrochloric acid and nitric acid did not interfere in the determination, but perchloric acid caused severe signal depression.

SLAVIN *et al* [2881] investigated the influence of various tube materials on the determination of strontium. They obtained a characteristic mass of 2 pg 1%.

In 0.2% nitric acid, strontium can be thermally pretreated up to about 1200°C without losses. Using pyrocoated graphite tubes and an ultrafast heating rate, the optimum atomization temperature is 2500°C.

10.47. Sulfur

The direct determination of sulfur by atomic absorption spectrometry employing commercial instruments has not yet been possible because the main resonance line is at 180.7 nm in the vacuum UV range. KIRKBRIGHT [654] succeeded in determining sulfur at this line in a modified atomic absorption spectrometer with a flushed monochromator, a nitrous oxide/acetylene flame shielded with nitrogen and an electrodeless discharge lamp. The characteristic concentration was 9 mg/L 1% and the detection limit about 5 mg/L. KIRKBRIGHT found no influence from the bonding state of sulfur and he used the procedure for the direct determination of sulfur in crude oil.

Using a similar instrument with an electrodeless discharge lamp, flushed optics and a vacuum monochromator, ADAMS and KIRKBRIGHT [2001] succeeded in determining sulfur in a graphite furnace. At the 180.7 nm spectral line, they obtained a characteristic mass of

0.42 ng 1%; the other lines at 182.0 nm and 182.6 nm were slightly less sensitive with 0.68 ng 1% and 1.5 ng 1%, respectively. The authors found identical reference curves for sulfate, thiocyanate, and thiourea. For sulfate, the authors presumed that atomization was via the dissociation of SO_2.

A number of indirect methods, which can be widely employed, have also been described. ROE *et al* [1042] converted organically bound sulfur into sulfate by means of the SCHÖNINGER or BENEDIKT digestion methods, precipitated this as barium sulfate and determined the barium in the precipitate after solution in EDTA solution. VARLEY and CHIN [1261] determined water-soluble sulfate in soil extracts by adding a known quantity of barium chloride and determining the excess barium by AAS. CAMPBELL and TIOH [2123] used the same procedure to determine sulfate in fertilizer. The precipitated barium sulfate can also be taken up in ammonia EDTA solution and the barium determined by AAS. ROSE and BOLTZ [1057] determined sulfur dioxide by oxidizing it to sulfate, precipitating as lead sulfate, and after centrifugation, determining the excess lead in the supernatant liquid by AAS. DUNK *et al* [315] added exactly measured quantities of barium chloride solution to sulfate solutions and determined the excess barium. CHRISTIAN and FELDMAN [232] utilized the occurrence of chemical interferences for the determination of a number of non-metals including sulfur. For example, they found a direct relationship between the absorption of calcium and the sulfate concentration in the solution under analysis. Naturally other ions interfere strongly in this determination and it is not specific.

FUWA and VALLEE [400] described a procedure for the determination of sulfur by molecular absorption spectrometry at an SO_2 absorption band. An air/hydrogen flame was directed into a 273 cm long heated Vycor tube and a hydrogen lamp served as the radiation source. The absorption was measured at 207 nm with a spectral bandpass of about 0.7 nm. They attained a characteristic concentration of about 10 mg/L 1%.

10.48. Tantalum

Tantalum can be determined in a nitrous oxide/acetylene flame at the 271.4 nm resonance line with a characteristic concentration of 20 mg/L 1% and a detection limit of about 1 mg/L. The absorption of tantalum is markedly increased in the presence of hydrofluoric acid and iron, so that sample and reference solutions should contain at least 0.1% hydrofluoric acid to eliminate this interference and to obtain the best sensitivity. BOND [141] preferred the addition of 0.1 M ammonium fluoride to increase the sensitivity.

VAN LUIPEN [3004] found that sulfuric acid, phosphoric acid, titanium and vanadium interfered in the determination of tantalum in the nitrous oxide/acetylene flame. The most probable explanation for these interferences is the varying volatilities of the compounds containing tantalum. Presumably, compounds such as $Ta_2O_5 \cdot Ti_2O_3$, $Ta_2O_5 \cdot V_2O_3$, or $Ta_2O_2(SO_4)_3 \cdot 5 H_2O$ are formed in the nitrous oxide/acetylene flame. WALLACE *et al* [3029] found improved linearity, sensitivity, precision, and detection limit for tantalum when they added an excess of aluminium. Aluminium may reduce the tendency to form refractory oxides and facilitate atomization. Reduction of lateral diffusion could be a further explanation of this improvement. In the presence of hydrofluoric acid and aluminium, these authors obtained a detection limit of 0.5 µg/L.

10.49. Technetium

HARELAND *et al* [468] described the first determination of technetium. In a fuel gas rich air/acetylene flame a characteristic concentration of 3.0 mg/L 1% can be achieved at the 261.4/261.6 nm doublet. Interferences from alkaline-earth metals can be eliminated by adding aluminium (about 100 mg/L per 50 mg/L interfering ions).

BAUDIN *et al* [87] determined technetium by the graphite furnace technique and found the best characteristic concentration of 1×10^{-8} g 1% at the 429.7 nm resonance line. The atomization temperature was about 2200°C. KAYE and BALLOU [2462] reported a markedly improved detection limit of 6×10^{-11} g measured at the non-resolved doublet at 261.4/261.6 nm. They used a demountable hollow cathode lamp as radiation source and found that an atomization temperature of 3300°C was the optimum compromise between sensitivity and tube lifetime. This value is, however, rather doubtful.

10.50. Tellurium

Tellurium can be easily determined free of interferences in an air/acetylene flame. At the 214.3 nm resonance line the characteristic concentration is 0.3 mg/L 1% and the detection limit is about 0.02 mg/L. In an air/hydrogen flame a somewhat improved sensitivity and a detection limit of 0.015 mg/L can be obtained.

WU *et al* [1358] examined the influence of 25 various ions on the absorption of tellurium and found no interferences worth mentioning. They also studied the extraction of tellurium with various complexing agents and organic solvents very thoroughly. They obtained very good results with ammonium pyrrolidine dithiocarbamate/pentan-2,4-dione.

Higher tellurium concentrations can be determined at the 225.9 nm resonance line with a characteristic concentration of about 5 mg/L 1%.

VAKAHARA and MUSHA [2686] determined tellurium in an argon (or nitrogen)/hydrogen diffusion flame with a characteristic concentration of 0.13 mg/L. They observed no interferences from 0.1 M to 2 M hydrochloric acid, hydrofluoric acid, nitric acid, sulfuric acid, or phosphoric acid; only perchloric acid caused signal depression. The addition of 2000 mg/L magnesium as the chloride very effectively eliminated the interferences due to various elements.

For a determination by the graphite furnace technique, KUNSELMAN and HUFF [2515] found that the tellurium signal was enhanced by hydrochloric acid, nitric acid, and sulfuric acid. They therefore used the analyte addition technique. BEATY [91] determined tellurium in rock samples after digestion in hydrofluoric acid and extraction with MIBK. SIGHINOLFI *et al* [2864] investigated geochemical samples using the same procedure, but included an additional back extraction of the tellurium into aqueous medium.

EDIGER [2217] showed that nickel is a suitable matrix modifier and stabilizes otherwise volatile tellurium compounds up to 1200°C. WEIBUST *et al* [3041] also investigated suitable matrix modifiers and found that cadmium, copper, palladium, platinum and zinc were all equally good and stabilized inorganically bound tellurium up to 1050°C. Organically bound tellurium, in contrast, is only stabilized by palladium, platinum and silver. Consequently, only palladium and platinum are suitable for both organically and inorganically bound tellurium. The optimum temperature for atomization off the L'vov platform is 1800°C.

Tellurium can be determined with high sensitivity using the hydride technique. The determination limit is about 0.5 ng absolute or 0.02 µg/L. FIORINO *et al* [2249] determined tellurium in foodstuffs and found that after the usual dilutions there were no acids or cations in interfering concentrations in the sample. They obtained excellent agreement with neutron activation analysis and other techniques. GREENLAND and CAMPBELL [2315] determined nanogram traces of tellurium in silicate rocks and found that in the concentrations normally present in such samples, neither iron nor copper interfered. BÉDARD and KERBYSON [2069] determined tellurium in copper and eliminated the interference through co-precipitation with lanthanum hydroxide.

FLEMING and IDE [2254] determined tellurium in steel in 3 M hydrochloric acid; they found that iron reduces the interferences due to copper, nickel, molybdenum, etc. They therefore maintained the iron content constant at 20 mg, and for highly alloyed steels they added iron correspondingly. The determination of tellurium in steel was also examined in detail in the author's laboratory [3054]. It was found that iron in concentrations up to 1 g/L did not interfere in a mixture of 7.5% hydrochloric acid and 7.5% nitric acid. The other metals present in low alloy steels did not interfere either, so that the determination can be performed directly against aqueous reference solutions that merely contain the same acid content.

Like selenium, the hexavalent oxidation state of tellurium gives no measurable signal using the hydride technique. Brief heating with concentrated hydrochloric acid is sufficient to effect reduction [2869]. SINEMUS *et al* [2869] first determined tellurium(IV) selectively in surface waters after acidifying with hydrochloric acid and nitric acid. Thereafter they reduced by heating with hydrochloric acid and determined the total tellurium content.

10.51. Thallium

Thallium can be determined easily and without interferences in an air/acetylene flame. At the 276.8 nm resonance line the characteristic concentration is 0.2 mg/L 1% and the detection limit is about 0.01 mg/L.

Higher thallium concentrations can be determined at the 377.6 nm resonance line which gives a characteristic concentration of about 3 mg/L 1%.

The determination of thallium in the graphite furnace is plagued by signal depression caused by halides. The volatility of thallium is too high to permit halides to be separated to any extent during thermal pretreatment. They are therefore volatilized during atomization and cause vapor-phase interferences. MANNING *et al* [2621] investigated the influence of magnesium chloride on conventional atomization off the tube wall and atomization from a tungsten wire in a furnace at constant temperature. In the latter case for atomization under isothermal conditions, virtually no signal depression could be observed. SLAVIN *et al* [2880] found that a similar improvement could be obtained when atomization was off the L'vov platform.

FULLER [2279] found severe influences from hydrochloric acid in perchloric acid for atomization from the tube wall. He attributed these to the formation of volatile chlorides. Sodium chloride suppressed the thallium signal completely, while sulfuric acid and nitric acid had very little influence. In the view of the author they formed a stable oxide. He found that the addition of 1% sulfuric acid to the sample solution permitted an interference-free determination of thallium in the presence of chlorides.

L'vov *et al* [2576] [2591] showed that lithium (as the nitrate) substantially reduced interferences due to sodium chloride. This is based on the formation of the relatively stable lithium chloride in the vapor phase, so that thallium is largely undisturbed.

KUJIRAL *et al* [2510] determined thallium in cobalt and nickel alloys; they found that tartaric acid and sulfuric acid permitted higher thermal pretreatment temperatures and reduced interferences.

KOIRTYOHANN *et al* [2481] found that the thallium signal was almost completely suppressed in the presence of 0.5 M perchloric acid. The effect remained even when the graphite tube was heated to above the boiling point of perchloric acid. The acid, or a decomposition product, reacts with graphite to form a thermally stable product that is subsequently volatilized and then interferes in the atomization. SLAVIN *et al* [2878] reported that this interference depends to a large extent on the graphite tube. In a perfectly coated tube, perchloric acid had only a minor influence on the thallium signal.

SLAVIN *et al* [2878] also found that sulfuric acid is a suitable matrix modifier for thallium, permitting loss-free thermal pretreatment up to 700 °C. They reported that using this modifier and atomizing off the L'vov platform in a stabilized temperature furnace, perchloric acid in concentrations up to 1 M had no influence on the thallium signal. No influences due to sodium chloride or other halides could be observed either [2882]. The optimum temperature for atomization off the L'vov platform is 1400 °C.

LANGMYHR *et al* [2532] determined thallium in silicate rocks directly in the solid samples with a precision of 10–20%. A similar procedure has been used in the author's laboratory for the determination of thallium in rock samples; good agreement was obtained with neutron activation analysis [1295].

10.52. Tin

Tin can be determined in air/hydrogen, fuel gas rich air/acetylene, and nitrous oxide/acetylene flames with comparable sensitivity. The corresponding values for the three most important resonance lines are compiled in table 38 (see also table 8, page 112). The best charac-

Table 38. Sn resonance lines

Wavelength nm	Energy level (K)	Characteristic concentration (mg/L 1%)		
		Air/H₂	Air/Acetylene	N₂O/Acetylene
224.6	0–44 509	1.1	2.0	3.0
286.3	0–34 914	1.8	3.5	5.4
235.4	1692–44 145	2.2	2.5	3.8

teristic concentration of 1.1 mg/L 1% can be achieved in an air/hydrogen flame at the 224.6 nm resonance line; the detection limit is about 0.02 mg/L.

ALLAN [42] and GATEHOUSE and WILLIS [407] employed a strongly reducing air/acetylene flame which gave a very favorable signal-to-noise ratio. After the introduction of the air/hydrogen flame by CAPACHO-DELGADO and MANNING [206] [208] and hollow cathode lamps containing molten tin by VOLLMER [1276], this element can be determined very well by atomic absorption spectrometry. Nevertheless, various interferences appear to influence the determination, so that the use of a nitrous oxide/acetylene flame is recommended for complex matrices. CAPACHO-DELGADO and MANNING [208] found a strong depression of the signal in the air/hydrogen flame caused by phosphoric and sulfuric acids and a slight influence from nitric acid. Nitric acid also had a slightly negative influence in the air/acetylene flame, while the other two acids enhanced the tin signal. Interferences from cations appear to be negligible. AGAZZI [20] found interferences from phosphate and pyrophosphate in an oxygen/hydrogen flame with a direct injection burner and also an increase of the signal from high sodium concentrations. On the other hand, AMOS and WILLIS [48] found a reduction of 15% in the tin signal by 5000 mg/L sodium in an air/hydrogen flame. LEVINE *et al* [742] found that even small concentrations of potassium enhanced the signal by 10% in a fuel gas rich air/acetylene flame. BOWMAN [148] discovered a slight increase in the tin signal in a nitrous oxide/acetylene flame due to ammonium iodide.

The best characteristic concentration of 0.6 mg/L 1% can be achieved in an argon/hydrogen diffusion flame, but numerous other elements cause interferences in this flame. NAKAHARA *et al* [903] found that the majority of these interferences could be effectively eliminated by the addition of 0.1% iron(III) chloride. RUBEŠKA and MIKŠOVOKY [1068] found that acids depressed the tin signal in this flame and that virtually all metallic elements enhanced the signal, in some cases considerably (up to 100%). These authors accounted these effects to an easier heat transfer due to the additional salts and thus to better atomization of tin.

KAHN and SCHALLIS [617] were unable to observe an absorption signal for 100 mg/L tin in MIBK when using an air/hydrogen flame. Thereupon HARRISON and JULIANO [472] investigated the influence of organic solvents on the absorption of tin. They found that even small concentrations of alcohols, ketones and organic acids reduce the absorption of tin in an air/hydrogen flame considerably. This effect grows with increasing chain length and branching of the solvents. For example, 10% methanol reduces the absorbance of 50 mg/L tin by about 25%, 10% ethanol reduces it by about 70% and 10% n-propanol by 80%. In the presence of 10% n-butanol, hardly any further tin signal can be observed. 10% acetone also suppresses the tin signal almost completely. These authors attributed this effect to a reduction of the concentration of hydrogen atoms in the flame by organic meterials; hydrogen appears to play an important role in the atomization of tin, either indirectly by reducing the monoxide to metal or directly through the intermediate formation of the hydride [1067].

These effects exclude the determination of tin in organic solvents in an air/hydrogen flame. Since the fuel gas rich air/acetylene flame is difficult to operate with organic solvents, only the nitrous oxide/acetylene flame remains. Nevertheless, ionization can be observed in this flame and this should be eliminated by the addition of alkali. MENSIK and SEIDEMANN [861] separated tin from rock and soil samples by sublimation with ammonium iodide

$(SnO_2 + 4NH_4I \rightarrow SnI_4 + 4NH_3 + 2H_2O)$. After solution in hydrochloric acid, the tin was extracted with trioctylphosphine oxide/MIBK and determined in a nitrous oxide/acetylene flame, free of interferences.

DONALDSON [2203] extracted tin from ores, iron, steel and non-ferrous alloys in sulfuric acid solution with toluene under the addition of potassium iodide, tartaric acid and ascorbic acid. The solution was evaporated to dryness and the residue taken up in hydrochloric acid under the addition of tartaric acid and potassium; the determination was performed in a nitrous oxide/acetylene flame.

GLADNEY and GOODE [2304] found that neutral solutions of tin could not be stored in glass containers at all, and only for up to 24 hours in plastic containers. For longer storage, the solutions should contain at least 0.1 M acid.

The determination of tin by the graphite furnace technique is, like the flame technique, subject to a number of difficulties. BARNETT and MCLAUGHLIN [2060] determined tin in iron, copper and zinc alloys using the analyte addition technique; they merely dissolved the samples in nitric acid. RATCLIFFE *et al* [2780] also determined tin in steels and found that it was only possible to work with nitric acid solutions. Hydrochloric acid suppressed the tin signal completely, while sulfuric acid and perchloric acid caused low and severely scattered signals. The authors also found interferences due to chromium, nickel, titanium and niobium.

DEL MONTE TAMBA and LUPERI [2196] determined tin in low-alloy and high-alloy steels with estimation against the same analytical curve. By careful selection of dissolution procedures, the authors were able to maintain the concentration of nitric acid virtually constant for all samples. Under these conditions, the determination of tin is free of the interferences caused by the other constituents generally present in steel. MARKS *et al* [2626] determined tin in complex nickel alloys directly in the solid samples. They calibrated directly against standard alloys and found that there was little background using this procedure, since no metal salts are formed.

THOMPSON *et al* [2966] found that calcium enhanced the tin signal. TOMINAGA and UMEZAKI [2979] reported interferences due to numerous chlorides and nitrates, and also several sulfates. They were able to substantially reduce, or even eliminate, these interferences by adding ascorbic acid. HOCQUELLET and LABEYRIE [2378] found that the addition of ammonium nitrate distinctly enhanced the sensitivity for tin and eliminated matrix interferences for the analysis of foodstuffs. This is in agreement with results obtained in the present author's laboratory [2321], where the best results for tin were obtained by adding ammonium hydroxide directly to nitric acid sample solutions in the graphite tube.

A number of authors have investigated the influence of the graphite surface on the determination of tin. HOCQUELLET and LABEYRIE [2378] found that coating the tube with tantalum or tantalum carbide brought about a marked signal enhancement. FRITZSCHE *et al* [2272] showed that impregnating the tube with molybdenum, tantalum, tungsten, or zirconium influenced the surface characteristics favorably for the atomization of tin. The use of impregnated tubes and gas stop during atomization minimized the interferences due to numerous ions. For the analysis of organotin compounds, VICKREY *et al* [3012] found that treatment with zirconium brought the best results with respect to the elimination of interferences. The determination could be performed directly against aqueous reference solutions.

L'vov *et al* [2579] explained the effect of surface treatment by surface kinetics. In normal, uncoated tubes, the sample solution penetrates the graphite, so that atomization "off the wall" is really atomization out of the graphite lattice. Only STURGEON and CHAKRABARTI [2933] found better sensitivity for tin with atomization in uncoated tubes.

This might be explained by the influence of the gas stream on the sensitivity. This influence is more pronounced in a pyrocoated tube than in an uncoated tube [2321]. The atomization of tin is in all probability via the thermal dissociation of the gaseous oxide, thus explaining the influence of the gas stream. In an uncoated tube, the reducing effect of graphite can, directly or indirectly, cause atomization, even when gas is flowing, while in a coated tube this is hardly possible. Using ammonium hydroxide as matrix modifier, tin can be thermally pretreated up to 800°C without losses; the optimum temperature for atomization off the L'vov platform is 2200°C.

Tin may also be determined by the hydride technique; in contrast to the other hydride-forming elements, the sensitivity is relatively strongly dependent on pH. The determination is best performed in a saturated boric acid solution; the determination limit is approximately 5.0 ng Sn absolute or 0.5 µg/L.

VIJAN and CHAN [3016] determined tin in particulate matter collected on filters and eliminated interferences due to copper, nickel, antimony, and arsenic by adding sodium oxalate or by prior co-precipitation with hydrated manganese dioxide. PYAN and FISHMAN [2768] found that EDTA was better than sodium oxalate for eliminating interferences in the determination of tin in water. The authors obtained even better results by using higher acid concentrations and diluted borohydride solution.

BRAMAN and TOMKINS [2103] determined inorganic tin and methyltin compounds in environmental samples through reaction with sodium borohydride at pH 6.5. The various hydrides were collected in a cold trap and then separated by fractional volatilization and determined individually.

BÉDARD and KERBYSON [2069] determined tin in metallic copper after prior co-precipitation with lanthanum hydroxide to avoid interferences. FLEMING and IDE [2254] determined tin in steel in the presence of tartaric acid. To avoid interferences, they maintained the iron content constant at 20 mg. In this way it was also possible to control the influences due to copper, nickel and molybdenum. In the author's laboratory it was found that the determination of tin in low-alloy steels was free of interferences when performed in an acid mixture of saturated boric acid solution and 0.6% nitric acid [3054]. Up to 0.1 g/L iron, as well as the other metals usually present in such samples, had no influence on the tin signal, so that the determination could be performed against aqueous reference solutions with matched acid content.

10.53. Titanium

Titanium can be determined in a nitrous oxide/acetylene flame at the 364.3 nm resonance line with a characteristic concentration of 2 mg/L 1% and a detection limit of about 0.05 mg/L. A multitude of other resonance lines were examined by CARTWRIGHT *et al* [214] (compare table 9, page 113), of which a number are compiled in table 39.

Table 39. Ti resonance lines

Wavelength nm	Energy level (K)	Characteristic concentration (mg/L 1%)
364.3	170–27 615	2
335.5	170–29 971	3
399.9	387–25 388	5
399.0	170–25 227	5
395.6	170–25 439	5
394.8	0–25 318	10

AMOS and WILLIS [48] found that the titanium signal is noticeably enhanced in the presence of iron or hydrofluoric acid. An increase in the concentration of hydrofluoric acid from 0.2% to 4% brought a 50% higher signal, and 0.2% iron doubled the absorption. It is therefore recommended to maintain the concentrations of these substances as constant as possible; with higher iron and hydrofluoric acid concentrations the effect appears to be largely controlled. BOND [141] found that 0.1 M ammonium fluoride solution increased the signal quite strongly and so he recommended this reagent.

RAO [1022] determined titanium, free of interferences, in ores after a lithium metaborate digestion in which he maintained concentrations of 2% lithium metaborate, 0.8% ammonium fluoride and 0.05–0.2% SiO_2 (corresponding to 12.5–50% SiO_2 in the original sample). Under these conditions the maximum sensitivity for titanium was achieved.

KOMÁREK *et al* [2490] made a detailed investigation on the influence of the valency state of titanium, of mineral acids and of various organic complexing ligands on the absorption of titanium in the nitrous oxide/acetylene flame. The thermal processes taking place in the dry aerosol particles appear to have greater significance for atomization than the complexing equilibria in solution. In the presence of chromotropic acid, the determination of titanium is more sensitive and virtually interference free.

The determination of titanium by the graphite furnace technique depends to a very large extent on the quality of the pyrolytic coating of the graphite tube and on the rate of heating for atomization [2615] [2881]. The determination is remarkably free of interferences. In 0.2% nitric acid, titanium can be thermally pretreated up to 1400°C without losses. The optimum atomization temperature, even using an ultrafast heating rate and pyrocoated graphite tubes, is above 2700°C.

10.54. Tungsten

Tungsten can be determined in a nitrous oxide/acetylene flame at the 400.9 nm resonance line with a characteristic concentration of 11 mg/L 1% and a detection limit of about 1 mg/L. Although AMOS and WILLIS [48] and MANNING [794] mention various resonance lines with better sensitivity, the 400.9 nm line is to be preferred because of its more favorable signal-to-noise ratio.

MUSIL and DOLEŽAL [2680] determined small amounts of tungsten in steel and zirconium alloys by reducing with tin(II) and extracting the thiocyanate complex with MIBK. The organic phase was nebulized directly into a nitrous oxide/acetylene flame. CHONG and BOLTZ [2156] described the indirect determination of tungsten via precipitation as lead(II) tungstate and the determination of lead in the supernatant liquid.

10.55. Uranium

Uranium can be determined in a nitrous oxide/acetylene flame at the 351.5 nm resonance line with a characteristic concentration of 50 mg/L 1% and a detection limit of about 30 mg/L. The resonance lines at 358.5 nm and 356.7 nm exhibit better sensitivity, but a less favorable signal-to-noise ratio. MANNING [794] examined further uranium lines and determined their characteristic concentrations.

Since uranium is markedly ionized in a nitrous oxide/acetylene flame, 0.1% potassium (as chloride) should be added to sample and reference solutions to suppress this effect.

KEIL [2463] found that a number of metals enhanced the uranium signal to varying degrees, gallium having the most pronounced influence. With an addition of 10 g/L Ga, the author obtained a characteristic concentration of 16 mg/L 1% and a detection limit of 30 mg/L at the 358.5 nm resonance line. The characteristic concentration is 26 mg/L 1% and the detection limit 10 mg/L at the 351.5 nm resonance line.

ALDER and DAS [2018–2020] described an indirect method for the determination of uranium based on the oxidation of uranium(IV) by copper(II). The resulting copper(I) is extracted as the neocuproine complex and determined.

Uranium can only be determined by the graphite furnace technique when the highest quality pyrocoated graphite tubes and an ultrafast heating rate are used. Even under these conditions, the optimum atomization temperature is above $2700\,°C$.

GOLEB [424] [425] determined the $^{235}U/^{238}U$ ratio employing a cooled hollow cathode lamp and a flameless atomization cell.

10.56. Vanadium

Vanadium can be determined in a nitrous oxide/acetylene flame at the triplet 318.3/318.4/318.5 nm with a characteristic concentration of about 2 mg/L 1% and a detection limit of 0.04 mg/L. CARTWRIGHT [214] examined numerous other resonance lines, a number of which are compiled in table 40.

CAPACHO-DELGADO and MANNING [207] found an enhancement of the vanadium signal in the presence of phosphoric acid. GOECKE [422] discovered that iron in concentrations over 1000 mg/L caused interferences, so he removed it from strong hydrochloric acid solutions by extraction with isopropyl ether. The high concentration of hydrochloric acid then had to be removed by evaporation. GOECKE found further that 200 mg/L aluminium enhanced the vanadium signal by a third, so he added this cation to all sample and reference solutions and was thereby able, with optimization of the nitrous oxide/acetylene flame, to obtain a characteristic concentration of 0.2 mg/L 1%.

Table 40. V resonance lines

Wavelength nm	Energy level (K)	Characteristic concentration (mg/L 1%)
318.3	137–31 541 ⎫	
318.4	323–31 722 ⎬	2
318.5	553–31 937 ⎭	
370.4	2425–29 418	5
437.9	2425–25 254	5
438.5	2311–25 112	6
370.5	2311–29 296	10
257.4	553–39 391	20
251.7	323–40 039	35
253.0	553–40 064	35
250.8	137–40 001	40

WEST *et al* [1306] attributed these increases in the absorption to a hindrance of the lateral diffusion in the flame. The additives delay the atomization of vanadium and thereby reduce the time available for lateral diffusion of the atoms in the flame. The atoms under study are concentrated in the center of the flame and hence in the optical beam.

The introduction of pyrolytically coated graphite tubes greatly facilitated the determination of vanadium by the graphite furnace technique [2615]. Nevertheless, the quality of the pyro-coating has a direct influence on the signal [2881]; in the author's laboratory it was found that with perfect tubes there was no drift in the sensitivity over 200–300 determinations.

THOMPSON *et al* [2966] obtained a detection limit of 1 μg/L by adding calcium and performing the determination in a pyrocoated tube; they did not observe any interferences. STUDNICKI [2928] reported that virtually all mineral acids depress the vanadium signal rather severely, and SCHWEIZER [1101] described an interference due to calcium in an uncoated graphite tube. He avoided this interference by using the analyte addition technique.

KRISHNAN *et al* [2509] determined vanadium in biological tissues. WEISS *et al* [3043] determined vanadium in sea-water after reduction to vanadium(IV) with ascorbic acid and pre-concentration on an ion exchanger. They found that both the graphite furnace technique and neutron activation analysis provided equally reliable results.

In 0.2% nitric acid, vanadium can be thermally pretreated up to about 1500°C without losses; using an ultrafast heating rate, the optimum atomization temperature is 2700°C.

10.57. Yttrium

Yttrium can be determined in a nitrous oxide/acetylene flame at the 410.2 nm resonance line with a characteristic concentration of 2 mg/L 1% and a detection limit of about 0.1 mg/L. The resonance lines at 407.7 nm, 412.8 nm and 414.3 nm exhibit only slightly lower sensitivity. To suppress ionization, 0.1–0.2% potassium (as chloride) should be added to sample and reference solutions.

WISE and SOLSKY [3073] determined yttrium in zirconium oxide after a fusion with LiBO$_2$/ H$_3$BO$_3$. They added a spectrochemical buffer containing potassium and EDTA to obtain maximum sensitivity in a stoichiometric nitrous oxide/acetylene flame. No interferences were caused by sulfate, phosphate, calcium or magnesium; silicon can be fumed off with hydrofluoric acid before the fusion.

For the determination of yttrium in rock samples, SEN GUPTA [2841] used the graphite furnace technique with pyrolytically coated tubes. To avoid interferences, yttrium and the lanthanides were separated from the major constituents by co-precipitation with calcium and iron.

WAHAB and CHAKRABARTI [3026] [3027] made a detailed investigation of the atomization of yttrium. They found that treating pyrolytically coated tubes with a carbide-forming metal (lanthanum, tantalum, zirconium) increased the sensitivity by a factor of around two. This effect can probably be traced back to the use of imperfectly coated tubes. Atomization from metal surfaces (tantalum or tungsten foil inserted into the tube) brought a further increase in the sensitivity, a lower atomization temperature and negligible memory effects. Under optimum conditions, the authors obtained a characteristic mass of 0.3 ng 1%. They proposed that atomization is via the thermal dissociation of the gaseous monoxide.

10.58. Zinc

Zinc is one of the metals most frequently determined by atomic absorption spectrometry. A characteristic concentration of 0.01 mg/L 1% and a detection limit of about 0.001 mg/L can be achieved in an air/acetylene flame at the 213.8 nm resonance line. An improved signal-to-noise ratio and a detection limit better than 0.001 mg/L are obtained in an air/hydrogen flame.

In the early years of atomic absorption spectrometry, DAVID [267] and ALLAN [41] engaged in thorough studies on the determination of zinc and found that AAS was superior to all other methods for the determination of this element.

The determination of zinc appears to be free of interferences in an air/acetylene flame as SPRAGUE and SLAVIN [1171] established for the analysis of biological materials. PLATTE and MARCY [989] verified this by finding no influence on the absorption of 1 mg/L zinc by 1000 mg/L sulfate, phosphate, nitrate, nitrite, bicarbonate, silicate, EDTA and nine other cations. The interferences quoted by GIDLEY and JONES [414] [415] can be traced to the use of an unsuitable lamp and an unsuitable burner.

High zinc concentrations can be advantageously determined at the 307.6 nm resonance line; the characteristic concentration at this line is about 100 mg/L 1%. Unnecessary dilutions can be avoided and high concentrations can be determined with good precision.

In principle, the determination of zinc by the graphite furnace technique is very simple and largely free of interferences. Since the dissociation constant of zinc chloride is lower than for the chlorides of most other elements, the halides should cause few interferences. The greatest problem for the determination of zinc is its very high sensitivity (characteristic mass 0.3 pg 1%) combined with its universal presence. Consequently, difficulties are encountered with high blank values and severe contamination [2987]. SOMMERFELD et al [2897] reported contamination of pipet tips by zinc; even repeated washing with hydrochloric acid could not eliminate the problem. The reader is also referred to section 6.4 (Problems of Trace Analysis).

CRUZ and VAN LOON [245] found that potassium nitrate and aluminium nitrate enhanced the zinc signal, while magnesium nitrate, iron nitrate and calcium chloride caused severe signal depression. CLARK *et al* [235] found that phosphoric acid, hydrochloric acid, sodium chloride, potassium chloride, and silicate also depressed the zinc signal. CAMPBELL and OTTOWAY [2128] found that sodium chloride and especially magnesium chloride depress the signal very severely. On the other hand, FERNANDEZ and MANNING [361] were unable to observe any interferences due to sodium chloride.

VIEIRA and HANSEN [3013] managed to get the problem of contamination under control for the determination of zinc in serum and urine. They worked on a microscale, using only 10 µL of sample. LANGMYHR *et al* [2526] determined zinc in teeth directly in the solid samples using the less sensitive resonance line at 307.6 nm. JULSHAMN [2441] found that 1 M perchloric acid only caused minor signal depression.

KINGSTON *et al* [2468] determined zinc in sea-water after prior separation of the major constituents on an ion exchange column. For the same determination, JAN and YOUNG [2432] preferred extraction with APDC in MIBK followed by back extraction in 4 M nitric acid. SMITH and WINDOM [2894] extracted with dithizone in chloroform and then back extracted with nitric acid.

CAMPBELL and OTTAWAY [2128] atomized diluted sea-water samples directly at 1490°C and so were able to separate the analyte signal from the background. GUEVREMONT [2325] also used a direct procedure for the determination of zinc in sea-water; he added citric acid as matrix modifier and obtained a detection limit of 0.1 µg/L zinc in sea-water.

MAIER *et al* [2608] determined zinc in surface waters and drinking water directly without interferences. VÖLLKOPF *et al* [3023] determined zinc in contaminated industrial waste waters and found that phosphoric acid was an excellent matrix modifier, permitting thermal pretreatment up to about 900°C. Primary ammonium phosphate can also be used as matrix modifier. For atomization off the L'VOV platform in a stabilized temperature furnace, no signal depression was caused by the matrix, even with severely contaminated waste water. Zeeman-effect background correction was required to correct the high background attenuation.

LUNDBERG and FRECH [2572] determined zinc in steel directly in solid samples. CARRONDO *et al* [2133] determined zinc in sewage sludge by dispensing a suspension directly into the graphite tube. The results were in good agreement with those obtained by the flame technique after digestion.

The optimum temperature for atomization off the L'VOV platform is 1600°C.

10.59. Zirconium

Zirconium can be determined in a nitrous oxide/acetylene flame at the 360.1 nm resonance line with a characteristic concentration of 10 mg/L 1% and a detection limit of about 1 mg/L. Further resonance lines are compiled in table 41.

Table 41. Zr resonance lines

Wavelength nm	Energy level (K)	Characteristic concentration (mg/L 1%)
360.1	1241–29 002	10
354.8	570–28 750	15
303.0	1241–34 240	15
301.2	570–33 764	17
298.5	0–33 487	17
362.4	570–28 157	20

Zirconium is ionized by about 10% in the nitrous oxide/acetylene flame [794]; to suppress this effect, 0.1% potassium (as chloride) should be added to sample and reference solutions [1152].

Amos and Willis [48] found that the absorption of zirconium is enhanced considerably in the presence of hydrofluoric acid and high iron concentrations; they therefore added 2% hydrofluoric acid to all solutions. Slavin *et al* [1152] verified this and also found that hydrochloric acid had a similar influence. 10% hydrochloric acid increased the signal of a zirconium reference solution by 400% compared to the acid-free solution. Bond [141] obtained the strongest effect by adding 0.1 M aluminium fluoride solution which gave an eight fold signal. Wallace *et al* [3029] found that the addition of aluminium in excess improved the linearity, sensitivity, precision and detection limit for zirconium. It is possible that aluminium reduces the tendency to form a refractory oxide and facilitates atomization. It is equally possible that aluminium reduces lateral diffusion and thus concentrates atoms more in the center of the flame. In the presence of 1% hydrofluoric acid and 0.2% aluminium, the authors obtained a detection limit of 0.3 mg/L. Bond *et al* [142] [143] found that the absorption of zirconium is enhanced in the presence of many nitrogen containing compounds, which can react as Lewis bases, and from this they developed a procedure for the determination of ammonium traces. Tyler [1247] extracted zirconium as the cupferon complex with a benzene/isopentanol (1:1) mixture and so eliminated a number of interferences. Foster [383] found that a thenoyltrifluoroacetone (TTA)/xylene mixture was ideally suited for the extraction of zirconium and that it enhanced the signal by a factor of at least five.

11. Specific Applications

Because atomic absorption spectrometry is specific and has great freedom from interferences, its range of applications is very wide. Although the actual analytical methods are often largely similar, the various problems posed by different branches of analysis justify a separate treatment of each area. Emphasis will be placed on the actual application of atomic absorption spectrometry, while sample preparation, such as dissolution or digestion of solid substances, can be assumed as largely known and will therefore be only briefly touched upon. This is in particular the case when the preparation of samples does not differ from known wet chemical procedures; only special preparation techniques will be described more fully. The interferences already mentioned under the individual elements will also be assumed as known and only special interferences will receive fuller treatment.

11.1. Body Fluids and Tissues

One of the most important applications of AAS is in the routine clinical-chemical laboratory for the determination of calcium and magnesium in various body fluids. The determination of sodium and potassium [956], which is largely carried out in simple flame photometers, is also occasionally performed by AAS. The determination of elements such as iron, copper and zinc in serum has become increasingly routine, because they cannot be determined with sufficient speed and reliability by other methods.

The determination of *calcium* and *magnesium* in serum and urine was first described by WILLIS in 1960 in various publications and it is now one of the most widely used applications of AAS. A survey of the extensive literature which has been published on the determination of these elements in biological materials is given in table 42.

The simplest procedure, which is employed most frequently, is the direct analysis of the 1:20 to 1:50 diluted sample in an air/acetylene flame. To suppress phosphate interference, 1% sodium EDTA, 0.5% lanthanum in hydrochloric acid, or 0.25% strontium must be added to sample and reference solutions. Lanthanum may only be added to diluted serum, as otherwise the protein will coagulate. For this reason EDTA is occasionally preferred. Although protein does not interfere in the determination of calcium and magnesium when using a good nebulizer/burner system at the high dilution and in the presence of lanthanum, various authors have found better reproducibility with deproteinized samples [1081] [1121]. It must be taken into consideration, however, that trichloroacetic acid and hydrochloric acid influence the determination of these two elements, i.e. the reference solutions must contain the same concentration of the corresponding acid. For the determination of calcium and magnesium in urine, BHATTACHARYA and WILLIAMS [2086] proposed the use of the nitrous oxide/acetylene flame, since it is then only necessary to dilute with a potassium buffer to suppress ionization; other interferences do not occur.

Numerous comparative measurements with conventional procedures, such as complexometric titrations, flame emission spectrometry, fluorimetry, etc., have shown that for the determination of calcium and magnesium in the most varying body fluids, atomic absorption

Table 42. Bibliographical survey: Application of atomic absorption spectrometry for the determination of calcium and magnesium in body fluids and tissues (reference numbers)

Element	Serum	Urine	Other body fluids, tissues, etc.
Ca	[416] [420] [484] [583] [671] [873] [956] [958] [998] [1041] [1081] [1113] [1245] [1323] [1324] [1384] [2079] [2332]	[416] [496] [673] [873] [1081] [1144] [1244] [1245] [1263] [1328] [1384] [2086]	[121] [283] [416] [462] [504] [673] [873] [910] [951] [2373] [2406] [2561] [2660]
Mg	[279] [416] [420] [438] [467] [505] [533] [542] [552] [554] [672] [782] [873] [956] [958] [1081] [1113] [1175] [1322] [1325] [2079] [2331]	[279] [416] [496] [505] [533] [554] [782] [873] [1081] [1175] [1244] [1328] [2086]	[121] [229] [279] [283] [416] [442] [462] [463] [476] [505] [542] [545] [554] [782] [873] [951] [969] [2373] [2406] [2561] [2660]

spectrometry is at least as good, and with respect to precision and speed is usually superior [438] [467] [1041] [1081] [1144] [1263].

Nevertheless, HANSEN [465] has drawn attention to the great care required in the determination of calcium, since this element is relatively sensitive to changes in the gas ratio of the flame. When all parameters of the nebulizer/burner system have been optimized, a precision of better than 1% for calcium and 0.6% for magnesium can be obtained.

The determination of *sodium* and *potassium* in serum has been described by WILLIS [1326], HERRMANN and LANG [504], PASCHEN [956] and BERNDT and JACKWERTH [2079]. It is best carried out on a 1:50 dilution of the sample in an air/hydrogen flame, and sodium should be measured at the secondary line at 330.2 nm to avoid too high dilutions.

If an air/acetylene flame is employed, the sodium content in the potassium reference solution must be matched approximately to that in the samples. However, the addition of a neutral spectrochemical buffer, such as cesium, to sample and reference solutions is preferable.

The micro method described by PASCHEN [956] for the determination of sodium, potassium, calcium and magnesium in a single serum dilution is particularly elegant. 100 µL serum are diluted 1:100, or better 1:50 [957], with 0.25% strontium chloride solution and analyzed directly for the four elements. The addition of strontium eliminates both the influence of phosphate on calcium and the ionization of sodium and potassium in the air/acetylene flame. The relative standard deviations for the determination lie between 0.3% and 0.5%.

GRUNBAUM and PACE [446] described a micro method for the determination of sodium, potassium, calcium and magnesium in urine. 0.1 mL urine is diluted 1:100 under addition of lanthanum and cesium and aspirated directly into the flame. HAZEBROUCQ [484] controlled the sodium, potassium, calcium and magnesium contents of solutions used for extrarenal dialysis (artificial kidney machines) and found that AAS is superior to FES. The precision for sodium is 1%. GUTTMANN [2331] developed a diagnostic method for latent cardiac insufficiency via the determination of the electrolytes sodium, potassium, calcium and magnesium. BERNDT and JACKWERTH [2079] determined the four serum electrolytes using the loop technique, a micromethod only requiring 5–20 µL serum.

However, not just serum and urine can be examined by AAS; HANKIEWICZ [463] for example analyzed gastric juices for magnesium by diluting them 1:10 with water and nebulizing directly into the flame.

The graphite furnace technique is not particularly suitable for the determination of sodium, potassium, calcium and magnesium in serum since very high dilutions are necessary and, further, the expected precision and accuracy cannot be achieved because of the high risk of contamination [836].

A number of authors investigated the determination of calcium and magnesium in liver tissue; techniques of solubilization were of major interest. MENDEN *et al* [2660] found that dry ashing followed by repeated treatment with hydrochloric acid, *aqua regia* and nitric acid gave the best results. LOCKE [2561] proposed low temperature ashing with subsequent dissolution in nitric acid. He found that sodium, potassium, phosphate and sulfuric acid contents had to be well matched in the reference solutions to avoid interferences on the determinations in the flame. IIDA *et al* [2406] preferred a pressure digestion with nitric acid and perchloric acid. HINNERS [2373] found that it was sufficient to lyophilize the liver sample and extract with 1% nitric acid to determine calcium and magnesium quantitatively.

The elements *iron, copper* and *zinc* are present in similar concentrations of about 1 mg/L ($= 100$ µg%) in serum and therefore count as trace elements. A survey of the literature published on the determinations of these three elements in various body fluids and tissues is given in table 43.

While tissue samples, for obvious reasons, must be wet or dry ashed before the analysis, iron, copper and zinc can frequently be determined directly in serum. Zinc is the easiest to determine because of its high sensitivity [1035]; it can be determined directly against aqueous reference solutions without interferences in 1:5 diluted serum and undiluted or 1:2 diluted urine [1030]. To prevent blockage of the burner slot, a three-slot burner head is used frequently with the air/acetylene flame. DAWSON and WALKER [280] preferred higher dilutions and scale expansion. The determination of copper and iron in serum presents greater problems because of the low sensitivities of these two elements. Owing to the relatively high viscosity of blood serum, it cannot be compared with aqueous reference solutions; on the other hand, higher dilutions considerably worsen the sensitivity and precision of the determination. For this reason, various authors have recommended deproteinization [110] [1218] [2130] or extraction [952] [1385], which however is somewhat time consuming for routine determinations. OLSEN and HAMLIN [930] recommended a simple deproteinization procedure with which 95% of the hemoglobin iron can be removed at the same time.

Deproteinization with trichloroacetic acid is by no means without problems, as established by FIELDING and RYALL [366], since it can have a considerable influence on the subsequent analysis. The precipitation of protein reduces the liquid phase by up to about 13% since the

Table 43. Bibliographical survey: Application of atomic absorption spectrometry for the determination of copper, iron and zinc in biological materials

Element	Serum, Plasma	Urine	Other body fluids, tissues, etc.
Cu	[110] [135] [294] [493] [506] [529] [836] [952] [1035] [1150] [1171] [1204] [1301] [2080] [2130] [2235] [2284] [2290] [2613] [2683] [3033] [3042]	[110] [1150] [1204] [2284] [2956]	[110] [230] [462] [476] [929] [1204] [2132] [2234] [2250] [2373] [2406] [2436] [2442] [2593] [2660] [2663] [2683] [2956] [3082]
Fe	[112] [135] [419] [493] [836] [927] [930] [1035] [1171] [1281] [1301] [1362] [1370] [1385] [2080] [2861]	[112] [135] [1161] [1380] [2956]	[135] [462] [476] [507] [930] [1370] [1381] [1382] [2026] [2406] [2561] [2660] [2956]
Zn	[280] [455] [482] [493] [784] [836] [952] [958] [1000] [1030] [1035] [1171] [2080] [2290] [3013]	[280] [784] [1000] [2594] [3013]	[121] [230] [280] [462] [476] [929] [951] [2050] [2234] [2250] [2373] [2406] [2436] [2526] [2561] [2593] [2594] [2660] [2663] [2896] [3082]

protein occupies about the same volume before and after the precipitation. Furthermore, the precipitate has a positive surface charge which can lead to adsorption of negatively charged ions, such as iron complexes, from the supernatant liquid, especially in the presence of EDTA. Further, SIERTSEMA [2861] reported a spectral interference in the determination of iron due to trichloroacetic acid. ALDRIGHETTI *et al* [2026] lyophilized blood serum, dissolved the residue in dilute hydrochloric acid and determined iron directly in the solution after centrifugation.

On the other hand, direct determinations also are not free of problems. While the determination of copper usually presents no difficulties, various points must be observed for the determination of iron. The viscosity problem can be very reliably solved by diluting the serum 1:2 with water and matching the reference solutions with a commercially available synthetic plasma expander [1035]. Nevertheless, this procedure can only be employed for iron when a reliable hemolysis can be carried out after the sample has been taken. Hemolytic sera can only be determined with reasonable reliability after deproteinization. Furthermore, the direct method is not suitable for the determination of iron in *serum,* since 40–70 µg Fe/ 100 mL too much are always found compared to other (e. g. colorimetric) procedures (figure

138). In plasma, the determined iron values agree well with other methods [493]. The direct injection method proposed by MANNING [2613] for the determination of copper in serum is particularly suitable. WEINSTOCK and UHLEMANN [3042] used this procedure to analyze routinely undiluted, untreated serum. After 500 determinations they observed no memory effects, no carryover and no blockage of the nebulizer or burner. Calibration was against pool serum. The reproducibility for copper was 1.8% for a series and 2.2% for day to day. Using the same procedure, BERNDT and JACKWERTH [2080] determined iron, copper, and zinc in serum.

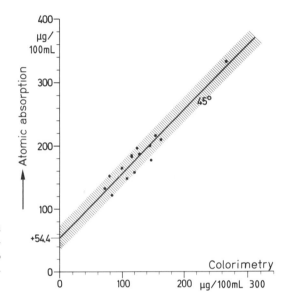

Figure 138. For the determination of iron by AAS in serum which has not been previously treated, 40–70 µg Fe/100 mL too much are usually found compared to colorimetric procedures or deproteinized sera.

The determination of iron and copper in urine offers no great problems. The direct determination of acidified (2 mL urine + 0.1 mL conc. sulfuric acid), undiluted urine has the necessary precision, sensitivity and specificity for clinical chemistry. There are no interferences due to protein and inorganic components, and the agreement with other, more time-consuming procedures is good [1161].

The greatest disadvantage for the determination of iron, copper and zinc in serum or plasma in the flame is the high sample requirement of 1 to 2 mL per determination if the injection method is not used. Thus, for these determinations the graphite furnace technique is frequently used, since only about 1–2 µL serum are required for each determination and the precision is better. The disadvantage of this micro procedure is the markedly greater time requirement for each determination.

The remarks made for the determination of iron in the flame naturally also apply here, i. e. a direct determination can only be made on plasma but not on serum. The majority of authors therefore prefer a micro deproteinization procedure recommended by OLSEN et al [927] which gives perfect results even with only 20 µL serum [1362].

Copper in serum, and iron and copper in urine can be determined either directly or after simple dilution with water in the graphite furnace [112] [1301] [2683] [3038]. The agreement with other methods is good, as shown by various authors [2235] [2290]. Because of the high risk of contamination, the determination of zinc by the graphite furnace technique requires special precautions [3013]. For this reason it is sometimes easier to use the flame technique [2593].

Atomic absorption spectrometry also serves ideally for the determination of the iron binding capacity [930] [1370] [1385] and iron in hemoglobin [507] [1381] [1382].

A micro procedure has also been worked out for determining the iron binding capacity using the graphite furnace technique [112] [1362], giving very good agreement with the macro method of OLSON and HAMLIN [930]. The sample requirement here is also around 20 μL serum.

Increasingly, tissue samples, especially liver tissue, are being examined by AAS for their content of iron, copper and zinc. JOHNSON [2436] performed a wet digestion in hydrochloric acid and nitric acid and determined copper and zinc by the flame technique. EVENSON and ANDERSON [2234] fumed off liver tissue with nitric acid and determined copper by the graphite furnace technique and zinc by the flame technique. MENDEN *et al* [2660] proposed a dry ashing procedure for the determination of eight elements.

HINNERS [2373] extracted lyophilized liver tissue with 1% nitric acid and obtained a quantitative recovery for copper and zinc, but not for iron. IIDA *et al* [2406] employed a pressure digestion in nitric acid/perchloric acid, while LOCKE [2561] proposed low temperature ashing. CARPENTER [2132] used enzymatic digestion for liver and kidney tissues.

JULSHAMN and ANDERSON [2442] examined tetraalkylammonium hydroxide as a solvent for muscle tissue and found that it was very suitable for the subsequent determination of copper. For the digestion of brain tissue, WUYTS *et al* [3082] investigated various methods and found that during dry ashing, alkaline phosphates leach copper out of porcelain crucibles. They therefore used a wet ashing procedure with success.

BAGLIANO *et al* [2050] developed a simple and rapid method for the determination of zinc in hair using a digestion in nitric acid/perchloric acid and the flame technique. SOHLER *et al* [2896] determined zinc in fingernails directly in the solid sample using the graphite furnace. LANGMYHR *et al* [2526] determined zinc in teeth also by solid sampling in the graphite furnace.

Even though virtually nothing was known about the other trace elements in the human body two decades ago, the publication of a number of review articles [278] [293] [1098] [1201] at the beginning of the 1970s showed the suddenly awakened interest in metallic elements which are often present in concentrations under 1 mg/L (100 μg%) in body fluids. There are at least three reasons for making a thorough investigation of these trace elements: Concentrations that are too low, due to dietry insufficiencies or to metabolic disturbances, can lead to deficiency diseases. Concentrations that are too high because of excessive environmental intake or metabolic disturbances can produce toxic effects. Abnormal concentrations of trace elements can also be caused by certain illnesses, so that a way is perhaps offered for a better understanding of these illnesses.

For a long time, the greatest obstacle for a routine determination of these elements was the large sample requirement and the necessity of a time consuming sample preparation proce-

Table 44. Bibliographical survey: Application of atomic absorption spectrometry for the determination of essential and non-essential trace elements in biological materials

Element	References
Al	[389] [707] [708] [2309] [2442] [2509] [2534] [2652] [2714] [2889] [2973]
B	[65]
Co	[294] [521] [1192] [1329] [2032] [2056] [2436] [2547] [2601]
Cr	[199] [200] [201] [277] [355] [429] [970] [983] [1059] [1085] [1383] [2050] [2147] [2314] [2329] [2372] [2436] [2461] [2513] [2543] [2738] [2803] [2837]
Mn	[24] [96] [230] [429] [462] [787] [941] [1301] [2032] [2073] [2250] [2373] [2406] [2436] [2442] [2525] [2526] [2541] [2561] [2660] [2802] [2848]
Mo	[983]
Ni	[294] [521] [523] [918] [970] [1086] [1136] [1200] [1201] [1203] [2000] [2003] [2036] [2066] [2112] [2336] [2372] [2747] [2947] [3022] [3094]
Rb	[1207] [2561]
Se	[1117] [1130] [2160] [2410] [2437] [2690]
Sr	[96] [246] [272] [294] [297] [876] [1287] [2364]
V	[2509]

dure for determinations in the flame. The graphite furnace technique has given the biochemist and the medical practitioner the means of collecting the quantity of data necessary for making a statistically reliable prognosis for the first time.

With trace elements, a distinction is made between essential, non-essential, and toxic elements. Here it must be taken into consideration that an essential element can be thoroughly toxic when it is present in excess, and that a toxic element may only become dangerous for humans above a certain concentration. It must further be considered that an element regarded as indifferent today may only be so because its importance for the human body has not yet been recognized. Nowadays, there is an increasing tendency not to attribute a particular effect to an element, but to refer to an essential or a toxic concentration of that element. Table 44 gives a literature survey on the determination of essential and indifferent trace elements in biological materials; and in table 45, the normal values for essential elements are given together with their effects on the human body.

The problem of "normal values" has been referred to by a number of authors [2946] [3010]. These are usually the mean values found for a large number of visibly healthy, non-exposed subjects. Particularly with trace elements, the values assumed as normal have decreased in recent years, even by several orders of magnitude. In part, the reason is to be found in the more sensitive analytical techniques that have become available. But to a major extent, sources of contamination have been recognized and eliminated. Nevertheless, the values presented in table 45 should only be regarded as a general guide [3010].

KRISHNAN *et al* [707] [708] made a detailed examination of the *aluminium* content in the brain in connection with neurofibrillitic degeneration. The brain tissue was wet ashed with nitric acid and the aluminium content determined in a nitrous oxide/acetylene flame. McDERMOTT and WHITEHILL [2652] found dry ashing better than wet ashing; they determined aluminium by the graphite furnace technique. They found a value of 1.3 µg Al/g

Table 45. Some essential trace elements: Normal values in human serum or plasma and their effect

Element	Normal value μg/L	Effect in the body
Aluminium	1–10	Influence on cell mitosis; increased values with neurofibrillitic degeneration.
Chromium	0.1–0.5	Essential for glucose metabolism; deficiency with arterial sclerosis and diabetes.
Cobalt	0.02–0.2	Vitamin B_{12} coupling; excess with cardiomyopathy.
Manganese	0.5 (mean)	Essential for glucose and lipid metabolism and a number of other important biochemical processes.
Molybdenum	0.6 (mean)	Essential for xanthine oxidase and aldehyde oxidase.
Nickel	2–3	Deficiency with liver cirrhosis and chronic renal insufficiency; increased values with myocardic infarction, etc.
Selenium	50–100	Connected with vitamin E.
Strontium	50	Hardens bones and teeth, possibly connected with osteoporosis.
Vanadium	0.02–1	Possibly essential for fatty acid metabolism.

dried weight in brain tissue. JULSHAMN and ANDERSEN [2442] found that for the determination of aluminium in muscle tissue, tetraalkylammonium hydroxide was not a suitable solvent. GORSKY and DIETZ [2309] found that aluminium can accumulate in toxic quantities in the brain as a result of rhenal failure. They found a mean value of 28 ± 9 μg/L in serum for a direct determination by the graphite furnace technique. FUCHS *et al* [389] also determined aluminium in serum with the graphite furnace. By using temperature programming they were able to thermally decompose 25 μL serum directly and determine the aluminium, free of interferences.

SMEYERS-VERBEKE *et al* [2889] determined aluminium in serum and urine; they also used temperature programming to remove the organic matrix. The authors found that external contamination was the biggest problem. The precision and accuracy were improved by the use of an autosampler.

OSTER [2714] diluted serum with Triton X-100 and nitric acid and found good agreement between the direct determination and the analyte addition technique using the graphite furnace. The detection limit for this procedure is 2.5 μg/L and the mean value in healthy subjects is quoted as greater than 4 μg/L (range 2.5–7 μg/L). TODA *et al* [2973] determined aluminium in serum samples from gel permeation chromatography by atomizing off the L'VOV platform and obtained a detection limit of 0.4 μg/L.

Chromium, even after prior dry [983] or wet ashing and extraction [355], can hardly be determined with sufficient sensitivity in the flame. The graphite furnace technique, on the other hand, offers sufficient sensitivity for a direct determination; DAVIDSON and SECREST [277] found that the various oxidation states of chromium exhibited the same sensitivity. ROSS *et al* [429] [1059] found that treatment with nitric acid for the determination of chromium in serum and urine brought no alterations and they thus decided on a direct determination, like other authors [970] [1085].

PEKAREK *et al* [2738] dispensed 50 µL serum directly into the graphite tube and attempted to avoid interferences by selective volatilization.

ROUTH [2803] found that background attenuation was reduced when hydrogen was mixed with the inert gas. KAYNE *et al* [2461] established that background correction was only complete when a halogen lamp is used as the continuum source. They obtained a mean value of 0.14 µg/L in serum. HINDERBERGER *et al* [2372] found that an interference-free determination of chromium in biological materials is only possible when a matrix modifier such as primary ammonium phosphate is used and atomization is off the L'VOV platform.

KUMPULAINEN [2513] determined chromium in human milk using the graphite furnace technique and obtained a mean value of 1 µg/L. JOHNSON [2436] determined chromium in liver by the flame technique after prior digestion in hydrochloric acid and nitric acid. BAGLIANO *et al* [2050] used the graphite furnace technique to determine chromium in hair.

Cobalt is the only essential element whose concentration is too low even for a direct determination by the graphite furnace technique. SULLIVAN *et al* [1192] described the determination of cobalt in tissue after wet ashing and DELVES determined cobalt in oxidized blood with the flame after extraction in MIBK. ALT and MASSMANN [2032] used the graphite furnace technique to determine cobalt in serum after prior wet digestion and extraction. LIDUMS [2547] preferred an ion exchanger for separation.

Manganese is present in blood serum in relatively high concentrations so that a direct determination in the flame can be made in a good atomic absorption spectrometer after 1:2 dilution with water [787]. For the determination of manganese in urine, an extraction is however to be preferred [24] [941]. Manganese in serum and other body fluids can be determined by the graphite furnace technique [2032] [2802]. The results are mostly in agreement with those from ashed [96] or extracted [429] samples. For the determination of manganese in urine, LEKEHAL and HANOCQ [2541] used a wet digestion and extraction prior to determination in the graphite furnace.

For the analysis of liver tissue, BELLING and JONES [2073] employed wet digestion with subsequent extraction, while IIDA *et al* [2406] preferred a pressure digestion with direct analysis of the digestion solution in the graphite furnace. MENDEN *et al* [2660] dry ashed tissue samples with repeated acid treatment and determined eight elements. HINNERS [2373] extracted lyophilized liver tissue with 1% nitric acid and was able to recover manganese quantitatively. For the determination of manganese, JULSHAMN and ANDERSEN [2442] dissolved muscle tissue in a tetraalkylammonium hydroxide. LANGMYHR *et al* [2525] determined manganese in bones directly by solid sampling in the graphite furnace. They found that excess chloride must be removed, so they added nitric acid to the sample in the tube. Using the same procedure, the authors also determined manganese in teeth [2526].

PIERCE and CHOLAK [983] determined *molybdenum* in blood and urine in a flame after ashing and extraction with APDC/MIBK.

Nickel can be determined in serum, whole blood or urine in a flame after acid treatment or deproteinization and subsequent extraction with APDC/MIBK [294] [521] [918] [1086] [1200]. Nickel can be determined directly in serum by the graphite furnace technique employing 50 µL samples [970] [1136], or with addition of lanthanum in nitric acid solution [1203].

The use of a temperature program for the thermal pretreatment of large serum volumes is indispensable. BEATY and COOKSEY [2066] found that by introducing air or oxygen briefly into the graphite furnace during thermal pretreatment, virtually residue-free ashing of the serum is possible. Thus, even with larger serum volumes, series determinations free of interferences are possible. VÖLLKOPF *et al* [3022] made an in-depth study on the significance of pyrolytic coating of graphite tubes on the determination of nickel.

The IUPAC Reference Method for the determination of nickel in serum and urine [2112] recommends an acid digestion followed by extraction with APDC in MIBK prior to determination by the graphite furnace technique. This method is in agreement with the findings of a number of working groups who have reported that the accuracy and precision of the nickel determination are better after an extraction step [2000] [2003] [2036] [3094]. Nevertheless, by the use of a suitable matrix modifier and atomization off the L'vov platform, a reliable and accurate direct determination should also be possible [2372].

CLINTON [2160] determined *selenium* in blood by the hydride technique after digestion in nitric acid/perchloric acid. He found a mean value of 29 ng/g in blood. NEVE and HANOCQ [2690] extracted selenium prior to determination by the graphite furnace technique. ISHIZAKI [2410] also used an extraction procedure and added nickel as a matrix modifier for the determination in the graphite furnace. For the determination of selenium, MANNING [2614] reported a spectral interference due to larger quantities of iron. This could interfere in the direct determination of selenium in blood and some tissues. This interference can be corrected by Zeeman-effect background correction.

Strontium can be determined in urine in a flame after dry ashing and addition of lanthanum [1287], or directly after 1:2 dilution with a hydrochloric acid/lanthanum solution [297]. BEK *et al* [96] determined strontium in serum directly by the graphite furnace technique and obtained good agreement with values from ashed samples. HELSBY [2364] determined strontium in tooth enamel by the graphite furnace technique. He was able to eliminate interferences due to the calcium and phosphate matrix by using the analyte addition technique. As a result of the high sensitivity of the graphite furnace technique and the low sample requirement, it is possible to make a strontium distribution profile over the tooth.

KRISHNAN *et al* [2509] investigated the determination of *vanadium* in biological tissues by the graphite furnace technique.

Besides these essential trace elements, other "indifferent" elements such as gold, lithium or rubidium are determined in serum and urine since they are introduced into the body therapeutically. The interest here is not the normal concentrations, which are very low, but solely the increased content in blood or serum and the amount excreted in the urine.

Gold is administered for the treatment of articular rheumatism. Gold in serum can be determined directly in a flame after 1:2 to 1:4 dilution. The viscosity of the reference solutions should be matched to that of the samples [302] [303] [313] [762]. For determination in blood, Triton X-100 is advantageously added for hemolysis [477]. Urine is treated with trichloroacetic acid, nitric acid/perchloric acid [314], or potassium permanganate [69] and the gold then extracted with MIBK. For all clinical requirements, the flame gives sufficient sensitivity and precision. Nevertheless, this determination is increasingly performed by the graphite furnace technique, in part due to the low sample requirement [2212] [2601] [2824]. WAWSCHINEK and RAINER [3039] recommended an extraction of the gold prior to determination in the graphite furnace. KAMEL *et al* [2451] [2452] determined gold in protein fractions separated by gel chromatography, and in tissue [2453] with good agreement between the flame technique, the graphite furnace technique, and neutron activation analysis.

Lithium is administered for the treatment of manic psychosis. Lithium in serum can be easily determined in the flame after 1:10 dilution [466] [744] [745] [1009]; in urine, greater dilutions may be necessary [1383]. FRAZER *et al* [385] determined lithium in erythrocytes after 1:50 dilution with water. By using the flame injection technique for the determination of lithium in serum, BERNDT and JACKWERTH [2079] were able to markedly reduce the sample requirement. The graphite furnace technique is also employed for determining very low lithium concentrations and for microanalysis [1301] [2601] [2906].

Platinum is administered as cis-dichlorodiammine-platinum(II) for tumor therapy. The determination of this element in body fluids by the graphite furnace technique has been described by several authors [2055] [2400] [2439] [2739]. *Palladium* is also determined in blood and urine by this technique [2439].

Rubidium is also employed in the therapy of psychoses. Like lithium, it is determined in serum, plasma, whole blood and urine employing 1:10 dilution with water for serum or blood and in undiluted urine [1207].

Besides the metallic elements, various non-metals or anions in body fluids are determined indirectly by AAS. Of especial mention is the method described by BARTELS [82] for the determination of *chloride* in serum in which an excess of silver nitrate solution is added, and after centrifugation, the silver in the supernatant liquid is determined directly. WOLLIN [1350] and also ROE *et al* [1042] determined *sulfate* in serum and urine by first oxidizing with H_2O_2 (SCHÖNINGER) or potassium chlorate (BENEDICT) and then adding barium, and after centrifugation, determining the barium in the supernatant liquid.

ZAUGG and KNOX [1372] [1373] described a method for the determination of *phosphate* in which ammonium molybdate is added after deproteinization, the free heteropoly acid is shaken with octan-2-ol and molybdenum determined in the organic extract. PARSONS *et al* [954] modified this procedure for automated analyses and the simultaneous determination of calcium in 0.2 mL plasma. The method is suitable for 0.2–20 mg P/100 mL and 1–20 mg Ca/100 mL and has a relative standard deviation of about 1%.

MANNING and FERNANDEZ [804] described the direct determination of phosphorus in urine and liver tissue by the graphite furnace technique. They employed the absorption of a PO_x radical which can be observed at a lead ion line at 220.3 nm.

The determination of metals of toxicological interest was described early in the history of AAS. In 1962, WILLIS reported the determination of *bismuth, cadmium, lead* and *mercury* in urine after extraction with an organic solvent. Similarly, most other toxic metals are extracted with APDC in MIBK over a wide pH range. A literature survey on the various publications in this area is compiled in table 46.

The fact that the "normal values" of these elements cannot be detected by direct nebulization into a flame, so that an enrichment by means of extraction is always required, somewhat hindered the expansion of atomic absorption spectrometry in this field. It is therefore hardly surprising that the considerably more sensitive flameless procedures were rapidly applied in this sector. The first successful attempt was the procedure of BRANDENBERGER and BADER [152–155] for the determination of mercury, later followed by the boat tech-

Table 46. Bibliographical survey: Application of atomic absorption spectrometry for the determination of toxic elements in biological materials

Element	References
Ag	[178] [199] [200] [201] [2396] [2436] [2526]
As	[298] [2171] [2252] [2437] [2560] [2960] [3051]
Be	[138] [2317] [2401]
Bi	[299] [644] [1329] [2800]
Cd	[111] [112] [199] [200] [201] [218] [247] [294] [319] [429] [483] [738] [739] [750] [784] [1058] [1329] [2031] [2115] [2132] [2197] [2198] [2221] [2234] [2289] [2372] [2373] [2422] [2442] [2525] [2533] [2569] [2660] [2663] [2693] [2740] [2758] [2766] [2918] [2992] [3079] [3099]
Hg	[111] [146] [564] [565] [566] [703] [711] [735] [753] [754] [786] [847] [863] [864] [871] [874] [920] [926] [1249] [1329] [2129] [2214] [2236] [2250] [2437] [2502] [2552] [2659] [2715] [2825] [2974] [3069]
Pb	[46] [53] [85] [109] [113] [199] [200] [201] [216] [217] [219] [292] [294] [308] [318] [333] [357] [359] [480] [482] [483] [492] [508] [509] [523] [591] [592] [593] [616] [618] [685] [695] [696] [697] [712] [783] [784] [869] [983] [1052] [1056] [1106] [1112] [1301] [1327] [1329] [1386] [2033] [2057] [2072] [2092] [2105] [2132] [2240] [2241] [2242] [2294] [2295] [2313] [2372] [2379] [2380] [2421] [2525] [2533] [2623] [2660] [2663] [2693] [2758] [2791] [2861] [2919] [2959] [3024] [3075]
Sn	[2983]
Te	[644] [1117] [2155]
Tl	[111] [112] [148] [482] [784] [1080] [1120] [2132] [2151] [2279] [2871]

nique [616] which can be applied to virtually all elements of toxicological interest. Urine undiluted and whole blood diluted 1:10 are introduced into a tantalum boat, dried near the flame, lightly ashed and then inserted into the flame of the atomic absorption spectrometer. The method is rapid and simple and permits routine detection of toxically increased values without special sample preparation. The technique has been described in detail for lead in urine [616], lead in whole blood [618] and thallium in various biological materials [248]. Thereafter followed the DELVES system, which was specially developed for the determination of lead in whole blood, and also the cold vapor technique for the determination of mercury. The conclusive breakthrough of AAS into toxicology came with the introduction of the graphite furnace technique, which can be universally employed for all the toxic elements mentioned here [1301].

The determination of *arsenic* can be performed by both the graphite furnace technique and the hydride technique. THIEX [2960] made an acid digestion of biological materials and extracted arsenic prior to determination by the graphite furnace technique. LO and COLEMAN [2560] preferred dry ashing for tissue samples; they took the residue up in hydrochloric acid and performed the determination directly in the graphite furnace. FITCHETT *et al* [2252] separated various arsenic species in urine by extraction and determined them in the graphite furnace. COX [2171] determined arsenic in urine by the hydride technique after digestion in nitric acid/sulfuric acid/perchloric acid. A direct determination is also possible when a suitable antifoaming agent is added [3051].

The fact that *lead* can only be determined in whole blood or urine after ashing and extraction [869] [985] has already been mentioned. KOPITO *et al* [696], however, further emphasize that despite all the advances, this determination remains relatively difficult and contains many sources of error; it must be carried out with the greatest care and requires experience. SCHLEBUSCH and NIEHOFF [1089] go a step further by maintaining that despite extraction and the allied enrichment, the flame does not give sufficient sensitivity for the determination of normal values (approx. 10 µg Pb/100 mL).

The DELVES procedure [292] for the determination of lead in whole blood has found wide distribution and application (see page 45). The results correspond well with the conventional flame technique [509] and colorimetry [359]. Besides the original method, in which the blood sample is dried in a nickel crucible and treated with hydrogen peroxide, numerous modifications have been described. EDIGER and COLEMAN [318] merely dried the blood sample on a hot-plate and then introduced it directly into the flame, while BARTHEL *et al* [85] replaced hydrogen peroxide by nitric acid and ROSE and WILDEN [1056] by *aqua regia* and were thus able to improve the precision of the determination. JOSELOW and BOGDEN [592] extracted the lead with APDC/MIBK before carrying out the determination in the DELVES system. MARCUS *et al* [2623] improved the procedure further and used it for routine determinations. BRATZEL and REED [2105] reported that one of the major sources of error for the determination of lead in blood is in the taking of the sample. JACKSON *et al* [2421] determined lead in urine without sample preparation directly against spiked urine samples. GRAEF [2313] determined lead in hair taken from beards directly by solid sampling.

An interesting variation has been described by CERNIK and SAYERS [219], who placed a droplet of blood on filter paper, and after drying, punched out a standard disk and analyzed it in the DELVES system. Even under very different conditions, the blood spread on the filter paper so that the lead values showed no significant variations [217]. JOSELOW and SINGH [593] found that the precision of this procedure could be markedly improved if the cup remained in the holder during a measurement series; nevertheless, the paper disk method appears to be less precise than that with liquid blood.

The initial attempts at determining lead in blood or urine by the graphite furnace technique gave only unsatisfactory results [200] [201]. For the determination using a carbon rod furnace, VOLOSIN *et al* [3024] found that neither the direct procedure nor diluting with Triton X-100 gave the correct results. They therefore used an extraction step for the determination of lead in blood. Other authors reported that simple pretreatment with Triton X-100 [712] or with nitric acid [482] facilitated the determination of lead in urine [784]. ALT and MASSMANN [2033] made a comparison between a direct determination under the addition of ammonium nitrate, dissolution in a quarternary ammonium hydroxide, and a pressure digestion; they found that all three procedures gave comparable results for the determination of lead in blood. STOEPPLER *et al* [2919] automated the procedure for the determination of lead in whole blood. They added nitric acid for deproteinization and to act as a matrix modifier, and determined lead directly in the supernatant liquid.

FERNANDEZ [2240] developed a procedure in which the sample is merely diluted with Triton X-100 and determined directly against aqueous reference solutions. The procedure was later automated [2241] and the accuracy verified by numerous interlaboratory tests.

HINDERBERGER *et al* [2372] showed that by adding primary ammonium phosphate as matrix modifier and atomizing off the L'VOV platform, all interferences for the determination of lead in blood and urine can be eliminated. Thereupon, FERNANDEZ [2242] modified his procedure and used Triton X-100 for dilution and primary ammonium phosphate for matrix modification. Atomization is from the L'VOV platform in a stabilized temperature furnace, and is free of interferences.

MÉRANGER *et al* [2663] have drawn attention to the fact that blood samples, even when stabilized with heparin, cannot be stored for any length of time. Independent of the container material and the temperature, the lead content is significantly lower within a week.

For the determination of lead in liver and kidney tissue, CARPENTER [2132] proposed enzymatic digestion, while BARLOW and KHERA [2057] dissolved tissue in a quarternary ammonium hydroxide. This procedure is excellent for the solubilization of liver and placenta tissue for subsequent determination directly in the graphite furnace. WITTMERS *et al* [3075] determined lead in bone ash after dissolution in nitric acid and the addition of lanthanum solution. LANGMYHR *et al* [2525] [2533] determined lead in bones and teeth directly by solid sampling. It is necessary to remove chlorides, but this can be done by adding nitric acid directly to the sample in the graphite tube.

For the determination of *cadmium* in blood, serum or urine with the flame, wet ashing with subsequent extraction with APDC/MIBK is most suitable [294] [738] [739]. Although a direct determination of cadmium in the DELVES system appears to be possible [218] [319], prior extraction is preferred for precision determinations [750]. For determinations by the graphite furnace technique, extraction [112] or cold ashing [784] is of advantage because of the high volatility. Nevertheless, with careful thermal pretreatment and atomization at lower temperatures (e. g. 1300 °C), direct determinations, which are in good agreement with extraction procedures, appear to be possible [429] [1058]. Using modern graphite furnaces, the determination of cadmium appears to be well controlled. STOEPPLER and BRANDT [2918] added nitric acid to whole blood for deproteinization and matrix modification and determined cadmium in the supernatant liquid. Urine is merely diluted and determined directly. ALT [2031] compared this procedure for blood with a procedure using a quarternary ammonium hydroxide, and an extraction procedure. The best values were obtained for the addition of nitric acid; the quarternary ammonium hydroxide caused severe background that could only be eliminated by using Zeeman-effect background correction. The extraction procedure is prone to contamination and cannot therefore be recommended. DELVES and WOODWARD [2198] found that a combination of matrix modification with ammonium phosphate and oxygen ashing in the graphite tube during thermal pretreatment gave excellent results. For the determination of cadmium in urine, BRUHN and NAVARRETE [2115] found that dilution with Triton X-100, matrix modification with ammonium phosphate, and atomization at 800 °C using an ultrafast heating rate is an excellent procedure. Nevertheless, HINDERBERGER *et al* [2372] established that an interference-free determination of cadmium in blood and urine is only possible using matrix modification and atomization off the L'VOV platform.

PRUSZKOWSKA *et al* [2766] made a detailed investigation on the determination of cadmium in urine and found that secondary ammonium phosphate is superior to primary as matrix modifier. They diluted urine 1:5, added secondary ammonium phosphate as modifier and atomized at 1500–1700°C off the L'vov platform in a stabilized temperature furnace. To the reference solutions they added 0.2% sodium chloride in addition. They corrected the high background attenuation using Zeeman-effect background correction. They obtained a characteristic concentration of 0.2 µg/L referred to undiluted urine. The determination is free of non-spectral interferences. It has been clearly demonstrated by various authors [2198] [2221] [3099] that the cadmium content in the blood of smokers is significantly higher than for non-smokers. MÉRANGER *et al* [2663] point out that blood samples in which cadmium is to be determined should not be stored for longer periods. Even in blood samples stabilized with heparin, the cadmium content is distinctly higher within a week, regardless of the temperature or container material.

EVENSON and ANDERSON [2234] determined cadmium in liver by first fuming off with nitric acid and then taking up the residue in dilute nitric acid for direct determination in the graphite furnace. HINNERS [2373] extracted lyophilized liver tissue with nitric acid for a subsequent determination by the graphite furnace technique. JULSHAMN and ANDERSEN [2442] investigated tetraalkylammonium hydroxide as a solvent for muscle tissue and found it excellently suited for the subsequent determination of cadmium. JACKSON and MITCHELL [2422] homogenized tissue under the addition of water and analyzed the emulsion directly. LANGMYHR *et al* [2525] [2533] determined cadmium in bones and teeth directly by solid sampling in the graphite furnace. The authors removed chlorides by adding nitric acid directly to the sample in the graphite tube.

In principle, even though *mercury* can be determined in the flame after extraction and respective enrichment [111] [864] or by the boat technique [863], nowadays the cold vapor technique is used almost exclusively (refer to page 76 *et seq*). The sole procedural differences are in the preparation of the sample. Urine is usually allowed to stand for several hours with sulfuric acid/potassium permanganate [711] [754]. The same procedure can also be used for blood, plasma [711] and hair [920]. BOUCHARD [146] recommended digestion in chromic acid because of the great speed and LINDSTEDT and SKARE [754] decomposed blood with nitric acid/perchloric acid at 70°C overnight. MAGOS and CERNIK [786] determined mercury in urine, blood and tissue without prior ashing; they were only able to determine the inorganic mercury, while after ashing total mercury can be determined. A good review on other flameless methods for the determination of mercury has been published by MANNING [797].

Sodium borohydride has become established as an alternative reductant to tin(II) chloride for the cold vapor technique [2659]. OSTER [2715] found that total mercury in urine could be determined very reliably using sodium borohydride as reductant after prior treatment of the urine with potassium permanganate solution and a mixture of nitric acid/sulfuric acid. SHIERLING and SCHALLER [2825] analyzed urine and blood directly, reducing with borohydride and collecting the mercury on a gold gauze. They were able to attain a sensitivity and specificity to permit even "normal" mercury contents in blood and urine to be detected.

Organomercury compounds are also reduced by sodium borohydride, although sensitivities vary somewhat [2974]. Several authors used weaker reductants to separate the organic mercury compounds from inorganic mercury to permit discrete determination [2129] [2214] [2552]. The special problems associated with the determination of mercury have been discussed in sections 8.4 and 10.27.

11.2. Foodstuffs and Drinks

In the 1970s the determination of trace elements in foodstuffs increased dramatically. The new possibilities offered by the graphite furnace, hydride, and cold vapor techniques played a significant role. The content of trace elements in foodstuffs is influenced by a number of sources (refer to figure 140, page 375). Influences on vegetable products include the nature of the soil, fertilizers, crop protection agents, insecticides, pesticides, etc., and on the proximity of roads and industrial installations. Animal products are influenced primarily by feedstuffs and the environment, while aquatic species are influenced by their habitat itself. All foodstuffs are further influenced by preparation, storage, packing etc. It is worth remembering that handling procedures can also cause a decrease and not just an increase in the trace element contents: For example, the removal of peel or external leaves, or washing or cooking can leach out soluble constituents. Problems can already begin with collecting the sample, a step over which the analyst often has no influence. The contents of trace elements can be so falsified that a subsequent analysis has little meaning. After sampling, a representative portion of the sample must be taken, shredded or crushed, mixed, homogenized, etc. A number of authors, for example, have dealt with the homogenization of canned products for the subsequent determination of lead [2440] [2942].

Thereafter, organic substances must be decomposed by digestion, separation and/or enrichment steps must be performed if necessary, followed by the actual determination of the trace elements. The procedure is similar for elements present in higher concentrations except that enrichment steps can be omitted. The determination is usually performed by the flame technique after suitable dilution of the digestion solution.

Metallic elements can be present in foodstuffs either bound chemically to functional organic groups or as inorganic salts. Generally, a digestion is required to decompose the organic material or to break the organometallic bond. Methods of sample preparation depend on the analyte element, the nature of the sample, and the facilities available to the analytical laboratory. It should consequently be realized that every method is a compromise between quantitative recovery of the analyte element, and speed, simplicity, reliability and demands of the procedure.

A widely used procedure is wet digestion in which the sample is oxidized with suitable acids at increased temperature. Very frequently, a mixture of nitric acid and perchloric acid is used [2286] [2389] [2747] [2866] [2867] [2953] [3001] [3019], often under the addition of sulfuric acid [2147] [2249] [2392] [2627] [2960]. The latter acid combination permits very rapid digestion that can also be automated. A number of authors follow this digestion step with an extraction step [2147] [2867] [2953], because perchloric acid, for example, attacks graphite. However, this is no longer a problem due to the availability of high quality pyrolytically coated graphite tubes.

A somewhat less efficient procedure, but definitely less dangerous, is digestion in a mixture of nitric acid and sulfuric acid [2010], frequently under the addition of an oxidizing agent such as potassium perchlorate [2134], hydrogen peroxide [2711], or vanadium pentoxide [2220] [2267]. REAMER and VEILLON [2782] found a wet digestion with phosphoric acid/nitric acid/hydrogen peroxide to be particularly suitable for the subsequent determination of selenium. AGEMIAN and CHEAM [2008] used a digestion in sulfuric acid/hydrogen peroxide, combined with oxidation with permanganate/persulfate for the determination of organic arsenic and mercury compounds in fish. SCHACHTER and BOYER [2822] performed extractions with nitric acid in a Soxhlet apparatus. SPERLING [2898] [2901] [2902], in particular, miniaturized the scale of nitric acid/sulfuric acid and nitric acid/perchloric acid digestions for the determination of cadmium in marine organisms and was thereby able to bring contamination problems under control.

KNAPP *et al* [2476] [2477] [2478] described a mechanized system for performing wet chemical digestion procedures especially with chloric acid/nitric acid. Particularly for volatile elements, such as cadmium, selenium and mercury, they found this acid mixture to be more suitable than all others.

Digestion procedures in PTFE-lined autoclaves (refer to figure 70, page 123) offer interesting possibilities. Such digestions are usually complete within about an hour. PAUS [966] reported the digestion of fish and seaweed for the subsequent determination of cadmium, copper, iron, lead, mercury and zinc. HOLAK [528] found good agreement with the official AOAC method for the determination of mercury in fish. Using radio tracers, TÖLG *et al* [700] established that even with volatile elements such as selenium and mercury, losses are less than 2%. ADRIAN [19] worked with larger sample quantities (5g) and lower temperatures, but at the cost of time; he allowed the sample to stand at ambient temperature overnight and then heated to 90°C for three hours.

Pressure digestions are particularly suitable prior to the determination of very volatile elements, since losses can be avoided [2917] [3050] and it is possible to work with nitric acid/sulfuric acid or nitric acid/hydrochloric acid without perchloric acid [2832] [2917] [3050]. The digestion solutions, after dilution, can generally be used directly for the determinations by AAS. Employing atomization off the L'VOV platform in a stabilized temperature furnace, trace determinations largely free of non-spectral interferences are possible. Such digestion procedures are also usually well suited for subsequent determinations by the hydride technique [3050].

BEHNE *et al* [2071] found that chromium in brewer's yeast could only be determined without losses after pressure digestion. In the meantime it has been shown, however, that other digestion procedures for chromium are free of losses [2147], but that this element can be retained to a high percent on acid-insoluble residues.

As well as wet digestion, oxidation with oxygen at increased temperature, dry ashing, is carried out, particularly for less volatile elements. The advantage of this procedure is that the organic constituents are completely destroyed free of residues. This is naturally advantageous for the following analysis. Relatively large sample aliquots can be ashed, although the procedure then takes somewhat longer. The major disadvantage of this procedure is that losses can occur, mostly through volatilization, but also occasionally through the formation of insoluble oxides or silicates.

Ashing can be performed directly [2308] [2359] [2706] [3035] or under the addition of a reagent such as magnesium nitrate [2232] [2269] [2560] [2782] [2858] at 500°C to 550°C. The addition of nitric acid gives a particularly clean and easily soluble ash [2316], while the addition of sulfuric acid helps to avoid losses. FEINBERG and DUCAUZE [2238] found that the addition of sulfuric acid permitted the ashing temperature to be raised to 980°C without significant losses of cadmium or lead taking place.

MENDEN *et al* [2660] found that not only the temperature program used for ashing, but also the dissolution of the ash are of major significance for the recovery of the analyte. They developed a multiashing procedure in which, after a preashing step, potassium sulfate and nitric acid are added, followed by nitric acid alone, and then *aqua regia*. FETTEROLF and SYTY [2247] reported that chewing gum can only be completely solubilized by a combined dry and wet digestion procedure.

Occasionally, the SCHÖNINGER procedure, in which the sample is ignited in an oxygen flask, is used [2382] [2986]. However, only small quantities of sample material can be treated by this procedure and the sample-to-wall ratio is very unfavorable. TÖLG [620] described a modified dry ashing procedure using activated oxygen that is free of losses. A digestion procedure using activated oxygen is only suitable for samples with very low inorganic salt content, however. Even a few percent of inorganic salts form a crust through which the oxygen plasma cannot pass. For ultratrace analysis, a compound technique developed by TÖLG's group is particularly suitable [2345]. The sample is incinerated in oxygen and the volatilized elements condense on a cold finger. They are then dissolved by refluxing with hydrochloric acid or nitric acid in the same apparatus and the solution used directly for the determination.

Mercury occupies a special position among the trace elements (refer to sections 8.4 and 10.27). Samples cannot be dry ashed, even with the addition of suitable additives, without losses [2269], and even homogenization [2137] or freeze drying [2771] are not without problems. Even wet digestions for mercury are not without risks, since losses can occur, depending on the digestion time and temperature [2137] [2477] [3008], especially if the oxidizing power of the digestion solution is not high enough.

Consequently, procedures permitting sample digestion at low temperature or in a very short time have been developed. BOUCHARD [146] digested fish and meat in chromic acid at 25 °C. The digestions were complete within ½ to 1½ hours. VELGHE *et al* [3007] added potassium permanganate and sulfuric acid to fish samples and warmed briefly. They found that the fish dissolved in less than a minute and that organic mercury compounds were quantitatively broken down. DUSCI and HACKETT [2213] dissolved fish within two to three minutes at 60 °C in nitric acid under the addition of sodium molybdate.

For the subsequent determination of mercury, pressure digestions in an autoclave are especially suitable, since they are free of losses if performed properly [528] [700] [966] [3050]. SEEGER [2836] described a simplified procedure using a normal screwtop container at low temperature and slight pressure.

ROOK *et al* [2798] and also GLADNEY and OWENS [2305] described a procedure of incineration in a stream of oxygen with freezing out of the mercury in a cold trap that is very good for the quantitative, loss-free determination of this element.

The possibilities of speciation of mercury have already been discussed in section 10.27. EGAAS and JULSHAMN [2220] have pointed out that the determination solely of the total mercury content can give a false picture. The content of antagonistic elements such as selenium must also be taken into account.

Selenium also occupies a special position among the trace elements. Numerous tests with animals have clearly shown that this element is essential, but up to the present no deficiency diseases have been reported for man. The concentration window between the toxicological threshold and the essential requirement is given as 1–2 powers of ten and is thus less than with any other trace element. Interactions between selenium and other trace elements are very pronounced, making physiological evaluation extremely difficult [2382].

As well as the oxidative digestion of biological materials, an extraction with dilute acid or an organic solvent has been investigated for various elements and samples. FREEMAN *et al* [2267] extracted inorganic and organically-bound arsenic from fish tissue and obtained the same results as with a wet digestion. MAURER [2641] extracted seven elements from foodstuffs with hydrochloric acid and nitric acid and found that this procedure was better, quicker and more simple than dry ashing. HINNERS *et al* [2375] extracted cadmium from rice with 1% nitric acid and obtained the same results as with a wet digestion in nitric acid and perchloric acid.

LOVE and PATTERSON [2564] extracted apples with acetone to investigate for residues of the pesticide tricyclohexyltin hydroxide. TRACHMAN *et al* [2983] determined tin in biological materials after dissolution in a quarternary ammonium hydroxide.

CRUZ *et al* [2178] have drawn attention to the importance of a speciation of lead. As well as total lead, which is determined after a nitric acid/sulfuric acid/perchloric acid digestion, the contents of extractable, volatile lead and tetra-alkyl lead are important. For the former the authors extracted with hexane. Volatile lead was separated by heating the sample to 150 °C and collecting the volatilized constituents in a cold trap. The determination was performed by warming the trap slowly and measuring directly in the graphite furnace. Tetra-alkyl lead is determined by the same procedure, except that a separating column is additionally used. For the determination of tetra-alkyl lead in fish tissue, SIROTA and UTHE [2873] homogenized the tissue and shook it with an aqueous solution of EDTA and benzene. For the separation of inorganic arsenic from fish BROOKE and EVANS [2109] distilled from 6.6 M hydrochloric acid solution and also used chelation and extraction, followed by back extraction. Using the hydride technique, both extraction procedures gave results in good agreement with each other. The authors preferred the distillation procedure because it is quicker and more effective.

Solid samples can be determined directly by the graphite furnace technique, since the sample can be thermally decomposed *in situ* in the graphite tube. One of the first applications of this type originated from PICKFORD and ROSSI [2741]. They analyzed the NBS standard reference material beef liver. Calibration was by means of the analyte addition technique by dispensing aqueous reference solutions directly onto the sample in the graphite tube. Agreement with the certificate values for copper, lead, manganese, and silver were within 10%. CHAKRABARTI *et al* [2142] later examined the same sample for six elements and found that problems due to contamination were considerably fewer with direct solid sampling.

LORD *et al* [2562] pressed pellets from freeze dried pulverized muscle tissue and dispensed them directly into the graphite tube. For aluminium, chromium and copper it was possible to calibrate against aqueous reference solutions, but not for lead and zinc since the appearance times for the sample and reference solution were different. Using a stabilized temperature furnace, it should be possible to eliminate this problem through peak area integration.

LANGMYHR *et al* [2522] [2523] [2527] made a detailed investigation of direct solid sampling of fish, liver, and botanical samples and determined cadmium, chromium, cobalt, copper, lead, manganese, nickel, and phosphorus. They used the analyte addition technique and also determinations directly against aqueous reference solutions and found good agreement with values obtained from digestion procedures and certificate values. Direct solid sampling is fast, requires little sample material, and exhibits few problems due to contamination or high blank values.

GROBENSKI *et al* [2320] combined direct solid sampling with *in situ* ashing by introducing oxygen briefly into the furnace during thermal pretreatment, thus destroying the organic constituents. They found far fewer spectral and non-spectral interferences using this technique than when analyzing digestion solutions.

Provided that the concentrations of the analyte elements are sufficiently high, **edible oils** and **liquid fats** can be nebulized directly into a flame after suitable dilution with an organic solvent. For the determination of trace elements by the flame technique, a prior ashing is unavoidable. Frequently, combustion in oxygen, such as the SCHÖNINGER technique [2986] or under reduced air supply in the presence of sulfuric acid [2985], with subsequent extraction is used.

Trace elements in edible oils and fats can be determined directly by the graphite furnace technique after slight dilution (to permit better dispensing). Frequently, multistep temperature programs are required to remove the organic constituents. In figure 139 the determination of phosphorus in soya oil is shown as an example [3050]. PRÉVÔT and GENTE-JAU-NIAUX [2760] determined phosphorus in edible oils directly after 1:1 dilution with MIBK and obtained a detection limit of about 0.5 mg/kg. SLIKKERVEER *et al* [2884] improved the procedure further and found that the addition of lanthanum (as lanthanum-2,4-pentadionate) gives improved precision and excellent recovery. These authors found that the ability to select complex temperature programs on the graphite furnace is particularly important.

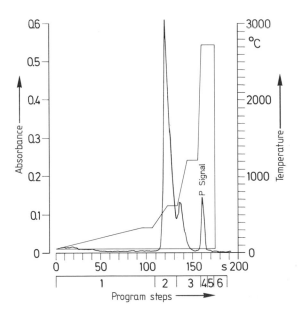

Figure 139. Determination of phosphorus in soya oil by the graphite furnace technique. Temperature program and the background and element signals.

Milk can be nebulized undiluted into the flame; however, the concentrations of the analyte elements are occasionally too low, so that they must be extracted after wet ashing or determined from the residue after dry ashing. BROOKS *et al* [166] preferred deproteinization of milk with trichloroacetic acid to be able to determine the main elements free of interferences. SULEK *et al* [2944] compared the determination of lead in dried milk by the flame technique, after dry ashing and subsequent chelation and extraction, and the graphite furnace technique. For the latter technique the sample was dissolved in nitric acid in a steam bath and determined directly. Both procedures gave good agreement with voltametric procedures; the graphite furnace technique is the simplest.

LAGATHU and DESIRANT [722] determined iron, cobalt, copper, manganese, and strontium in milk by the graphite furnace technique with relative standard deviations of 2–5%. MANNING and FERNANDEZ determined copper and strontium [802] and lead [803] directly using background correction.

Almost all **dairy products** must be wet or dry ashed before analysis. A sole exception is the determination of copper in butter and other fats; this is most conveniently carried out via a simple extraction of the dissolved or melted sample with nitric acid [15]. PRICE *et al* [1005] described a similar extraction for nickel in edible fats which is also carried out with 10% nitric acid in the presence of carbon tetrachloride.

Drinks, with the exception of fruit juices or concentrates still containing pulp, can be nebulized directly into a flame. The flame is used in effect for the *in situ* decomposition of the organic material. MCHARD *et al* [2655] [2656] described a hydrolysis procedure using nitric acid at 80 °C for the determination of eight trace elements in orange juice by the flame technique. Liqueurs and concentrates must be diluted since they have high viscosities, and carbon dioxide in drinks should be driven off before analysis, most suitably by warming. BORIELLO and SCIAUDONE [2093] determined copper, iron, lead and zinc in canned and bottled beer after warming with nitric acid and hydrogen peroxide. They found that canned beer had significantly higher concentrations of lead and zinc, and comparable concentrations of iron and copper. Drinks containing high or changeable sugar or alcohol contents should be analyzed by the analyte addition technique if possible, since a high alcohol content in particular can influence the individual elements to varying degees. VARJU [1258] reported that a direct determination is also possible with alcoholic drinks if the alcohol content is brought to 50% and not allowed to vary by more than 5%.

WATKINS *et al* [3035] examined **packaging materials,** especially colored paper, for lead and found very high contents in some cases. The samples were solubilized by wet or dry ashing. TRACHMAN *et al* [2983] determined tin in packing materials after extraction with water, acetic acid, and organic solvents. Tin is added to plastic foils as a stabilizer.

A bibliographical survey on the analysis of foodstuffs and drinks is presented in table 47. A detailed review with references up to 1978 has been published by FRICKE *et al* [2268].

11.3. Soils, Fertilizers and Plants

Agricultural chemistry encompasses three main branches; the analysis of soil extracts, fertilizers, and plants. Besides the main elements sodium, potassium, calcium and magnesium,

Table 47. Bibliographical survey: Application of atomic absorption spectrometry in foodstuffs chemistry

Element	Animal Products	Vegetable Products	Oils, Fats	Drinks, Milk, Dairy Products
Ag	[2316] [2420] [2748]	[2420]		
Al	[129] [2562]	[525]		
As	[2008] [2109] [2232] [2249] [2253] [2267] [2389] [2405] [2560] [2711] [2857] [2858] [2960] [3050]	[2232] [2405] [2711] [2857] [2858] [2960] [3050]		[2858] [3050]
B	[526]	[526]		
Bi	[2797]	[2797]		
Ca	[121] [489] [951] [1129] [2641] [2660]	[594] [951] [2641]	[23] [382] [448]	[166] [386] [594] [722] [2655] [2656] [3050]
Cd	[49] [129] [527] [2010] [2142] [2233] [2238] [2389] [2522] [2660] [2706] [2747] [2753] [2832] [2874] [2892] [2898] [2901] [2902]	[751] [2090] [2233] [2238] [2286] [2375] [2479] [2747] [2832] [2892] [3050]	[2985]	
Cl				[405]
Co	[598] [2420] [2527] [2874]	[2420] [2866]		[387] [722] [3050]
Cr	[49] [129] [2010] [2316] [2420] [2527] [2543] [2562] [2706] [2874]	[594] [2071] [2134] [2420] [2984]		[387] [594]
Cu	[49] [129] [203] [489] [490] [491] [674] [741] [1129] [2010] [2231] [2238] [2443] [2522] [2562] [2641] [2660] [2748] [2874] [2892]	[39] [594] [2231] [2238] [2359] [2641] [2867] [2892] [3050]	[1143] [2985]	[15] [180] [386] [387] [594] [722] [802] [897] [1046] [1184] [1258] [1331] [1378] [2093] [2656]
Fe	[129] [489] [674] [1129] [2231] [2308] [2443] [2641] [2660] [2874]	[273] [525] [594] [2231] [2641]	[1143]	[386] [387] [594] [722] [1003] [1046] [1258] [2093] [2656]
Hg	[57] [146] [528] [531] [576] [754] [895] [934] [949] [1225] [1229] [1251] [2008] [2044] [2087]			

Table 47. Continued

Element	Animal Products	Vegetable Products	Oils, Fats	Drinks, Milk, Dairy Products
	[2137] [2186] [2213] [2220] [2445] [2771] [2798] [2852] [2925] [3007] [3008] [3050]	[531] [576] [948] [2512] [2798] [2836] [3050] [3053]		[2986] [3050]
K	[121] [490] [1226] [2641]	[273] [594] [2641]	[1001]	[166] [386] [512] [594] [722]
Li		[594]		[114] [594]
Mg	[121] [489] [674] [951] [2641] [2660]	[36] [273] [594] [951] [2641]	[23] [448]	[116] [387] [549] [722] [1258] [2656]
Mn	[489] [674] [2231] [2443] [2522] [2641] [2660] [2748] [2874]	[174] [273] [594] [2231] [2359] [2641]	[1143] [2985]	[387] [594] [722] [2656]
Mo				[1046]
Na	[121] [490] [867] [2641]	[273] [594] [2641]	[130] [1001]	[166] [386] [512] [594] [722]
Ni	[2010] [2233] [2527] [2747] [2874]	[2233] [2747]	[1005] [1143]	[180] [387]
P	[654] [2523]	[2523]	[1143] [2760] [2884] [3050] [2985]	[654] [804]
Pb	[129] [144] [527] [590] [1055] [1179] [2010] [2178] [2233] [2237] [2238] [2389] [2392] [2440] [2443] [2522] [2660] [2706] [2747] [2748] [2753] [2873] [2874] [3050]	[352] [2170] [2178] [2225] [2233] [2238] [2247] [2392] [2440] [2627] [2747] [2984]		[369] [456] [724] [803] [1055] [1377] [1378] [2093] [2392] [2944] [3050]
Rb		[594] [1185]		[594]
Sb	[2232] [2249]	[2034] [2232]		[2034]
Se	[2220] [2249] [2253] [2345] [2404] [2405] [2778] [2851] [3050]	[2160] [2283] [2345] [2382] [2404] [2405] [2778] [2782] [2953] [3019] [3050]		
Sn	[276] [1054] [1129] [2232] [2378]	[2232] [2378] [2564] [3050]		[1003] [1093] [3050]
Sr	[275] [1242]	[270] [594] [1242]		[722] [802] [1242]
Te	[2249]			
Zn	[49] [121] [129] [489] [674] [951] [1129] [1246] [2010] [2231] [2316] [2443] [2641] [2660] [2874]	[174] [267] [273] [594] [951] [2231] [2359] [2641]	[23] [2985]	[387] [594] [1378] [2093]

Table 48. Bibliographical survey: Application of atomic absorption spectrometry to the analysis of soils, fertilizers and plants

Element	Soils and Soil Extracts	Fertilizers	Plants
Al	[173] [720] [968] [1365] [3088]		
As	[2005] [2256] [2963] [3017] [3037]		[2318] [2511] [2735] [2893] [2963] [3037]
B	[464] [979]	[461] [980] [1288]	[859] [963] [2227] [2283]
Ca	[173] [269] [968]		[117] [268] [274] [418] [1259] [2376]
Cd	[612] [2208] [2270] [2419] [2746] [2775] [2792] [2828] [2994] [3088]		[2994]
Co	[2270] [2419] [2746] [3088]		[1127] [2283]
Cr	[173] [2661] [3088]		
Cu	[39] [64] [173] [464] [1260] [2270] [2419] [2746] [2792] [2830] [3088]	[461] [490] [846]	[22] [39] [418] [1040] [1259] [2283]
Fe	[37] [64] [173] [968] [1365] [2830] [3088]	[461] [846]	[22] [37] [273] [274] [418] [1040] [1111] [1259] [2283]
Hg	[3087]		
K	[173] [269] [968] [1221]	[490] [989] [1230]	[274] [418]
Li	[173]		[961]
Mg	[269] [441] [968] [1321]	[846]	[117] [274] [418] [1259] [2283]
Mn	[64] [173] [464] [535] [2746] [2830] [3088]	[461] [846]	[22] [174] [273] [274] [418] [1040] [1111] [1259] [2283]
Mo	[464] [1250] [2465]	[271] [530]	[503] [2283] [2689]
Na	[269] [968]	[490] [2169]	[274] [418]
Ni	[173] [2270] [2419] [2661] [2746] [2792] [3088]		
P		[2383]	
Pb	[612] [2270] [2419] [2746] [2792] [3088]		[532] [612] [1111] [2225] [2363]
Rb	[173]		
SO₄	[401] [1261]	[2123]	
Sb			[2034]
Se	[2005]	[553]	[1017] [2283]
Si	[721] [1365] [2369]		
Sn			[1319]
Sr	[173] [272]		[272] [1242]
Zn	[41] [64] [173] [464] [2746] [2792] [3088]	[461] [846]	[22] [41] [174] [274] [418] [1040] [1259] [2283] [2376]

the trace elements copper, manganese, zinc, iron, molybdenum and boron are of especial interest because they play an important role in the life cycle of plants. If these elements are not contained in the soil in sufficient quantity, damage to plants or poor harvests can result; the soil should be enriched with fertilizer. On the other hand, over-fertilization can also lead to damage, so that a regular control is necessary. The aluminium content of soils is also frequently examined nowadays since it is of importance for regeneration and the acidity of the soil. Toxic elements, such as arsenic, cadmium, lead, etc., are also of great importance with respect to environmental protection. A literature survey on the individual elements determined in soil extracts, fertilizers, and plants is compiled in table 48.

Flame atomic absorption spectrometry is very suitable for the direct analysis of **soil extracts** since the often complex and concentrated matrix rarely gives greater problems. For trace analyses, the extract can be directly nebulized with ammonium acetate or other extracting agents using a 3-slot burner head. The results are in good agreement with neutron activation analysis [173] and frequently are better than with conventional procedures [464]. The freedom from interferences and speed of AAS permit easier automation when large numbers of samples must be handled [532]. Boron is the only element which frequently cannot be determined directly with the required sensitivity in soil and soil extracts. A special extraction procedure with an organic solvent brings an increase in the sensitivity by a factor of about 25 and is particularly useful for this element [979] [1288].

KRISHNAMURTI *et al* [2508] found that the digestion of soils with nitric acid and hydrogen peroxide at 100 °C was especially advantageous for subsequent analysis by the flame technique. Organic material is readily oxidized and contamination is no problem. Naturally, the much more sensitive graphite furnace and hydride techniques play an important role for the determination of trace elements in soil extracts. Most authors [2208] [2419] [2737] [2746] [2828] [2994] prefer an extraction of the complexed metallic trace elements with an organic solvent such as MIBK or chloroform to avoid interferences. YAMASAKI *et al* [3088] determined a large number of trace elements in soil extracts directly in the graphite furnace, but found numcrous interferences. It is probable that such interferences can be eliminated by atomizing off the L'VOV platform in a stabilized temperature furnace.

The determination of arsenic in soil extracts by the hydride technique appears to be largely free of interferences [2005] [2963] [3017] [3037]. Compared to other techniques, it has the advantage of speed, sensitivity and accuracy. The hydride technique is particularly suitable for routine analyses [2963] and has been automated for the determination of arsenic and selenium [2005].

The analysis of **fertilizers** mostly requires a prior digestion; however, for the analysis of the most important trace elements, fuming off once with concentrated hydrochloric acid and dissolving the residue in dilute hydrochloric acid is frequently sufficient. The results obtained for manganese, copper, zinc, and iron using this technique are in agreement with those obtained after a potassium hydrogen sulfate fusion or a hydrofluoric acid/nitric acid digestion [461]. MCBRIDE [846] made a critical investigation on the determination of copper, iron, manganese, magnesium, and zinc in fertilizers by AAS [1363] and additionally carried out comparative measurements. The results were very satisfactory and the procedure was adopted as the official method of the AOAC.

COROMINAS *et al* [2169] compared various methods for the determination of sodium in fertilizers and obtained the best accuracy and precision with AAS. CAMPBELL and TIOH [2123] determined sulfate in fertilizers by precipitating it as barium sulfate, taking up the precipitate with ammoniacal EDTA solution and determining barium. WOODIS *et al* [3078] described an indirect procedure for determining biuret in mixed fertilizers and urea. An alcoholic solution of biuret and copper is treated with a strong base so that a biuret/copper complex is formed and the excess copper is precipitated. The dissolved copper, which corresponds to the concentration of the biuret, is determined.

MANNING and FERNANDEZ [804] reported the direct determination of phosphorus in fertilizers using the graphite furnace technique. They utilized the absorption of a PO_x radical at a lead ion line at 220.3 nm which is specific for phosphorus. HOFT *et al* [2383] determined phosphorus in aqueous solutions of fertilizers using a nitrous oxide/acetylene flame at the 213.6 nm resonance line. They found that the method was fast, specific, and free of interferences.

The analysis of **plant materials** practically always requires ashing of the dried material; atomic absorption spectrometry offers the advantage that the ash, mostly dissolved in dilute hydrochloric acid, can be analyzed directly. Only very rarely are further enrichment steps such as extraction in an organic solvent required.

A number of authors examined the relationship between sample preparation and analytical results for plant analysis. GIRON [418] found no significant differences between dry and wet ashing with nitric acid/perchloric acid for the determination of eight elements; only the precision was better with dry ashing. VARJU [1259] found that pretreatment of the plant samples with nitric acid before the dry ashing accelerated this and increased the solubility of the ash. Copper and iron as their nitrates are not bound to silicic acid and can therefore also be quantitatively determined. An extraction of plant materials with 6 M hydrochloric acid at 110°C in an autoclave dissolves iron, copper, manganese and zinc quantitatively, but calcium and magnesium to less than 50%.

ROBLES and LACHICA [1040] compared wet ashing (with nitric acid) with three simple extraction procedures employing nitric acid/hydrochloric acid, hydrochloric acid alone, and hydrochloric acid/hydrofluoric acid, respectively, to remove the silicic acid. The last procedure gave the highest values for iron, copper, manganese and zinc. AGUILAR *et al* [22] examined the relationship between the conventional analysis of the sap, and found good agreement for copper, manganese and zinc, while the results for iron were up to ten times higher for the conventional method.

ELTON-BOTT [2227] determined boron in plants by treating the ashed sample with concentrated sulfuric acid to convert borate to boric acid, then reacting with methanol and conducting the volatile boric acid methyl ester formed into a nitrous oxide/acetylene flame.

FURR *et al* [2283] determined boron, cobalt, copper, iron, magnesium, manganese, molybdenum, selenium, and tin in apples, millet and vegetables grown on soils fertilized with fly ash. They found high contents of all these elements; fly ash can thus be used to correct certain soil deficiencies, but care must be taken to avoid an enrichment of toxic elements such as selenium. ELFVING *et al* [2225] investigated fertilization using waste paper and found somewhat increased lead levels in vegetables.

SEGAR *et al* [1111] determined lead, iron and manganese in an aqueous suspension of leaves by the graphite furnace technique and found good agreement with other procedures. HEN-NING and JACKSON [503] determined molybdenum in plants by the graphite furnace technique after dry ashing and dissolution of the residue in hydrochloric acid; they obtained good agreement with colorimetric procedures. NEUMANN and MUNSHOWER [2689] determined molybdenum in plants after wet digestion in nitric acid. They found that nitric acid reduced various interferences in the graphite furnace distinctly.

URE and MITCHELL [2994] determined cadmium in plant materials in hydrochloric acid solution by the graphite furnace technique after prior dry ashing at 450 °C. They found no noticeable interferences and obtained an excellent detection limit.

Numerous papers have been published on the determination of arsenic in plant materials by the hydride technique [2318] [2511] [2735] [2893] [2963] [3017] [3037]. Generally, a wet digestion in nitric acid/sulfuric acid [2511] [2963], or nitric acid/perchloric acid [2318] [2735] [3017], or sulfuric acid/hydrogen peroxide [2893] is performed prior to the determination. All authors are in agreement that this technique is sensitive, selective, fast, simple and economical. Interferences due to transition elements or acids are not observed, so that this technique is well suited to automation [3017].

An in-depth study was made by one group on the determination of antimony in plants such as cabbage and grass [2034]. This group found that only the hydride-generation AAS technique provided sufficient sensitivity and a quantitative recovery.

11.4. Water

Water analysis includes the investigation of drinking water, fresh waters and sea-water, and also sediments. Waste waters and other polluted or endangered waters are handled in the following section (11.5, Environment).

Atomic absorption spectrometry is particularly suited for use in water analysis. The analyte metals are generally already present in solution and it is thus only necessary to select a technique of sufficient sensitivity for the analysis. The concentrations of dissolved substances in fresh waters is generally so low that apart from ionization, no major interferences for the determination of the alkali metals are to be expected. As long as the concentrations of the analyte elements are sufficiently high, or the sensitivity of flame AAS adequate, the metal salts dissolved in water can be determined directly without preparation. For natural waters this is principally valid for *sodium, potassium, calcium* and *magnesium* and also frequently for *zinc*. The most important trace elements, copper, iron, cobalt, nickel, manganese and chromium, on the other hand, usually cannot be determined with the required sensitivity by the direct method. Generally a preconcentration procedure or the use of the graphite furnace technique is necessary. A bibliographical survey on the use of atomic absorption spectrometry in water analysis is compiled in table 49.

The most frequently employed preconcentration procedure in water analysis for trace elements is complexing the metal ions with ammonium pyrrolidine dithiocarbamate (APDC) and subsequently extracting into methyl isobutyl ketone (MIBK). This extraction system has the advantage that relatively stable chelates are formed with numerous metals over a wide pH range, so that with a single extraction the majority of trace elements can be simul-

Table 49. Bibliographical survey: Application of atomic absorption spectrometry in water analysis

Element	References
Ag	[225] [1109] [1308] [2432]
Al	[373] [1302] [2339] [2608]
As	[2038] [2154] [2168] [2171] [2251] [2252] [2448] [2515] [2688] [2721] [2751] [2752] [2838] [2843] [2844] [2869] [2908] [3037]
Au	[1387]
B	[2608] [2954] [3000]
Ba	[2066] [2435] [3072]
Be	[510]
Bi	[2869]
Ca	[108] [125] [136] [192] [242] [371] [374] [386] [510] [989] [2891] [3072]
Cd	[72] [177] [510] [718] [770] [771] [965] [967] [988] [1015] [1109] [1168] [2100] [2116] [2128] [2255] [2324] [2326] [2339] [2360] [2432] [2468] [2608] [2664] [2779] [2788] [2838] [2894] [2902] [2903] [2905]
Cl	[404]
Co	[162] [168] [177] [182] [321] [337] [375] [510] [597] [815] [916] [965] [1077] [1109] [1168] [2468] [2608]
Cr	[177] [225] [289] [321] [361] [510] [815] [865] [883] [916] [989] [1302] [2172] [2189] [2339] [2360] [2432] [2608] [2838]
Cu	[72] [136] [162] [167] [177] [192] [320] [321] [337] [361] [371] [374] [386] [434] [510] [597] [704] [718] [815] [900] [916] [965] [967] [988] [989] [1015] [1109] [1168] [1237] [1302] [2100] [2116] [2255] [2339] [2360] [2432] [2468] [2779] [2894]
Cs	[2271]
Fe	[136] [162] [168] [177] [192] [320] [321] [337] [386] [510] [588] [596] [704] [916] [967] [988] [989] [1077] [1109] [1111] [1168] [1300] [1302] [2318] [2339] [2360] [2432] [2468] [2608]
Ge	[2806]
Hg	[372] [621] [798] [926] [1235] [1302] [2007] [2009] [2635] [2644] [2921] [3087]
K	[125] [192] [371] [374] [386] [3072]
Li	[56] [371] [374] [3072]
Mg	[125] [136] [192] [371] [374] [989] [2891] [3072]
Mn	[72] [136] [177] [320] [321] [361] [371] [374] [596] [597] [815] [916] [928] [967] [989] [1077] [1109] [1111] [1168] [1300] [1302] [2131] [2339] [2360] [2468] [2475] [2608] [2649]
Mo	[195] [289] [815] [899] [2256] [2498] [2684]
Na	[125] [192] [371] [374] [386] [510] [3072]
Ni	[136] [162] [168] [177] [182] [321] [337] [375] [588] [597] [815] [916] [965] [989] [1077] [1109] [1168] [2100] [2116] [2339] [2432] [2468] [2608] [2779] [2894]
P	[1371]
Pb	[72] [136] [162] [168] [177] [320] [361] [375] [597] [916] [965] [967] [988] [1109] [1111] [1168] [1302] [2122] [2178] [2255] [2293] [2341] [2360] [2377] [2432] [2468] [2608] [2669] [2779] [2783] [2788] [2838] [3018]
Rb	[125] [3072]
Sb	[2448] [2515] [2869]

Table 49. Continued

Element	References
Se	[699] [2168] [2180] [2366] [2448] [2515] [2535] [2628] [2767] [2751] [2838] [2869] [2907]
Sn	[2768]
Sr	[125] [371] [374] [3072]
Te	[2515] [2869]
V	[815] [901] [1109] [2498] [3043]
W	[2501]
Zn	[136] [162] [168] [177] [337] [361] [371] [374] [421] [434] [597] [916] [967] [988] [989] [1015] [1109] [2116] [2128] [2255] [2325] [2360] [2432] [2468] [2540] [2608] [2779] [2894]

taneously extracted. Using this procedure, many authors successfully extracted and subsequently determined *copper, zinc, iron, cobalt, nickel, lead, manganese* and other elements. JOYNER and FINLEY [596] complexed iron and manganese in sea-water with sodium diethyl dithiocarbamate and extracted the complex with MIBK. CHAU *et al* [228] found that *chromium* can best be extracted as the acetyl acetonate complex, and DELAUGHTER [289] employed diphenyl dithiocarbazone for chromium and dithiol for *molybdenum*, respectively, with extraction into MIBK.

In contrast, other authors found that these elements could be optimally extracted with APDC/MIBK. MIDGETT and FISHMAN [865] ascertained that chromium must be oxidized to the hexavalent state before extraction with this system. Nevertheless, the extraction of trace elements with organic solvents is not completely free of problems. OLSEN and SOMMERFELD [928] draw attention to the relative instability of various metal chelates. Manganese in particular begins to precipitate directly after extraction. These authors therefore evaporated the extract to dryness and dissolved the residue in 1 M hydrochloric acid/acetone (1:1). KOIRTYOHANN and WEN [691] found that the sensitivity of various extracted trace elements was dependent upon the pH of the aqueous solution and they eliminated this dependence by adding 0.2 M perchloric acid to the aqueous solution. FISHMAN [373] extracted *aluminium* from water with oxine/MIBK at pH 8; the extraction must be finished within three minutes however, otherwise magnesium oxinate comes out of solution and takes the aluminium with it. HICKS *et al* [510] draw attention to a possible interference due to the presence of water-miscible organic solvents in waste water.

MULFORD [894] described the APDC/MIBK extraction system in detail and BURRELL [184] treated the technique of solvent extraction in combination with atomic absorption spectrometry for the determination of trace elements in sea-water. Comprehensive instructions and methodology on the application of AAS in water analysis have been published by FISHMAN and DOWNS [374], BREWER, SPENCER and SMITH [162], and EDIGER [317].

JOYNER *et al* [597] published a comprehensive report on the various procedures for sample enrichment which are of interest for the trace analysis of sea-water. The authors deal critically with evaporation, co-precipitation, solvent extraction and other preconcentration procedures, and give a comprehensive bibliographical survey on this topic.

KORKISCH and KRIVANEC [2498] determined *molybdenum* and *vanadium* in a nitrous oxide/acetylene flame after separation as the citrate complex on an ion exchange column. VANDERBORGHT and VAN GRIEKEN [2999] combined chelation with 8-hydroxyquinoline and preconcentration on activated charcoal for the determination of trace elements in natural water. HALL and GODINHO [2339] found that freeze drying of natural waters for the determination of *aluminium, cadmium, chromium, copper, iron, manganese,* and *nickel* was just as effective as extraction or enrichment on an ion exchange column.

During the 1970s the graphite furnace technique found rapid acceptance for water analysis. With this technique, the various trace elements in natural surface waters can be determined directly without sample preparation [1294]. The major inorganic constituents in water cause a certain degree of signal depression [361] [1302], but this remains constant above a given total salt concentration. It is thus possible to avoid this interference by matching the matrix, for example by making up reference solutions in tap water [1302]. Other authors preferred the analyte addition technique [72] [321]; elements such as chromium and nickel exhibit hardly any matrix influences [321] [883]. Many authors [2117] [2360] [2608] [2838] report on the successful application of the graphite furnace technique for the determination of a large number of trace elements. For the analysis of drinking water, for example, this technique offers high sensitivity, good precision and selectivity, and requires little time since virtually no sample preparation is required [2838]. Compared to neutron activation analysis (after prior freeze drying) or to X-ray fluorescence (after prior enrichment on a cellulose exchanger), graphite furnace AAS offers the same results, but with the major advantage that no sample preparation is required [2117].

The determination of *lead* in water is frequently performed by the graphite furnace technique. MITCHAM [2669] reported severe depression of the signal for the direct determination, so he included an extraction step. VIJAN and SADANA [3018] separated lead by co-precipitation with manganese dioxide and determined the lead using the hydride technique. For the determination of lead by the graphite furnace technique, CALLIO [2122] found that the addition of phosphoric acid largely eliminated the interferences. REGAN and WARREN [2783] used ascorbic acid for the same purpose. As well as total lead, CRUZ *et al* [2178] also determined extractable, volatile and tetra-alkyl lead. GARDNER and HUNT [2293] have drawn to attention the risk of loss of lead during filtration which can cause substantial errors.

MÉRANGER and SUBRAMANIAN [2664] examined four different methods for determining *cadmium* in drinking water. They found that the determination is free of interferences, and thus preferred a direct determination using the graphite furnace technique. Using a suitable matrix modifier and atomization off the L'vov platform in a stabilized temperature furnace, the determination of lead in natural waters is also virtually free of interferences, so that extraction steps are not required.

SZYDLOWSKI [2954] and VAN DER GEUGTEN [3000] determined *boron* in natural waters by the graphite furnace technique. They found that the addition of barium or calcium and magnesium improved the sensitivity substantially. CRANSTON and MURRAY [2172] reported the speciation of *chromium* in natural waters. FLORENCE and BATLEY [2255] made a critical examination of speciation and the problems of trace analysis in water.

A large number of publications deal with the element *arsenic*. A number of authors employ the graphite furnace technique; the addition of nickel as matrix modifier permits calibration directly against aqueous reference solutions [2908]. OWENS and GLADNEY [2721] found that the addition of nickel could be omitted if the concentration of nitric acid is held constant. POLDOSKI [2752] reported a spectral interference in the determination of arsenic by the graphite furnace technique due to silicate in suspended matter. This interference is not observed using the hydride technique. SHAIKH and TALLMAN [2843] reduced arsenic with sodium borohydride, collected the arsine and dispensed it into the graphite furnace in dissolved form.

The majority of authors determine arsenic free of interferences by the hydride technique [2168] [2171] [2251] [2312] [2751] [2869] [3037]. It is necessary to reduce pentavalent arsenic to trivalent; potassium iodide is used preferentially [2168] [2869] and the reaction is complete within about 30 minutes at ambient temperature. FISHMAN and SPENCER [2251] mention that samples containing organically bound arsenic should be pretreated either with UV radiation or a sulfuric acid/potassium persulfate digestion. A number of authors have reported the speciation of arsenic in water [2038] [2154] [2252] [2688] [2844]. It is possible to exploit the fact that arsenic(III) can be reduced to hydride in neutral solution, while arsenic(V) can only be reduced in strongly acidified solution [2688] [2844]. Organic arsenic compounds are usually determined after selective extraction or volatilization.

Selenium is also determined relatively frequently in water. An interference-free determination by the graphite furnace technique can be expected when nickel is added as matrix modifier [2628] [2907] and atomization is off the L'VOV platform in a stabilized temperature furnace. For a determination by the hydride technique it is necessary to reduce selenium(VI) to selenium(IV), since the higher oxidation state exhibits no measurable signal. A reduction with iodide requires great care since selenium is easily reduced to the elemental state [2168]. Reduction with boiling semiconcentrated hydrochloric acid is more suitable [2535], especially in an autoclave [2869]. SINEMUS *et al* [2869] utilized the fact that selenium(VI) gives no measurable signal for the selective determination of both valency states.

GOULDEN and BROOKSBANK [2312] and also SINEMUS *et al* [2869] determined *antimony* in water by the hydride technique. The reduction of antimony(V) to antimony(III) with potassium iodide is virtually spontaneous. SINEMUS *et al* also determined *bismuth* and *tellurium* by the hydride technique. Bismuth requires no preparation, but tellurium requires brief heating with semiconcentrated hydrochloric acid to reduce it to the tetravalent state. PYEN and FISHMAN [2768] reported the determination of *tin* in water by the hydride technique; the acid and borohydride concentrations were critical.

Nowadays, the determination of *mercury* in water is performed almost exclusively by the cold vapor technique. Particularly for the determination of the "natural" content an amalgamation on gold is used. Permanganate or persulfate are frequently used for the digestion of organomercury compounds. However, AGEMIAN *et al* [2007] [2009] draw attention to the fact that this digestion procedure cannot be employed in the presence of high chloride contents, since chloride is immediately oxidized to chlorine gas. These authors thus used UV radiation, which permitted an excellent recovery for numerous organomercury compounds and showed no susceptibility to interference by chlorides. The technique only fails when the sample has a high content of suspended matter since UV radiation cannot penetrate turbid water. In such cases the authors used a sulfuric acid/dichromate digestion, which extracts mercury very effectively from particulate matter, followed by UV photo-oxidation.

MATSUNAGA *et al* [2635] established that a mercury concentration of 0.5 μg/L, even in acidified solution, rapidly decreases. The presence of sodium chloride prevents adsorption losses on the walls of the container, so that sea-water samples are less endangered. The mercury content in a sea-water sample containing 0.2 M sulfuric acid remains unchanged for 60 days. Polyethylene containers are unreliable since they are often contaminated; glass containers are easier to clean. STOEPPLER *et al* [2644] [2921] reported similar findings and also the fact that mercury diffuses through the walls of polyethylene containers. These authors also reported that the ratio of methylmercury to ionic mercury changes with storage.

The special problems associated with the determination of mercury and its speciation are discussed in sections 8.4 and 10.27.

The graphite furnace and hydride techniques are used almost exclusively nowadays to determine trace elements in **sea-water** whose concentrations, at least in the open oceans, are much lower than in fresh water. The procedure used most frequently is extraction with an organic solvent, followed by back extraction into acid medium and subsequent determination in the graphite furnace. Following upon the introduction of matrix modification, atomization off the L'VOV platform in a stabilized temperature furnace and Zeeman-effect background correction, there are even prospects that direct determinations will be possible [2875]. This is particularly valid for the higher trace element concentrations in coastal waters and estuaries.

SEGAR and GONZALEZ [1107] [1109] were the first to make a comprehensive examination of the analysis of sea-water with the graphite furnace technique and found that volatile elements such as lead, cadmium, silver and zinc could not be determined. More difficultly atomized elements such as iron, cobalt, manganese, nickel and vanadium could be freed sufficiently from the sodium chloride matrix, but owing to co-volatilization the sensitivity is in part strongly reduced. Only iron was still sensitive enough for a direct determination; for this reason the authors recommended extraction of the trace elements with APDC/MIBK before determination in the graphite furnace [1111]. This method has also been successfully used by other authors [704] [967]. An especial advantage of the graphite furnace technique for sea-water analysis is that the extraction can be performed on a micro scale [704] and the determination carried out "on site", so that samples do not need to be transported [185].

Extraction with an organic solvent followed by back extraction into an acid is the procedure used most frequently for enrichment. As well as APDC/MIBK [2432], dithizone/chloroform [2377] [2894], 8-hydroxyquinoline/chloroform [2475] and other extracting agents are used. DE JONG and BRINKMAN [2189] determined chromium(II) and chromium(VI) separately in the graphite furnace after selective extraction.

SPERLING [2902] [2903] [2905] made a detailed examination of the determination of cadmium in sea-water. He especially drew attention to the fact that cadmium in natural waters is often bound, so that only a portion is complexed by even such strong chelating agents as APDC. Consequently a digestion is required before extraction. He also mentioned the problem of contamination when handling the samples that can cause substantial discrepancies.

RASMUSSEN [2779] compared extraction with APDC/DEDC and back extraction into nitric acid with enrichment on a Chelex-100 ion exchange column for cadmium, copper, lead, nickel and zinc. He only found good agreement for cadmium, while for the other elements the results were falsified due to high reagent blanks. BRULAND *et al* [2116] performed the same comparison and found similar values for cadmium and zinc. Copper and nickel apparently were not removed completely from sea-water by the ion exchanger. KINGSTON *et al* [2468], on the other hand, found that a Chelex-100 column could be used very successfully, and under suitable conditions permits quantitative separation down to the sub-nanogram range.

MUZZARELLI and ROCCHETTI [899] [900] [901] determined copper, molybdenum, and vanadium in sea-water after treating it with persulfate and running it down a Chitosan column to separate sodium chloride. WEISS *et al* [3043] preconcentrated vanadium on an ion exchange column after reduction with ascorbic acid and found that the results obtained by the graphite furnace technique were just as reliable as neutron activation analysis.

LUNDGREN *et al* [771] were able to determine cadmium in sea-water directly by the graphite furnace technique using an ultrafast rate of heating and temperature control. They atomized cadmium at 820°C, a temperature at which sodium chloride is not noticeably volatile, and obtained a detection limit of 0.03 µg/L. Using the same technique, RIANDY *et al* [2788] determined cadmium and lead in sea-water. CAMPBELL and OTTAWAY [2128] reported the direct determination of cadmium and zinc.

In 1974, EDIGER *et al* [320] reported that the addition of ammonium nitrate facilitated the analysis of sea-water. The reaction

$$NaCl + NH_4NO_3 \rightarrow NaNO_3 + NH_4Cl$$

takes place and the end products, sodium nitrate and ammonium chloride, can be separated at temperatures around 400°C, while sodium chloride is not noticeably volatile until temperatures above 1100°C. They successfully determined lead, iron, copper and manganese with detection limits around 1 µg/L.

Other authors also used this matrix modifier successfully for the determination of elements such as lead [2341] or manganese [2131] [2649]. STEIN *et al* [2907] [2908] determined arsenic and selenium in river estuaries using nickel as a matrix modifier. This modifier had also originally been proposed by EDIGER [320]. SZYDLOWSKI [2954] determined boron in sea-water under the addition of barium to enhance the sensitivity.

GUEVREMONT *et al* [2324] [2325] [2326] investigated organic matrix modifiers and found that EDTA or citric acid was very suitable for cadmium and citric acid for zinc. Using these modifiers, the authors attempted to make the analyte elements more volatile than the concomitants. The atomization temperature fell well below the volatilization temperatures of other matrix constituents, so that exact temperature control or background correction were not necessary.

CARNRICK *et al* [2131] reported the successful direct determination of manganese in sea-water using matrix modification and atomization off the L'VOV platform in a stabilized temperature furnace. Calibration was against aqueous reference solutions.

STURGEON *et al* [2931] compared various methods for the determination of trace elements in sea-water. They found that ICP-AES was not sensitive enough and that isotope mass spectrometry was too expensive and too slow. In contrast, graphite furnace AAS is rapid and sufficiently sensitive, except for the lowest contents of certain elements. For these elements, an extraction or enrichment on an ion exchange column can be employed. The accuracy attainable with the graphite furnace technique is usually very good.

Finally, the analysis of **particulate matter** and **sediments** will be discussed briefly as it is occasionally associated with water analysis. The sediment fraction is usually digested in one of the common acid mixtures and, after dilution, is analyzed by the flame technique [50] [177] [380]. SCHOCK and MERCER [2830] found that using less sensitive resonance lines for iron, copper, and manganese saved a good deal of time compared to making the usual dilutions. They found no deterioration in the accuracy and precision, while there were fewer errors and less risk of contamination.

HONGVE and HOLTH-LARSEN [2391] determined calcium in sediments from inland lakes and were only able to obtain correct results using the nitrous oxide/acetylene flame. LOSSER [2463] determined iron in nitric acid extracts of sediments from the continental shelf. The addition of 0.3 M ammonium chloride was necessary to obtain correct values. KIM *et al* [2465] determined molybdenum by the flame technique after extraction with "Aliquat 336", a quarternary aliphatic amine.

RANTALA and LORING [2773] [2774] determined a number of trace elements in suspended matter and sediments after digestion in hydrofluoric acid and *aqua regia*. Using a series of chemical extractions and digestions, FILIPEK and OWEN [2248] determined the distribution of heavy metals throughout the various mineralogical constituents of sediments. SCOTT [2834] reported losses of chromium due to adsorption on the silicate residue if heating was prolonged during digestion procedures, so he controlled the digestion time very carefully. SINEX *et al* [2870] found that extraction of NBS sediment standards with HNO_3/HCl (9:1) gave a very high recovery and good precision. However, the procedure failed for real samples and cannot therefore be recommended.

AGEMIAN and CHAU [2006] determined 20 elements in sediments after pressure digestion in nitric acid, perchloric acid and hydrofluoric acid at 140°C. The authors found that organic materials are fully decomposed and silicates are completely dissolved, so that a quantitatives recovery by AAS is guaranteed.

FISHMAN and SPENCER [2251] determined arsenic in sediments using the hydride technique. Samples containing organically bound arsenic require pretreatment with a sulfuric acid/potassium persulfate digestion. Without a digestion, or with a digestion in nitric acid/sulfuric acid, the results are too low. PYEN and FISHMAN [2767] determined selenium in sediments by the hydride technique after a prior persulfate digestion. In this way all organic selenium compounds are included in the determination; this is not the case with an acid/permanganate digestion.

For the determination of mercury in sediments, AGEMIAN and DA SILVA [2009] used a sulfuric acid/dichromate digestion, which extracts mercury very effectively from particulate matter, followed by a photo-oxidation of organomercury compounds by UV radiation. The graphite furnace technique is very well suited for the determination of trace elements in sediments [151]. Under given prerequisites, sediments can also be dispensed into the graphite furnace and analyzed [380].

11.5. Environment

The general term environment covers a whole range of factors that influence our lives. It is hardly possible to pick out any specific area and regard this in isolation since all factors influence each other mutually. In figure 140 the complex environmental relationships are depicted in simplified form and possible interactions indicated. The analysis of foodstuffs, water, biological materials, plants, and soil has been treated in some depth in earlier sections. It remains in this section to discuss air, waste water, sewage sludge, and refuse and garbage.

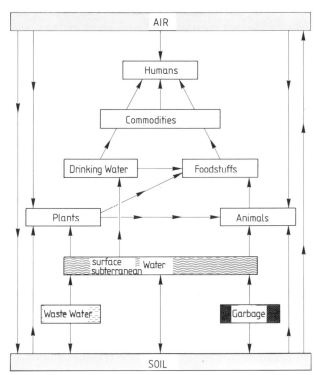

Figure 140. Possible interactions within the elemental cycle between humans and their environment and mutual interactions between subdivisions of the environment (according to [2043]).

Metals and their compounds are among the most insidious pollutants of our environment because they are not biologically degradable. Only a very few metals are nontoxic, even at high concentration; a large number are very active and even at very low concentrations cause changes in humans, animals, and plants. Removal of metals from natural cycles is usually only possible via the formation of insoluble or inactive compounds and sediments. Even in these forms they are a potential source of danger, since they can be mobilized through the action of microorganisms or a change of pH, for example. Metals in the environment can originate from natural sources or as a result of human activity.

For the examination of **air,** a distinction is made between gaseous metals or metallic compounds and particulate matter such as dust, fly ash, and aerosols. The various forms of aggregation require differing collection and sample preparation techniques. Characterization of the chemical and toxicological properties of dust and aerosols has become one of the most important tasks assigned to the analytical chemist. While the physical chemical examination of gases down to the trace range is nowadays generally well under control, the analysis of airborne dust and aerosols, dust from emissions, and smoke or fog presents considerable difficulties [2185].

On a global scale, the majority of particulate matter in the air originates from natural sources; only 10–20% can be ascribed to anthropological sources [2209] [2923], i.e. directly from human activity. However, these sources are not evenly distributed over mainland areas. In typical conurbations, in which human activity is distinctly most pronounced, the highest emissions from many sources occur and the larger part of the population is subjected to pollutants [2185]. Dust and aerosols up to about 5 μm can enter the lungs and thus influence people directly. But even larger particles are not harmless, since their influence on natural waters and plants [2639] [2829] can have an indirect influence on humans. Representative sample collecting of particulate emissions still has many problems that must be overcome. DANNECKER [2185] has drawn attention to the difficulty of collecting samples representative for the whole mass stream from diffuse sources such as stack gases with outputs of several hundred thousand cubic meters per hour at temperatures up to 300°C and fluctuating dust concentrations lying in the milligram to gram per cubic meter range. The constituents in flue gases are frequently in gaseous or vapor form and thus evade collection.

DANNECKER [2185] made a detailed examination of the enrichment behavior of metals in mass streams from large firing plants (power stations, etc.); initially, various digestion procedures were investigated. The most reliable results were obtained using a combined, multistep digestion with nitric acid, perchloric acid, and hydrofluoric acid in open PTFE tubes in a temperature-programmed heating block. The author indicated that deviations from the true value were small when the most suitable parameters were selected and when the analytical technique operated in the optimum concentration range. This signifies the use of the graphite furnace and hydride techniques for trace and nanotrace analyses.

DANNECKER established that the quantities of heavy metals emitted from large coal-fired plants represent substantial mass streams which can cause serious pollution, especially in conurbations. With the exception of manganese, all of the elements investigated are more or less enriched in the flue dust of treated gases. Similar effects are also observed with garbage incineration plants where a large portion of the ecologically toxic elements reach the environment. The output of heavy metals in stack gas dust is calculated as 270 g/h lead and 11 g/h cadmium. It is of considerable concern that exactly these elements are concentrated in the fine grain dust and especially in the airborne dust. Highly toxic elements like cadmium, lead, and vanadium are up to eight times more concentrated in suspended dust than in coarse grain dust; this is likely to be of significance with regard to toxicology and epidemiology.

LAW and GORDON [2538] also investigated the distribution of elements between the bottom ash, fly ash and atmospheric particles of a garbage incineration plant. GARDNER [2292] described the digestion of coaldust in nitric acid and perchloric acid under the addition of sulfuric acid for the subsequent determination of mercury. MAY and STOEPPLER [2645] used a similar digestion procedure, with the addition of dichromate as required, for a variety of environmental samples. OWENS and GLADNEY [2720] described the digestion of fly ash in nitric acid, sulfuric acid, hydrofluoric acid, and perchloric acid for subsequent determination of beryllium by the graphite furnace technique. LIKAITS *et al* [2549] investigated digestion procedures for flue dust for the determination of arsenic and found that both nitric acid/sulfuric acid and nitric acid/perchloric acid gave values that were too low. A hydrochloric acid/nitric acid digestion, under the addition of hydrofluoric acid when silicate contents are high, gave the correct results. The determination of arsenic in the nitrous oxide/acetylene flame is rapid and gives accurate results with good precision. SILBERMAN and FISHER [2865] dissolved flue ash in hydrofluoric acid at ambient temperature and determined 19 elements by the flame and graphite furnace techniques. Only selenium could not be determined by this procedure since it remained up to 75% in the residue.

BELCHER and BROOKS [99] determined Na, K, Mg, Ca and Sr in coalash in an air/acetylene flame. The samples were dissolved in hydrofluoric acid and diluted with a lanthanum solution to avoid interferences. OBERMILLER and FREEDMAN [922] dissolved coalash in perchloric acid and hydrofluoric acid and determined Na, K, Ca and Fe. They also added lanthanum for the determination of the alkaline earth metals to avoid interferences. STOEPPLER and BACKHAUS [2917] examined a pressure digestion with hydrochloric acid/nitric acid for various environmental samples. For coal and coalash, depending on the elements to be determined and the analytical technique, POLLOCK [2754] applied an oxygen bomb, low temperature ashing, a fusion, or a number of special digestion techniques. A number of authors used fusion, with lithium metaborate for example, for the subsequent determination of trace elements in fly ash by the flame technique [2125] [2682] and the graphite furnace technique [2700] [2722]. Mercury, lead and tin are volatilized during these procedures [2682], but it is possible to avoid losses of lead through careful temperature control [2700]. HANSEN and FISHER [2348] investigated the distribution of elements between various phases of coal fly ash. MILLS and BELCHER [2668] have published an overview on the preparation, digestion, and techniques of analysis of coalash.

Dust particles in air are generally collected by filtration techniques with a twenty-four hour collection time. Coarse dust particles that settle out are collected in containers over a 4-week period and measured by weighing. Standard procedures have been published for both techniques, but these will not be further treated here. DANNECKER [2185] has made a detailed investigation of the analysis of airborne dust.

SACHDEV [1077] reported the determination of manganese, iron, cobalt, and nickel in air and water using a modified atomic absorption spectrometer. KETTNER [640] compared the determination of lead in dust deposits by the dithizone method and by AAS, and found the latter just as accurate and also faster and more reliable. Various other authors determined lead [181] [295] [548] [559] [1094] [1227], cadmium [1094] [1227], iron [534] [559] [1227], copper [534] [1094], zinc [559] [1094] and other elements [1227] by extracting the filters with acid, or by low temperature or oxygen plasma ashing of the filters and taking up the residue in acid, and nebulizing the solution into the flame.

DUMAREY *et al* [2211] determined mercury bound to particles by pyrolysis and enrichment on gold wool after collection on glass fiber filters. Only a small portion of the mercury is determined, however, since the majority is gaseous and not bound to particles [2449].

BETTGER *et al* [2085] also used glass fiber filters and determined beryllium in air by the graphite furnace technique after dissolving the sample with nitric acid/sulfuric acid/hydrofluoric acid. The authors observed losses of BeO when the samples were evaporated to dryness. For the determination of lead, cadmium, and zinc by the flame technique and arsenic by the hydride technique, HUBERT *et al* [2399] dissolved the samples from glass fiber filters with nitric acid/hydrochloric acid.

DANNECKER [2185] found that glass fiber filters permit a large air throughput, but because of the highly fluctuating element blank values, this type of filter is more suitable for the gravimetric determination of the airborne dust concentration rather than for the elemental analysis of airborne dust. RANWEILER and MOYERS [2776] found polystyrene filters especially useful since they can be totally ashed at 400 °C and have relatively low blank values compared to glass fiber filters.

Due to their low and constant metal blank values, cellulose membrane filters are excellently suited for multielement determinations using the most varied analytical techniques. To determine the element contents of airborne dust samples, the membrane filter is digested oxidatively in a PTFE vessel with nitric acid under the addition of a small amount of hydrofluoric acid and perchloric acid [2185]. GELADI and ADAMS [2297] found that the presence of perchloric acid during the determination of beryllium or manganese by the graphite furnace technique leads to errors. They therefore evaporated to dryness, or used low temperature ashing, and took up the residue in hydrofluoric acid/nitric acid. According to the latest knowledge, however, the interference caused by perchloric acid should hardly be observed for atomization off the L'VOV platform in a stabilized temperature furnace [2878].

STOLZENBURG and ANDREN [2922] exposed filters to the vapors of nitric acid and hydrofluoric acid and found the same values for ten elements using this extraction procedure as for complete dissolution in nitric acid and perchloric acid. DE JONGHE and ADAMS [2190] found that particulate lead is trapped on cellulose filters, but not organically bound lead; this they collected in iodine monochloride solution downstream of the filter.

PACHUTA and LOVE [2724] punched out disks from cellulose filters and placed them in nickel microcrucibles for the determination of lead by the flame technique. THOMASSEN *et al* [2962] used a similar procedure for the determination of cadmium, copper, lead, and manganese by the graphite furnace technique. TORSI *et al* [2981] collected airborne dust electrostatically on a graphite tube and determined lead. SIEMER and WOODRIFF [2860] and NOLLER and BLOOM [2695] sucked air through graphite tubes or graphite crucibles and thus used the porous graphite directly as a filter. NOLLER *et al* [2696] have published an overview on conventional methods of collection and on the direct collection and determination of metals in airborne dust by the graphite furnace technique.

BEGNOCHE and RISBY [2070] have dealt with the problem of pore size for filters used for the subsequent determination of metals in airborne dust. SEIFERT and DREWS [2839] discuss the problems of sampling, sample preparation and contamination for the analysis of dust.

OMANG [932] and also JANSSENS and DAMS [572] [573] determined cadmium and lead in air by the graphite furnace technique and found excellent sensitivity and accuracy. BRODIE and MATOUŠEK [2108] added phosphoric acid to prevent losses during ashing for the determination of cadmium in airborne dust. RANTALA and LORING [2775] dissolved samples in *aqua regia* and hydrofluoric acid and determined cadmium free of interferences by atomizing at 950°C using an ultrafast heating rate. SLAVIN [2876] showed that graphite furnace AAS is the most suitable technique for the determination of chromium in environmental samples. It exhibits the best detection limits and is cheaper and more simple than other procedures. VIJAN and CHAN [3016] reported the determination of tin and VIJAN and WOOD [1271] the determination of arsenic in airborne dust samples by the hydride technique. SHEAFFER *et al* [2847] determined silver in precipitation down to 10^{-6} mg/L by the graphite furnace technique.

Relatively little is reported in the literature on the direct determination of metal traces in air by AAS. WHITE [1317] described an atomic absorption spectrometer that permitted the direct examination of air for cadmium and lead vapors. EDWARDS [322] determined silver in air with a detection limit of 3 µg/m³ by conducting the air directly into the primary air supply of a flame. THILLIEZ [1224] used AAS to determine lead alkyls in the vicinity of a manufacturing plant. CHAU *et al* [2153] and ROBINSON *et al* [2796] combined gas chromatography with AAS to determine lead alkyls in air by the graphite furnace technique. SHENDRIKAR and WEST [2849] found that selenium is not retained very effectively in filters, so they collected it successfully in water.

Mercury is largely present in air as the vapor or in the form of gaseous organic compounds and is not bound to particulate matter. It is therefore frequently determined by the cold vapor technique, either directly or after preconcentration [70] [71] [703]. LINDNER [752] determined mercury in indoor air by simply drawing the air through an absorption cell mounted in an atomic absorption spectrometer. SCHIERLING and SCHALLER [2826] [2827] collected mercury by amalgamation on a gold gauze and were thereby able to achieve high sensitivity even after a short collection period. ALDRIGHETTI *et al* [2027] made a thorough examination of the influence of flowrate and collection time for this technique.

GARDNER [2291] collected mercury in a trap containing potassium permanganate and sulfuric acid. JANSSEN *et al* [2434] draw attention to the fact that this technique cannot be used for particulate matter. They also found that portable detectors do not recover mercury compounds, so that they employed adsorption on charcoal-filled papers for the determination of mercury in air. The filters are boiled in nitric acid for the subsequent determination. POLLOCK [2755] used a collection tube filled with activated charcoal and silver or gold for mercury in air; the tube is subsequently heated to 500–600°C. ROSE and BOLTZ [1057] determined sulfur dioxide in air indirectly via lead sulfate and the determination of lead by AAS.

Toxic heavy metals generally only pass into surface waters in low concentrations. Because of their pronounced tendency of accumulating in constituents of the foodstuffs cycle, however, in the medium and long term they can become dangerous to human health. Their determination in **waste waters** is therefore of the utmost importance with respect to environmental control.

The composition of waste waters can vary substantially, both in respect to their heavy metal contents and also their total content of inorganic and organic materials. While household waste water hardly contains larger quantities of toxic heavy metals, industrial waste waters can be heavily contaminated. The nature of the determination and any observed interferences can depend markedly on the type of waste water to be analyzed.

GUILLAUMIN [2327] determined nine elements in power station waste water by the graphite furnace technique directly against aqueous reference solutions and observed no major interferences. KUNSELMAN and HUFF [2515] determined antimony, arsenic, selenium and tellurium by graphite furnace AAS using the analyte addition technique and found good agreement with the hydride technique. MARTIN and KOPP [2628] determined selenium in industrial waste waters by the graphite furnace technique under the addition of 1 % nickel as matrix modifier.

SMITH *et al* [2893] determined arsenic in environmental samples and found few interferences. In the present author's laboratory [3060], however, it was found that the high concentrations of transition elements, such as iron, copper or nickel, that can be expected in industrial waste waters cause almost complete suppression of the arsenic signal. Pentavalent arsenic is also more susceptible to interferences than trivalent. Waste water must be treated with *aqua regia* or nitric acid/hydrogen peroxide before analysis by the hydride technique to permit detection of any elemental or organically bound arsenic. The next logical step is then a reduction of the trivalent state with potassium iodide in semiconcentrated hydrochloric acid. Interferences are relatively minor in this medium. To prevent signal depression due to higher nickel concentrations, iron should be added to sample and reference solutions. Any interferences due to the presence of selenium can be controlled by adding copper to all solutions. In this way it is possible to determine arsenic free of interferences in industrial waste waters containing up to 500 mg/L iron and copper, and 100 mg/L nickel.

BRAMAN and TOMKINS [2103] determined inorganic tin and methyltin compounds in environmental samples by the hydride technique. They generated the hydrides with sodium borohydride, collected them in a cold trap and then volatilized them one after the other by slow warming.

VÖLLKOPF *et al* [3023] determined lead, cadmium, chromium, and zinc in industrial waste waters and in garbage seepage water by the graphite furnace technique. To prevent severe signal depression due to the high total salt contents, they found that the addition of ammonium phosphate as matrix modifier for the determination of lead, cadmium, and zinc, or phosphoric acid for zinc, and magnesium nitrate for chromium was indispensable. Atomization off the L'VOV platform in a stabilized temperature furnace eliminated non-spectral interferences completely, so that the determinations could be performed against aqueous reference solutions that merely contained the same concentrations of matrix modifiers. The authors applied Zeeman-effect background correction to correct the very high non-specific attenuation.

CRUZ *et al* [2178] have drawn attention to the fact that as well as total lead, the concentrations of extractable, volatile and tetra-alkyl lead are of importance. They have published methods for their determination by the graphite furnace technique. PARKER [2731] determined organic silicon compounds in water in the nitrous oxide/acetylene flame after extraction with pentan-1-ol/MIBK. CASSIDY *et al* [2135] used porous polymers to collect organic silicon compounds which they separated on a molecular sieve. CRISP *et al* [2176] determined anionic detergents indirectly by extracting with bis-ethylenediamine-copper(II) in chloroform and determining the copper by the graphite furnace technique. PAKALNS and FARRAR [2726] investigated the influence of surfactants on the extraction of various metals.

Interferences could be eliminated by the addition of aluminium nitrate and higher concentrations of APDC.

The use of **sewage sludge** as an agricultural fertilizer is a convenient way of reintroducing valuable materials back to the natural cycles and also an economic way of disposing of the sludge. Comprehensive analytical control is nevertheless necessary to prevent detrimental changes to the soil, for example the accumulation of heavy metals. Analytical chemistry plays an important role in decision making processes on the agricultural use of sewage sludge; it can decide the fertilizing value of the sludge and thus allow a reduction in the use of mineral fertilizers, and it can also monitor the accumulation of heavy metals in the soil by objective controls [2094].

RITTER *et al* [2792] investigated a number of digestion procedures for sewage sludge for the subsequent determination of cadmium, copper, lead, nickel, and zinc. They found that dry ashing at 550°C is the best procedure; the precision is good and the procedure is faster and easier to perform than a complete digestion in which, after dry ashing, the sample must be fumed off repeatedly with hydrochloric and hydrofluoric acids. Extraction with 6 M nitric acid gives results that are too low for most elements and samples. Wet ashing with nitric acid and perchloric acid is also suitable and gives results that are in good agreement with the complete digestion procedure, except that scatter is greater due to contamination. Digestion with nitric acid and hydrochloric acid gives values for all elements that are significantly too low. Regrettably these authors did not include the widely employed digestion with nitric acid and hydrogen peroxide in their examinations. THOMPSON and WAGSTAFF [2969] found that a simple nitric acid digestion of sewage sludge in calibrated glass containers for the subsequent determination of cadmium, chromium, copper, lead, nickel, and zinc is easy to perform and provides correct results. MARTIN and KOPP [2628] used a digestion in nitric acid and hydrogen peroxide for the subsequent determination of selenium by the graphite furnace technique. Under the addition of 1% nickel as matrix modifier, this technique provided good sensitivity, accuracy and precision.

For the determination of cadmium, chromium, copper, lead, nickel, and zinc in sewage sludge, CARRONDO *et al* [2133] compared a direct determination in the graphite furnace with a determination in the flame after various digestion procedures. Ashing at 450°C, digestion in sulfuric acid and nitric acid, or in nitric acid and hydrogen peroxide are relatively time consuming. In the graphite furnace, a diluted suspension of sludge in 1% nitric acid can be analyzed directly against aqueous reference solutions. With respect to precision and accuracy, this rapid direct method is comparable to various digestion procedures and analysis in the flame. STERRITT and LESTER [2916] used the same direct procedure for homogenized samples and found it not only quicker but also more precise than the analysis of wet or dry ashed samples in the flame because there are fewer sources of error.

HAYNES [2356] determined antimony and arsenic in incinerable household **garbage.** The samples were ashed at 450–500°C under the addition of magnesium nitrate as ashing additive and nickel nitrate to prevent losses. The determination in the graphite furnace was in good agreement with the certificate values of standard reference materials. Due to the inhomogeneity of the samples, scattering of the results by up to 20% was observed.

The analysis of **sediments** has gained in importance with the increasing awareness of environmental control. Generally, a given sediment fraction is digested in one of the common acid mixtures and the determinations performed by the flame technique [50] [177] [380]. KRISHNAMURTI *et al* [2508], for example, used nitric acid/hydrogen peroxide. GABRIELLI *et al* [2285] determined lead in suspended matter by introducing it directly into the flame after ashing under an infrared lamp. The graphite furnace technique can be used equally well for the analysis of sediments, especially when the concentrations are low [151].

SCHMIDT and DIETL [2828] determined cadmium in a zirconium lined graphite furnace after extraction with APDC/MIBK and back extraction into dilute nitric acid. BRIESE and GIESY [2106] determined cadmium in suspended matter directly in the graphite furnace under the addition of ammonium nitrate or ammonium sulfate. Under given prerequisites, sediments can be determined directly in suspension by the graphite furnace technique [380].

TERASHIMA [2958] determined arsenic in sediments by the hydride technique, after prior digestion in nitric acid, perchloric acid and hydrofluoric acid under the addition of potassium permanganate, in good agreement with other methods. AGEMIAN and BEDEK [2005] used the same digestion mixture for the determination of arsenic and selenium. They found that both elements were quantitatively extracted from all phases, including the silicate phase. CHAU *et al* [2152] determined dimethylselenide and dimethyldiselenide, the most important selenium metabolites in the biological system, in the atmospheres of sea-water sediment systems. MAHER [2607] determined inorganic and methylated arsenic species in estuarine sediments by the hydride technique after solvent extraction and separation by ion exchange chromatography.

MATSUNAGA and TAKAHASHI [2636] determined organically bound mercury in sediments; the compounds can be reduced directly by tin(II) chloride in the presence of sodium hydroxide and copper(II). The mercury is collected on gold granules and subsequently released by heating rapidly to 500°C.

MAY and PRESLEY [2646] determined iron, nickel, and vanadium in coastal tar by dissolving in an organic solvent and analyzing directly by the graphite furnace technique. The ratio of nickel to vanadium is characteristic for the origin of crude oil and is suitable for its identification. Later, these authors employed a digestion of crude oil in nitric acid, perchloric acid and sulfuric acid [2647].

11.6. Rocks, Minerals and Ores

Because of its high specificity and freedom from interferences, which make separation of the main constituents even with trace analyses superfluous, atomic absorption spectrometry has found rapid acceptance in the geochemical laboratory. It not only replaces conventional wet chemical analyses, but also stands on an equal footing with other more time-consuming procedures such as X-ray fluorescence or even neutron activation analysis.

At the start of the 1960s, the elements calcium, magnesium, potassium, iron, manganese, copper and zinc in various rock samples and silver and gold in ores were principally determined. After the introduction of the nitrous oxide/acetylene flame, elements such as aluminium, silicon and titanium were added in increasing measure. Thus, with atomic absorption spectrometry it is possible to determine the most important metallic elements of interest in rocks, usually with sufficient sensitivity and very good precision. The graphite furnace and hydride techniques have made a substantial contribution to the determination of trace elements. Additionally, various enrichment procedures have been developed to permit low-interference determinations of trace elements by the flame technique. A survey of literature published up to the present on the application of AAS to the analysis of rocks, minerals and ores is compiled in table 50. Geochemistry is also the first specialized area to which atomic absorption spectrometry was applied and for which a comprehensive monograph [2] containing much useful information was published.

The most difficult part of a rock or ore analysis is frequently the digestion procedure. This is particularly the case when the sample to be analyzed consists of silicate material. A number of digestion procedures which have been tested in connection with AAS will therefore be briefly discussed.

The most widely known and customary procedure, which is also very suitable for a subsequent analysis by AAS, is digestion with hydrofluoric acid/sulfuric acid [45]: 5 mL 40% hydrofluoric acid are added to 100 mg substance in a platinum dish and then 3 mL concentrated sulfuric acid are added drop by drop. The dish is warmed on a water bath overnight and then carefully evaporated until the sulfuric acid fumes. If the digestion is not complete it must be evaporated to dryness, repeated and then finally made up in a volumetric flask. Various variations of this simple HF/H_2SO_4 digestion have been described. For example, GALLE [402] used less sulfuric acid and then made a potassium pyrosulfate digestion on the residue. Other authors prefer hydrochloric acid or perchloric acid with hydrofluoric acid since sulfuric acid can occasionally cause certain interferences [1241] [1243]. BILLINGS and ADAMS [124] dissolved various silicate and carbonate rocks, such as mica and feldspar, in hydrofluoric acid/nitric acid and obtained good agreement with other analytical procedures, such as flame emission spectrometry, X-ray fluorescence, etc. [122]. SANZOLONE *et al* [2820] used a digestion in hydrofluoric acid/nitric acid for various geological samples and dissolved the residue in hydrochloric acid. KANE [2454] used a hydrofluoric acid/perchloric acid digestion mixture. ARMANNSSON [2042] described a digestion in hydrofluoric acid, sulfuric acid and perchloric acid. SUTCLIFFE [2949] fumed the samples off with nitric acid/sulfuric acid and took the residue up in hydrochloric acid.

LANGMYHR and PAUS [725] [728] described a digestion procedure employing solely hydrofluoric acid with which the silicon can be removed, or also quantitatively retained. The authors checked this procedure for the analysis of silicate rocks [725], silicate materials such as glass, quartzite and sand [726], bauxite [727], feldspar [2529], and also various standard rocks, ores and cements [728]. Depending on whether silicon is to be retained or not, the sample is heated with hydrofluoric acid in either a sealed or an open plastic vessel. If the temperature is maintained below the boiling point of the azeotropic mixture (38.26% HF) at 112°C, thick walled vessels, preferably of PTFE, are sufficient for the digestion. For digestions at higher temperatures a sealed system is preferably used [964]. These PTFE autoclaves (figure 70, page 123) reduce the digestion times considerably—to between ½ to 1½ hours, depending on the nature of the rock—and operate virtually free of losses and contamination; this is of great advantage especially for trace analyses [444]. Similar digestion systems have also been described by BERNAS [115] and RANTALA and LORING [1019].

Table 50. Bibliographical survey: Application of atomic absorption spectrometry to the analysis of rocks, minerals and ores

Element	References
Ag	[158] [329] [376] [439] [494] [538] [715] [730] [766] [768] [913] [1072] [1164] [1231] [1232] [1373] [2159] [2275] [2347] [2532] [2682] [3031]
Al	[66] [115] [131] [150] [402] [495] [496] [725] [726] [727] [728] [729] [765] [855] [1153] [1257] [1366] [2323] [2682] [2924] [2980]
As	[2159] [2328] [2489] [2667] [2809] [2958]
Au	[158] [198] [211] [439] [931] [1028] [1115] [1126] [1132] [1165] [1182] [1231] [1232] [2159] [2275] [2332] [2340] [2347] [2493] [2596] [2682] [2732] [2810] [2863] [2943]
B	[765]
Ba	[765] [1066] [2403] [2565] [2682] [2980] [3034]
Be	[765] [1153] [2499] [2565]
Bi	[496] [546] [730] [2159] [2347] [2454]
Ca	[44] [45] [115] [122] [124] [126] [131] [164] [290] [402] [725] [726] [727] [728] [729] [765] [768] [855] [898] [1036] [1241] [1243] [1257] [1349] [1366] [2306] [2682] [2924] [2980]
Cd	[245] [341] [496] [730] [766] [768] [1274] [1295] [2042] [2159] [2307] [2347] [2497] [2532] [2909]
Co	[124] [126] [164] [183] [768] [1101] [1142] [2042] [2347] [2479] [2565] [2682] [2820] [2980] [3034] [3070]
Cr	[45] [93] [727] [728] [768] [855] [1101] [1142] [2347] [2565] [2682] [2980] [3034]
Cs	[435] [436] [1109] [2566]
Cu	[66] [106] [107] [164] [264] [340] [768] [855] [1101] [1180] [1190] [1243] [1295] [2042] [2159] [2347] [2497] [2565] [2682] [2820] [2980] [3034] [3070]
Fe	[44] [45] [66] [115] [122] [124] [126] [131] [164] [183] [402] [495] [496] [725] [726] [727] [728] [729] [768] [860] [1036] [1142] [1190] [1241] [1243] [1257] [1349] [1366] [2045] [2682] [2924] [2980]
Ga	[991] [2159] [2347] [2530] [2682]
Ge	[991]
Hg	[496] [768] [1010] [1233] [1262] [1294] [1295] [2104] [2159] [2682] [2862] [2982]
In	[2159] [2530] [2682]
Ir	[443] [769]
K	[44] [45] [115] [122] [124] [126] [131] [164] [290] [725] [726] [727] [728] [729] [768] [855] [898] [1036] [1241] [1349] [1366] [2566] [2682] [2789] [2790] [2924] [2980]
La, Lanthanides	[642] [643] [1255] [2648] [2840] [2841]
Li	[435] [1176] [2565] [2566]
Mg	[44] [45] [115] [124] [126] [131] [164] [290] [341] [402] [725] [726] [727] [728] [729] [765] [855] [898] [1036] [1190] [1241] [1243] [1257] [1349] [1366] [2682] [2924] [2980]
Mn	[45] [164] [202] [290] [402] [495] [496] [725] [727] [728] [729] [768] [1036] [1241] [1257] [1349] [1366] [2347] [2497] [2682] [2924] [2980] [3070]

Table 50. Continued

Element	References
Mo	[1021] [1101] [1142] [1254] [2347] [2682] [2949]
Na	[44] [45] [115] [122] [124] [126] [131] [164] [290] [725] [726] [727] [728] [729] [855] [898] [1036] [1241] [1349] [1366] [2566] [2682] [2924] [2980]
Ni	[45] [93] [124] [126] [164] [183] [245] [768] [1101] [1142] [2042] [2347] [2565] [2682] [2820] [2980] [3034] [3070]
Pb	[89] [264] [329] [340] [435] [496] [766] [768] [1142] [2127] [2157] [2159] [2347] [2363] [2497] [2532] [2682] [2700] [2859] [2980] [3070]
Pd	[2159] [2275] [2732] [2810]
Pt	[1115] [1208] [2159] [2275] [2957]
Rb	[122] [124] [126] [164] [435] [731] [1043] [2566]
Re	[2226]
Rh	[1095] [1115] [2275]
Sb	[914] [1183] [2146] [2159] [2347] [2682] [2980] [3044]
Se	[1118] [1167] [2159]
Si	[66] [115] [131] [150] [437] [725] [727] [728] [729] [855] [1257] [1366] [2045] [2118] [2299] [2682] [2980] [3030]
Sn	[911] [2159] [2682] [3045]
Sr	[45] [93] [122] [124] [126] [765] [855] [1243] [2565]
Te	[91] [1167] [2159] [2167] [2315] [2864]
Ti	[115] [131] [150] [402] [495] [496] [725] [726] [727] [728] [729] [765] [855] [1022] [1256] [1257] [1366] [2565] [2682] [2924] [2980]
Tl	[384] [496] [768] [1124] [1294] [1295] [2159] [2347] [2532]
U	[2018]
V	[115] [234] [728] [765] [1101] [2488] [2565] [2924] [2980] [3034]
W	[631] [1020] [2466]
Y	[2840] [2841]
Zn	[66] [106] [107] [150] [202] [245] [264] [329] [335] [340] [435] [766] [768] [855] [1142] [1190] [2042] [2159] [2347] [2497] [2532] [2565] [2682] [2980] [3070]
Zr	[765] [2924]

The latter is only suitable for small excess pressures, however, since it is not contained in a pressure jacket.

TOMLJANOVIC and GROBENSKI [2980] used a digestion under pressure prior to the determination of the main and trace constituents in iron ores. GILL and KRONBERG [2299] found for the determination of silicon that a digestion with hydrofluoric acid under the addition of nitric acid and hydrochloric acid in an autoclave gave the best compromise between speed, accuracy, and precision. In its speed, the procedure is comparable with manual X-ray fluorescence analysis.

For volatile elements, such as mercury [2104] [2862] [2982], pressure digestions in a sealed PTFE container have the advantage that losses due to volatilization cannot occur (when the apparatus is used correctly). PRICE and WHITESIDE [2763] described a modified digestion procedure that is generally applicable for silicate materials. They weigh 0.5 g sample together with 5 mL water, 2 mL *aqua regia,* and 1 mL hydrofluoric acid into a PTFE auto-clave and heat for 30 minutes at 160°C. After cooling, 10 mL 4% boric acid are added rapidly and then the autoclave is again heated to 160°C for 20 minutes. The authors emphasize that the second heating step is essential for complete dissolution of the sample.

AYRANCI [2045] [2046] described the digestion of silicate materials with hydrofluoric acid, sulfuric acid, and phosphoric acid in a PTFE cylinder under a nitrogen atmosphere for the subsequent determination of main, minor, and trace constituents from a single solution. The author laid stress on the importance of a non-oxidizing environment for this digestion procedure.

The use of common alkaline digestion procedures, such as fusions with sodium carbonate/potassium carbonate, has proved of little value for trace determinations since the carbonates are frequently not available in the required degree of purity and the marked increase in the total salt concentration can lead to interferences.

Nevertheless, a number of fusion procedures are reported in the literature that are worth mentioning. BURDO and WISE [2118] fused the sample with sodium carbonate and sodium borate and then dissolved the melt in acid molybdate solution for subsequent determination of silicon. A silico-molybdate complex is formed that prevents the polymerization and precipitation of silicate.

CHOW [2157] determined lead in columbite after prior fusion with pyrosulfate and solubilization in tartaric acid. GUEST and MACPHERSON [2323] found a fusion with sodium peroxide the most suitable for the determination of aluminium in sulfidic and silicate materials and ores.

SUHR and INGAMELLS recommended a rapid and simple procedure employing a lithium metaborate ($LiBO_2$) digestion with which silicon as well as other metallic elements in silicate rocks and minerals can be digested and maintained in solution. This procedure was originally recommended for emission spectrometry [1191] and colorimetry [553], and was only later applied to atomic absorption spectrometry [855] [1190].

According to the original directions of SUHR and co-workers, 0.100 g of the finely powdered sample is mixed with 0.500 g $LiBO_2$ and heated in a baked-out graphite crucible in a muffle furnace at 900–1000°C for 10–15 minutes. The clear melt is agitated in the hot crucible so that small pellets are formed which do not adhere to the walls and then it is rapidly poured into 40 mL nitric acid contained in a 200 mL PTFE or polyethylene beaker. The beaker is covered and the contents stirred with a magnetic stirrer to obtain complete solution. From this solution—with further dilution if required—silicon, aluminium, magnesium, calcium, potassium, sodium, manganese, chromium, strontium, titanium, copper, zinc and other elements can be determined. YULE and SWANSON [1366] checked this procedure for nine elements with various standard rocks and found good agreement with the certificate values. Good agreement was also found with wet chemical procedures (which take five times as long) for both the main constituents (up to 80% SiO_2) and for the trace elements. GOVINDARAJU and L'HOMEL [437] semi-automated this procedure for the subsequent determination of silicon in silicate rocks, either directly or after passage through a cation exchanger. The latter method gives a precision of better than 1% and excellent accuracy. INGAMELLS [556] has published a good overview on the use of lithium metaborate digestions for geological samples.

VAN LOON and PARISSIS [1257] modified the lithium metaborate digestion independently from SUHR and adapted it especially for subsequent atomic absorption analyses. Silicon, aluminium, calcium, iron, potassium, magnesium, manganese, sodium and titanium were determined in 12 standard silicate rocks. Exact procedures for the preparation required for precise and interference-free atomic absorption analyses are given for every element. The results obtained compare very well with values from three other analytical procedures. BOAR and INGRAM [131] also give detailed procedures for the analysis of silicate rocks with this digestion procedure.

For the determination of potassium, RICE [2790] compared a fusion with lithium metaborate with a wet digestion in hydrofluoric acid and found no differences in the precision; no losses of potassium were observed during heating at 900 °C for 15 minutes. NUHFER and ROMANOSKY [2700] determined lead after a prior lithium metaborate fusion and found that there were no losses of lead at fusion temperatures below 1100 °C. A temperature of at least 950 °C is nevertheless necessary, otherwise the digestion is not quantitative. For the fusion of silicate materials with lithium metaborate, MUTER and NICE [2682] found that mercury and tin are volatilized.

The application of atomic absorption spectrometry to geochemical analyses usually brings no special problems. The tendency to use hotter flames is recommended, with the nitrous oxide/acetylene flame for the determination of the alkaline-earths, for example [764]. This can be recommended the more when the concentrations of silicon and aluminium are higher, of if sulfuric acid has been used for the digestion. Under certain circumstances the analyte addition technique must be employed for the determination of very low concentrations in complex or concentrated matrices in order to avoid errors: frequently, preconcentration or extraction procedures, such as the APDC/MIBK system, are employed for enrichment.

CLARK and VIETS [2159] described a multielement extraction system for the determination of 18 trace elements in geochemical samples. HANNAKER and HUGHES [2347] extracted 15 trace elements as their chloro complexes from geological materials with DEDC and 8-hydroxyquinoline in MIBK and n-butyl acetate. SANZOLONE [2820] extracted cobalt, copper, and nickel with DEDC in MIBK. For the determination of cadmium and zinc, ARMANNSSON [2042] found that back extraction from the organic solvent into dilute hydrochloric acid was necessary.

KIM et al [2466] used long chain alkylamines in chloroform for the preconcentration of tungsten in geological samples. KORKISCH and SORIO [2499] extracted beryllium as its acetylacetonate with chloroform and removed other co-extracted elements, particularly aluminium, on a cation exchanger. WELSCH and CHAO [3044] volatilized antimony as its tri-iodide by heating with ammonium iodide. After absorbing in 10% hydrochloric acid, the antimony was extracted with tri-n-octyl-phosphine oxide (TOPO) in MIBK. These authors used the same procedure to volatilize tin as its tetraiodide [3045], which, after extraction, was determined in a nitrous oxide/acetylene flame. MICHAEL [2667] determined arsenic indirectly in sulfidic samples also containing antimony by reducing to the hydride, oxidizing in iodine solution, and complexing with ammonium molybdate. After extraction of the heteropoly acid with MIBK, the molybdenum was determined in a nitrous oxide/acetylene flame. Various authors used anion exchangers to separate and preconcentrate cadmium [2909], vanadium [2488] and other elements [2497]. STRELOW *et al* [2924] described the use of a single cation exchange column for the separation of 10 main and minor constituents from silicate rocks. By the use of various eluents the elements can be eluted after each other from the column; solely Ti(IV) and Zr are eluted together.

Nevertheless, a number of authors determine numerous elements directly from rock digestions with good success. LUECKE [2566] determined the alkali metals in a variety of rocks and minerals and found that high aluminium and phosphate contents interfered with the determination of rubidium and cesium in particular. The addition of EDTA and the use of the nitrous oxide/acetylene flame eliminated these interferences. In later work, LUECKE [2565] investigated matrix influences on the determination of trace elements in rock samples of varying geochemical composition. Eleven elements were determined in synthetic matrices corresponding to ultrabasic, basic, intermediate, and acid rocks. The author found that the analytical curves for reference solutions made up in water or hydrochloric acid never agreed with the curves obtained from samples containing matrix. The author recommended measuring against a synthetic matrix (or silicon), or taking a rock sample of similar composition but with the lowest possible analyte content and preparing an analytical curve using the addition technique. For the analysis of silicate rocks, WARREN and CARTER [3034] also recommended performing the determination, after prior digestion in a low-pressure PTFE autoclave at 100°C, against reference solutions containing the main constituents in similar concentrations.

GLIKSMAN *et al* [2306] determined calcium with high precision in phosphate rocks in a nitrous oxide/acetylene flame. They found that using a modern atomic absorption spectrometer, this determination could be performed with the same accuracy and precision as other established procedures, e.g. gravimetrically as the sulfate. The sample throughput is three times higher with AAS and the costs are up to 70% lower owing to the economical use of reagents.

WALSH [3030] has drawn attention to the necessity of high accuracy and precision for the determination of high silicate contents in rocks. Generally, results are given to two decimal places. To obtain the necessary accuracy in a relatively short space of time, the author used a combined gravimetric and AAS procedure. After a sodium carbonate fusion, the sample is fumed off with 20% hydrochloric acid to make SiO_2 insoluble. The residue is taken up in 20% hydrochloric acid and the remaining SiO_2 determined gravimetrically. A few milligrams SiO_2 are present in solution and are determined by AAS and the result added to the gravimetric value.

SEN GUPTA [2840] determined yttrium and the lanthanides in rocks and minerals; he found that the sensitivity was better by a factor of 2–5 when absolute ethanol was used as solvent instead of water. The addition of lanthanum eliminated interferences and occasionally led to a further increase in sensitivity. RICE [2789] reported an interlaboratory investigation on the determination of potassium in rocks and minerals. The greatest scattering of the results was caused by contamination or losses. When these problems were eliminated, AAS with the air/acetylene flame and cesium as ionization buffer always gave correct results, while AES with an air/propane flame and lithium as reference element gave results that were too high, at least for low potassium concentrations.

DALL'AGLIO [264] compared atomic absorption with optical emission spectrometry and LANGMYHR and PAUS [729] carried out an interlaboratory test in which AAS stood in very good agreement with X-ray spectrometry, emission spectrometry, conventional wet chemical techniques and other methods. It is therefore hardly surprising that atomic absorption spectrometry has found rapid acceptance in the routine geochemical laboratory [748].

GOVINDARAJU [435] [436] described a procedure for the direct introduction of solid samples into a flame. Powdered rock samples are mixed with an ionization buffer (e. g. sodium carbonate), brought onto a threaded iron rod wetted with acetone and introduced directly into the flame. Lithium, rubidium, cesium, lead, and zinc can be determined with detection limits down to $1 \mu g/g$. WILLIS [3070] examined the effectiveness of atomization for lead, cobalt, copper, manganese, nickel, and zinc when suspensions of geological samples were nebulized directly into the flame. Only particles smaller than $12 \mu m$ make any substantial contribution to the observed absorption signal. The efficiency of atomization increases rapidly with decreasing particle size. The efficiency of atomization varied by a factor of two for various rock samples at a particle size of less than $44 \mu m$.

LUECKE *et al* [329] [766] described the use of the boat technique for the determination of very low contents of lead, silver and zinc in soil samples for geochemical prospecting. They mentioned the high speed and sensitivity of the determinations. FRATTA [384] also used the boat technique for the determination of thallium in silicate rocks in the ppb range.

Various authors employed the graphite furnace technique for the determination of trace elements in acid digestions. SCHWEIZER [1101] used a modified hydrofluoric acid/perchloric acid digestion, repeatedly fumed off with hydrochloric acid at the end for the determination of the main and minor constituents in carbonaceous rock materials in the flame, and the trace elements chromium, cobalt, copper, molybdenum, nickel and vanadium in the graphite furnace. Chromium and cobalt could be determined against aqueous reference solutions, but for the other elements the analyte addition technique had to be employed; the relative standard deviations were between 2.5% and 4%. Other authors used normal hydrofluoric acid/perchloric acid digestions directly for the determination of beryllium [1123], vanadium [234], cadmium, nickel, zinc [245], aluminium, iron, manganese and titanium [495] [496] and cobalt [768]. In some cases the perchloric acid appears to reduce the non-specific attenuation [245], but it depresses the signal for cobalt very strongly [768], and also interferes with a number of other elements, so that a separation appears quite desirable [234] [1123]. To reduce the interferences further, several authors extracted such elements as tellurium [91] or thallium [496] [1124] from the sample solution before determination in the graphite furnace.

For the determination of tellurium, SIGHINOLFI *et al* [2864] employed an extraction with MIBK followed by a back extraction into the aqueous phase. CORBETT and GOODBEER [2167] separated tellurium by precipitation as the element and were thus able to perform determinations in the $0.01-1 \mu g/g$ range without interferences. KANE [2454] determined nanogram quantities of bismuth in rocks after prior digestion in hydrofluoric acid/perchloric acid. The bismuth is extracted as its iodide into MIBK and then transferred to the aqueous phase with EDTA and acid. SEN GUPTA [2841] determined yttrium and the rare earth elements in rocks after prior separation by co-precipitation with calcium and iron.

Many of the interferences that have been reported for the determination of trace elements by the graphite furnace technique, and which make a prior separation necessary, can be directly attributed to unsuitable instruments or unfavorable conditions. The severe systematic errors reported by HEINRICHS [2363] for the determination of lead in geological samples, for example, can be clearly assigned to vapor-phase interferences. These interferences should no longer be evident for atomization off the L'VOV platform in a stabilized temperature furnace. As shown by SLAVIN *et al* [2878] for aluminium and thallium, the interferences caused by perchloric acid are eliminated under optimum conditions. The application of Zeeman-effect background correction can also aid interference-free trace element determinations, especially in the presence of high salt concentrations [2700].

A number of authors have made detailed investigations on the direct analysis of solid samples by the graphite furnace technique. Time-consuming digestion procedures can thereby be avoided and the highest sensitivity achieved. The direct determination of easily atomized elements such as bismuth, cadmium, lead, mercury, silver, thallium and zinc [496] [730] [768] [1294] [1295] and also cesium and rubidium [731] has been reported. For most of these elements, good separation was achieved through selective volatilization by atomizing at relatively low temperatures (e. g. 1400–1700°C). The rock samples are generally molten at such temperatures, but the vapor pressures are low so that only minor background attenuation occurs. The analyte element is atomized with sufficient speed from the molten rock sample. By using a good background corrector, this procedure can also be applied at higher temperatures, so that copper, for example, can be determined in solid samples [1295].

LANGMYHR *et al* [2530–2532] used a graphite furnace heated by a high frequency induction generator for the direct analysis of solids. They determined cadmium, lead, silver, thallium, and zinc in silicate rocks [2532], gallium and indium in bauxite, sulfidic ores and other materials [2530], and 13 trace elements in phosphate rocks [2531]. The agreement with the conventional flame or graphite furnace technique and with certificate values was good. The relative standard deviations were between 5% and 20%, depending on the concentration of the analyte element.

GONG and SUHR [2307] determined cadmium in geological samples and found that for low cadmium contents the sample had to be ground more finely since the cadmium is more deeply embedded in the matrix. A less finely ground sample is adequate for higher contents since the cadmium is more adsorbed on the surface. The determinations were performed against aqueous reference solutions with good accuracy. SIEMER and WEI [2859] determined lead in silicate rocks by atomizing at 1200–1500°C, calibration being directly against aqueous reference solutions. For a number of standard reference materials they found excellent agreement with the certificate values.

As well as the advantages mentioned above, there are also a number of distinct disadvantages with direct solids analysis. Since the maximum usable sample size is in the order of a few milligrams, the particle size should not exceed 1 μm to ensure that the dispensed aliquot is representative of the whole. Nevertheless, the relative standard deviation for direct solids analysis is typically 10–20%, which is sometimes insufficient for the requirements. Calibration also occasionally causes difficulties, and the time and effort required to make precise weighings in the region of 1 mg should not be underestimated. Nevertheless, this procedure is ideally suited for the rapid, semiquantitative examination of a sample for trace elements.

Little has been reported in the literature on the application of the hydride technique to the analysis of geological samples. The work reported is nevertheless very successful and few interferences were observed. TERASHIMA [2958] determined arsenic in rocks, minerals and sediments after prior digestion in hydrofluoric acid, perchloric acid and nitric acid; arsenic was reduced with potassium iodide before the determination. The values agreed very well with those obtained by other methods. KOKOT [2489] modified the procedure so that he could determine 300 samples per day. GUIMONT *et al* [2328] only found interferences due to cobalt and nickel; interferences due to cobalt were only evident at concentrations a 100 times higher than those normally found in rocks and sediments. EDTA, cyanide and thiocyanate were used successfully to eliminate the interference due to nickel; a 5% solution of potassium thiocyanate was the most effective. RUBEŠKA and HLAVINKOVÁ [2809] draw attention to the fact that EDTA or thiocyanate can only be used to eliminate interferences when the final pH after addition of sodium borohydride solution is alkaline or when the sample solution is sufficiently alkaline in the vicinity of sodium borohydride pellets. Neither EDTA nor thiocyanate have any effect in 1 M hydrochloric acid. The authors performed a digestion with *aqua regia* and reduced arsenic(V) to arsenic(III) with potassium iodide and ascorbic acid. They found that interferences were caused by nickel above 10 µg, cobalt above 100 µg, and copper and iron above 1000 µg. Interferences should then only be expected with samples containing more than 10 µg/g nickel, and these are rare. CHAN and VIJAN [2146] sought a specific, sensitive, precise, and accurate method for the determination of antimony in rocks. The method should also be robust and economical, requiring little time and effort. They found that the hydride technique was very well suited, since the main constituents caused no interferences and interfering transition elements are only present in low concentrations. Sulfides and sulfites are decomposed during sample pretreatment. The interference due to iron is practically eliminated by potassium iodide, which is added to reduce antimony(V) to antimony(III). GREENLAND and CAMPBELL [2315] determined nanogram quantities of tellurium in silicate rocks. They found that the concentrations of iron and copper normally present in such rocks caused no interferences. The procedure is rapid, sensitive, and has sufficient precision and accuracy; 5 ng/g tellurium in rocks can be determined.

The cold vapor technique is most suitable for the determination of mercury, preferably after prior sample digestion in a PTFE autoclave. BRANDVOLD and MARSON [2104] determined mercury in ores and concentrates after a pressure digestion in sulfuric acid/hydrofluoric acid under the addition of potassium permanganate. TOTH and INGLE [2982] determined mercury in manganese nodules after a pressure digestion with hydrofluoric acid and nitric acid. In the Pacific basin they found values between 20 and 50 ng/g, in active "mountain crests" 95–290 ng/g, and in rapidly forming hydrothermic crusts up to 1300 ng/g. SIGHINOLFI *et al* [2862] determined mercury in various silicate materials after digestion with nitric acid/sulfuric acid in an autoclave. They found that the detection limit depended on the purity of the reagents; using ultrapure acids they attained approximately 10 ng/g.

The elements *silver* and *gold* hold a special position in geochemical analysis. A docimastic analysis, which is usually employed for the determination of these elements, is difficult and very time-consuming; atomic absorption spectrometry offers speed and sensitivity. Only the solubilization of the sample occasionally causes difficulties and frequently requires certain modifications for each raw material. FIXMAN and BOUGHTON [376] dissolved silver-containing minerals in nitric acid under the addition of hydrogen peroxide. RAWLING *et al* [1025] [1026] dissolved the sample in hydrochloric acid/nitric acid and maintained the silver in solution by always using 6 M hydrochloric acid.

RUBEŠKA *et al* [1072] give three digestion procedures for sulfidic silver ores employing (1) nitric acid/tartaric acid under addition of mercury(II) nitrate, or (2) nitric acid/sulfuric acid, or (3) sulfuric acid, under certain circumstances with the addition of tartaric acid. HUFFMANN *et al* [538] found, on the other hand, that simply heating the sample with concentrated nitric acid for about 15 minutes brings all silver into solution. The authors examined 44 ore samples with contents between 1 and 8000 µg/g silver and found no mentionable differences between the values obtained from the simple nitric acid extraction and those from complete solution of the sample. The agreement with conventional methods of determination was also good. SPITZER and TESIK determined silver in copper ironstone [494] and in calcine [1165] from 7 M hydrochloric acid solution, from which the iron had previously been extracted. The agreement with docimastic analyses was good.

For the determination of silver in ores and minerals, WALTON [3031] treated the samples with an acid mixture containing hydrofluoric acid and fumed off to dryness. The residue was treated with strong ammonia solution and the extract nebulized directly into the flame. The recovery was virtually quantitative for sulfidic ores, but about 3% too low for others.

The determination of gold does not appear to be completely interference-free and acid solutions especially cannot be compared to cyanide ones. Normally, however, gold is not determined directly from solution, but is extracted because the natural concentrations are too low for a direct analysis and the requirements for accuracy are high. BUTLER *et al* [198] complexed gold with dimethylglyoxime and extracted with MIBK. STRELOW *et al* [1182] examined the problem of gold analysis in detail and found that gold can be quantitatively extracted directly from 3 M hydrochloric acid with MIBK without a complexing agent. With a mixture of aqueous solution to MIBK in the ratio 10:1, 99% of the gold can be extracted in one stage. SPITZER and TESIK [1165] used the same procedure to extract gold from calcine. The co-extracted iron was removed from the organic phase in a second stage. These authors, like TINDALL [1231] [1232], found good agreement with values obtained by docimastic analyses. GREAVES [439] extracted gold with MIBK in the presence of hydrobromic acid instead of hydrochloric acid and obtained satisfactory results.

RUBEŠKA *et al* [2810] found that gold and palladium can be extracted from geological samples in one step with dibutylsulfide and toluene. The detection limits for both elements are 20 ng/g in the flame and 1 ng/g in the graphite furnace; this is below the mean concentration of gold in rocks and sufficient for palladium in basic rocks. PARKES and MURRAY-SMITH [2732] used the same procedure after roasting the samples at 600°C and dissolving in hydrochloric acid/nitric acid.

Owing to its high sensitivity, the graphite furnace technique has been increasingly used in recent years for the determination of gold and other noble metals. MACHIROUX and AHN [2596] determined gold in lead and zinc sulfide ores after dissolving the samples in hydrochloric acid/nitric acid under the addition of bromine and extracting with MIBK. SIGHINOLFI and SANTOS [2863] extracted gold as the bromoaurate with MIBK from hydrobromic acid solution and dispensed the organic phase directly into the graphite furnace. They attained a detection limit of 0.6–0.8 ng/g. HADDON and PANTONY [2332] shook ore samples suspended in hydrochloric acid with pentyl acetate containing bromine for three hours. They found that the extraction was virtually 100% with good reproducibility.

Since the procedure is specific for gold, no interferences occur in the graphite furnace. KONTAS [2493] described a rapid procedure for the determination of gold using the graphite furnace technique without extraction with an organic solvent. The samples are stood with hydrochloric acid and nitric acid overnight and then shaken under the addition of water and centrifuged. From the supernatant liquid, gold is precipitated as its amalgam by adding tin(II) chloride and mercury(I) nitrate. After dissolution in hydrochloric acid/hydrogen peroxide, the determination is performed in the graphite furnace. As well as gold, FREYER and KERRICH [2275] also determined silver, palladium, platinum, and rhodium in rock samples. The samples are digested in *aqua regia* and then fumed off with hydrofluoric acid to free them from silicon. After taking up the residue in hydrochloric acid, tellurium chloride solution is added and then reduced to the metal with tin(II) chloride. Gold, palladium, platinum, rhodium, and silver are precipitated with the tellurium, but otherwise nothing else. The precipitate is taken up in *aqua regia* and is then ideal for determination by the graphite furnace technique since virtually all concomitants have been separated and tellurium itself is very volatile. No matrix interferences are observed and even a background corrector is not required.

11.7. Metallurgy and Plating

In the analytical laboratories of iron and steel works, direct reading spectrometers (known as quantometers) are used nowadays principally for production control. Alongside these, atomic absorption spectrometry has also secured its position, so that metallurgy now belongs to one of the main fields of application. A literature survey on the determinations carried out on irons, steels and alloys is compiled in table 51.

For **iron** and the majority of **steels,** the dissolution procedure presents no major problems; frequently they can be dissolved in a hydrochloric acid/nitric acid mixture, which is ideally suited for subsequent atomic absorption analysis. BEYER [119] warmed about 500 mg sample with 10 mL 1+1 hydrochloric acid and then added concentrated nitric acid drop by drop until all soluble material had been taken up. KINSON [645] [646] and BELCHER [100] found that tungsten steels can be better dissolved in phosphoric acid/hydrofluoric acid. HUSLER [547] dissolved tool steel in nitric acid/hydrofluoric acid, and SCHILLER [1088] employed the same acid mixture to dissolve niobium in low alloy steels. HEADRIDGE and SOWERBUTTS [488] used a pressure digestion in a PTFE autoclave for the determination of total aluminium in steel. DAMIANI *et al* [2184] dissolved ferrosilicon alloys by fuming off the silicon with hydrofluoric acid/nitric acid, taking up the residue in nitric acid/perchloric acid and making up to volume with hydrochloric acid.

Table 51. Bibliographical survey: Application of atomic absorption spectrometry to the analysis of iron and steel

Element	References
Ag	[522] [2048] [2210] [2572]
Al	[47] [145] [175] [488] [679] [680] [791] [2184] [2742] [2846]
As	[513] [2059] [2060] [2254] [2780] [3049] [3054]
Bi	[2041] [2060] [2210] [2254] [2264] [2572] [2846] [3049] [3054]
Ca	[487] [743] [1220] [2184]
Cd	[919] [2060] [2572]
Co	[675] [851] [852] [1169]
Cr	[119] [547] [675] [680] [743] [945] [947] [1045] [1053] [1299] [2729]
Cu	[119] [646] [675] [680] [743] [1169]
Fe	[743] [919]
Mg	[98] [119] [851] [919] [1169] [1220] [2846]
Mn	[100] [119] [537] [547] [675] [680] [743] [919] [947] [1169] [2184] [2846]
Mo	[271] [547] [663] [675] [887] [990] [1045]
Nb	[1088]
Ni	[119] [236] [645] [675] [743] [919] [947] [1169] [1299] [2846]
P	[2353] [3061] [3068]
Pb	[328] [743] [851] [919] [2040] [2059] [2060] [2210] [2254] [2260] [2261] [2264] [2265]
Sb	[2059] [2060] [2254] [2259] [2780] [3049] [3052] [3054] [3083]
Se	[972] [2210] [2244] [2254] [3049] [3054]
Si	[656] [675] [845] [1004]
Sn	[903] [2059] [2060] [2169] [2203] [2254] [2321] [2780] [3049] [3054]
Te	[75] [819] [2163] [2210] [2254] [3049] [3054]
Ti	[486] [905] [2184]
Tl	[2119] [2210]
V	[207] [457] [547] [675] [2846] [2872]
W	[547] [2680] [2972]
Zn	[632] [743] [2572] [2846]

The analytical procedure is usually adjusted to suit the nature of the analysis to be carried out and the required accuracy. Interferences of a chemical nature are rarely observed in steel analyses. In the presence of higher silicon contents, interferences occur in the determination of manganese, but these can be avoided by adding calcium and using a 3-slot burner head. Chromium is influenced by iron in the air/acetylene flame, so that either the iron content must be matched in the reference solutions or a nitrous oxide/acetylene flame must be employed. KÖNIG, SCHMITZ and THIEMANN [680] found that the determination of copper, manganese and aluminium is influenced both by the acid content and the iron concentration, so they matched the reference solutions correspondingly. KNIGHT and PYZYNA [675] solved this problem for the analysis of nine elements in tool steel by calibrating with standard steels of the NBS and found very good agreement with emission spectrometry and X-ray fluorescence. With the latter, phase problems which lead to errors can occur, as examples have shown. HUSLER [547] used a nitrous oxide/acetylene flame for the determination of chromium, manganese, molybdenum, vanadium and tungsten in tool steel. After the addition of iron as the main element to the reference solutions and potassium as ionization buffer, the determinations in this flame were free of interferences.

The use of standard steels for calibration has proved of especial value for high precision analyses. If high accuracy is necessary, such as in the determination of the main alloying constituents in chrome-nickel steels, small interferences can have a marked effect. As WELZ and SEBESTYEN [1299] were able to show, even with matching of the acid and iron contents, an accuracy of only 1% could be achieved for chromium and nickel. With standard steels for calibration and employing baseline suppression and scale expansion, an accuracy of 0.3% relative can be achieved. An amazing fact is that the standard steels do not particularly have to resemble the sample in their composition; no special selection of standard steels must be made.

QURESHI *et al* [2769] mention the fact that iron can be relatively easily extracted from steel solutions prior to the determination of trace elements. The partition of iron(III) between the aqueous phase and 0.1 M 2-hexylpyridine in benzene is a function of the hydrochloric acid concentration. In 7 M hydrochloric acid iron can be largely separated from numerous elements, so that the subsequent trace determinations are free of interferences.

Magnesium, which has considerable effect on the properties of cast iron, can only be determined with great difficulty by conventional methods. Atomic absorption spectrometry, because of its simplicity and high sensitivity, offers the ideal method for this element. BELCHER and BRAY [98] developed a procedure in 1962 which is still frequently used nowadays as the standard method. 1 g sample is dissolved in 30 mL 1 + 1 hydrochloric acid and oxidized with 5 mL nitric acid. The solution is evaporated to dryness, mildly heated for a further five minutes, taken up in 10 mL hydrochloric acid and then diluted with strontium chloride solution. The use of a nitrous oxide/acetylene flame offers advantages for both magnesium and calcium [1220]. SHAW and OTTAWAY [2846] successfully applied the graphite furnace technique to the determination of magnesium in iron and steels.

The determination of *molybdenum* in steel is preferably carried out in a nitrous oxide/acetylene flame since only slight interferences are to be expected. KIRKBRIGHT [663] only added the corresponding amount of iron to the reference solutions and found otherwise no interferences. Nevertheless, POLLOCK [990] frequently found values that were too low for the determination of molybdenum in slags, even in the nitrous oxide/acetylene flame, and was only able to obtain the theoretical values by adding 1000 mg/L aluminium. In an air/acetylene flame, which must be run fuel gas rich, numerous interferences occur, but these can be eliminated by the addition of either a large excess of aluminium [271] or 2% ammonium chloride [887].

The determination of *aluminium* [175] [488] [791], *vanadium* [207] [457], *titanium* [486], *niobium* [1088] and other refractory elements in steel can be carried out in a nitrous oxide/acetylene flame with no interferences worth mentioning, only the iron concentration should be matched. BOSCH *et al* [145] reported a rapid determination of acid-soluble aluminium in steel in the important concentration range 0.002–0.060% Al. According to this procedure, 20 samples can be analyzed in 11 minutes with a standard deviation of about ±0.001% Al. SHAW and OTTAWAY [2846] determined aluminium in steel by the graphite furnace technique. The acid-soluble aluminium was determined by dissolving the sample in nitric acid and dispensing directly into the graphite furnace. To determine the acid-insoluble aluminium, the residue was fused with sodium carbonate/sodium tetraborate and dissolved in nitric acid. PERSSON *et al* [2742] also employed the graphite furnace technique to determine aluminium in various steels. They dissolved the samples in hydrochloric acid/nitric acid and added ammonium sulfate to eliminate the influence of hydrochloric acid. The iron content in the sample reduced the sensitivity by about 20% and was thus matched in the reference solutions. The varying iron contents in different steels and the influence of chromium, molybdenum or nickel in stainless steels could be disregarded.

COLLIN *et al* [239] analyzed the various *non-metallic phases* in steel which have considerable influence on the physical and mechanical properties. The steel was dissolved in hydrofluoric acid and the residue (sulfides, carbides, nitrides, oxides, aluminosilicates of calcium, iron, magnesium and manganese) was fused with sodium carbonate/sodium tetraborate. After dissolution in sulfuric or hydrochloric acid and dilution, the individual elements were determined in the flame.

McAULIFFE [845] has described a procedure for the direct determination of *silicon* in cast iron. 0.50 g sample is warmed with 25 mL 1.5 M sulfuric acid until solution is complete. Then 10 mL 12% ammonium persulfate solution are added and the solution is heated to boiling for about one minute until it is clear. After cooling, the solution is diluted, filtered and nebulized into the flame directly. The values obtained by this procedure agree very well with those of other methods and with certificate values. KIRKBRIGHT *et al* [656] determined silicon in low alloyed steels by atomic fluorescence spectrometry and found no interferences.

For the determination of *tungsten* in metallic samples, TINDALL [2972] has drawn attention to the importance of not heating digestion solutions to above 70°C. He pulverized the samples and dissolved them slowly in concentrated hydrochloric acid in a waterbath at 60–70°C. MUSIL and DOLEŽAL [2680] reduced tungsten with tin(II) chloride and extracted the thiocyanate complex into MIBK. The organic phase was nebulized directly into a nitrous oxide/acetylene flame. No interferences due to other constituents in the alloys were observed.

HOFTON and HUBBARD [522] determined *silver* in steel after extraction of the dithizonate in MIBK. This represents a considerable simplification compared to conventional methods. PETERSON [972] determined *selenium* free of interferences in stainless steel in the range 0.2–0.5% by dissolving the sample in hydrochloric acid/nitric acid and then treating it with sulfuric acid/phosphoric acid.

MARČEC *et al* [819] determined traces of *tellurium* in steels and iron by precipitating with tin(II) chloride and extracting the diethyl dithiocarbamate complex with pentyl acetate. The subsequent atomic absorption analysis is interference-free. BARNETT and KAHN [75] described a procedure for the determination of tellurium in steel without prior extraction. 2 g steel are dissolved in 25 mL of an acid mixture consisting of two parts concentrated nitric acid and one part 1 + 1 sulfuric acid and then diluted to 100 mL. Since marked non-specific attenuation occurs with this solution at the 214.3 nm resonance line, a background corrector must be used.

For a large number of trace elements in iron and steel, the hydride technique and the graphite furnace technique have opened up substantial possibilities, both in respect to the attainable determination limits and the speed, simplicity and reliability of the determinations. FLEMING and IDE [2254] reported the determination of *antimony, arsenic, bismuth, lead, selenium, tellurium* and *tin* in steels by the hydride technique in 1976. At that time the authors were still using an argon/hydrogen diffusion flame, so the sensitivities were not optimum. They nevertheless recognized the importance of the acid concentration in the analysis solution for the control of interferences and also found that iron reduces the interferences due to nickel and other transition elements. The determination of the hydride-forming elements in low-alloy steels was later investigated in the present author's laboratory [3049] [3054]. An acid mixture could be found for each element to permit an interference-free determination directly against aqueous reference solutions. The influence of nickel was also investigated and it was found that increasing the acid concentration, or more especially adding iron, reduced this interference drastically. The interference-free determination range for arsenic, for example, could be increased by three orders of magnitude [3049] (refer to figure 120, page 239). While FLEMING and IDE [2254] reduced antimony(V) to antimony(III) with potassium iodide before the determination, in the present work the determination was performed directly against antimony(V) reference solutions [3052].

BARNETT and MCLAUGHLIN [2060] determined a number of trace elements in iron alloys by the graphite furnace technique after dissolution of the samples in nitric acid. Antimony, arsenic, and lead could be determined directly against synthetic reference solutions containing the main alloy constituents in similar concentrations; bismuth, cadmium, and tin had to be determined by the analyte addition technique. DULSKI and BIXLER [2210] determined bismuth, lead, selenium, silver, tellurium, and thallium in iron by five different methods; they found that a direct determination by the graphite furnace technique brought a substantial saving of time and provided satisfactory results. The authors used spiked samples for calibration.

FRECH [2259] made a detailed investigation on the determination of *antimony* in steel by the graphite furnace technique. The samples were dissolved in *aqua regia*. He found that in the presence of oxygen, which had diffused into the graphite tube, the antimony signal was dependent on both the hydrochloric acid and nitric acid concentrations. He attributed the severe signal depressions to losses in the form of antimony chloride. If great care is taken to exclude oxygen, the two acids do not interfere in the determination of antimony. Chromium enhances the signal, but above 200 mg/L the enhancement is constant, so that this quantity of chromium was always added. At a heating rate of 450 °C/s nickel enhanced the antimony signal while iron depressed it. At heating rates of 900 °C/s or faster, no influence was observed.

FRECH and CEDERGREN [2260] [2261] made a thorough practical and theoretical examination of the determination of *lead* in steel by the graphite furnace technique. They found that in the presence of a sufficient quantity of hydrogen the sample must be thermally pretreated at 900 K to eliminate chlorine according to:

$$FeCl_2 + H_2 \rightarrow Fe + 2\,HCl$$

Otherwise volatile lead chloride is formed, leading to losses and thus signal depression. In a graphite furnace in which oxygen is largely excluded, hydrogen is formed at high temperature through reaction of the graphite with water that remains in the tube after the drying step. FRECH and CEDERGREN [2261] also made a comparison of various furnace types for this determination and found substantial differences in the reaction behavior. Based on this work, FRECH et al [2265] described a routine procedure for the determination of lead in steel that gives good agreement with polarography and flame AAS with extraction. As in many fields, steel analysis clearly indicates that severe interferences occur for the graphite furnace technique when inadequate instruments are used and the operating conditions are not optimum. Chloride interferences during the determination of lead are virtually eliminated when atomization is off the L'VOV platform in a stabilized temperature furnace [2882].

The work of RATCLIFFE et al [2780] also clearly indicates this tendency. For the determination of *tin* in steel, these authors reported total signal suppression due to hydrochloric acid and severe scattering caused by sulfuric acid and perchloric acid. Chromium, nickel, niobium, and titanium also caused interferences. By carefully selecting the dissolution procedure, DEL MONTE TAMBA and LUPERI [2196] succeeded in maintaining a constant hydrochloric acid concentration for all steel types. In this way, tin in all samples could be determined against the same reference curve. The authors found that in the graphite furnace the constituents normally present in steel caused no interferences to the determination. The procedure is suitable for the determination of tin in the range 0.002–0.04% in steel, and since the procedure is rapid it is ideal for routine analyses. GROBENSKI et al [2321] investigated the determination of tin in steel using a stabilized temperature furnace and atomization off the L'VOV platform. They observed no interferences and were able to perform the determinations directly against aqueous reference solutions.

For the determination of selenium in steels and alloys, FERNANDEZ et al [2244] reported a spectral interference due to iron; this interference could be eliminated by applying Zeeman-effect background correction. YAMADA et al [3083] reported similar spectral interferences due to copper, iron, and lead on the determination of antimony.

A determination that first became possible with the introduction of the graphite furnace technique, and especially the L'VOV platform, the stabilized temperature furnace and Zeeman-effect background correction, is the determination of *phosphorus* in steel. This determination had been reported by WHITESIDE and PRICE [3068] and also HAVEZOV et al [2353], but under closer scrutiny the methods described by these authors could not be verified. On the basis of detailed theoretical and practical investigations, PERSSON and FRECH [2741] showed that reliable results for phosphorus could only be obtained for atomization under isothermal conditions. This was verified by comprehensive work performed in the present author's laboratory on the determination of phosphorus in steel [3061]. It is essential that atomization is off the L'VOV platform in a stabilized temperature furnace. The most suitable combination is a massive pyrolytic graphite platform in an uncoated graphite tube. The sample can thus hardly react with the graphite, but the very low partial pressure of oxygen favoring this determination is present. An 0.2% lanthanum solution is the most suitable matrix modifier, and should be added to both sample and reference solutions. Under these conditions the determination of phosphorus in steel can be performed directly against aqueous reference solutions. Using a continuum source background corrector, a severe spectral interference that depends on the bandpass is observed; this leads to overcompensation [3061]. This interference is eliminated using transverse AC Zeeman-effect background correction. Approximately 0.002% phosphorus in steel can then be determined free of interferences.

The possibility of direct **solids** analysis offered by the graphite furnace technique can also be of interest to steel analysis. The most important arguments include speed of analysis, avoidance of digestion procedures, and the best detection limits. HEADRIDGE *et al* used an induction furnace heated to a constant temperature of 700–2200 °C into which the samples were inserted. They determined bismuth [2041], silver [2048], and lead [2040] in steel. The authors used preanalyzed steels for calibration; good results were also obtained using aqueous reference solutions.

Using a similar procedure, FRECH *et al* [2264] [2572] determined bismuth, cadmium, lead, silver and zinc in steels. They established that at 2100 °C the recovery was only about 80%, so that aqueous reference solutions should not be used for calibration. The authors emphasized that for direct solids analysis only a furnace that is in temperature equilibrium has any practical value.

Atomic absorption spectrometry is equally suitable for the determination of trace elements and main constituents in **alloys.** Only the techniques and the accuracy requirements are different. SATTUR [1079] examined numerous non-ferrous alloys routinely by AAS. He determined copper, iron, manganese and nickel in *copper alloys* in which the copper was present to more than 80% in some cases. He similarly determined alloys containing aluminium, lead, tin and zinc. GIDLEY and JONES [414] [415] analyzed copper alloys for zinc, ELWELL and GIDLEY for lead [328] and manganese [7], and CAPACHO-DELGADO and MANNING [206] [208] determined tin in brass and lead alloys using an air/acetylene flame because chemical interferences were stronger in an air/hydrogen flame. SPRAGUE *et al* [1167] determined selenium and tellurium in various copper alloys. NAKAHARA *et al* [904] determined beryllium in copper alloys with a nitrous oxide/acetylene flame and found no interferences other than those from palladium and silicon.

ROSALES [2801] determined copper in flotation products containing up to 50% copper by using less sensitive resonance lines with maximum speed and accuracy. ALDER *et al* [2016] determined copper in copper alloys after releasing it electrographically from the surface and recovering it on moist filter paper.

LIKAITS *et al* [2549] determined arsenic in blister copper after prior digestion in hydrochloric acid/nitric acid. The authors draw attention to the importance of digesting the samples in the presence of an oxidizing acid, since arsenic trichloride is volatile at temperatures a little above 100 °C. Digestion in sulfuric acid/nitric acid or perchloric acid/nitric acid results in values for arsenic that are too low. TSUKAHARA and TANAKA [2988] determined silver in copper alloys after extraction as the tri-n-octylmethylammonium silver bromide complex. NAKAHARA and MUSHA [2686] determined tellurium in copper smelting sludges in an argon/hydrogen diffusion flame and observed no interferences due to acids except perchloric acid. Magnesium chloride effectively eliminated interferences caused by various elements. CELIS *et al* [2139] determined aluminium in dispersion-hardened copper/aluminium oxide alloys. They dissolved the alloys in sulfuric acid/hydrogen peroxide, removed the copper electrolytically, evaporated the solution to dryness, and fused the residue with sodium carbonate.

BARNETT and McLAUGHLIN [2060] used the graphite furnace technique to determine a number of trace elements in copper alloys. They were able to determine antimony and lead against reference solutions that contained the main alloy constituents; arsenic, bismuth, cadmium, and tin had to be determined by the analyte addition technique. SHAW and OTTAWAY [2845] determined lead in copper alloys using the graphite furnace technique after prior dissolution of the samples in nitric acid. MEYER *et al* [2665] volatilized selenium from molten copper alloys at 1100–1150 °C in a stream of oxygen. The selenium was collected in a cold trap, dissolved in nitric acid and determined in the graphite furnace. The detection limit for this procedure is 3×10^{-11} g Se or 0.1 ng/g Se in copper or copper alloys. FRITZSCHE *et al* [2272] found for tin that reactions on the graphite surface hindered the determination. They thus impregnated the graphite tubes with tantalum, tungsten, or zirconium.

BAKER and HEADRIDGE [2052] determined bismuth, lead, and tellurium in copper alloys directly in the solid samples in an induction furnace. Sample quantities were in the range 2–30 mg; calibration was against preanalyzed standards.

Similar to the analysis of steel, the main constituents in alloys can only be determined with the required accuracy when a spectrometer suitable for precision analysis is employed. The use of a less sensitive resonance line to avoid excessive dilutions [1299] [3062] and making comparative analyses against standard alloys of similar composition are also frequently recommended.

LEWIS, OTT and SINE [747] recommended atomic absorption spectrometry as the standard method for the determination of lead in *nickel alloys*. They determined lead in concentrations of 0.0001% downwards and found no chemical interferences. DYCK [316] determined chromium, magnesium and manganese in nickel alloys and found that AAS was rapid and reliable. ANDREW and NICHOLS [54] determined magnesium in high purity nickel and in nickel alloys with a standard deviation of 0.00005%. BURKE [179] found that the concentrations of antimony, lead, bismuth and tin in nickel are too low for a precise direct determination, so he employed co-precipitation with manganese(IV) oxide for enrichment [179]. BURKE [2119] found that traces of thallium in nickel alloys could be quantitatively extracted out of a 10% hydrochloric acid solution with 2% ascorbic acid and 9% potassium iodide into 5% tri-n-octyl-phosphine oxide (TOPO) in MIBK. The following determination in the flame is interference free. For the determination of traces of bismuth, cadmium, lead, silver, and zinc in nickel alloys, KIRK *et al* [2469] used an ion exchanger to separate the matrix. ARENDT [60] described two procedures for the determination of niobium in nickel alloys and NEWLAND [911] extracted bismuth from nickel alloys with MIBK before the determination.

WELCHER and KRIEGE [1289] developed a special procedure for the precision determination of Al, Cr, Co, Fe, Mo, Nb, Ta, Ti, V and W in high temperature nickel alloys. The agreement with values obtained by the tedious wet chemical method was good. Furthermore, the accuracy and precision were checked by repeated determinations on a standard alloy. CARPER [212] determined the three main constituents in iron-nickel-cobalt alloys. The composition of the alloy is particularly critical with respect to its magnetic properties. 150 mg sample are dissolved in 5 mL *aqua regia* and diluted to 500 mL for the determination of cobalt. The solution is further diluted 1:4 for iron and 1:20 for nickel. The measurement is against aqueous reference solutions, which in each case contain the other two alloy constituents in the corresponding concentrations. The relative standard deviation for cobalt (content about 3%) is 1.5%, for iron (content about 17%) 0.28% and for nickel (content about 80%) 0.21%. The agreement with classical chemical analysis, which is more time consuming, is excellent. No chemical interferences were observed when the reference solutions contained Ni—Fe—Co in the ratio 80:17:3. WELCHER and KRIEGE [1290] determined the main elements in high temperature cobalt alloys after dissolution in hydrochloric acid/nitric acid/hydrofluoric acid. The references did not have to be matched exactly to the sample solution; only slight interferences occurred and the agreement with certificate values was good.

Since hydrochloric acid suppresses the signals for bismuth and thallium completely and depresses the signal for lead, and sulfuric acid influences the sensitivity for tellurium, WELCHER et al [1291] [2625] dissolved nickel alloys in a mixture of equal parts of nitric acid, hydrofluoric acid and water for subsequent analysis by the graphite furnace technique. For the determination of arsenic, however, the hydrofluoric acid must be removed. The authors emphasized that in the presence of nickel, no losses of selenium take place during thermal pretreatment in the graphite furnace. The results of this method are in good agreement with emission spectrometry and X-ray fluorescence. KUJIRAI et al [2510] also described a rapid procedure for the determination of thallium in nickel and cobalt alloys. DULSKI and BIXLER [2210] investigated five various procedures for the determination of traces of bismuth, lead, selenium, silver, tellurium, and thallium in nickel alloys; the direct method using the graphite furnace technique was the fastest. Calibration was against spiked sample solutions.

MARKS et al [2626] determined bismuth, lead, selenium, tellurium, thallium, and tin in complex nickel alloys directly from the solid sample by the graphite furnace technique. A chip of about 1 mg was volatilized without pretreatment. Calibration was against preanalyzed alloys. Since no metal salts are formed as in solution, much lower background attenuation was observed. The authors found that this procedure was sufficiently accurate for the requirements; for best precision and accuracy they nevertheless recommend that the samples should be taken into solution. Direct solids analysis can also provide valuable information on the distribution of trace elements within larger samples. HEADRIDGE and THOMPSON [2358] used an induction furnace to determine bismuth directly from solid nickel alloy samples.

The fact that so few interferences are observed for the determination of arsenic and selenium in nickel alloys by the graphite furnace technique derives from the matrix modifying influence of nickel. EDIGER [2217] described this influence clearly in his first publication on matrix modification. The advantage brought to the graphite furnace technique through the formation of stable compounds between nickel and elements such as arsenic and selenium is a major source of interference in the hydride technique, however. Nevertheless, DRINKWATER [2206] managed to perform an interference-free determination of bismuth in nickel alloys by the hydride technique by adding EDTA solution as buffer. KIRKBRIGHT and TADDIA [2472] eliminated the influence of nickel on the determination of selenium by adding tellurium. The constants of stability of tellurides are lower than those of the corresponding selenides. The influence of nickel on arsenic can be substantially reduced through the addition of iron in acid solution [3049] [3058], so that it is possible to determine arsenic in nickel alloys directly. For the determination of traces of selenium in nickel alloys and nickel oxide, separation of the nickel by precipitation as the hydroxide is recommended [3059]. YOUNG [3092] has published an overview on the determination of 28 elements in nickel alloys, refinery sludges and residues.

QUARRELL *et al* [1011] determined antimony and tin in *lead alloys* used for cable screening. They dissolved the sample in a mixture of fluoroboric acid and hydrogen peroxide and determined antimony in the air/acetylene flame and tin in the nitrous oxide/acetylene flame directly from this solution. BELL [104] has drawn attention to the necessity of a rapid determination of the constituents in lead alloys used in the electrical industry and also that lead and tin are not soluble in the same medium. Mixtures of fluoroboric acid, hydrogen peroxide and di-sodium EDTA dissolve lead alloys rapidly at room temperature, and after dilution, the solutions can be aspirated directly into the flame. SKERRY and CHAPMAN [1131] dissolved tin/lead solders in hydrochloric acid/nitric acid by heating the solder in the form of fine turnings to 195°C and adding to the acid under stirring. HWANG and SANDONATO [550] determined numerous trace elements in a tin/lead solder after dissolving in fluoroboric acid/nitric acid. During investigations performed in the present author's laboratory [3062], it was found that the procedure recommended by HWANG and SANDONATO for the dissolution of soft solders was the only workable one. It is important that the mixing ratio of 3 parts 65% nitric acid, 2 parts 35% fluoroboric acid and 5 parts deionized water is strictly maintained, otherwise the sample will not be completely dissolved, or metastannic acid precipitates on dilution. By choosing suitable analytical lines, it is possible to perform numerous determinations from a single digestion solution; namely the main constituents lead (368.4 nm) and tin (266.1 nm), and the trace elements arsenic (graphite furnace), antimony, copper, nickel (low concentrations in the graphite furnace), silver and bismuth.

GUERRA [2322] dissolved tin/lead solders in hydrochloric acid/nitric acid for the subsequent determination of gold and copper. TSUKAHARA and TANAKA [2988] extracted silver as the tri-n-octylmethylammonium silver bromide complex from lead alloys and determined it directly in the flame. MACHATA and BINDER [2595] determined the main and trace elements in buckshot for the identification of ammunition. They found that the graphite furnace technique was more suitable than the flame. BERNDT *et al* [2084] determined arsenic in lead and lead alloys by the hydride technique and found that the range of interference-free determination could be extended by almost four orders of magnitude when 6 M hydrochloric acid was used instead of the usual 0.5 M. This was verified by independent work performed in the present author's laboratory.

BARTON *et al* [86] determined calcium in lead/calcium alloys with a relative standard deviation of 0.3% and a mean recovery of 100%.

BELL [102] [103] investigated the application of atomic absorption spectrometry for the analysis of *aluminium alloys* in detail and found that Cd, Cr, Cu, Fe, Mg, Mn, Ni and Zn could be determined free of interferences. The sample is simply dissolved in 1 + 1 hydrochloric acid, with the addition of a few drops of hydrogen peroxide if required, and correspondingly diluted. For the determination of alkaline-earth metals the addition of lanthanum or the use of a nitrous oxide/acetylene flame is recommended. WILSON [1339] determined silver traces in aluminium alloys and compared the results with the conventional method of VOLHARD [1274]; the agreement was very good. MANSELL et al [816] likewise found good agreement with wet chemical procedures for the determination of Ca, Cu, Mn and Zn in *magnesium alloys*. Since the majority of alloys also contain aluminium, they added lanthanum to the samples. Besides other elements, WILSON [1340] determined chromium and zirconium in aluminium alloys and found good agreement with titrimetric or gravimetric values. ELROD and EZELL [327] made a more comprehensive study on the determination of chromium in aluminium and aluminium oxide materials and found that potassium peroxydisulfate eliminated all the interferences due to larger amounts of aluminium, iron or titanium in the air/acetylene flame. The samples were dissolved in $1+1$ sulfuric acid and then sufficient potassium peroxydisulfate was added before dilution so that the end solution contained 1%. The determination could be made against aqueous chromium reference solutions. The agreement with certificate values was very good and the day-to-day precision of AAS was better than the diphenylcarbazide method by about a factor of three.

CAMPELL [204] determined silicon in aluminium alloys; he dissolved 0.5 g sample in 10 mL 20% sodium hydroxide solution with subsequent oxidation of the silicon with hydrogen peroxide. The solution was acidified, made up to 100 mL and directly aspirated into the flame. The solution was further diluted for higher silicon contents. After the reference solutions had been matched with the main constituents, an agreement of 1% relative with wet chemical procedures could be achieved. NAKAHARA et al [904] determined beryllium in aluminium and magnesium alloys in a nitrous oxide/acetylene flame and found no interferences other than those caused by palladium and silicon.

JANOUŠEK [2433] made a detailed examination of the determination of beryllium in aluminium alloys and found that both aluminium and silicon interfere severely. Thermally stable compounds are formed that do not decompose even in the nitrous oxide/acetylene flame. Procedures proposed by other authors using fluoride [1016], 8-hydroxyquinoline [377], or diethyleneglycol-monobutyl [904] are only effective up to aluminium concentrations of around 4000 mg/L. JANOUŠEK therefore proposed a digestion in nitric acid/hydrofluoric acid/water in the ratio 2:1:4 in a PTFE autoclave. This eliminates the influence of aluminium up to 10 000 mg/L. A fluoride complex of aluminium is formed that prevents the formation of aluminium-oxygen compounds. BURKE [2119] determined traces of thallium in aluminium alloys in the flame after prior extraction with TOPO in MIBK. FRITZSCHE et al [2272] determined tin in aluminium alloys by the graphite furnace technique. The authors found that the atomization of tin is hindered by reactions with the graphite surface, so they impregnated the graphite tubes with tungsten, zirconium or tantalum solutions. The impregnated tubes are more resistant than untreated tubes against the acid concentrations required to prevent the hydrolysis of tin. Investigations in the present author's laboratory have shown that graphite tubes coated with an impervious layer of pyrolytic carbon are at least as good as tubes impregnated with metal salts. ISHIZUKA et al [2411] atomized aluminium alloys with a ruby laser and determined aluminium, chromium, copper, iron, manganese, molybdenum, nickel, and vanadium by AAS.

MYERS [538] determined aluminium, vanadium and iron in *titanium alloys,* and iron in *zirconium alloys*. The samples were dissolved in hydrofluoric acid and oxidized with a little nitric acid. The mean deviation over several months was $\pm 0.10\%$ for aluminium and vanadium and ±0.002% for iron in the titanium alloys. CAPACHO-DELGARDO and MANNING [206] [208] determined tin in zirconium by dissolving the metal in a mixture of HBF_4 and HCl.

SCHLEWITZ and SHIELDS [1090] determined chromium, copper, iron, nickel and tin in zirconium alloys and found that the matrix must be matched for an exact analysis. For the determination of tantanlum in zirconium alloys, they recommended an extraction with MIBK [1091].

THORMAHLEN and FRANK [1228] determined silicon in *niobium alloys,* which they dissolved in sulfuric acid/hydrogen peroxide. PABALKAR *et al* [2723] determined iron, copper, and nickel in *tungsten alloys* containing 7.5% nickel and 2.5% copper. The alloy was dissolved in hydrofluoric acid/nitric acid under the addition of ammonium citrate to maintain tungsten in solution. ORTNER *et al* [2712] [2713] determined arsenic, lead, and silicon in tungsten alloys by the graphite furnace technique. They impregnated the graphite tube with sodium tungstate solution to obtain better reproducibility and longer tube lifetime.

LIEBER [749] determined the main constituents in *silver alloys* after dissolving them in hydrochloric acid/nitric acid, and SCHAEFER and VOMHOF [1084] determined gold in jewelry alloys after solution in *aqua regia* in good agreement with gravimetry and X-ray fluorescence, and with high precision.

Besides the analysis of alloys, atomic absorption spectrometry is also suitable for the determination of traces in **high purity metals.** In general, two precautionary measures must be taken. First, slight background effects can occur when high total salt concentrations are aspirated into a flame that must be taken into account; and second, the analyte addition technique should be employed to avoid the influences of viscosity (transport interferences). Further, attention must be drawn to the risk of aspirating high concentrations of copper and silver into a premix burner with acetylene as the fuel gas since acetylide formation can take place, leading to an explosion in the burner system. In such cases, the removal of the main constituent or dilution is recommended if a fuel gas other than acetylene cannot be used.

MCCRACKEN *et al* [848] reported the determination of traces of tin, cadmium and zinc in copper after prior separation of the copper by means of ion exchangers, and GOODWIN [432] determined silver in copper without separation and found that only nitrose gases interfere and must therefore be boiled off before the determination. WUNDERLICH and BURGHARDT [3080] determined traces of silver in pure copper after prior extraction with dithizone in carbon tetrachloride and back extraction into the aqueous phase. WUNDERLICH and HAEDELER [3081] determined traces of bismuth, cadmium, and zinc in pure copper. These authors draw attention to a potential spectral interference due to overlapping of the copper line at 213.85 nm with the zinc line at 213.86 nm. Copper was thus separated either as its sulfide with hydrogen sulfide or electrolytically.

BÉDARD and KERBYSON [2068] [2069] determined traces of arsenic, bismuth, selenium, tellurium, and tin in copper by the hydride technique. To prevent interferences during hydride generation, they separated the trace elements beforehand by precipitation with iron oxide or lanthanum oxide. For each element, the detection limits are between 0.001 and 0.006 µg/g in copper. ALDUAN *et al* [2028] determined antimony in copper by the hydride technique and eliminated interferences by adding potassium thiocyanate. For the same determination,

LINDSJÖ [2550] added thiourea to complex the copper. Frequently, using 6 M hydrochloric acid is sufficient to reduce the interference due to copper on the hydride-forming elements by several orders of magnitude [3058].

HAYNES [2357] used the graphite furnace technique to determine antimony, arsenic, selenium, and tellurium in ultrapure copper. No interferences were observed and calibration could be performed against aqueous reference solutions containing the same quantities of copper and nitric acid as the sample solutions. Copper acts as a matrix modifier for these elements, so that relatively high pretreatment temperatures can be applied in the graphite furnace. MEYER, HOFER and TÖLG [2665] determined selenium in copper, silver, gold, lead, and bismuth after volatilization in a stream of oxygen at 1100–1150 °C. The selenium is collected in a cold trap, dissolved in nitric acid and determined by the graphite furnace technique. The detection limits are 0.1 ng/g in copper, 0.05 ng/g in silver and gold, 0.2 ng/g in bismuth, and 10 ng/g in lead.

FORRESTER *et al* [2257] determined arsenic, bismuth, cadmium, lead, selenium, silver, tellurium, and thallium in high purity nickel. The nickel was dissolved in ultrapure nitric acid and the solution used directly for determinations in the graphite furnace. The results were in good to excellent agreement with certificate values; the detection limits are between 0.01 and 0.5 µg/g for each element, referred to solid nickel.

HAMNER *et al* [2344] determined antimony, arsenic, bismuth, selenium, silver, and tellurium in metallic chromium by the graphite furnace technique. A sodium peroxide/sodium carbonate fusion had sufficient oxidizing power to prevent all losses of the trace elements. The chromium matrix caused interfering background attenuation, but this could be controlled by using 10 µL aliquots and continuum source background correction. Nickel stabilized most of the analyte elements and improved the sensitivity. Remaining interferences were eliminated by using the analyte addition technique.

LINDSJÖ [2550] determined bismuth in pure cobalt by the hydride technique. He oxidized the cobalt to cobalt(III) with hydrogen peroxide in sodium hydroxide solution and complexed with EDTA. This complex is still stable in 5 M hydrochloric acid, while the bismuth complex decomposes.

MCELHANEY [2654] determined gold, cobalt, and silver in pure aluminium by the graphite furnace technique, but did not have complete success with a dual-wavelength procedure for background correction. LANGMYHR and RASMUSSEN [2530] used an induction furnace to determine gallium and indium directly in solid aluminium samples. NORVAL and GRIES [2699] determined thallium in metallic cadmium by direct solid sampling in the graphite furnace. They found this procedure to be more sensitive than working with solutions; no time-consuming extraction procedures are required, the risk of contamination is minimal, and very small sample weights can be used.

NEUMANN determined chromium, iron, cobalt, manganese and nickel in high purity molybdenum and tungsten and found no interferences [908], while for the determination of calcium, either the addition of cesium and strontium was necessary [907] or, as for the determination of aluminium, the use of a nitrous oxide/acetylene flame [909]. In all cases prior separation was not required.

SCARBOROUGH *et al* [1083] described a procedure for the determination of chromium, cobalt, iron and nickel in concentrations under 10 µg/g in high purity sodium metal. The trace elements under study are precipitated together with added lanthanum at pH 9 and thus separated from sodium. The precipitate is dissolved in hydrochloric acid and determined by AAS. HUBER *et al* [2398] determined chromium, iron, manganese, and nickel in sodium metal by the graphite furnace technique; the sodium was distilled off *in vacuo* beforehand and the residue taken up in hydrochloric acid. GARBETT *et al* [2287] [2288] also investigated the determination of chromium, iron, and nickel in sodium by the graphite furnace technique. They found that the usual sodium salts are more volatile than the analyte elements. These authors volatilized sodium chloride selectively at 1100°C under the addition of ammonia and determined the analyte elements free of interferences.

PANDAY and GANGULY [947] determined chromium, manganese and nickel in metallic bismuth and tellurium and found marked background attenuation which they eliminated by removing the main constituents. ROTH, BOHL and SELLERS [1061] determined traces of tellurium in high purity bismuth by extracting with MIBK from 6 M hydrochloric acid solution and aspirating the organic layer directly into the flame. The method is accurate and rapid; the effectiveness of the extraction is 98–99%.

YUDELEVICH *et al* [3093] determined 13 elements in ultrapure rhenium and 15 elements in ultrapure gallium by the graphite furnace technique. DITTRICH *et al* [2201] investigated the determination of selenium and tellurium in various semiconductor materials by the graphite furnace and hydride techniques. They found that the semiconductor materials caused interferences in both techniques, but that these could largely be brought under control under optimum conditions. For the graphite furnace technique this signifies atomization from the L'VOV platform in a stabilized temperature furnace and for the hydride technique the use of higher acid concentrations.

DOOLAN and BELCHER [2204] have published a good overview on the analysis of non-ferrous alloys and ultrapure metals using atomic spectroscopic methods with particular reference to sample preparation procedures.

The determination of trace *noble metals* in other noble metals does not appear to be completely free of interferences; various authors use additions of 0.5% copper and 0.5% cadmium [1114], 0.5–1% lithium [485] [1234] or 1% sodium and 1% copper [57] to eliminate various interferences. ADRIAENSSENS and VERBEEK [17] [18] found that in a 2% potassium cyanide solution, most of the interelemental interferences observed with noble metals in acid solutions did not occur and they determined gold, palladium, platinum and silver without interferences and with good precision in cupellation granules. KALLMANN and HOBART [623] also determined gold, palladium and silver in noble metal beads from smelting processes in cyanide solution with an accuracy of 1%. HEINEMANN [2361] [2362] made a detailed investigation on the mutual influence of the platinum metals in an air/acetylene flame and found that lanthanum or uranium is an effective spectrochemical buffer. The graphite furnace technique also appears to be ideally suited for noble metal analyses since interelemental interferences hardly occur [16]. KRAGTEN and REYNAERT [2507] used the graphite furnace technique to determine traces of iron in gold and silver; gold was reduced to the metal and silver was precipitated as the chloride before the determination. Iron is not precipitated and can be determined from the solution. ROWSTON and OTTAWAY [2804] made an exhaustive investigation on the atomization of noble metals in the graphite furnace and established recommended conditions for their determination. Various authors have drawn attention to the risk of background attenuation during noble metal analyses in the flame which must be taken into consideration [324] [687]. Two good reviews on the analysis of noble metals have been published by SEN GUPTA [1116] and BEAMISH and VAN LOON [90].

Of interest in connection with the analysis of high purity metals is the preconcentration procedure for trace metals on activated charcoal or mercury described by JACKWERTH and co-workers. Activated charcoal can act as a trace adsorber for enrichment of numerous metals from high purity aluminium, silver, tungsten [563], cadmium or magnesium [562]. For the subsequent determination, the activated charcoal suspended in dilute nitric acid can be aspirated directly into the flame, or better, the activated charcoal is leached out by evaporating to dryness with nitric acid and the residue is taken up in a small quantity of acid.

In a similar manner, voluminous precipitates or trace chelating agents (insoluble in acid) can be used. For example, trace elements in manganese alloys were preconcentrated on silver halides [2078] or in pure copper by precipitating the matrix as copper sulfide [2428]. Gold and palladium in various ultrapure metals were preconcentrated as the dithizone complex by adding excess dithizone [2429].

When two metals are in electrical contact, the more base of the two determines the potential and will therefore alone be attacked by oxidizing agents. If a piece of metal coated with a layer of mercury is dissolved in acid, the trace impurities are enriched while the main constituent passes into the amalgam layer. The "end point" of the solution can be clearly identified potentiometrically. This enrichment procedure has been successfully employed for the determination of numerous trace elements in high purity aluminium [519], cadmium [561] [2385] and zinc [560]. JACKWERTH [2424] has published a summary on the preconcentration of trace elements (refer also to section 6.3.2).

Plating baths are mostly suitable for direct analysis by atomic absorption spectrometry after dilution; however, since the determinations are mostly simple, little has been published on this application of the technique. SHAFTO [1119] reported the determination of copper, iron, lead and zinc in nickel baths, and WHITTINGTON and WILLIS [1318] analyzed various plating baths for trace elements. KOMETAIN [694] determined the concentrations of silver, cobalt, copper, iron, nickel and lead in gold baths; he further employed AAS for the determination of the composition of plating baths, diffusion data and the thickness of the electro-plated coatings, etc. KRAFT [701] described the application of AAS for determining cobalt in gold baths, silver, copper and iron in chromium, nickel and tin baths, and copper in copper cyanide baths.

VAN LOON [1253] determined iridium in a noble metal multicomponent concentrate with satisfactory precision, and EZELL [336] reported the indirect determination of chloride in liquid industrial wastes via precipitation with silver nitrate. EZELL found that the precision with AAS is better than with titrimetric methods. KUNISHI and OHNO [717] determined sulfate in a rhodium bath. They exploited the suppressing effect of sulfate on the signal for iron and its elimination by lanthanum for the determination of sulfate.

KAPETAN [624] reported the determination of cobalt and impurities in gold baths. Before the introduction of AAS, cobalt was determined polarographically, and copper, iron and lead by optical emission. The replacement of both these methods by AAS saves a considerable amount of time. The trace elements are determined directly from the undiluted bath, and for cobalt the bath is diluted 1:20.

11.8. Coal, Oil and Petrochemistry

Resulting from rising oil prices and the slow growth of nuclear energy, the prospects of a marked increase in the use of coal have risen in recent years. This fact is of significance to analytical chemistry, since coal-fired power stations, for example, release substantial mass streams of heavy metals, leading to a strain on the environment (refer to DANNECKER [2185]). The analysis of air, airborne dust, etc. has been fully discussed in section 11.5. It remains in this section to discuss the analysis of coal and its products.

For the analysis of *coal,* the solubilization of the sample is relatively difficult. This is one of the reasons why the graphite furnace technique has found ready acceptance in coal analysis, since it offers the possibility of analyzing solids and slurries directly. MILLS and BELCHER [2668] have published a good overview on the analysis of coal, coke and similar products, including sample preparation and subsequent determination of the analyte elements by atomic spectroscopy.

DITTRICH and LIESCH [306] determined chromium in coke used for making electrodes. They ashed the samples, fused the ash with potassium bisulfate, and took up the fusion mixture in 0.5 M sulfuric acid. The procedure is suitable for 0.1–2 µg/g chromium in carbon. HARTSTEIN *et al* [479] determined 10 elements in coal by the flame technique after prior digestion under pressure in fuming nitric acid (2½ hours at 150°C) and then in hydrofluoric acid (15 minutes at 150°C).

POLLOCK [2754] determined 12 trace elements in coal by the flame technique after dry ashing and dissolution in acid, and 11 elements by the graphite furnace technique. For the determination of arsenic and antimony he used low temperature ashing and the hydride technique, while for bismuth, germanium, tellurium, and tin he applied high temperature ashing followed by the hydride technique. Selenium was likewise determined by the hydride technique, but after sample digestion in an oxygen bomb. Mercury was determined by the cold vapor technique after prior digestion in an oxygen bomb. GLADFELTER and DICKERHOOF [2300] determined iron in coal after prior extraction in hydrochloric acid/nitric acid.

OWENS and GLADNEY [2720] determined beryllium in coal by the graphite furnace technique after prior digestion in nitric acid/hydrofluoric acid/sulfuric acid under the addition of perchloric acid. GLADNEY [2301] mentions the significance of the distribution of beryllium in coal purification, fluidization and gasification processes. The chemical analysis of coal takes a long time, especially because solubilization is very time consuming. High temperature ashing is relatively quick, but leads to losses of organically-bound trace elements. The author therefore applied direct solids analysis by the graphite furnace technique and found that for a sample weight of about 1 mg the scatter was less than 10%. The detection limit for beryllium was 0.005 ng and correct results were obtained both for calibration against aqueous reference solutions and for the analyte addition technique.

LANGMYHR and AADALEN [2521] also employed direct solids analysis for the determination of copper, nickel, and vanadium in coal and petroleum coke. The results also agreed with certificate values for both calibration against aqueous reference solutions and for the analyte addition technique.

GARDNER [2292] determined total mercury in coal using the cold vapor technique with an absolute detection limit of 5 ng. The pulverized coal was stood overnight with nitric acid and sulfuric acid, and then heated under reflux until the solution was light yellow. The solution was then distilled under the addition of perchloric acid and glycine, and mercury was determined from the fraction boiling between 130°C and 340°C. MURPHY [2679] used a digestion in sulfuric acid and hydrogen peroxide for the subsequent determination of mercury and increased the sensitivity by accumulation on silver wool.

The application of AAS in **petrochemistry** ranges from the analysis of crude oil through the control of additives and impurities in the most varying intermediate and end products to the determination of wear metals in used lubricating oil. AAS offers the great advantage that the samples frequently can be diluted with a suitable solvent, and then directly aspirated into the flame. The reference material must frequently be chosen with great care, otherwise false measurements can occur. For the majority of oil analyses, oil-soluble metal organic salts, mostly cyclohexane-butyrates, etc., which are offered by various manufacturers, are suitable.

LANG *et al* [2518] [2519] found, however, that for nickel and vanadium, at least, various organometallic compounds gave different absorbance values both in the air/acetylene flame and the nitrous oxide/acetylene flame. They therefore recommended that the analyte addition technique be used to prevent systematic errors. For the determination of zinc in lubricating oils and additives, LUKASIEWICZ and BUELL [2568] also found deviations of up to 10% when the same reference solutions were used for all types of oil. The analyte addition technique, or better matching of the reference solution with a similar oil, proved the remedy in this case.

As well as the use of organometallic compounds to prepare reference solutions, various authors also checked the applicability of aqueous solutions. Thus, for example, oil or petroleum/water emulsions that could be nebulized directly into the flame were recommended [2371] [2756]. HON *et al* [2390] found that aqueous inorganic reference solutions diluted with iso-butyric acid could be used for the determination of metals in lubricating oils. For the determination of metals of wear in lubricating oils, various authors [2110] [2814] [2989] mixed the oil samples with a small amount of acid and performed the determinations directly against similarly prepared references. A number of digestion procedures for oils have also been reported. These include acids [2647] [3028] and combustion [2786] [3014]. After such "mineralization" procedures it is possible to perform determinations directly against aqueous reference solutions.

The determination of nickel, which is a strong catalyst poison, and vanadium, which leads to corrosion problems, in *crude oil* was reported by various authors after the introduction of the graphite furnace technique [222] [923] [933] [2517]. While the majority introduce the sample into the graphite tube either directly or after dilution and then carry out thermal pretreatment, ESKAMANI *et al* [332] preferred ashing with a magnesium sulfonate at 650°C before introducing the sample into the graphite tube.

The determination of catalyst poisons in crude oil and distillation residues was one of the first applications of AAS in petrochemistry. BARRAS [79] [80] examined oil sump for nickel, copper, and iron and found that the results with AAS compared favorably with other analytical procedures. KERBER [633] determined nickel in *gas oil* diluted 1:10 and examined a large number of solvents for their suitability and combustion properties in the flame. He found that numerous aliphatic hydrocarbons, alcohols and ketones did not dissolve the sample completely, and that although aromatic hydrocarbons were good solvents, they burnt with yellow, sooty flames which lead to poor results. p-Xylene was finally established as a good solvent with acceptable combustion properties. TRENT and SLAVIN [1243] likewise determined nickel in gas oil and similar substances and found that in the trace range stray light must be corrected since values which are too high are otherwise obtained. These authors also employed p-xylene as solvent, but diluted 1:5, and used the analyte addition technique.

MOORE *et al* [878] used dioxan to dilute oil samples. Other authors [127] had difficulty with this solvent as it has a tendency to freeze on the nebulizer jet. ROBINSON [1037] determined sodium in gas oil and found no interferences from potassium. CAPACHO-DELGADO and MANNING [207] determined vanadium in gas oil using a nitrous oxide/acetylene flame. By employing a 1:3 dilution and the analyte addition technique, they were able to detect 0.05 mg/L vanadium in the oil.

Frequently, barium, calcium and zinc salts are employed as additives in *lubricating oils*. Since they are always present in relatively high concentrations, their determination by AAS presents no difficulties; the sample must only be correspondingly diluted. MOSTYN and CUNNINGHAM [889] determined zinc in lubricating oils with an air/acetylene flame and n-heptane as the diluting solvent.

For these analyses, SLAVIN [12] recommended MIBK as solvent and the use of a nitrous oxide/acetylene flame for the determination of calcium and barium additives. The determination is free of chemical interferences in this flame; 0.2% sodium should merely be added to sample and reference solutions to suppress ionization.

BINDING and GAWLICK [127] examined various solvents for the analysis of barium, calcium and zinc additives. n-Hexane gave an unsteady display and cyclohexane froze behind the nebulizer jet. Barium additives were either poorly soluble or insoluble in toluene and MIBK. n-Heptane proved to be suitable, but the authors recommended iso-octane as the ideal solvent for this application. Barium and calcium were determined in a nitrous oxide/acetylene flame and the sodium salt of petroleum sulfonic acid was added to suppress ionization. The authors made a detailed investigation on the emission of the nitrous oxide/acetylene flame in the regions of the calcium and barium resonance lines and found that increasing the lamp current improved the signal-to-noise ratio for the barium determination considerably. The authors employed two nebulizers to be able to work with only a single dilution of the oil samples. One nebulizer adjusted for optimum performance served for the less sensitive barium and a second nebulizer having only about $1/10^{th}$ of the aspiration performance served for the more sensitive calcium and zinc.

PETERSON and KAHN [974] diluted the samples with xylene and also determined these three elements in a nitrous oxide/acetylene flame, but reduced the sensitivity for zinc by turning the burner head through 90°. SALAMA [1078] employed less sensitive resonance lines for all three elements and was also able to work with a single dilution. According to the procedure recommended by the German Standards Association (DIN) for the determination of barium, calcium, and zinc in lubricating oils, the oil sample is diluted with petroleum and nebulized directly into the flame (air/acetylene for zinc; nitrous oxide/acetylene for barium and calcium). The determinations are performed against reference solutions containing the analyte elements as oil-soluble compounds. An alkali metal is added to all solutions to suppress ionization.

SUPP [1205] determined antimony in *additives* by dissolving them in MIBK under addition of 2-ethylhexanoic acid and aspirating directly into the flame. He found good agreement with the wet chemical method and no interferences from zinc or barium. Antimony dialkyldithiocarbamate and dialkylphosphorodithionate give oils excellent anti-wear, high pressure and anti-oxidation properties. The author also determined lead in lead dipentyldithiocarbamate, a much used additive for lubricating oils and greases, by this method [1206].

NORWITZ and GORDON [921] determined lithium in sebacate-based lubricants which, because of their high and low temperature properties, gain continuously in importance. Such lubricants typically contain 8% lithium stearate and 1.5% barium petroleum sulfonate. The sample is treated with hydrochloric acid, extracted with ether and the aqueous phase fumed off with perchloric acid and then analyzed by AAS.

MOSTYN and CUNNINGHAM [889] determined molybdenum, which is in the form of the disulfide in *greases,* after prior ashing in a quartz crucible. Unpublished investigations in the present author's laboratory have shown that molybdenum disulfide suspensions in oil can be directly determined in a nitrous oxide/acetylene flame, but that the height of the signal is strongly dependent upon the particle size. Further, after dilution and during aspiration the suspension must be magnetically stirred since separation otherwise occurs. KÄGLER [601] [602] determined lithium, sodium, and calcium in greases, and sodium in *fuel oil* and found that atomic absorption spectrometry is considerably simpler than X-ray fluorescence and flame emission spectrometry. 10–200 mg of the grease are homogenized with 4–8 mL butanol, made up to 100–200 g with 1 M hydrochloric acid and well shaken. Lithium, sodium, and calcium are determined directly from the HCl extract. Fuel oil is diluted with either toluene or a toluene/methanol mixture (8:2) and aspirated directly. For trace determinations of sodium or vanadium in light fraction fuel oils, the author's own researches have shown that this can be directly aspirated either undiluted or diluted 1:2.

KÄGLER [603] made a detailed examination of the determination of sodium in various mineral oil products with respect to instrument parameters and solvent influences.

A relatively difficult analytical problem is the determination of trace metals in light fraction fuel oils used for gas turbines since they must satisfy stringent purity requirements. Particularly vanadium must be reliably determined in concentrations below 0.1 µg/g. The graphite furnace technique is particularly suitable for these analyses; nevertheless, the furnace must be fitted with a graphite tube which prevents the oil from creeping during thermal pretreatment. Also, a temperature program with a slow rate of temperature increase is important for the removal of the matrix. When these requirements are met, 0.01 µg/g vanadium can be determined in fuel oil [1296].

The graphite furnace technique has been used for a relatively long period for the determination of trace elements in *oil* and *petroleum products.* For the determination of aluminium, beryllium, chromium, and manganese, RUNNELS *et al* [2813] found it advantageous to treat the graphite tubes with lanthanum or zirconium. The sensitivity, reproducibility, and lifetime of the tubes were improved by this treatment. ROBBINS *et al* [2794] compared the determination of beryllium using two different furnace systems; they found that a tube furnace with treated graphite tubes gave a ten times better detection limit. ROBBINS and WALKER

[2795] determined traces of cadmium in petroleum and found that next to the higher sensitivity the low sample requirement of the graphite furnace technique was an advantage, since less sample had to be ashed.

For the determination of copper in oil, KUNDU and PRÉVÔT [2514] recommended mixing oxygen with the purge gas during thermal pretreatment to accelerate destruction of the organic matrix. ROBBINS [2793] determined manganese directly in petroleum products in the range 10–300 ng/g and eliminated interferences by using the analyte addition technique. ESKAMANI et al [331] determined selenium in petrol (gasoline) and other petrochemical products after prior digestion in a PTFE autoclave.

WALKER et al [3028] used the hydride technique for the determination of selenium in petroleum and petroleum products. They digested the samples in fuming nitric acid, sulfuric acid and perchloric acid in Kjeldahl flasks and observed no interferences due to the other elements present in petroleum products. The agreement between two independent laboratories and a third laboratory using neutron activation analysis was excellent.

KNAUER and MILLIMAN [2478] determined mercury in petroleum products by the cold vapor technique. The authors examined two digestion procedures; combustion of the sample in a WICKBOLD apparatus in oxygen/hydrogen with collection of mercury in acidified permanganate solution, and an acid digestion in nitric acid/sulfuric acid. The authors found that both procedures require a great deal of experience.

MILLER et al [866] determined silicon in traces of *silicone grease* and found AAS to be ideal.

In 1961, ROBINSON [1038] reported the determination of lead in *petrol* (gasoline) by AAS and found it rapid, interference-free and selective. He diluted petrol with iso-octane to obtain a more favorable measurement range. Later, a number of authors found differences in the absorbance for tetraethyl lead and tetramethyl lead which brought difficulties, especially with mixtures. TRENT [1239] examined the determination and found that tetramethyl lead in heptane, MIBK or toluene gave a signal two to three times higher than tetraethyl lead. The length of the nebulizer aspirating capillary was also of importance. Finally, TRENT employed a 25 cm long capillary and iso-octane as diluent, since no differences between the two additives could then be observed. Tetraethyl lead required 30 seconds to give a stable signal, while the full signal height is obtained immediately with tetramethyl lead.

MOSTYN and CUNNINGHAM [889] therefore used an accurately measured aspiration time of 15 seconds to avoid waiting until equilibrium had been reached. With this procedure the authors achieved a relative standard deviation of 1.6%.

KASHIKI et al [628] found that tetramethyl lead, tetraethyl lead, and lead alkyls could be determined by calibration against standard lead alkyls when 3 mg iodine were added to 1 mL petrol sample and diluted to 50 mL with MIBK before aspiration. Later, the same authors described this procedure for the graphite furnace technique to facilitate determinations at the lower concentration range (0.02–0.1 mg Pb/L). Also in this case, all the lead compounds examined exhibited the same sensitivity after the addition of iodine [627].

McCORRISTON and RITCHIE [2650] observed that there was no difference between the various alkyl lead compounds and that the signal immediately reached full sensitivity when a direct injection burner was used instead of a premix burner. LUKASIEWICZ et al [2567] employed a nitrous oxide/hydrogen flame for the determination of lead in petrol and found that the composition of the petrol had virtually no influence. POLO-DIEZ *et al* [2756] nebulized petrol/water emulsions directly into the flame and observed no influences due to the various matrices or alkyl lead compounds.

KOLB *et al* [692] used a combination of gas chromatography/atomic absorption spectrometry for the separation and element specific determination of both the lead additives in petrol. In this case AAS served as a highly specific detector for gas chromatography by only responding to the metallic elements.

Later, SEGAR [1108] [1110] described this procedure for the graphite furnace technique. He conducted the eluate from the gas chromatograph directly into a continuously heated graphite tube at 1500–1700°C and was able to achieve good separation of five lead alkyls. In the following years, a number of reports were published describing the combination of gas chromatography [2165] [2796] or liquid chromatography [2096] [2191] with the graphite furnace technique for speciation of alkyl lead compounds in petrol. BYE *et al* [2121] made a comparison between the flame and the furnace techniques as specific detectors for tetraalkyl lead compounds. The results from both techniques are generally in good agreement and the recovery is similar; the graphite furnace technique has the advantage that it is two orders of magnitude more sensitive than the flame technique.

MOORE *et al* [878] determined nickel and copper and MOSTYN and CUNNINGHAM [889] determined zinc in petrol by aspirating it undiluted. BARTELS and WILSON [84] examined jet aviation fuels for manganese, which is added to these fuels as methyl-cyclopentadienyl-manganesetricarbonyl to eliminate the typical vapor trails. When undiluted fuel is aspirated, additives in the range 0.02–0.3 weight percent can be directly determined by AAS. Higher concentrations can be determined after dilution.

Within the last few years the determination of *wear metals in lubricating oils* has become an important application of atomic absorption spectrometry. The rapidity and high precision of these determinations permit continuous monitoring of engines, gears and other machine parts. Frequently, the element specific examination of a lubricating oil enables early recognition and often localization of incipient damage to be made.

Usually the lubricating oil is diluted 1:10 with MIBK and aspirated directly. Calibration is mainly against oil-soluble standards (cyclohexanebutyrates, etc.) or against naphthenates [853]. SPRAGUE and SLAVIN [1138] [1173] developed an automated procedure for the determination of Fe, Ni, Cr, Pb, Cu, Ag, Sn, Mg and Al in aircraft lubricating oils. With a sample frequency of 14 seconds a reproducibility of 3% could be achieved. BURROWS *et al* [187] compared atomic absorption spectrometry and colorimetry for the examination of machine oils, which were ashed before the determination. The agreement was good. Similar results could also be obtained by AAS without ashing, however. This is because suspended metallic components in lubricating oils, as well as dissolved ones, can be determined by the flame technique so long as the particles are not too large.

KRISS and BARTELS [83] [709] thoroughly examined the effectiveness of the nebulizer/burner system for the determination of suspended particles in oil. For iron particles of 3.0 μm and 1.2 μm, the recovery was only in the order of a few percent. Nevertheless, the authors state that such particle sizes are hardly present in genuine oil. Further, they found that the effectiveness of atomization is independent of the flame and that problems with the reproducibility come from the oil and not from the flame [710]. The authors have reported a procedure with which the large particles can also be determined by AAS. The oil is diluted with 14 mL MIBK and 5 mL of a mixture consisting of 80% methanol, 5% water and 15% concentrated hydrochloric acid. For the 1.2 μm particles the recovery is 100% after having stood an hour, but for the 3 μm particles the waiting time is 16 hours [709]. Based on this procedure, various authors reported acid treatment of lubricating oils. SABA and EISENTRAUT [2814] diluted aircraft lubricating oil with 4-methylpentan-2-one and shook the sample with hydrochloric acid/hydrofluoric acid for ten seconds for the subsequent determination of titanium. The entire analysis only requires 1–2 minutes and the attainable detection limit is 0.03 μg/g titanium. The authors found no differences between reference solutions prepared from organo-titanium compounds and from oil and metallic titanium treated in the same way as the samples. For the determination of chromium, copper, iron, magnesium, molybdenum, and vanadium in used aircraft lubricating oils, TUELL *et al* [2989] treated the samples with a mixture of hydrofluoric acid, hydrochloric acid and nitric acid in the ratio 2:3:3. These authors also found no differences between reference solutions prepared from organo-metallic compounds and those treated in the same way as the samples. BROWN *et al* [2110] determined aluminium, copper, iron, magnesium, molybdenum, nickel, thallium, and tin in aircraft lubricating oils after prior treatment with a little hydrofluoric acid and *aqua regia* in an ultrasonic bath at 65°C. They found that the procedure is very rapid and independent of the particle size. The recovery was 97–103% and the relative standard deviation, in dependence on the analyte element and its concentration, was 4–10%.

With lubricating oil analyses, it is usually not the absolute determination of the metal concentration that is of importance, but only typical increases in the relative concentration which can indicate incipient damage or damage that has already occurred. A widely distributed monitoring network for a heavy contractors plant, for example, has been developed according to this system and has proved of great use [1279]. KAHN *et al* [615] have also shown that no statistical deviation can be observed between AAS and AES for oil analyses.

The analysis of wear metals in lubricating oil by the graphite furnace technique has been described by two groups of workers [165] [233]. However, this does not appear to be a typical application since it is neither a trace nor a micro determination and the flame is in all cases more rapid.

MAY and PRESLEY [2646] investigated the analysis of *coastal crude oil residues* and the identification of the origin. A trace element determination is the most suitable; the ratio Ni:V is typical for many crudes. Neutron activation analysis and graphite furnace AAS are the most suitable analytical methods; graphite furnace AAS is naturally much more economical and simple. The crude oil sample can be dissolved and analyzed for iron, nickel, and vanadium directly in the graphite furnace. Better agreement can be obtained with neutron activation analysis, however, after wet digestion of the sample in nitric acid/perchloric acid/sulfuric acid in the ratio 5:2:2 [2647].

MENZ and CONRADI [2662] used the determination of trace elements in crude oils as a method for their geochemical classification. They found that xylene is not a particularly good diluent for the graphite furnace technique, since layering and precipitation occur after a longer standing time. A mixture of 80% 100–140°C petroleum and 20% toluene proved to be more suitable; no separation occurred even after several day's standing. The authors also established that reproducible results could only be obtained with automatic sample dispensing into the graphite furnace.

LIEU and WOO [2548] reported that the existing pressure in an oil well was only sufficient to permit low recovery of oil. Pumping water into oil wells is the most important and widely used secondary method for recovery of oil. The water pumped in under pressure must not contain more than a minimum of suspended matter, otherwise the pumps will block and too high a pressure will be generated. The authors determined a large number of elements in water and in suspended particles by both the flame and graphite furnace techniques; the recovery was 91% or better. These procedures were faster and required less sample material than other techniques.

MANNING *et al* [2616] determined arsenic in synthetic *process waters* used for crude oil recovery by four different techniques. The production of fuels from oil shale and bituminous sand has gained substantial importance in recent years. Particularly for the on-site treatment of oil shale there is at least as much water as oil, so that there is a considerable waste removal problem. The process waters frequently contain high and fluctuating organic and inorganic contents, including toxic metals such as arsenic. Higher arsenic contents can be determined directly by the flame technique. For a determination in the graphite furnace the sample is diluted and nickel is added as matrix modifier; the analyte addition technique is used. For a determination by the hydride technique, the sample must be treated with nitric acid and sulfuric acid, otherwise low results are obtained. The results obtained by the three AAS techniques and by ICP-AES were in excellent agreement with each other and also with the mean result obtained from an interlaboratory analysis.

11.9. Glass, Ceramics, Cement

With glass, and more especially with ceramic materials, it is not the actual analysis which is the problem, but solubilization of the samples. Glass is normally dissolved in hydrofluoric acid and the silicon removed since it is usually not of interest. The subsequent analysis by AAS is very simple since the total salt concentration is low. Ceramics are frequently dissolved in hydrofluoric acid, also often in combination with perchloric acid, etc. Occasionally small PTFE autoclaves can be of use [964] [2039]. Failing this, alkaline fusions must be used. However, under certain circumstances, these can make the AAS analysis difficult.

ADAMS and PASSMORE [13] [14] determined magnesium, calcium and strontium in *glass* and *ceramics* and also the alkali metals in ceramics. The chemical interferences which are expected for the alkaline-earth elements in the air/acetylene flame were eliminated with a mixture of lanthanum and EDTA. JONES [584] determined Li, Na, K, Ca, Mg, Ba, Cu, Mn, Fe, Co, Ni, Pb and Zn in various glasses and found good agreement with both certificate values and photometric results. PASSMORE and ADAMS determined iron, zinc [959] and copper [960] in numerous glasses and found good agreement with colorimetric results. SLAVIN *et al* [1148] determined arsenic in glass, and BOBER and MILLS [132] analyzed lead glasses for lead in good agreement with chemical and spectrochemical procedures. PRAGER and GRAVES [2759] determined lead in glasses at the secondary resonance line at 216.4 nm after prior dissolution in hydrofluoric acid/nitric acid. There was no significant difference between the results obtained by flame AAS, X-ray fluorescence, and gravimetry; the AAS method was faster and more simple. ANDREW [2039] determined boron in glass after dissolving a larger sample quantity in a minimum of hydrofluoric acid under the addition of hydrochloric acid and nitric acid in a closed PTFE vessel. In this way a high boron concentration in the solution was obtained. Excess hydrofluoric acid was removed by adding quartz powder. WISE and SOLSKY [3074] determined aluminium in glass, under the addition of 1000 mg/L calcium and some perchloric acid as spectrochemical buffers, in good agreement with certificate values. CHARLIER and MAYENCE [227] determined chromium, cobalt, copper, manganese and nickel in glasses against matched reference solutions without interferences after fuming off with hydrofluoric acid/perchloric acid and taking the residue up in 5% hydrochloric acid. The detection limits were around 0.5% metal in glass. HANSEN and HALL [2349] determined 14 elements in glasses. They prepared sample and reference solutions in a common matrix containing 0.4% cesium chloride and 2% hydrochloric acid, and thereby eliminated numerous analytical errors and the need for special sample preparation techniques or the analyte addition technique. BURDO and WISE [2118] fused glasses with sodium carbonate/sodium borate and took up the residue in acidified molybdate solution for the subsequent determination of silicon. A silico-molybdate complex is formed that prevents polymerization of the silicon and has a buffering action against possible interferences. The authors achieved an accuracy of 0.2–0.3% absolute for samples containing 14–94% SiO_2 and obtained a substantial saving of time compared to gravimetric methods.

Various authors have examined the application of the graphite furnace technique for glass analyses. FULLER [392] dissolved high purity quartz in hydrofluoric acid, volatilized off the silicon tetrafluoride and the acid in the graphite tube and determined iron and copper down to 0.01 µg/g in good agreement with colorimetry. He also determined chromium, cobalt, copper, iron, manganese, nickel and vanadium in glasses by fuming them off to dryness with hydrofluoric acid/perchloric acid, dissolving the residue in warm perchloric acid and extracting the trace elements with MIBK for subsequent determination in the graphite furnace [396]. WOOLLEY [1357] determined iron and copper in the high purity glasses required for the production of fiber optics in the graphite furnace. The sample preparation is of particular interest: The sample is pulverized by an electro-hydraulic pressure wave which is generated in the water contained in the sample by a spark. The sample is then dissolved in a PTFE autoclave and the silicon is separated. FULLER [2278] determined cobalt, copper, iron, and nickel in glasses using the graphite furnace technique after prior digestion of the sample in hydrofluoric acid/nitric acid and extraction of the analyte elements as their o-phenanthroline complexes in MIBK. Lead is determined after a digestion in hydrofluoric acid/sulfuric acid/hydrochloric acid by separation as the sulfate, dissolution in hydrochloric acid and extraction as the diethyldithiocarbamate complex in MIBK. SIEMER and WEI [2859] determined lead in glasses directly by solids analysis in a graphite furnace. Finely ground glass is mixed with graphite powder and heated in the furnace to 1350°C. The lead is atomized, while the majority of the matrix remains in the furnace. Using peak area integration, these authors found excellent agreement with certificate values compared to aqueous reference solutions. HERMANN [2370] determined selenium in glass by the hydride technique after prior digestion in hydrofluoric acid, hydrochloric acid and nitric acid.

A determination carried out frequently nowadays in connection with environmental control is the testing of *glazed ceramic surfaces* for their release of toxic heavy metals. The vessel is allowed to stand in 4% acetic acid for 24 hours at $22 \pm 2\,°C$ and the metals of interest then determined. KRINITZ and FRANCO [706] compared induction periods of half an hour and 24 hours and found an asymptotic increase in the metal concentration. In an interdisciplinary test, relative standard deviations of 4–5% were obtained for lead and cadmium.

MUNTZ [896] determined yttrium in *hafnium* and *zirconium oxides* in good agreement with other methods after the samples had been dissolved in hydrofluoric acid/sulfuric acid. WISE and SOLSKY [3073] determined yttrium in zirconium oxide after a fusion with lithium metaborate and boric acid, and dissolution in hot perchloric acid. To attain maximum sensitivity they added potassium and EDTA as buffers. Sulfate, phosphate, calcium, and magnesium did not cause any interferences; any silicon that might be present can be fumed off with hydrofluoric acid before the fusion.

HAVEZOV and TAMNEV [2354] determined beryllium in *β-aluminium oxide ceramics* in the nitrous oxide/acetylene flame. The sample is taken up rapidly at $300\,°C$ in strong phosphoric acid; this reduces chemical interferences and at the same time facilitates atomization.

VAN LOON [1252] has described a procedure for the determination of aluminium in materials with high silicon contents. He found good agreement with certificate values. BELCHER [97] dissolved tungsten carbide in orthophosphoric acid/hydrochloric acid under addition of nitric acid and determined iron in these solutions in good agreement with X-ray fluorescence analysis.

The determination of secondary constituents and trace elements in **cement** was carried out at an early stage by AAS. CAPACHO-DELGADO and MANNING [209] determined Fe, Mg, Mn, K, Li, Na, Sr, Al and Ti in cement against aqueous reference solutions in which the calcium concentration had been matched. For Mg, Sr, Al and Ti the nitrous oxide/acetylene flame was employed. MONTAGUT-BUSCAS *et al* [875] determined cobalt, after extraction with MIBK, in cement to be used for the construction of reactors.

LANGMYHR and PAUS [2528] described a digestion procedure for cements, clinker, raw materials, and silicious limestone using hydrochloric acid under the addition of hydrofluoric acid. For the determination of Al, Ca, Fe, Mg, Mn, K, Na, Si, and Sr in cements and the associated raw materials, CROW and CONOLLY [2177] proposed a lithium metaborate fusion. The residue is taken up in nitric acid in the normal manner. The fusion takes one-and-a-half hours per sample.

While the routine accuracy of AAS of approximately 1–3% is sufficient for the determination of the secondary constituents in cement, calcium must be determined with the greatest accuracy. CROW, HIME and CONNOLLY [244] found after a comprehensive study that it is possible to achieve an accuracy of 0.2% by this method when using standard cements for calibration. The authors give complete methodology for the analysis of all the important elements in cement by AAS.

TENOUTASSE [1222] determined ten elements including calcium in cement after digestion in hydrofluoric acid/hydrochloric acid in a PTFE autoclave. No foreign ions (except boric acid) were added and all elements could be determined from one solution. FIERENS and

DEGRE [367] have described an automatic procedure for the determination of 11 elements in cement, raw materials and clinker products.

11.10. Plastics, Textiles, Paper

SLAVIN [1141] reported the determination of manganese in *plastics* after prior wet ashing with concentrated sulfuric acid. To avoid matrix influences, he employed the analyte addition technique. FARMER [339] determined iron in nylon 66; he dry ashed the sample, dissolved the residue in hydrochloric acid and extracted with APDC/MIBK. DRUCKMAN [311] analyzed polypropylene for titanium, aluminium and iron and found that atomic absorption spectrometry had decisive advantages over earlier procedures. A sample of the polymer is ashed at 800 °C and taken up in molten sodium carbonate. This is dissolved in 1.5 M sulfuric acid and aspirated directly into the flame. This procedure can also be extended to other plastics (polyethylene, polystyrene, PVC) and other metals such as magnesium, copper and zinc. PARALUSZ [950] determined silicon in various polymers and made a comprehensive study of various references and hollow cathode lamps; the most varied analytical methods were also compared and atomic absorption spectrometry proved to be especially suitable.

OLIVIER [924] employed AAS for the determination of 20 metallic elements in a multitude of polymers and he gives detailed methodology for the sample preparation. Soluble polymers are dissolved in a solvent suitable for atomic absorption spectrometry, such as MIBK, DMF [925], cyclohexanone, formic acid, etc., and aspirated directly. Insoluble polymers are digested in sulfuric acid/hydrogen peroxide; volatile metals are precipitated twice with subsequent enrichment. The author emphasizes that atomic absorption spectrometry permits rapid and accurate determination of metals in polymers both in the μg/g and % ranges.

PRICE [1002] determined antimony, lead, iron, cobalt, copper, manganese, tin and zinc in raw materials and in polymers such as rayon, nylon and polyesters directly in the flame rapidly and accurately after dissolution in sulfuric acid/hydrogen peroxide. KERBER determined gold in polyester fibers [634], aluminium and iron in polypropylene, and copper and iron in fluorinated hydrocarbon polymers [637] by directly introducing the samples into the graphite furnace; he obtained good agreement with flame AAS, colorimetry and emission spectrometry.

TRACHMAN *et al* [2983] determined tin in packing materials made of plastic to which organotin compounds are added as stabilizers. The plastic sample is extracted with 3% acetic acid/8% heptane/50% ethanol and the solution used directly after acidification for a determination in the graphite furnace.

KORENAGA [2495] used the hydride technique to determine traces of arsenic in acrylic fibers containing antimony. After digestion in nitric acid/perchloric acid/sulfuric acid, arsenic and antimony are reduced to the trivalent state. In contrast to antimony(III), arsenic(III) can be extracted from hydrochloric acid and sulfuric acid solution. The determination is performed by the hydride technique after back extraction into the aqueous phase.

DELLA MONICA and MCDOWELL [291] determined chromium in *leather* and found that compared to alkaline fusions, AAS is reliable and brings a considerable saving in time.

HARTLEY and INGLIS [478] described the determination of aluminium in *wool* by AAS. The sample is dissolved in boiling hydrochloric acid and the extract is aspirated directly into the flame. The detection limit is 0.02% aluminium in the wool. SIMONIAN [1128] determined copper in textiles using the same procedure and found good agreement with colorimetry and electrolysis. OLIVIER [924] decomposed wool in 5% sodium hydroxide solution or in concentrated hydrochloric acid, and cotton in 72% sulfuric acid and determined numerous elements in the resulting solutions.

BETHGE and RÅDESTRÖM [118] employed AAS to determine copper, iron, manganese, calcium, magnesium and sodium in *cellulose* and found the accuracy and precision were comparable to those obtained by conventional methods. If a large number of samples are to be determined, or several metals in one sample, atomic absorption spectrometry saves a great deal of time. The samples were dry ashed, taken up in 6 M hydrochloric acid and aspirated directly. ANT-WUORINEN and VISAPÄÄ [59] have described a procedure for the determination of Na, K, Mg, Ca, Fe, Cu and Zn in various cellulose samples. The influence of various methods of pretreatment was examined in detail and the procedure was checked with X-ray fluorescence analysis. PERSSON [2745] determined aluminium in pulp by the graphite furnace technique.

PARALUSZ [950] investigated the content of organic silicon compounds in *paper* by extracting them with petrol, evaporating the solvent, taking up the residue in a minimum of MIBK and aspirating directly. LANGMYHR *et al* [732] determined cadmium, copper, lead and manganese in paper and paper pulp directly in the graphite furnace by introducing the solid or slurried sample and employing a background corrector. KERBER *et al* [637] found good agreement between the flame and graphite furnace techniques for the determination of iron, copper, manganese and silicon in paper. WATKINS *et al* [3035] determined lead in paper used for packing foodstuffs after prior wet and dry ashing. In some papers the authors found substantial lead contents. SIMON *et al* [2868] attempted to identify and classify various types of paper on the basis of their trace element contents. They found that the majority of papers could be identified by six characteristics. The contents of antimony, chromium, cobalt, copper, and manganese were the most important. Lead exhibits a high correlation with chromium, cobalt and manganese. The concentrations of cadmium, iron and magnesium are often markedly scattered within a single paper sort.

11.11. Radioactive Materials, Pharmaceuticals, and Miscellaneous Industrial Products

Atomic absorption spectrometry is suitable for numerous analytical problems in industry, and because of its simplicity, sensitivity and specificity it can be universally applied. Frequently, however, these numerous "secondary tasks" of AAS are either not reported or only reported in passing.

Elements with a high effective neutron capture efficiency retard radiative processes. Thus their presence in *nuclear fuels* and other materials, in which they could influence the process, is extremely undesirable, even in very low concentrations (10^{-5} to 10^{-7}%). The necessity of analyses in the microtrace and nanotrace ranges for numerous elements is evident.

Optical emission methods are frequently used, since many elements can be determined simultaneously or in rapid sequence. X-ray fluorescence analysis is also an important method.

Atomic absorption spectrometry also found application in the analysis of uranium at an early stage. It was found that the air/acetylene flame was not hot enough to atomize the uranium matrix so that a separation was necessary. The nitrous oxide/acetylene flame is more suitable, but even here the matrix is frequently separated to reduce problems due to the activity of the nebulized solution. TAKAHASHI and URUNO [1215] determined magnesium after separating the bulk of the uranium. HUMPHREY [541] performed the same determination, but without separating the uranium, and employed the analyte addition technique to avoid interferences. JURISK [600] determined aluminium, iron, and nickel in good agreement with colorimetric procedures. SCARBOROUGH [1082] analyzed uranium alloys for molybdenum, ruthenium, palladium, and rhodium. There were no mutual interferences among these four elements when the corresponding quantity of uranium was added to the reference solutions. SHEPHERD and JOHNSON [2850] separated uranium on an ion exchange column prior to the determination of iron and nickel; they performed the determinations both directly and by the analyte addition technique. The latter technique permits the determination of smaller contents and is very precise.

KORKISCH *et al* [2496] [2500] determined copper and lead in uranium oxide and yellow cake samples after prior separation of uranium and other ions on a strongly basic anion exchanger. WALKER and VITA [1280] determined Al, Ca, Cd, Co, Cr, Cu, Fe, Mg, Mn, Na, Ni, K, Pb, and Zn in uranium and uranium compounds. The sample is dissolved in 6–8 M nitric acid and the uranium is selectively separated with tributylphosphate. The aqueous layer is evaporated to dryness, the residue taken up in 0.2 M hydrochloric acid and the solution used directly for determinations by the flame technique. AAS is about as precise as colorimetry and much more precise than AES. PAGLIAI and POZZI [946] separated uranium by reversed phase chromatography (solid phase: KEL-F; stationary phase: TBP) and determined 10 elements by AAS without any problems.

PICHOTIN and CHASSEUR [975] determined magnesium, calcium, lithium, and potassium in plutonium metal after prior extraction of the plutonium with tri-n-octylphosphine and dibutylphosphoric acid. FRANZ *et al* [2258] investigated the composition of fuel rods containing thorium oxide and additions of aluminium oxide and silicon dioxide. They determined aluminium by AAS after prior pressure digestion in nitric acid/hydrofluoric acid.

The graphite furnace technique is particularly useful in the analysis of nuclear raw materials and recycling products. The ability to obtain high sensitivity with very small sample quantities meets not only the task at hand but also to a large extent the problems of analyzing radioactive materials. The reduction of the sample volume to a few microliters opens up interesting possibilities since less safety shielding is required. The inert purge gas with a relatively slow, regulated flowrate also aids in controlling radioactive radiation. PATEL *et al* [2733] determined cadmium, chromium, and cobalt in uranium directly in the graphite furnace without prior separation; the reference solutions contained the same quantity of uranium. BAGLIANO *et al* [2049] used the graphite furnace technique to determine chromium, cobalt, iron, and manganese in uranium oxide in the range 0.05–0.005%. The authors found that for samples with a uranium content of up to 0.8 g/L reference solutions made up in nitric acid could be used; nevertheless, it is preferable to match the uranium content in the reference solutions. PAGE *et al* [2725] determined cadmium and lithium in uranium oxide by dispensing the pulverized U_3O_8 samples directly into the graphite tube in quantities of about 1 mg. The working range was 0.05-2 µg/g for cadmium and 1-25 µg/g for lithium; the precision was better than 10%. The authors prepared the references by mixing the high purity oxides. HENN *et al* [2368] used the graphite furnace technique in a laboratory for recycling nuclear fuels principally for the analysis of uranium nitrate and plutonium nitrate solutions. An essential requirement for recycling is to maintain the purity of the solutions against numerous inactive and radioactive contaminants. Due to the high availability, the good determination limits, and a favorable analysis time, the authors preferred this technique to other analytical methods.

BAUDIN [2064] has published a good overview on the analysis of nuclear materials such as uranium, plutonium, and zirconium, and also on steels, cooling water, sodium coolant, etc. and discusses the special problems associated with radioactive materials.

A relatively frequent application of AAS is the analysis of traces in *ultrapure chemicals*. AGAZZI [20] determined zinc in hydrogen peroxide to which it is added as a stabilizer. The range of interest is 2-50 mg/L in hydrogen peroxide. LUNDY and WATJE [772] examined fuming nitric acid for Al, Fe, Cr, Ni and Zn. The sample was evaporated almost to dryness and then taken up in water. The results with AAS were in satisfactory agreement with the time-consuming colorimetric method. YANAGISAWA *et al* [1360] have described a procedure for the determination of calcium in phosphoric acid and its salts. Calcium is complexed with hydroxyquinoline and extracted with 3-methylbutan-1-ol. The extract is aspirated directly and the calcium concentration determined by the analyte addition technique. BEDROSIAN and LERNER [94] determined Na, K, Ca, Mg, Fe, Mn, Cu, Ni and Cr in elemental boron and obtained a relative standard deviation of 4% or better. Atomic absorption spectrometry compared well with emission techniques and the relative error of the latter was frequently the greater. SPITZER [1146] analyzed zinc oxide for Cu, Pb, Fe, Mn, Ni and Ca. The results agreed well with those for chemical analysis. After the reference solutions had been matched to the zinc concentration (40 g/L), all the elements could be determined free of interferences. This method is reliable and convenient.

CRISP *et al* [2175] determined non-ionic surfactants in the concentration range 0.05-2 mg/L indirectly by extracting with potassium tetracyanatozincate(II) in 1,2-dichlorobenzene and determining the zinc by AAS. JUNG and CLARKE [599] determined cadmium in fungicides at the secondary spectral line at 326.1 nm using the flame technique.

For the determination of selenium in technical sulfuric acid, HOLEN *et al* [2388] performed electrochemical preconcentration on a platinum wire. Atomization was in an argon/hydrogen diffusion flame with simultaneous heating of the wire. KATO [2459] determined silicon in silicon carbide after a prior fusion with sodium hydroxide and sodium peroxide and dissolution in dilute hydrochloric acid. ALDER and BUCKLOW [2017] determined chromium, copper, and zinc in carbon fibers. These metals influence the quality of the fibers since they interfere in the pyrolysis process. For the analysis, the carbon fibers are pulverized, mixed with nitric acid and sodium hexametaphosphate, and the solution nebulized directly into the flame. The authors found that neither sample quantity nor element concentration was a problem, and they attained good accuracy and precision with this method. If it is necessary to determine lower element concentrations, the method can also be transferred to the graphite furnace technique.

JACKWERTH an WILLMER [2430] used the flame injection technique to determine numerous elements in ultrapure barium and strontium salts. The main constituents, barium and strontium, are precipitated as their nitrates by fuming off with nitric acid. The solubility of barium nitrate and strontium nitrate in the resulting concentrated nitric acid solution is so low that the preconcentrated trace elements can be determined without interferences from the remaining solution.

HARRINGTON [470] determined ruthenium in hydrated ruthenium chloride and titanium in tetra-orthobutyltitanate in good agreement with gravimetric methods. OLIVIER [926] examined caustic soda for mercury using a flameless procedure. REVERSAT [1033] determined calcium, chromium, iron, copper, magnesium and nickel in single crystals of lithium hydroxide using the graphite furnace technique. MÜLLER-VOGT and WENDL [2676] determined silicon in single crystals of $LiNbO_3$ by the graphite furnace technique after prior fusion with sodium carbonate/potassium carbonate and extraction as the molybdate complex with pentanol. For the determination of copper, iron, and zinc the authors dissolved the crystals in hydrofluoric acid/nitric acid and nebulized directly into the air/acetylene flame. KOMETANI [2491] determined ultratraces of chromium, cobalt, copper, iron, manganese, nickel, and zinc in silicon tetrachloride used for the manufacture of glass fiber. The silicon tetrachloride is volatilized at low temperature, the residue dissolved in ultrapure hydrofluoric acid and analyzed by the graphite furnace technique. LANGMYHR and HAKEDAL [2524] determined cadmium, copper, iron, lead, manganese, and zinc in technical and analytical grade hydrofluoric acid and sulfuric acid, and also in ammonia solution. The acids were fumed off as far as possible in vitreous carbon boats remote from the graphite furnace and then the boat introduced into the furnace for atomization. Ammonia solution was dispensed directly into the furnace. KOWALSKA and KEDZIORA [2505] determined lead in high purity graphite by direct solids analysis in the graphite furnace. The precision of the determination is strongly dependent on the homogeneity of the sample; the detection limit was under 0.01 µg/g.

MAY and GREENLAND [2643] used the hydride technique to determine arsenic in phosphoric acid. They found that there were no interferences when arsenic was reduced to arsenic(III) with potassium iodide. Even antimony in the expected concentration range had no influence on the determination.

The analysis of *catalysts* on support materials without prior dissolution has been reported by various authors. KASHIKI and OSHIMA [626] aspirated a suspension of aluminium oxide catalysts in methanol directly into the flame and determined cobalt and molybdenum in good agreement with a digestion method. COUDERT and VERGNAUD [240] determined platinum on activated charcoal or aluminium oxide by introducing the powder evenly into the gas mixture with a screw feed, so guaranteeing pneumatic transport to the flame. JANOUŠKOVÁ *et al* [570] also determined platinum on aluminium oxide, but carried out the determination in the graphite furnace and only found slight interferences. HARRINGTON and BRAMSTEDT [2351] examined coatings of antimony, tantalum and tin oxides which are used together with noble metal oxides on a titanium substrate as electrochemical catalysts. The oxides are removed from the substrate by an alkali fusion and taken into solution by acid treatment. Antimony and tin are determined directly and tantalum after extraction in MIBK by the flame technique. KALLMANN and BLUMBERG [2450] determined platinum metal in catalysts for automobile exhaust gases and LABRECQUE [2518] determined cobalt, molybdenum, and nickel in petrochemical catalysts. SUPP and GIBBS [2948] determined selenium and tellurium in accelerators used for the vulcanization of natural and synthetic rubber in an air/acetylene flame after prior digestion in hydrochloric acid/nitric acid.

DALRYMPLE and KENNER [265] determined calcium in various *pharmaceutical preparations* and found good agreement between AAS and complexometric titrations. Zinc in crystalline insulin and insulin preparations was determined by SPIELHOLTZ and TORALBALLA [1162] by AAS and various other methods. AAS is more rapid and simpler for the same or better precision of the results.

LEATON [736] reported the application of AAS for the analysis of pharmaceutical products. As, Ca, Co, Cu, Fe, Mg, K and Hg were determined in solutions, powders and tablets with good accuracy. To reduce sample preparation to a minimum, the single substances were only dissolved in acid and aspirated directly into the flame.

KARAKHANIS and ANFINSEN [625] determined aluminium in pharmaceuticals by dissolving or leaching the samples with concentrated hydrochloric acid and aspirating directly after dilution. The method was free of interferences and in good agreement with complicated procedures. TARLIN and BATSCHELDER [1216] determined iron in tablets and capsules in agreement with colorimetric, volumetric and gravimetric procedures, and MOODY and TAYLOR [877] determined zinc in numerous pharmaceutical preparations with good accuracy and precision. SUZUKI *et al* [2950] used the flame technique to determine cobalt in vitamin B12 and proteins, while PECK [2736] used the graphite furnace technique for the same analysis. The samples were dissolved in hydrochloric acid and correspondingly diluted to obtain the optimum measurement range for cobalt. SZYDLOWSKI and VIANZON [2955] determined barium in solutions for infusion and injection directly in the graphite furnace after corresponding dilution. Barium can enter such solutions from glass containers or rubber closures. For the determination, stabilizers such as sodium chloride or sodium bisulfide must be matched in the reference solutions. THOMPSON and ALLEN [2964] described the determination of selenium in selenium tablets or foodstuffs containing selenium that are offered in the U.S.A. as supplements to the diet. The samples are suspended in acidified emulsions of mixed, non-ionic wetting agents and dispensed directly into the graphite furnace under the addition of nickel as matrix modifier.

A good review on the determination of 15 elements in the most varied pharmaceutical products has been published by SMITH [1157].

SEARLE *et al* [1103] employed AAS to determine lead in old *paints* and obtained a standard deviation of 0.05% lead with this method, while with gravimetry the standard deviation was 0.3% with a five times higher sample requirement. HENN [502] used the DELVES system for the same determination and found good agreement between this procedure and prior ashing.

MOTEN [890] determined chromium with a detection limit of 5 ppm in triphenylmethane paint additives. BRANDT [156] has reported the universal applicability of AAS to the analysis of raw materials, pigments, polymers and additives, and EIDER [323] found that the main advantage of AAS is the ability to directly analyze organometallic compounds and complex mixtures without prior separation. This signifies a considerable saving of time compared to wet chemical analyses for the same precision. For the determination of total chromium in chromium oxide paints NOGA [2694] proposed a digestion with potassium permanganate in sulfuric acid in a PTFE autoclave. Frequently an alkaline fusion is otherwise necessary for such samples. The determination of chromium is performed in the nitrous oxide/acetylene flame.

The determination of lead in paints by AAS is widely reported. PORTER [2757] digested latex paints in nitric acid and performed the determination in the air/acetylene flame. LAU and LI [2537] and also MITCHELL *et al* [2670] employed the DELVES system for this determination and found good agreement with conventional procedures. CASTELLANI *et al* [2136] performed the determination by the graphite furnace technique and obtained the same results for direct solids analysis and a digestion procedure. VANDEBERG *et al* [2996] reported an interlaboratory investigation on the determination of lead in paints by AAS.

FULLER [392] employed the graphite furnace for the determination of aluminium, iron, copper and manganese in titanium dioxide pigments. The sample was dissolved in hydrofluoric acid and introduced directly into the graphite tube; the agreement with flame AAS and X-ray fluorescence analysis was very good. Later, FULLER [2281] investigated the introduction of aqueous slurries of titanium dioxide pigments directly into the flame or the graphite furnace for the determination of copper, iron, lead, and manganese. An advantage of the samples is that all pigments have a uniform grain size of 10 ± 0.3 µm. The injection method was used for flame analysis since the nebulizer would otherwise clog too quickly. The results were in good agreement with other procedures. The direct procedure offers the greatest speed and simplicity, and also the lowest blank values. The determination limits for the four elements were 2 µg/g for the flame technique and 0.1 µg/g for the graphite furnace technique.

MATZ [841] determined Pb, Mg, Fe, K, Na, Zn, Cu and Cd in *rubber* seals by AAS; the samples were dry ashed at 650 °C, the ash dissolved in 6 M hydrochloric acid and then aspirated directly into the flame. The determination is rapid, simple and interference free.

DEILY [287] employed atomic absorption spectrometry for the determination of aluminium in *aluminium trialkyls*. The sample is hydrolysed in a 1% solution of hydrochloric acid in ethyleneglycol-monomethylether and the solution is aspirated directly into a nitrous oxide/acetylene flame.

PERKINS [971] determined sodium in *phosphorescent compounds* by dissolving about 250 mg in hydrochloric acid and diluting to 10 mL. AAS was superior to AES for this application.

SCOTT [1102] examined yttrium phosphors for europium by either dissolving the sample in hydrochloric acid or fusing with potassium carbonate. Europium-activated yttrium orthovanadate, yttrium oxide and yttrium oxysulfide find application in the color television industry and a rapid and simple method had to be found. Atomic absorption spectrometry completely meets all the requirements.

WOOLLEY [1356] has described the application of atomic absorption spectrometry in the *electronics industry* for the analysis of various raw materials and finished products. Synthetic quartz crystals and semiconductor materials, glasses, fluorescent lamp materials, plating baths, magnetic materials, thermistors, alloys and plastics are all analyzed.

Bibliography

Monographs and Books

[1] Analytical Methods for Atomic Absorption Spectrophotometry. Perkin-Elmer Corporation, Norwalk, Connecticut 1982.
[2] Angino, E. E., and Billings, G. K.: Atomic Absorption Spectrometry in Geology. Elsevier Publishing Company, Amsterdam/London/New York 1967.
[3] Atomic Absorption Spectroscopy. ASTM Special Technical Publication 443, Amer. Soc. Test. Mater., Philadelphia 1969.
[4] Christian, G. D., and Feldman, F. J.: Atomic Absorption Spectroscopy: Applications in Agriculture, Biology, and Medicine. Interscience Publishers, New York/London/Sydney 1970.
[5] Dean, J. A., and Rains, T. C.: Flame Emission and Atomic Absorption Spectrometry, Vol. 1, Theory. Marcel Dekker Inc., New York/London 1969.
[6] Dean, J. A., and Rains, T. C.: Flame Emission and Atomic Absorption Spectrometry, Vol. 2, Components and Techniques. Marcel Dekker Inc., New York 1971.
[7] Elwell, W. T., and Gidley, J. A. F.: Atomic Absorption Spectrophotometry, 2nd edition. Pergamon Press, Oxford/London/Edinburgh/New York/Toronto/Sydney/Paris/Braunschweig 1966.
[8] L'vov, B. V.: Atomic Absorption Spectrochemical Analysis. English translation: Adam Hilger, London 1970.
[9] Price, W. J.: Analytical Atomic Absorption Spectrometry. Heyden & Son Ltd., London/New York/Rheine 1972.
[10] Ramirez-Muñoz, J.: Atomic Absorption Spectroscopy. Elsevier Publishing Company, Amsterdam/London/New York 1968.
[11] Rubeška, I., and Moldan, B.: Atomic Absorption Spectrophotometry. English translation: Iliffe Books Ltd., London 1969.
[12] Slavin, W.: Atomic Absorption Spectroscopy. Interscience Publishers, New York/London/Sydney 1968.

Miscellaneous References

[13] Adams, P. B.: in Standard Methods of Chemical Analysis, Vol. III B, 6th edition. Edited by F. J. Welcher, Van Nostrand, Princeton, N. J. 1966.
[14] Adams, P. B., and Passmore, W. O.: Anal. Chem. **38,** 630 (1966).
[15] Adda, J., Rousselet, F., and Mocquat, C.: Rev. Latière Française **231,** 227 (1966).
[16] Adriaenssens, E., and Knoop, P.: Anal. Chim. Acta **68,** 37 (1973).
[17] Adriaenssens, E., and Verbeek, F.: At. Absorption Newslett. **12,** 57 (1973).
[18] Adriaenssens, E., and Verbeek, F.: At. Absorption Newslett. **13,** 41 (1974).
[19] Adrian, W. J.: At. Absorption Newslett. **10,** 96 (1971).
[20] Agazzi, E. J.: Anal. Chem. **37,** 364 (1965).
[21] Agazzi, E. J.: Anal. Chem. **39,** 233 (1967).
[22] Aguilar, A., Jaime, S., and Lachica, M.: 3rd CISAFA, Paris 1971, p. 545.
[23] Aitzetmüller, K.: 3rd CISAFA, Paris 1971, p. 487.
[24] Ajemian, R. S., and Whitman, N. E.: Ann. Ind. Hyg. Assoc. J. **30,** 52 (1969).
[25] Al Ani, M. J., Dagnall, R. M., and West, T. S.: Analyst **92,** 597 (1967).
[26] Aldous, K. M., Browner, R. F., Dagnall, R. M., and West, T. S.: Anal. Chem. **42,** 939 (1970).
[27] Aldous, K. M., Dagnall, R. M., and West, T. S.: Anal. Chim. Acta **44,** 457 (1969).
[28] Aldous, K. M., Mitchell, D. G., and Jackson, K. W.: 4th ICAS, Toronto 1973, p. 32.
[29] Alkemade, C. Th. J.: Anal. Chem. **38,** 1252 (1966).
[30] Alkemade, C. Th. J.: Appl. Optics **7,** 1261 (1968).
[31] Alkemade, C. Th. J.: in [5], p. 101.

[32] Alkemade, C. Th. J., Hoomayers, H. P., and Lijnse, P. L.: Spectrochim. Acta **27 B**, 149 (1972).
[33] Alkemade, C. Th. J., and Milatz, J. M. W.: Appl. Sci. Res. Sect. B **4**, 289 (1955).
[34] Alkemade, C. Th. J., and Milatz, J. M. W.: J. Opt. Soc. Am. **45**, 583 (1955).
[35] Alkemade, C. Th. J., and Voorhuis, M. H.: Z. Anal. Chem. **163**, 91 (1958).
[36] Allan, J. E.: Analyst **83**, 466 (1958).
[37] Allan, J. E.: Spectrochim. Acta **10**, 800 (1959).
[38] Allan, J. E.: Nature **187**, 1110 (1960).
[39] Allan, J. E.: Spectrochim. Acta **17**, 459 (1961).
[40] Allan, J. E.: Spectrochim. Acta **17**, 467 (1961).
[41] Allan, J. E.: Analyst **86**, 530 (1961).
[42] Allan, J. E.: Spectrochim. Acta **18**, 259 (1962).
[43] Allan, J. E.: 4th Australian Spectrosc. Conf. 1963.
[44] Althaus, E.: Analysentechn. Berichte **3**, 1964 (Bodenseewerk Perkin-Elmer & Co. GmbH, Über-lingen).
[45] Althaus, E.: N. Jahrb. Miner. Mh. **9**, 259 (1966).
[46] Amos, M. D., Bennett, P. A., Brodie, K. G., Lung, P. W. Y., and Matoušek, J. P.: Anal. Chem. **43**, 211 (1971).
[47] Amos, M. D., and Thomas, P. E.: Anal. Chim. Acta **32**, 139 (1965).
[48] Amos, M. D., and Willis, J. B.: Spectrochim. Acta **22**, 1325 (1966).
[49] Anderson, J.: At. Absorption Newslett. **11**, 88 (1972).
[50] Anderson, J.: At. Absorption Newslett. **13**, 31 (1974).
[51] Anderson, R. G., Maines, I. S., and West, T. S.: IAASC, Sheffield 1969, A 7.
[52] Anderson, R. G., Maines, I. S., and West, T. S.: Int. Sympos. Microchem., Graz 1970, Vol. B, p. 17.
[53] Anderson, W. N., Broughton, P. M. G., Dawson, J. B., and Fisher, G. W.: Clin. Chim. Acta **50**, 129 (1974).
[54] Andrew, T. R., and Nichols, P. N. R.: Analyst **87**, 25 (1962).
[55] Andrew, T. R., and Nichols, P. N. R.: Analyst **92**, 156 (1967).
[56] Angino, E. E., and Billings, G. K.: Geochim. Cosmochim. Acta **30**, 153 (1966).
[57] Antonacopoulos, N.: Chem. Mikrobiol. Technol. Lebensm. **3**, 8 (1974).
[58] Antonetti, A., Amiel, C., Mulher, R., and Rousselet, F.: 3rd CISAFA, Paris 1971, p. 589.
[59] Ant-Wuorinen, O., and Visapää, A.: Paperi Ja Puu **48**, 649 (1966).
[60] Arendt, D. H.: At. Absorption Newslett. **11**, 63 (1972).
[61] Armstrong, F. A. J., and Uthe, J. F.: At. Absorption Newslett. **10**, 101 (1971).
[62] Atkinson, R. J., Chapman, G. D., and Krause, L.: J. Opt. Soc. Am. **55**, 1269 (1965).
[63] Atwell, M. G., and Hebert, J. Y.: Appl Spectrosc. **23**, 480 (1969).
[64] Bachler, W.: Die Bodenkultur **20**, 17 (1969).
[65] Bader, H., and Brandenberger, H.: At. Absorption Newslett. **7**, 1 (1968).
[66] Bailey, N. T., and Wood, S. J.: Anal. Chim. Acta **69**, 19 (1974).
[67] Baker, M. R., and Vallee, B. L.: J. Opt. Soc. Am. **45**, 775 (1955).
[68] Baker, R. A., Hartshorne, D. J., and Wilshire, A. G.: At. Absorption Newslett. **8**, 21 (1969).
[69] Balazs, N. D. H., Pole, D. J., and Masarei, J. R.: Clin. Chim. Acta **40**, 213 (1972).
[70] Ballard, A. E., Stewart, D. W., Kamm, W. O., and Zuehlke, C. W.: Anal. Chem. **26**, 921 (1954).
[71] Ballard, A. E., and Thornton, C. W. D.: Ind. Eng. Chem. Anal. Ed. **13**, 893 (1941).
[72] Barnard, W. M., and Fishman, M. J.: At. Absorption Newslett. **12**, 118 (1973).
[73] Barnett, W. B.: Anal. Chem. **44**, 695 (1972).
[74] Barnett, W. B.: At. Absorption Newslet. **12**, 142 (1973).
[75] Barnett, W. B., and Kahn, H. L.: At. Absorption Newslett. **8**, 21 (1969).
[76] Barnett, W. B., and Kahn, H. L.: 3rd CISAFA, Paris, 1971.
[77] Barnett, W. B., and Kahn, H. L.: Clin. Chem. **18**, 923 (1972).
[78] Barnett, W. B., Kahn, H. L., and Manning, D. C.: At. Absorption Newslett. **8**, 46 (1969).
[79] Barras, R. C.: Jarrell-Ash Newslett. **13**, (1962).
[80] Barras, R. C., and Helwig, J. D.: Am. Petrol. Inst. Abstr. Refining Lit. **5** (1963).
[81] Barringer, A. R.: Inst. Mining & Metal. Bull. 714, **75 B**, 120 (1966).
[82] Bartels, H.: At. Absorption Newslett. **6**, 132 (1967).

[83] Bartels, T. T., and Slater, M. P.: At. Absorption Newslett. **9**, 75 (1970).

[84] Bartels, T. T., and Wilson, C. E.: At. Absorption Newslett. **8**, 3 (1969).

[85] Barthel, W. F., Smrek, A. L., Angel, G. P., Liddle, J. A., Landrigan, P. J., Gehlbach, S. H., and Chisolm, J. J.: J. Ass. Offic. Agr. Chem. **56**, 1252 (1973).

[86] Barton, H. N., Johnson, A. J., and Shepherd, G. A.: Dow Chem. Corp. RFP-**978** (1967).

[87] Baudin, G., Bonne, R., Chaput, M., and Feve, L.: 3rd CISAFA, Paris 1971, p. 853.

[88] Baudin, G., Chaput, M., and Feve, L.: Spectrochim. Acta **26 B**, 425 (1971).

[89] Bazhov, A. S.: J. Anal. Chem. USSR **23**, 1446 (1968).

[90] Beamish, F. E., and Van Loon, J. C.: Recent Advances in the Analytical Chemistry of the Noble Metals. Pergamon Press, Oxford/New York/Toronto/Sydney/Braunschweig 1972.

[91] Beaty, R. D.: At. Absorption Newslett. **13**, 38 (1974).

[92] Beaty, R. D.: At. Absorption Newslett. **13**, 44 (1974).

[93] Beccaluva, L., and Venturelli, G.: At. Absorption Newslett. **10**, 50 (1971).

[94] Bedrosian, A. J., and Lerner, M. W.: Anal. Chem. **40**, 1104 (1968).

[95] Beer, A.: Ann. Physik **86**, 78 (1852).

[96] Bek, F., Janoušková, J., and Moldan, B.: At. Absorption Newslett. **13**, 47 (1974).

[97] Belcher, C. B.: Anal. Chim. Acta **29**, 340 (1963).

[98] Belcher, C. B., and Bray, H. M.: Anal. Chim. Acta **26**, 322 (1962).

[99] Belcher, C. B., and Brooks, K. A.: Anal. Chim. Acta **29**, 202 (1963).

[100] Belcher, C. B., and Kinson, K.: Anal. Chim. Acta **30**, 483 (1964).

[101] Belcher, R., Dagnall, R. M., and West, T. S.: Talanta **11**, 1257 (1964).

[102] Bell, G. F.: At. Absorption Newslett. **5**, 73 (1966).

[103] Bell, G. F.: At. Absorption Newslett. **6**, 18 (1967).

[104] Bell, H. F.: Anal. Chem. **45**, 2296 (1973).

[105] Bell, W. E., Bloom, A. L., and Lynch, J.: Rev. Sci. Int. **32**, 688 (1961).

[106] Belt, C. B.: At. Absorption Newslett. **3**, 23 (1964).

[107] Belt, C. B.: Econ. Geol. **59**, 240 (1964).

[108] Bentley, E. M., and Lee, G. F.: Environ. Sci. Technol. **1**, 721 (1967).

[109] Berman, E.: At. Absorption Newslett. **3**, 111 (1964).

[110] Berman, E.: At. Absorption Newslett. **4**, 296 (1965).

[111] Berman, E.: At. Absorption Newslett. **6**, 57 (1967).

[112] Berman, E.: 4th ICAS, Toronto 1973.

[113] Berman, E., Valavanis, V., and Dubin, A.: Clin. Chem. **14**, 239 (1968).

[114] Bermejo-Martinez, F., and Baluja-Santos, C.: CSI XVII, Florence 1973, p. 577.

[115] Bernas, B.: Anal. Chem. **40**, 1682 (1968).

[116] Berry, J. W., Chappell, D. G., and Barnes, R. B.: Ind. Eng. Chem. Anal. Ed. **18**, 19 (1946).

[117] Berry, W. L., and Johnson, C. M.: Appl. Spectrosc. **20**, 209 (1966).

[118] Bethge, P. O., and Rådeström, R.: Svk. Papperstidn. **69**, 772 (1966).

[119] Beyer, M.: At. Absorption Newslett. **4**, 212 (1965).

[120] Beyer, M.: At. Absorption Newslett. **8**, 23 (1969).

[121] Bianchi, C. P.: Cell Calcium. Butterworth, London 1969.

[122] Billings, G. K.: At. Absorption Newslett. **4**, 312 (1965).

[123] Billings, G. K.: At. Absorption Newslett. **4**, 357 (1965).

[124] Billings, G. K., and Adams, J. A. S.: At. Absorption Newslett. **3**, 65 (1964).

[125] Billings, G. K., and Harriss, R. C.: Texas J. Sci. **17**, 129 (1965).

[126] Billings, G. K., Ragland, P. C., and Harriss, R. C.: Texas J. Sci. **18**, 277 (1966).

[127] Binding, U., and Gawlick, H.: Analysentechn. Berichte **18** (1969).

[128] Binnewies, M., and Schäfer, H.: Z. Anorg. Allg. Chem. **395**, 77 (1973).

[129] Bishop, J. R., Dunn, J. W., and Hill, W. H.: IAASC, Sheffield, 1969, B 6.

[130] Black, L. T.: J. Am. Oil Chem. Soc. **47**, 313 (1970).

[131] Boar, P. L., and Ingram, L. K.: Analyst **95**, 124 (1970).

[132] Bober, A., and Mills, A. L.: Appl. Spectrosc. **22**, 62 (1968).

[133] Bode, H., and Fabian, H.: Z. Anal. Chem. **162**, 328 (1958).

[134] Bode, H., and Fabian, H.: Z. Anal. Chem. **163**, 187 (1958).

[135] Böhmer, M., Auer, E., and Bartels, H.: Ärztl. Lab. **13,** 258 (1967).
[136] Boettner, E. A., and Grunder, F. I.: in Trace Inorganics in Water. Amer. Chem. Soc. Publ., Advan. Chem. Ser. **73,** 236 (1968).
[137] Bokowski, D. L.: At. Absorption Newslett. **6,** 97 (1967).
[138] Bokowski, D. L.: Am. Indust. Hygn. Assn. J. **29,** 474 (1968).
[139] Boling, A. E.: Spectrochim. Acta **22,** 425 (1966).
[140] Boling, F. A.: Spectrochim. Acta **23 B,** 495 (1968).
[141] Bond, A. M.: Anal. Chem. **42,** 932 (1970).
[142] Bond, A. M., and O'Donnell, T. A.: Anal. Chem. **40,** 560 (1968).
[143] Bond, A. M., and Willis, J. B.: Anal. Chem. **40,** 2087 (1968).
[144] Boppel, B.: Z. Anal. Chem. **268,** 114 (1974).
[145] Bosch, H., Büchel, E., and Lohau, K. H.: Analysentechn. Berichte **20** (1970).
[146] Bouchard, A.: At. Absorption Newslett. **12,** 115 (1973).
[147] Bouguer, P.: Essai d'Optique sur la gradation de la lumière. Paris 1729.
[148] Bowman, J. A.: Anal. Chim. Acta **42,** 285 (1968).
[149] Bowman, J. A., Sullivan, J. V., and Walsh, A.: Spectrochim. Acta **22,** 205 (1966).
[150] Bowman, J. A., and Willis, J. B.: Anal. Chem. **39,** 1210 (1967).
[151] Brady, D. V., Montalvo, J. G., Glowacki, G., and Pisciotta, A.: Anal. Chim. Acta **70,** 448 (1974).
[152] Brandenberger, H., and Bader, H.: Helv. Chim. Acta **50,** 1409 (1967).
[153] Brandenberger, H., and Bader, H.: At. Absorption Newslett. **6,** 101 (1967).
[154] Brandenberger, H., and Bader, H.: At. Absorption Newslett. **7,** 53 (1968).
[155] Brandenberger, H., and Bader, H.: Chimia **21,** 597 (1967).
[156] Brandt, J.: Am. Paint J. **57,** 28 (1973).
[157] Bratzel, M. P., and Chakrabarti, Ch. L.: Anal. Chim. Acta **63,** 1 (1973).
[158] Bratzel, M. P., Chakrabarti, Ch. L., and Sturgeon, R. E.: Anal. Chem. **44,** 372 (1972).
[159] Bratzel, M. P., Dagnall, R. M., and Winefordner, J. D.: Anal. Chem. **41,** 713 (1969).
[160] Bratzel, M. P., Dagnall, R. M., and Winefordner, J. D.: Anal. Chem. **41,** 1527 (1969).
[161] Bratzel, M. P., Winefordner, J. D., and Dagnall, R. M.: IAASC, Sheffield 1969, C3.
[162] Brewer, P. G., Spencer, D. W., and Smith, C. L.: in [3], 70 (1969).
[163] Brimhall, W. H.: Anal. Chem. **41,** 1349 (1969).
[164] Brimhall, W. H., and Adams, J.: Geochim. Cosmochim. Acta **33,** 1308 (1969).
[165] Brodie, K. G., and Matoušek, J. P.: Anal. Chem. **43,** 1557 (1971).
[166] Brooks, I. B., Luster, G. A., and Easterly, D. G.: At. Absorption Newslett. **9,** 93 (1970).
[167] Brooks, R. R., Presley, P. J., and Kaplan, I. R.: Anal. Chim. Acta **38,** 321 (1967).
[168] Brooks, R. R., Presley, P. J., and Kaplan, I. R.: Talanta **14,** 809 (1967).
[169] Browner, R. F., Dagnall, R. M., and West, T. S.: Anal. Chim. Acta **45,** 163 (1969).
[170] Browner, R. F., Dagnall, R. M., and West, T. S.: IAASC, Sheffield 1969, C4.
[171] Browner, R. F., Dagnall, R. M., and West, T. S.: Talanta **16,** 75 (1969).
[172] Browner, R. F., Patel, B. M., and Winefordner, J. D.: CSI XVII, Florence 1973, p. 227.
[173] Brunelle, R. L., Hoffman, C. M., Snow, K. B., and Pro, M. J.: J. Ass. Offic. Agr. Chem. **52,** 911 (1969).
[174] Buchanan, J. R., and Muraoka, T. T.: At. Absorption Newslett. **3,** 79 (1964).
[175] Buck, L.: Chimie Analytique **47,** 10 (1965).
[176] Buell, B. E.: in [5], p. 267.
[177] Bukenberger, U., Lodemann, C. K. W., and Loeschke, J.: Oberrhein. Geol. Abh. **21,** 43 (1972).
[178] Buneaux, F., and Fabiani, P.: Ann. Biol. Clin. **28,** 273 (1970).
[179] Burke, K. E.: Anal. Chem. **42,** 1536 (1970).
[180] Burke, K. E., and Albright, C. H.: J. Ass. Offic. Agr. Chem. **53,** 531 (1970).
[181] Burnham, C. D., Moore, C. E., Kowalski, T., and Krasniewski, J.: Appl. Spectrosc. **24,** 4 (1970).
[182] Burrell, D. C.: At. Absorption Newslett. **4,** 309 (1965).
[183] Burrell, D. C.: At. Absorption Newslett. **4,** 328 (1965).
[184] Burrell, D. C.: Anal. Chim. Acta **38,** 447 (1967).
[185] Burrell, D. C.: 3rd CISAFA, Paris 1971, p. 409.
[186] Burrell, D. C., and Wood, G. G.: Anal. Chim. Acta **48,** 45 (1969).

[187] Burrows, J. A., Heerdt, J. C., and Willis, J. B.: Anal. Chem. **37**, 579 (1965).

[188] Busch, K. W., and Morrison, G. H.: Anal. Chem. **45**, 712 A (1973).

[189] Busch, K. W., Morrison, G. H., and Feldman, M.: 4th ICAS, Toronto 1973, p. 78.

[190] Butler, L. R. P.: J. S. Afr. Inst. Mining Met. **62**, 786 (1962).

[191] Butler, L. R. P.: At. Absorption Newslett. **5**, 99 (1966).

[192] Butler, L. R. P., and Brink, D.: S. Afr. Ind. Chem. **17**, 152 (1963).

[193] Butler, L. R. P., and Brink, D.: IAASC, Sheffield 1969, A 3.

[194] Butler, L. R. P., and Brink, D.: in [6], p. 21.

[195] Butler, L. R. P., and Mathews, P. M.: Anal. Chim. Acta **36**, 319 (1966).

[196] Butler, L. R. P., and Schroeder, W. W.: IAASC, Sheffield 1969, A 6.

[197] Butler, L. R. P., and Strasheim, A.: Spectrochim. Acta **21**, 1207 (1965).

[198] Butler, L. R. P., Strasheim, A., Strelow, F. W. E., Mathews, P., and Feast, E. C.: CSI XII, Exeter 1965.

[199] Buttgereit, G.: Analysentechn. Berichte **26** (1972).

[200] Buttgereit, G.: Arbeitsmed. Sozialmed. Arbeitshyg. **10**, 286 (1972).

[201] Buttgereit, G.: Z. Anal. Chem. **267**, 81 (1973).

[202] Calkins, R. C.: Appl. Spectrosc. **20**, 146 (1966).

[203] Cameron, A. G., and Hackett, D. R.: J. Sci. Fd. Agric. **21**, 535 (1970).

[204] Campbell, D. E.: XVI Anachem Conf., Detroit 1968.

[205] Campbell, W. C., Ottaway, J. M., and Strong, B.: Talanta **21**, 837 (1974).

[206] Capacho-Delgado, L., and Manning, D. C.: At. Absorption Newslett. **4**, 317 (1965).

[207] Capacho-Delgado, L., and Manning, D. C.: At. Absorption Newslett. **5**, 1 (1966).

[208] Capacho-Delgado, L., and Manning, D. C.: Spectrochim Acta **22**, 1505 (1966).

[209] Capacho-Delgado, L., and Manning, D. C.: Analyst **92**, 553 (1967).

[210] Capacho-Delgado, L., and Sprague, S.: At. Absorption Newslett. **4**, 363 (1965).

[211] Carlson, G. G., and Van Loon, J. C.: At. Absorption Newslett. **9**, 90 (1970).

[212] Carper, J. L.: At. Absorption Newslett. **9**, 48 (1970).

[213] Cartwright, J. S., and Manning, D. C.: At. Absorption Newslett. **5**, 114 (1966).

[214] Cartwright, J. S., Sebens, C., and Manning, D. C.: At. Absorption Newslett. **5**, 91 (1966).

[215] Cartwright, J. S., Sebens, C., and Slavin, W.: At. Absorption Newslett. **5**, 22 (1966).

[216] Cernik, A. A.: 4th ICAS, Toronto 1973, p. 5.

[217] Cernik, A. A.: At. Absorption Newslett. **22**, 42 (1973).

[218] Cernik, A. A.: At. Absorption Newslett. **12**, 163 (1973).

[219] Cernik, A. A., and Sayers, M. H. P.: Brit. J. Industr. Med. **28**, 392 (1971).

[220] Chakrabarti, Ch. L.: Appl. Spectrosc. **21**, 160 (1967).

[221] Chakrabarti, Ch. L.: Anal. Chim. Acta **42**, 379 (1968).

[222] Chakrabarti, Ch. L., and Hall, G.: Spectrosc. Letters **6**, 385 (1973).

[223] Chakrabarti, Ch. L., Lyles, G. R., and Dowling, F. B.: Anal. Chim. Acta **29**, 489 (1963).

[224] Chakrabarti, Ch. L., Robinson, J. W., and West, P. W.: Anal. Chim. Acta **34**, 269 (1966).

[225] Chao, T. T., Fishman, M. J., and Ball, J. W.: Anal. Chim. Acta **47**, 189 (1969).

[226] Chapman-Andresen, C., and Christensen, S.: Compt. Rend. Trav. Lab. Carlsberg **38**, No. 2, 19 (1970).

[227] Charlier, H., and Mayence, R.: 3rd CISAFA, Paris 1971, p. 799.

[228] Chau, Y-K., Sim, S-S., and Wong, Y-H.: Anal. Chim. Acta **43**, 13 (1968).

[229] Cheek, D. B., Graystone, J. E., Willis, J. B., and Holt, A. B.: Clin. Sci. **23**, 169 (1962).

[230] Cheek, D. B., Powell, G. K., Reba, R., and Feldman, M.: Bull. Johns Hopkins Hosp. **118**, 338 (1966).

[231] Christensen, S.: At. Absorption Newslett. **11**, 51 (1972).

[232] Christian, G. D., and Feldman, F. J.: Anal. Chim. Acta **40**, 173 (1968).

[233] Chuang, F. S., and Winefordner, J. D.: Appl. Spectrosc. **28**, 215 (1974).

[234] Cioni, R., Innocenti, F., and Mazzuoli, R.: At. Absorption Newslett. **11**, 102 (1972).

[235] Clark, D., Dagnall, R. M., and West, T. S.: Anal. Chim. Acta **63**, 11 (1973).

[236] Clarke, W. E.: IAASC, Sheffield 1969, E 7.

[237] Clinton, O. E.: Spectrochim. Acta **16**, 985 (1960).

[238] Cobb, W. D., and Harrison, T. S.: Joint Symp. acc. Meth. Anal. Maj. Const., London 1970.

[239] Collin, J., Sire, J., and Merklen, J.: 3rd CISAFA, Paris 1971, p. 713.

[240] Coudert, M., and Vergnaud, J. M.: 3rd CISAFA, Paris 1971, p. 757.

[241] Cowley, T. G., Fassel, V. A., and Kniseley, R. N.: Spectrochim. Acta **23 B**, 771 (1968).

[242] Cragin, J. H., and Herron, M. M.: At. Absorption Newslett. **12**, 37 (1973).

[243] Creeser, M. S., and West, T. S.: Spectrochim. Acta **25 B**, 61 (1970).

[244] Crow, R. F., Hime, W. G., and Connolly, J. D.: J. Res. Develop. Lab. Portland Cement Ass. **9**, No. 2, 60 (1967).

[245] Cruz, R., and Van Loon, J. C.: Anal. Chim. Acta **72**, 231 (1974).

[246] Curnow, D. H., Gutteridge, D. H., and Horgan, E. D.: At. Absorption Newslett. **7**, 45 (1968).

[247] Curry, A. S., and Knott, A. R.: Clin. Chim. Acta **30**, 115 (1970).

[248] Curry, A. S., Read, J. F., and Knott, A. R.: Analyst **94**, 744 (1969).

[249] Dagnall, R. M.: II ČS. Conf. Flame Spectrosc., Zvikov 1973.

[250] Dagnall, R. M., Kirkbright, G. F., West, T. S., and Wood, R.: Anal. Chim. Acta **47**, 407 (1969).

[251] Dagnall, R. M., Taylor, M. R. G., and West, T. S.: Spectrosc. Letters **1**, 397 (1968).

[252] Dagnall, R. M., Thompson, K. C., and West, T. S.: Anal. Chim. Acta **36**, 269 (1966).

[253] Dagnall, R. M., Thompson, K. C., and West, T. S.: At. Absorption Newslett. **6**, 117 (1967).

[254] Dagnall, R. M., Thompson, K. C., and West, T. S.: Analyst **92**, 506 (1967).

[255] Dagnall, R. M., Thompson, K. C., and West, T. S.: Talanta **14**, 551 (1967).

[256] Dagnall, R. M., Thompson, K. C., and West, T. S.: Talanta **14**, 557 (1967).

[257] Dagnall, R. M., Thompson, K. C., and West, T. S.: Talanta **14**, 1151 (1967).

[258] Dagnall, R. M., Thompson, K. C., and West, T. S.: Talanta **14**, 1467 (1967).

[259] Dagnall, R. M., Thompson, K. C., and West, T. S.: Analyst **93**, 72 (1968).

[260] Dagnall, R. M., Thompson, K. C., and West, T. S.: Analyst **93**, 153 (1968).

[261] Dagnall, R. M., Thompson, K. C., and West, T. S.: Analyst **94**, 643 (1969).

[262] Dagnall, R. M., and West, T. S.: Talanta **11**, 1553 (1964).

[263] Dagnall, R. M., and West, T. S.: Appl. Opt. **7**, 1287 (1968).

[264] Dall'Aglio, M., Gragnani, R., and Visibelli, D.: Rend. Soc. Ital. Mineral. Petrol. Cia. Soc. Mines. Italy **24**, 188 (1968).

[265] Dalrymple, B. A., and Kenner, C. T.: J. Pharm. Sci. **58**, 604 (1969).

[266] Dalton, E. F., and Malanoski, A. J.: At. Absorption Newslett. **10**, 92 (1971).

[267] David, D. J.: Analyst **83**, 655 (1958).

[268] David, D. J.: Analyst **84**, 536 (1959).

[269] David, D. J.: Analyst **85**, 459 (1960).

[270] David, D. J.: Nature **187**, 1109 (1960).

[271] David, D. J.: Analyst **86**, 730 (1961).

[272] David, D. J.: Analyst **87**, 576 (1962).

[273] David, D. J.: At. Absorption Newslett. **1**, 45 (1962).

[274] David, D. J.: in Modern Methods of Plant Analysis, Vol. 5. Springer Verlag, Berlin/Göttingen/ Heidelberg 1962.

[275] David, D. J.: Analyst **89**, 747 (1964).

[276] David, D. J.: Spectrochim. Acta **20**, 1185 (1964).

[277] Davidson, I. W. F., and Secrest, W. L.: Anal. Chem. **44**, 1808 (1972).

[278] Dawson, J. B.: Proc. Soc. Anal. Chem. **1970**, 195.

[279] Dawson, J. B., and Heaton, F. W.: Biochem. J. **80**, 99 (1961).

[280] Dawson, J. B., and Walker, B. E.: Clin. Chim. Acta **26**, 465 (1969).

[281] Dean, J. A.: Analyst **85**, 621 (1960).

[282] Dean, J. A., and Carnes, W. J.: Anal. Chem. **34**, 192 (1962).

[283] Decker, C. F., Aras, A., and Decker, L. R.: Anal. Biochem. **8** 344 (1964).

[284] de Galan, L.: Spectrosc. Letters **3**, 123 (1970).

[285] de Galan, L., and Winefordner, J. D.: J. Quant. Spectrosc. Radiat. Transfer **7**, 251 (1967).

[286] de Galan, L., and Samaey, G. F.: Spectrochim. Acta **24 B**, 679 (1969).

[287] Deily, J. R.: At. Absorption Newslett. **5**, 119 (1966).

[288] Deily, J. R.: At. Absorption Newslett. **6**, 65 (1967).

[289] Delaughter, B.: At. Absorption Newslett. **4**, 273 (1965).
[290] Delfino, J. J., Bortleson, G. C., and Lee, G. F.: Environm. Sci. Technol. **3**, 1189 (1969).
[291] Della Monica, E. S., and McDowell, P. E.: J. Am. Leather Chem. Ass. **1971**, 21.
[292] Delves, H. T.: Analyst **95**, 431 (1970).
[293] Delves, H. T.: At. Absorption Newslett. **12**, 50 (1973).
[294] Delves, H. T., Shepherd, G., and Vinter, P.: Analyst **96**, 260 (1971).
[295] Den Tonkelaar, W. A. M., and Bikker, M. A.: Atmospher. Environm. **1971**, 353.
[296] Derschau, H. A. V., and Prugger, H.: Z. Anal. Chem. **247**, 8 (1969).
[297] Descube, J., Roques, N., Rousselet, F., and Girard, M. L.: Ann. Biol. Clin. **35**, 1011 (1967).
[298] Devoto, G.: Boll. Soc. Ital. Biol. Sper. **44**, 425 (1968).
[299] Devoto, G.: Boll. Soc. Ital. Biol. Sper. **44**, 1253 (1968).
[300] Dickinson, G. W., and Fassel, V. A.: Anal. Chem. **41**, 1021 (1969).
[301] Dickson, R. E., and Johnson, C. M.: Appl. Spectrosc. **20**, 214 (1966).
[302] Dietz, A. A., and Rubinstein, H. M.: Clin. Chem. **15**, 787 (1969).
[303] Dietz, A. A., and Rubinstein, H. M.: Ann. Rheum. Dis. **32**, 124 (1973).
[304] Dinnin, J. I., and Helz, A. W.: Anal. Chem. **39**, 1489 (1967).
[305] Dinnin, J. I.: Anal. Chem. **39**, 1491 (1967).
[306] Dittrich, K., and Liesch, G.: Talanta **20**, 691 (1973).
[307] Dodson, R., Forney, F., and Swift, E.: J. Am. Chem. Soc. **58**, 2573 (1936).
[308] Döllefeld, E.: Ärztl. Lab. **17**, 369 (1971).
[309] Donega, H. M., and Burgess, T. E.: Anal. Chem. **42**, 1521 (1970).
[310] Dowling, F. B., Chakrabarti, Ch. L., and Lyles, G. R.: Anal. Chim. Acta **28**, 392 (1963).
[311] Druckman, D.: At. Absorption Newslett. **6**, 113 (1967).
[312] D'Silva, A. P., Kniseley, R. N., and Fassel, V. A.: Anal. Chem. **36**, 1287 (1964).
[313] Dunckley, J. V.: Clin. Chem. **17**, 992 (1971).
[314] Dunckley, J. V.: Clin. Chem. **19**, 1081 (1973).
[315] Dunk, R., Mostyn, R. A., and Hoare, H. C.: At. Absorption Newslett. **8**, 79 (1969).
[316] Dyck, R.: At. Absorption Newslett. **4**, 170 (1965).
[317] Ediger, R. D.: At. Absorption Newslett. **12**, 151 (1973).
[318] Ediger, R. D., and Coleman, R. L.: At. Absorption Newslett. **11**, 33 (1972).
[319] Ediger, R. D., and Coleman, R. L.: At. Absorption Newslett. **12**, 3 (1973).
[320] Ediger, R. D., Peterson, G. E., and Kerber, J. D.: At. Absorption Newslett. **13**, 61 (1974).
[321] Edmunds, W. M., Giddings, D. R., and Morgan-Jones, M.: At. Absorption Newslett. **12**, 45 (1973).
[322] Edwards, H. W.: Anal. Chem. **41**, 1172 (1969).
[323] Eider, N. G.. Appl. Spectrosc. **25**, 313 (1971).
[324] Elliott, E. V., and Stever, K. R.: At. Absorption Newslett. **12**, 60 (1973).
[325] Ellis, D. W., and Demers, D. R.: Anal. Chem. **38**, 1943 (1966).
[326] Ellis, D. W., and Demers, D. R.: in Trace Inorganics in Water. Amer. Chem. Soc. Publ., Advan. Chem. Ser. **73**, 326 (1968).
[327] Elrod, B. B., and Ezell, J. B.: At. Absorption Newslett. **8**, 129 (1969).
[328] Elwell, W. T., and Gidley, J. A. F.: Anal. Chim. Acta **24**, 71 (1961).
[329] Emmermann, R., and Lücke, W.: Z. Anal. Chem. **248**, 325 (1969).
[330] Erinc, G., and Mangee, R. J.: Anal. Chim. Acta **31**, 197 (1964).
[331] Eskamani, A., Strecker, H. A., and Vigler, M. S.: 4th ICAS, Toronto 1973, p. 136.
[332] Eskamani, A., Vigler, M. S., Strecker, H. A., and Anthony, N. R.: 4th ICAS, Toronto 1973, p. 138.
[333] Evenson, M. A., and Pendergast, D. D.: Clin. Chem. **20**, 163 (1974).
[334] Everson, R. T., and Schrenk, W. G.: 4th ICAS, Toronto 1973, p. 55.
[335] Ezell, J. B.: At. Absorption Newslett. **5**, 122 (1966).
[336] Ezell, J. B.: At. Absorption Newslett. **6**, 84 (1967).
[337] Fabricand, B. P., Sawyer, R. R., Ungar, S. G., and Adler, S.: Geochim. Cosmochim. Acta **26**, 1023 (1962).
[338] Fallgatter, K., Svoboda, V., and Winefordner, J. D.: Appl. Spectrosc. **25**, 347 (1971).
[339] Farmer, M. H.: At. Absorption Newslett. **6**, 121 (1967).

[340] Farrar, B.: At. Absorption Newslett. **4,** 325 (1965).

[341] Farrar, B.: At. Absorption Newslett. **5,** 62 (1966).

[342] Fassel, V. A.: IAASC, Sheffield 1969, P7.

[343] Fassel, V. A.: CSI XVI, Heidelberg 1971, p. 63.

[344] Fassel, V. A., and Becker, D. A.: Anal. Chem. **41,** 1522 (1969).

[345] Fassel, V. A., and Golightly, D. W.: Anal. Chem. **39,** 466 (1967).

[346] Fassel, V. A., and Mossotti, V. G.: Anal. Chem. **35,** 252 (1963).

[347] Fassel, V. A., Mossotti, V. G., Grossman, W. E. L., and Kniseley, R. N.: XII CSI, Exeter, 1965.

[348] Fassel, V. A., Mossotti, V. G., Grossman, W. E. L., and Kniseley, R. N.: Spectrochim. Acta **22,** 347 (1966).

[349] Fassel, V. A., Myers, R. B., and Kniseley, R. N.: Spectrochim. Acta **19,** 1187 (1963).

[350] Fassel, V. A., Rasmuson, J. O., Kniseley, R. N., and Cowley, T. G.: Spectrochim. Acta **25 B,** 559 (1970).

[351] Fassel, V. A., Slack, R. W., and Kniseley, R. N.: Anal. Chem. **43,** 186 (1971).

[352] Favretto, L., Pertoldi Marletta, G., and Favretto Gabrielli, L.: At. Absorption Newslett. **12,** 101 (1973).

[353] Feldman, F. J.: Anal. Chem. **42,** 719 (1970).

[354] Feldman, F. J., Bosshart, R. E., and Christian, G. D.: Anal. Chem. **39,** 1175 (1967).

[355] Feldman, F. J., Knoblock, E. C., and Purdy, W. C.: Anal. Chim. Acta **38,** 489 (1967).

[356] Fernandez, F. J.: At. Absorption Newslett. **8,** 90 (1969).

[357] Fernandez, F. J.: At. Absorption Newslett. **12,** 70 (1973).

[358] Fernandez, F. J.: At. Absorption Newslett. **12,** 93 (1973).

[359] Fernandez, F. J., and Kahn, H. L.: At. Absorption Newslett. **10,** 1 (1971).

[360] Fernandez, F. J., and Manning, D. C.: At. Absorption Newslett **7,** 57 (1968).

[361] Fernandez, F. J., and Manning, D. C.: At. Absorption Newslett. **10,** 65 (1971).

[362] Fernandez, F. J., and Manning, D. C.: At. Absorption Newslett. **10,** 86 (1971).

[363] Fernandez, F. J., and Manning, D. C.: At. Absorption Newslett. **11,** 67 (1972).

[364] Fernandez, F. J., Manning, D. C., and Vollmer, J.: At. Absorption Newslett. **8,** 117 (1969).

[365] Ferris, A. P., Jepson, W. B., and Shapland, R. C.: Analyst **95,** 574 (1970).

[366] Fielding, J., and Ryall, R. G.: Clin. Chim. Acta **33,** 235 (1971).

[367] Fierens, P., and Degre, J. P.: CSI XVII, Florence 1973, p. 558.

[368] Fiorino, J., Kniseley, R. N., and Fassel, V. A.: Spectrochim. Acta **23 B,** 413 (1968).

[369] Fiorino, J. A., Moffitt, R. A., Woodson, A. L., Gajan, R. J., Huskey, G. E., and Scholz, R. G.: J. Ass. Offic. Agr. Chem. **56,** 1246 (1973).

[370] Firman, R. J.: Spectrochim. Acta **21,** 341 (1965).

[371] Fishman, M. J.: At. Absorption Newslett. **5,** 102 (1966).

[372] Fishman, M. J.: Anal. Chem. **42,** 1462 (1970).

[373] Fishman, M. J.: At. Absorption Newslett. **11,** 46 (1972).

[374] Fishman, M. J., and Downs, S. C.: US Geol. Surv. Water-Supply Papers 1540-C. US Gov. Print. Office, Washington, D. C. 1966.

[375] Fishman, M. J., and Midgett, M. R.: in Trace Inorganics in Water. Amer. Chem. Soc. Publ., Advan. Chem. Ser. **73,** 230 (1968).

[376] Fixman, M., and Boughton, L.: At. Absorption Newslett. **5,** 33 (1966).

[377] Fleet, B., Liberty, K. V., and West, T. S.: Anal. Chim. Acta **45,** 205 (1969).

[378] Fleming, H. D.: Spectrochim. Acta. **23 B,** 207 (1967).

[379] Fleming, L. W., and Stewart, W. K.: Clin. Chim. Acta **14,** 134 (1966).

[380] Förstner, U., and Müller, G.: Schwermetalle in Flüssen und Seen. Springer-Verlag, Berlin/Heidelberg/New York 1974.

[381] Forrester, A. T., Gudmundsen, R. A., and Johnson, P. O.: J. Opt. Soc. Am. **46,** 339 (1956).

[382] Foss, R. A., and Houston, D. M.: At. Absorption Newslett. **8,** 82 (1969).

[383] Foster, R. R.: At. Absorption Newslett. **7,** 110 (1968).

[384] Fratta, M.: 4th ICAS, Toronto 1973, p. 152.

[385] Frazer, A., Secunda, S. K., and Mendels, J.: Clin. Chim. Acta **36,** 499 (1972).

[386] Frey, S. W.: At. Absorption Newslett. **3,** 127 (1964).

[387] Frey, S. W., de Witt, W. G., and Bellomy, B. R.: Prc. Am. Soc. Brew. Chem. **1966,** 172.
[388] Fritze, K., and Stuart, C.: 4ᵗʰ ICAS, Toronto 1973, p. 7.
[389] Fuchs, Ch., Brasche, M., Paschen, K., Nordbeck, H., and Quellhorst, E.: Clin. Chim. Acta **52,** 71 (1974).
[390] Fuhrman, D. L.: At. Absorption Newslett. **8,** 105 (1969).
[391] Fukushima, S.: Microchim. Acta **1959,** 596.
[392] Fuller, C. W.: Anal. Chim. Acta **62,** 261 (1972).
[393] Fuller, C. W.: Anal. Chim. Acta **62,** 442 (1972).
[394] Fuller, C. W.: At. Absorption Newslett. **11,** 65 (1972).
[395] Fuller, C. W.: At. Absorption Newslett. **12,** 40 (1973).
[396] Fuller, C. W., and Whiteheard, J.: Anal. Chim. Acta **68,** 407 (1974).
[397] Fulton, A., and Butler, L. R. P.: Spectrosc. Letters **1,** 317 (1968).
[398] Fuwa, K., Pulido, P., McKay, R., and Vallee, B. L.: Anal. Chem. **36,** 2407 (1964).
[399] Fuwa, K., and Vallee, B. L.: Anal. Chem. **35,** 942 (1963).
[400] Fuwa, K., and Vallee, B. L.: Anal. Chem. **41,** 188 (1969).
[401] Galindo, G. G., Appelt, H., and Schalscha, E. B.: Soil Sci. Soc. Amer. Proceed. **33,** 974 (1969).
[402] Galle, O. K.: Appl. Spectrosc. **22,** 404 (1968).
[403] Gambrell, J. W.: At. Absorption Newslett. **10,** 81 (1971).
[404] Gambrell, J. W.: At. Absorption Newslett. **11,** 125 (1972).
[405] Garrido, M. D., Llaguno, C., and Garrido, J.: Am. J. Enology Viticult. **1971,** 44.
[406] Gatehouse, B. M., and Walsh, A.: Spectrochim. Acta **16,** 602 (1960).
[407] Gatehouse, B. M., and Willis, J. B.: Spectrochim. Acta **17,** 710 (1961).
[408] Gaumer, M. W., Sprague, S., and Slavin, W.: At. Absorption Newslett. **5,** 58 (1966).
[409] Gelder, Z. V.: Spectrochim. Acta **25 B,** 669 (1970).
[410] Giammarise, Λ.: Λt. Absorption Newslett. **5,** 113 (1966).
[411] Gibson, J. H., Grossman, W. E. L., and Cooke, W. D.: in Analytical Chemistry. Elsevier Publishers, Amsterdam 1962.
[412] Gibson, J. H., Grossman, W. E. L., and Cooke, W. D.: Anal. Chem. **35,** 266 (1963).
[413] Gidley, J. A. F.: CSI IX, Paris 1962, p. 263.
[414] Gidley, J. A. F., and Jones, J. T.: Analyst **85,** 249 (1960).
[415] Gidley, J. A. F., and Jones, J. T.: Analyst **86,** 271 (1961).
[416] Gimblet, E. G., Marney, A. F., and Bonsnes, R. W.: Clin. Chem. **13,** 204 (1967).
[417] Ginzburg, V. L., Livshits, D. M., and Satariana, G. I.: Zh. Analit. Khim. **19,** 1089 (1964).
[418] Giron, H. C.: At. Absorption Newslett. **12,** 28 (1973).
[419] Glenn, M. T., Savory, J., Fein, S. A., Reevers, R. D., Molnar, C. J., and Winefordner, J. D.: Anal. Chem. **45,** 203 (1973).
[420] Gochman, N., and Givelber, H.: Clin. Chem. **16,** 229 (1970).
[421] Goebgen, H. G., and Brockmann, J.: Wasser, Luft, Betrieb **12,** 11 (1968).
[422] Goecke, R.: Talanta **15,** 871 (1968).
[423] Goguel, R.: Spectrochim. Acta **26 B,** 313 (1971).
[424] Goleb, J. A.: Anal. Chem. **35,** 1978 (1963).
[425] Goleb, J. A.: Anal. Chim. Acta **34,** 135 (1966).
[426] Goleb, J. A.: Anal. Chim. Acta **36,** 130 (1966).
[427] Goleb, J. A., and Yokoyama, Y.: Anal. Chim. Acta **30,** 213 (1964).
[428] Gonzalez, J. G., and Ross, R. T.: Anal. Letters **5,** 683 (1972).
[429] Gonzalez, J. G., Ross, R. T., and Segar, D. A.: 4ᵗʰ ICAS, Toronto 1973, p. 37.
[430] Goodfellow, G. I.: Anal. Chim. Acta **36,** 1491 (1967).
[431] Goodfellow, G. I.: Appl. Spectrosc. **21,** 39 (1967).
[432] Goodwin, E.: At. Absorption Newslett. **9,** 95 (1970).
[433] Gough, D. S., Hannaford, P., and Walsh, A.: Spectrochim. Acta **28 B,** 197 (1973).
[434] Goulden, P. D., Brooksbank, P., and Ryan, J. F.: Int. Lab. **1973,** No. 5, 31.
[435] Govindaraju, K., Mevelle, G., and Chouard, C.: Anal. Chem **46,** 1672 (1974).
[436] Govindaraju, K., Hermann, R., Mevelle, G., and Chouard, C.: At. Absorption Newslett. **12,** 73 (1973).

[437] Govindaraju, K., and L'homel, N.: At. Absorption Newslett. **11,** 115 (1972).

[438] Gray, R., and Pruden, E. L.: Am. J. Med. Technol. **33,** 349 (1967).

[439] Greaves, M. C.: Nature **199,** 552 (1963).

[440] Greenfield, S., Jones, I. L., and Berry, C. T.: Analyst **89,** 713 (1964).

[441] Griffin, G. F.: Soil. Sci. Soc. Amer. Proc. **32,** 803 (1968).

[442] Griffith, F. D., Parker, H. E., and Rogler, J. C.: J. Nutr. **83,** 15 (1964).

[443] Grimaldi, F. S., and Schnepfe, M. M.: Talanta **17,** 617 (1970).

[444] Grobenski, Z.: Analysentechn. Berichte **31** (1974).

[445] Groenewald, T.: Anal. Chem. **40,** 863 (1968).

[446] Grunbaum, B. W., and Pace, N.: Microchem. J. **15,** 666 (1970).

[447] Güçer, S., and Massmann, H.: CSI XVII, Florence 1973, p. 51.

[448] Guillaumin, R.: At. Absorption Newslett. **5,** 19 (1966).

[449] Gupta, H. K. L., Amore, F. J., and Boltz, D. F.: At. Absorption Newslett. **7,** 107 (1968).

[450] Gutsche, B., and Herrmann, R.: Analyst **95,** 805 (1970).

[451] Gutsche, B., and Herrmann, R.: Naunyn-Schmiedebergs Arch. Pharm. **270,** 94 (1971).

[452] Gutsche, B., and Herrmann, R.: Z. Anal. Chem. **253,** 257 (1971).

[453] Gutsche, B., and Herrmann, R.: Z. Anal. Chem. **258,** 277 (1972).

[454] Haarsma, J. P. S., deJong, G. J., and Agterdenbos, J.: Spectrochim. Acta **29 B,** 1 (1974).

[455] Haas, T., Lehnert, G., and Schaller, K. H.: Z. Klin. Chem. Klin. Biochem. **5,** 27 (1967).

[456] Haelen, P., Cooper, G., and Pampel, C.: At. Absorption Newslett. **13,** 1 (1974).

[457] Hall, G., Cochrane, I. G., and Dorman, R. W.: IAASC, Shefield 1969, E 4.

[458] Hall, J. M., and Woodward, C.: Spectry. Letters **2,** 113 (1969).

[459] Halls, D. J., and Townshend, A.: Anal. Chim. Acta **36,** 278 (1966).

[460] Hambly, A. N., and Rann, C. S.: in [5], p. 241.

[461] Hammar, H. E., and Page, N. R.: At. Absorption Newslett. **6,** 33 (1967).

[462] Hanig, R. C., and Aprison, M. H.: Anal. Biochem. **21,** 169 (1967).

[463] Hankiewicz, J.: Polish Med. J. **VIII,** 779 (1969).

[464] Hanna, W. J.: Ag. Chem. **III,** 23 (1967).

[465] Hansen, A. C.: At. Absorption Newslett. **12,** 125 (1973).

[466] Hansen, J. L.: Am. J. Med. Technol. **34,** 625 (1968).

[467] Hansen, J. L., and Freier, E. F.: Am. J. Med. Technol. **33,** 217 (1967).

[468] Hareland, W. A., Ebersole, E. R., and Ramachandran, T. P.: Anal. Chem. **44,** 520 (1972).

[469] Harrington, D. E.: At. Absorption Newslett. **9,** 106 (1970).

[470] Harrington, D. E.: At. Absorption Newslett. **11,** 107 (1972).

[471] Harrison, W. W.: Anal. Chem. **37,** 1168 (1965).

[472] Harrison, W. W., and Juliano, P. O.: Anal. Chem. **41,** 1016 (1969).

[473] Harrison, W. W., and Juliano, P. O.: Anal. Chem. **43,** 248 (1971).

[474] Harrison, W. W., and Tyree, A. B.: Clin. Chim. Acta **31,** 63 (1971).

[475] Harrison, W. W., and Wadlin, W. H.: Anal. Chem. **41,** 374 (1969).

[476] Harrison, W. W., Yurachek, J. P., and Benson, C. A.: Clin. Chim. Acta **23,** 83 (1969).

[477] Harth, M., Haines, D. S. M., and Bondy, D. C.: Am. J. Clin. Pathol. **59,** 423 (1973).

[478] Hartley, F. R., and Inglis, A. S.: Analyst **92,** 622 (1967).

[479] Hartstein, A. M., Freedman, R. W., and Platter, D. W.: Anal. Chem. **45,** 611 (1973).

[480] Hasegawa, N., Hirai, A., Sugino, H., and Kashiwagi, T.: 3rd CISAFA, Paris 1971, p. 665.

[481] Hatch, W. R., and Ott, W. L.: Anal. Chem. **40,** 2085 (1968).

[482] Hauck, G.: Z. Anal. Chem. **267,** 337 (1973).

[483] Hauser, T. R., Hinners, T. A., and Kent, J. L.: Anal. Chem. **44,** 1819 (1972).

[484] Hazebroucq, G. F.: 3rd CISAFA, Paris 1971, p. 577.

[485] Headridge, J. B., and Ashy, M. A.: 4th ICAS, Toronto 1973, p. 65.

[486] Headridge, J. B., and Hubbard, D. P.: Anal. Chim. Acta **37,** 151 (1967).

[487] Headridge, J. B., and Richardson, J.: Analyst **94,** 968 (1969).

[488] Headridge, J. B., and Sowerbutts, A.: Analyst **98,** 57 (1973).

[489] Heckman, M.: J. Ass. Offic. Agr. Chem. **50,** 45 (1967).

[490] Heckman, M.: J. Ass. Offic. Agr. Chem. **53,** 923 (1970).

[491] Heckman, M.: J. Ass. Offic. Agr. Chem. **54,** 666 (1971).
[492] Hein, H.: Analysentechn. Berichte **29** (1973).
[493] Heinemann, G.: Z. Klin. Chem. Klin. Biochem. **10,** 467 (1972).
[494] Heinrich, G., Spitzer, H., and Tesik, G.: Z. Anal. Chem. **226,** 124 (1967).
[495] Heinrichs, H., and Lange, J.: Fortschr. Miner. **50,** Bh. 1, 35 (1972).
[496] Heinrichs, H., and Lange, J.: Z. Anal. Chem. **265,** 256 (1973).
[497] Hell, A., Ulrich, W. F., Shifrin, N., and Ramirez-Muñoz, J.: Appl. Opt. **7,** 1317 (1968).
[498] Helsby, C. A.: Talanta **20,** 779 (1973).
[499] Helsby, C. A.: Anal. Chim. Acta **69,** 259 (1974).
[500] Heneage, P.: At. Absorption Newslett. **5,** 64 (1966).
[501] Heneage, P.: At. Absorption Newslett. **5,** 67 (1966).
[502] Henn, E. L.: At. Absorption Newslett. **12,** 109 (1973).
[503] Henning, S., and Jackson, T. L.: At. Absorption Newslett. **12,** 100 (1973).
[504] Herrmann, R., and Lang, W.: Z. Ges. Exptl. Med. **134,** 268 (1961).
[505] Herrmann, R., and Lang, W.: Z. Ges. Exptl. Med. **135,** 569 (1962).
[506] Herrmann, R., and Lang, W.: Z. Klin. Chem. **1,** 182 (1963).
[507] Herrmann, R., Lang, W., and Stamm, D.: Blut **11,** 135 (1965).
[508] Hessel, D. W.: At. Absorption Newslett. **7,** 55 (1968).
[509] Hicks, J. M., Gutierrez, A. N., and Worthy, B. E.: Clin. Chem. **19,** 322 (1973).
[510] Hicks, J. E., McPherson, R. T., and Salyer, J. W.: Anal. Chim. Acta **61,** 441 (1972).
[511] Hieftje, G. M., and Malmstadt, H. V.: Anal. Chem. **40,** 1860 (1968).
[512] Hill, G. L., and Caputi, A.: Am. J. Enology Viticult. **20,** 227 (1969).
[513] Hill, U. T.: in [3], 83 (1969).
[514] Hinkle, M. E., and Learned, R. E.: US Geol. Survey Prof. Paper **650-D,** D 251 (1969).
[515] Hoare, H. C., Mostyn, R. A., and Newland, B. T. N.: Anal. Chim. Acta **40,** 181 (1968).
[516] Hoare, H. C., Mostyn, R. A., and Newland, B. T. N.: IAASC, Sheffield 1969, D6.
[517] Hobbs, R. S., Kirkbright, G. F., Sargent, M., and West, T. S.: Talanta **15,** 997 (1968).
[518] Hobbs, R. S., Kirkbright, G. F., and West, T. S.: Analyst **94,** 554 (1969).
[519] Höhn, R., Jackwerth, E., and Koos, K.: Spectrochim. Acta **29 B,** 225 (1974).
[520] Höhn, R., and Umland, F.: Z. Anal. Chem. **258,** 100 (1972).
[521] Hoffmann, H-D., and Fiedler, H.: Z. Ges. Inn. Med. Ihre Grenzgeb. **25,** 1065 (1970).
[522] Hofton, M. E., and Hubbard, D. P.: 3rd CISAFA, Paris 1971, p. 743.
[523] Hohnadel, D. C., Sunderman, F. W., Nechay, M. W., and McNeely, M. D.: Clin. Chem. **19,** 1288 (1973).
[524] Holak, W.: Anal. Chem. **41,** 1712 (1969).
[525] Holak, W.: J. Ass. Offic. Agr. Chem. **53,** 877 (1970)
[526] Holak, W.: J. Ass. Offic. Agr. Chem. **54,** 1138 (1971).
[527] Holak, W.: At. Absorption Newslett. **12,** 63 (1973).
[528] Holak, W., Krinitz, B., and Williams, J. C.: J. Ass. Offic. Agr. Chem. **55,** 741 (1972).
[529] Holtzman, N. A., Elliott, D. A., and Heller, R. H.: New England J. Med. **275,** 347 (1966).
[530] Hoover, W. L., and Duren, S. C.: J. Ass. Offic. Agr. Chem. **52,** 708 (1969).
[531] Hoover, W. L., Melton, J. R., and Howard, P. A.: J. Ass. Offic. Agr. Chem. **54,** 860 (1971).
[532] Hoover, W. L., Reagor, J. C., and Garner, J. C.: J. Ass. Offic. Agr. Chem. **52,** 708 (1969).
[533] Horn, D. B., and Latner, A. L.: Clin Chim. Acta **8,** 974 (1963).
[534] Hoschler, M. E., Kanabrocki, E. L., Moore, C. E., and Hattori, D. M.: Appl. Spectrosc. **27,** 185 (1973).
[535] Hossner, L. R., and Ferrara, L. W.: At. Absorption Newslett. **6,** 71 (1967).
[536] Hossner, L. R., Weger, S. J., and Ferrara, L. W.: ACS Meet., Atlantic City 1968.
[537] Hubbard, D. P., and Monks, H. H.: Anal. Chim. Acta **47,** 197 (1969).
[538] Huffman, C., Mensik, J. D., and Rader, L. F.: US Geol. Survey Prof. Paper **550-B,** B 189 (1966).
[539] Human, H. G. C., Butler, L. R. P., and Strasheim, A.: Analyst **94,** 81 (1969).
[540] Human, H. G. C., Strasheim, A., and Butler, L. R. P.: IAASC, Sheffield 1969, F1.
[541] Humphrey, J. R.: Anal. Chem. **37,** 1604 (1965).

[542] Hunt, B. J.: Clin. Chem. **15,** 979 (1969).
[543] Hunter, R. E., Kelsall, M. A., Bishop, W. J., and Woodworth, P. F.: 4th ICAS, Toronto 1973, p. 36.
[544] Hurford, T. R., and Boltz, D. F.: Anal. Chem. **40,** 379 (1968).
[545] Husdan, H., and Rapoport, A.: Clin. Chem. **15,** 669 (1969).
[546] Husler, J. W.: At. Absorption Newslett. **9,** 31 (1970).
[547] Husler, J. W.: At. Absorption Newslett. **10,** 60 (1971).
[548] Hwang, J. Y.: Canadian Spectrosc. March **1971,** 43.
[549] Hwang, J. Y., Mokeler, Ch. J., and Ullucci, P. A.: Anal. Chem. **44,** 2018 (1972).
[550] Hwang, J. Y., and Sandonato, L. M.: Anal. Chem. **42,** 744 (1970).
[551] Hwang, J. Y., Ullucci, P. A., and Mokeler, Ch. J.: Anal. Chem. **45,** 795 (1973).
[552] Hyatt, K. H., Levy, L., Nichaman, N., and Oscherwitz, M.: Appl. Spectrosc. **20,** 142 (1966).
[553] Ihnat, M.: Anal. Chim. Acta **82,** 293 (1976).
[554] Iida, C., Fuwa, K., and Wacker, W. E. C.: Anal. Biochem. **18,** 18 (1967).
[555] Ingamells, C. O.: Anal. Chem. **38,** 1228 (1966).
[556] Ingamells, C. O.: Anal. Chim. Acta **52,** 323 (1970).
[557] Intonti, R., and Stacchini, A.: Spectrochim. Acta **23 B,** 437 (1968).
[558] Jackson, A. S., Michael, L. M., and Schumacher, H. S.: Anal. Chem. **44,** 1064 (1972).
[559] Jackson, B., and Myrick, H. N.: Internat. Labor. May/June **1971,** 41.
[560] Jackwerth, E.: Z. Anal. Chem. **256,** 128 (1971).
[561] Jackwerth, E., Höhn, R., and Koos, K.: Z. Anal. Chem. **264,** 1 (1973).
[562] Jackwerth, E., Lohmar, J., and Wittler, G.: Z. Anal. Chem. **266,** 1 (1973).
[563] Jackwerth, E., Lohmar, J., and Wittler, G.: Z. Anal. Chem. **270,** 6 (1974).
[564] Jacobs, M. B., Goldwater, L. J., and Gilbert, H.: Am. Ind. Hyg. Assoc. J. **22,** 276 (1961).
[565] Jacobs, M. B., and Singerman, A.: J. Lab. Clin. Med. **59,** 871 (1962).
[566] Jacobs, M. B., Yamaguchi, S., Goldwater, L. J., and Gilbert, H.: Am. Ind. Hyg. Assoc. J. **21,** 475 (1960).
[567] Jacobsen, E., and Harrison, G. R.: J. Opt. Soc. Amer. **39,** 1054 (1969).
[568] James, C. H., and Webb, J. S.: Trans. Instn. Min. Metall. **73,** 633 (1964).
[569] Janauer, G. E., Smith, F. E., and Mangan, J.: At. Absorption Newslett. **6,** 3 (1967).
[570] Janoušková, J., Nehasilova, M., and Sychra, V.: At. Absorption Newslett. **12,** 161 (1973).
[571] Janssen, A., and Umland, F.: Z. Anal. Chem. **251,** 101 (1970).
[572] Janssens, M., and Dams, R.: Anal. Chim. Acta **65,** 41 (1973).
[573] Janssens, M., and Dams, R.: Anal. Chim. Acta **70,** 25 (1974).
[574] Jaworowski, R. J., and Weberling, R. P.: At. Absorption Newslett. **5,** 125 (1966).
[575] Jaworowski, R. J., Weberling, R. P., and Bracco, D. J.: Anal. Chim. Acta **37,** 284 (1967).
[576] Jeffus, M. T., Elkins, J. S., and Kenner, C. T.: J. Ass. Offic. Agr. Chem. **53,** 1172 (1970).
[577] Jenkins, D. R.: Spectrochim. Acta **23 B,** 167 (1967).
[578] Jenkins, D. R.: IAASC, Sheffield 1969, C 1.
[579] Jenkins, D. R.: Spectrochim. Acta **25 B,** 47 (1970).
[580] Jenkins, D. R., and Sudgen, T. M.: in [5], p. 151.
[581] Johnson, F. J., Woodis, T. C., and Cummings, J. M.: At. Absorption Newslett. **11,** 118 (1972).
[582] Johnson, H. N., Kirkbright, G. F., and Whitehouse, R. J.: Anal. Chem. **45,** 1603 (1973).
[583] Johnson, J. R. K., and Riechman, G. C.: Clin. Chem. **14,** 1218 (1968).
[584] Jones, A. H.: Anal. Chem. **37,** 1761 (1965).
[585] Jones, A. H.: At. Absorption Newslett. **9,** 1 (1970).
[586] Jones, D. I. H., and Thomas, T. A.: Hilger J. **9,** 39 (1965).
[587] Jones, G. W., Lewis, B., and Seaman, H.: J. Amer. Chem. Soc. **53,** 3992 (1931).
[588] Jones, J. L., and Eddy, R. D.: Anal. Chim. Acta **43,** 165 (1968).
[589] Jones, W. G., and Walsh, A.: Spectrochim. Acta **16,** 249 (1960).
[590] Jordan, J.: At. Absorption Newslett. **7,** 48 (1968).
[591] Joselow, M. M., and Bogden, J. D.: At. Absorption Newslett. **11,** 99 (1972).
[592] Joselow, M. M., and Bogden, J. D.: At. Absorption Newslett. **11,** 127 (1972).
[593] Joselow, M. M., and Singh, N. P.: At. Absorption Newslett. **12,** 128 (1973).

[594] Joseph, K. T., Panday, V. K., Raut, S. J., and Soman, S. D.: At. Absorption Newslett. **7,** 25 (1968).
[595] Joseph, K. T., Parameswaran, M., and Soman, S. D.: At. Absorption Newslett. **8,** 127 (1969).
[596] Joyner, T., and Finley, J. S.: At. Absorption Newslett. **5,** 4 (1966).
[597] Joyner, T., Healy, M. L., Chakrabarti, D., and Koyanagi, T.: Environm. Sci. Technol. **1,** 417 (1967).
[598] Julshamn, K., and Braekkan, O. R.: At. Absorption Newslett. **12,** 139 (1973).
[599] Jung, P. D., and Clarke, D.: J. Ass. Offic. Agr. Chem. **57,** 379 (1974).
[600] Jursik, M. L.: At. Absorption Newslett. **6,** 21 (1967).
[601] Kägler, S. H.: Analysentechn. Berichte **6** (1965).
[602] Kägler, S. H.: Erdöl, Kohle, Erdgas, Petrochemie **19,** 879 (1966).
[603] Kägler, S. H.: Erdöl, Kohle, Erdgas, Petrochemie **24,** 13 (1971).
[604] Kahn, H. L.: At. Absorption Newslett. **2,** 35 (1963).
[605] Kahn, H. L.: J. Chem. Education **43,** A 7 & A 103 (1966).
[606] Kahn, H. L.: J. Metals, **1966,** 1101.
[607] Kahn, H. L.: At. Absorption Newslett. **6,** 51 (1967).
[608] Kahn, H. L.: At. Absorption Newslett, **7,** 40 (1968).
[609] Kahn, H. L.: in Trace Inorganics in Water. Advan. Chem. Ser. **73,** 183 (1968).
[610] Kahn, H. L.: Amer. Laboratory **1969,** 52.
[611] Kahn, H. L.: At. Absorption Newslett. **10,** 58 (1971).
[612] Kahn, H. L., Fernandez, F. J., and Slavin, S.: At. Absorption Newslett. **11,** 42 (1972).
[613] Kahn, H. L., and Manning, D. C.: At. Absorption Newslett. **4,** 264 (1965).
[614] Kahn, H. L., and Manning, D. C.: Amer. Laboratory **1972,** 8.
[615] Kahn, H. L., Peterson, G. E., and Manning, D. C.: At. Absorption Newslett. **9,** 79 (1970).
[616] Kahn, H. L., Peterson, G. E., and Schallis, J. E.: At. Absorption Newslett. **7,** 35 (1968).
[617] Kahn, H. L., and Schallis, J. E.: At. Absorption Newslett. **7,** 5 (1968).
[618] Kahn, H. L., and Sebestyen, J. E.: At. Absorption Newslett. **9,** 33 (1970).
[619] Kahn, H. L., and Slavin, S.: At. Absorption Newslett. **10,** 125 (1971).
[620] Kaiser, G., Tschöpel, P., and Tölg, G.: Z. Anal. Chem. **253,** 177 (1971).
[621] Kalb, G. W.: At. Absorption Newslett. **9,** 84 (1970).
[622] Kallmann, S., and Hobart, E. W.: Anal. Chim. Acta **51,** 120 (1970).
[623] Kallmann, S., and Hobart, E. W.: Talanta **17,** 845 (1970).
[624] Kapetan, J. P.: in [3], 78 (1969).
[625] Karkhanis, P. P., and Anfinsen, J. R.: J. Ass. Offic. Agr. Chem. **56,** 358 (1973).
[626] Kashiki, M., and Oshima, S.: Anal. Chim. Acta **51,** 387 (1970).
[627] Kashiki, M., Yamazoe, S., Ikeda, N., and Oshima, S.: Anal. Letters **7,** 53 (1974).
[628] Kashiki, M., Yamazoe, S., and Oshima, S.: Anal. Chim. Acta **53,** 95 (1971).
[629] Kawamura, H., Tanaka, G., and Ohyagi, Y.: Spectrochim. Acta **28 B,** 309 (1973).
[630] Keats, G. H.: At. Absorption Newslett. **4,** 319 (1965).
[631] Keller, E., and Parsons, M. L.: At. Absorption Newslett. **9,** 92 (1970).
[632] Kelly, W. R., and Moore, C. B.: Anal. Chem. **45,** 1274 (1973).
[633] Kerber, J. D.: Appl. Spectrosc. **20,** 212 (1966).
[634] Kerber, J. D.: At. Absorption Newslett. **10,** 104 (1971).
[635] Kerber, J. D., and Barnett, W. B.: At. Absorption Newslett. **8,** 113 (1969).
[636] Kerber, J. D., Barnett, W. B., and Kahn, H. L.: At. Absorption Newslett. **9,** 39 (1970).
[637] Kerber, J. D., Koch, A., and Peterson, G. E.: At. Absorption Newslett. **12,** 104 (1973).
[638] Kerber, J. D., Russo, A. J., Peterson, G. E., and Ediger, R. D.: At. Absorption Newslett. **12,** 106 (1973).
[639] Kerbyson, J. D., and Ratzkowski, C.: Canad. Spectrosc. **13,** 102 (1968).
[640] Kettner, H.: Schriftenr. V. Wasser-, Boden-, Lufthyg. **29,** 55 (1969).
[641] King, A. S.: Astrophys. **28,** 300 (1908).
[642] Kinnunen, J., and Lindsjö, O.: Chemist-Analyst **56,** 25 (1967).
[643] Kinnunen, J., and Lindsjö, O.: Chemist-Analyst **56,** 76 (1967).
[644] Kinser, R. E.: Am. Ind. Hyg. Ass. J. **27,** 260 (1966).

[645] Kinson, K., and Belcher, C. B.: Anal. Chim. Acta **30,** 64 (1964).
[646] Kinson, K., and Belcher, C. B.: Anal. Chim. Acta **31,** 180 (1964).
[647] Kirchhof, H.: Spectrochim. Acta **24 B,** 235 (1969).
[648] Kirchhoff, G.: Pogg. Annalen, **109,** 275 (1860).
[649] Kirchhoff, G.: Phil. Mag. (4), **20,** 1 (1860).
[650] Kirchhoff, G., and Bunsen, R.: Phil. Mag. (4), **20,** 89 (1860).
[651] Kirchhoff, G., and Bunsen, R.: Phil. Mag. (4), **22,** 329 (1861).
[652] Kirkbright, G. F.: IAASC, Sheffield 1969, A5.
[653] Kirkbright, G. F.: Analyst **96,** 609 (1971).
[654] Kirkbright, G. F.: II ČS. Conf. Flame Spectrosc., Zvikov 1973.
[655] Kirkbright, G. F., Peters, M. K., and West, T. S.: Analyst **91,** 705 (1966).
[656] Kirkbright, G. F., Rao, A., and West, T. S.: Anal. Letters **2,** 465 (1969).
[657] Kirkbright, G. F., Sargent, M., and West, T. S.: At. Absorption Newslett. **8,** 34 (1969).
[658] Kirkbright, G. F., Sargent, M., and West, T. S.: Talanta **16,** 245 (1969).
[659] Kirkbright, G. F., Sargent, M., and West, T. S.: Talanta **16,** 1467 (1969).
[660] Kirkbright, G. F., Semb, A., and West, T. S.: Talanta **14,** 1011 (1967).
[661] Kirkbright, G. F., Semb, A., and West, T. S.: Talanta **15,** 441 (1968).
[662] Kirkbright, G. F., Semb, A., and West, T. S.: Spectrosc. Letters **1,** 7 (1968).
[663] Kirkbright, G. F., Smith, A. M., and West, T. S.: Analyst **91,** 700 (1966).
[664] Kirkbright, G. F., and Troccoli, O. E.: Spectrochim. Acta **28 B,** 33 (1973).
[665] Kirkbright, G. F., Troccoli, O. E., and Vetter, S.: Spectrochim. Acta **28 B,** 1 (1973).
[666] Kirkbright, G. F., and Vetter, S.: Spectrochim. Acta **26 B,** 505 (1971).
[667] Kirkbright, G. F., Ward, A. F., and West, T. S.: Anal. Chim. Acta **64,** 353 (1973).
[668] Kirkbright, G. F., and West, T. S.: Appl. Optics **7,** 1305 (1968).
[669] Kirkbright, G. F., West, T. S., and Wilson, P. J.: At. Absorption Newslett. **11,** 53 (1972).
[670] Kirkbright, G. F., West, T. S., and Wilson, P. J.: At. Absorption Newslett. **11,** 113 (1972).
[671] Klein, B., Kaufman, J. H., and Morgenstern, S.: Clin. Chem. **13,** 388 (1967).
[672] Klein, B., Kaufman, J. H., and Oklander, M.: Clin. Chem. **13,** 788 (1967).
[673] Klein, B., Kaufman, J. H., and Oklander, M.: Clin. Chem. **13,** 79 (1967).
[674] Knauer, G. A.: Analyst **95,** 476 (1970).
[675] Knight, D. M., and Pyzyna, M. K.: At. Absorption Newslett. **8,** 129 (1969).
[676] Kniseley, R. N.: in [5], p. 189.
[677] Kniseley, R. N., D'Silva, A. P., and Fassel, V. A.: Anal. Chem. **35,** 911 (1963).
[678] Knudson, E. J., and Christian, G. D.: Anal. Letters **6,** 1039 (1973).
[679] König, P., Schmitz, K. H., and Thiemann, E.: Z. Anal. Chem. **244,** 232 (1969).
[680] König, P., Schmitz, K. H., and Thiemann, E.: Arch. Eisenhüttenwesen **40,** 53 (1969).
[681] Koirtyohann, S. R.: Anal. Chem. **37,** 601 (1965).
[682] Koirtyohann, S. R.: At. Absorption Newslett. **6,** 77 (1967).
[683] Koirtyohann, S. R.: in [5], p. 295.
[684] Koirtyohann, S. R.: 4th ICAS, Toronto 1973, p. 88.
[685] Koirtyohann, S. R., and Feldman, C.: in Developments in Applied Spectroscopy, Vol. 3. Plenum Press, New York 1964, p. 180.
[686] Koirtyohann, S. R., and Pickett, E. E.: Anal. Chem. **37,** 601 (1965).
[687] Koirtyohann, S. R., and Pickett, E. E.: Anal. Chem. **38,** 585 (1966).
[688] Koirtyohann, S. R., and Pickett, E. E.: Anal. Chem. **40,** 2068 (1968).
[689] Koirtyohann, S. R., and Pickett, E. E.: Spectrochim. Acta **23 B,** 673 (1968).
[690] Koirtyohann, S. R., and Pickett, E. E.: Spectrochim. Acta **26 B,** 349 (1971).
[691] Koirtyohann, S. R., and Wen, J. W.: Anal. Chem. **45,** 1986 (1973).
[692] Kolb, B., Kemmner, G., Schleser, F. H., and Wiedeking, E.: Z. Anal. Chem. **221,** 166 (1966).
[693] Kolihová, D., and Sychra, V.: Anal. Chim. Acta **59,** 477 (1972).
[694] Kometain, T. Y.: Plating **56,** 1251 (1969).
[695] Kopito, L., Byers, R. K., and Schwachman, H.: New England J. Med. **276,** 949 (1967).
[696] Kopito, L., Davis, M. A., and Schwachman, H.: Clin. Chem. **20,** 205 (1974).
[697] Kopito, L., and Schwachman, H.: J. Lab. & Clin. Med. **70,** 326 (1967).

[698] Kornblum, G. R., and de Galan, L.: Spectrochim. Acta **28 B**, 139 (1973).

[699] Kopp, J. F.: 4th ICAS, Toronto 1973, p. 71.

[700] Kotz, L., Kaiser, G., Tschöpel, P., and Tölg, G.: Z. Anal. Chem. **260**, 207 (1972).

[701] Kraft, E. A.: Amer. Laboratory, August **1969**, 8.

[702] Kranz, E.: Emissionsspektroskopie. Akademie-Verlag, Berlin 1964, p. 160.

[703] Krauser, L. A., Henderson, R., Shotwell, H. P., and Culp, D. A.: Amer. Ind. Hyg. Assn. J. **32**, 331 (1971).

[704] Kremling, K., and Petersen, H.: Anal. Chim. Acta **70**, 35 (1974).

[705] Kriege, O. H., and Welcher, G. G.: Talanta **15**, 781 (1968).

[706] Krinitz, B., and Franco, V.: J. Ass. Offic. Agr. Chem. **56**, 869 (1973).

[707] Krishnan, S. S., and Crapper, D. R.: 4th ICAS, Toronto 1973, p. 4.

[708] Krishnan, S. S., Gillespie, K. A., and Crapper, D. R.: Anal. Chem. **44**, 1469 (1972).

[709] Kriss, R. H., and Bartels, T. T.: At. Absorption Newslett. **9**, 78 (1970).

[710] Kriss, R. H., and Bartels, T. T.: At. Absorption Newslett. **11**, 110 (1972).

[711] Kubasik, N. P., Sine, H. E., and Volosin, M. T.: Clin. Chem. **18**, 1326 (1972).

[712] Kubasik, N. P., and Volosin, M. T.: Clin. Chem. **20**, 300 (1974).

[713] Kumamaru, T.: Anal. Chim. Acta **43**, 19 (1968).

[714] Kumamaru, T., Hayashi, Y., Okamoto, N., Tao, E., and Yamamoto, Y.: Anal. Chim. Acta **35**, 524 (1966).

[715] Kumamaru, T., Otani, Y., and Yuroka, Y.: Bull. Chem. Soc. Jap. **40**, 429 (1967).

[716] Kumamaru, T., Tao, E., Okamoto, N., and Yamamoto, Y.: Bull. Chem. Soc. Jap. **38**, 2204 (1966).

[717] Kunishi, M., and Ohno, S.: At. Absorption Newslett. **13**, 29 (1974).

[718] Kuwata, K., Hisatomi, K., and Hasegawa, T.: At. Absorption Newslett. **10**, 111 (1971).

[719] Kuzovlev, I. A., Kuznetsov, Y. N., and Sverdlina, O. A.: Zaw. Lab. **39**, 428 (1973).

[720] Laflamme, Y.: At. Absorption Newslett. **6**, 70 (1967).

[721] Laflamme, Y.: At. Absorption Newslett. **7**, 101 (1968).

[722] Lagathu, J., and Desirant, J.: Rev. Franc. Corps Gras **19**, 169 (1972).

[723] Lambert, H.: Photometria, sive de mesura et gradibus luminis colorum et umbrae. 1760.

[724] Lamm, S., Cole, B., Glynn, K., and Ullmann, W.: New England J. Med. **289**, 574 (1973).

[725] Langmyhr, F. J., and Paus, P. E.: Anal. Chim. Acta **43**, 397 (1968).

[726] Langmyhr, F. J., and Paus, P. E.: Anal. Chim. Acta **43**, 506 (1968).

[727] Langmyhr, F. J., and Paus, P. E.: Anal. Chim. Acta **43**, 508 (1968).

[728] Langmyhr, F. J., and Paus, P. E.: At. Absorption Newslett. **7**, 103 (1968).

[729] Langmyhr, F. J., and Paus, P. E.: At. Absorption Newslett. **8**, 131 (1969).

[730] Langmyhr, F. J., Solberg, R., and Wold, L. T.: Anal. Chim. Acta **69**, 267 (1974).

[731] Langmyhr, F. J., and Thomassen, Y.: Z. Anal. Chem. **264**, 122 (1973).

[732] Langmyhr, F. J., Thomassen, Y., and Massoumi, A.: Anal. Chim. Acta **68**, 305 (1974).

[733] Larkins, P. L.: Spectrochim. Acta **26 B**, 477 (1971).

[734] Larkins, P. L., and Willis, J. B.: Spectrochim. Acta **26 B**, 491 (1971).

[735] Least, C. J., Rejent, T. A., and Lees, H.: At. Absorption Newslett. **13**, 4 (1974).

[736] Leaton, J. R.: J. Ass. Offic. Agr. Chem. **53**, 237 (1970).

[737] Lebedev, V. I., and Dolidze, L. D.: IAASC, Sheffield 1969, F5.

[738] Lehnert, G., Klavis, G., Schaller, K. H., and Haas, T.: Brit. J. Industr. Med. **26**, 156 (1969).

[739] Lehnert, G., Schaller, K. H., and Haas, T.: Z. Klin. Chem. **6**, 174 (1968).

[740] Lemonds, A. J., and McClellan, B. E.: Anal. Chem. **45**, 1455 (1973).

[741] Leonard, E. N.: At. Absorption Newslett. **10**, 84 (1971).

[742] Levine, J. R., Moore, S. G., and Levine, S. L.: Anal. Chem. **42**, 412 (1970).

[743] Levine, S. L.: Anal. Chem. **40**, 1376 (1968).

[744] Levy, A. L., and Katz, E. M.: Clin. Chem. **15**, 787 (1969).

[745] Levy, A. L., and Katz, E. M.: Clin. Chem. **16**, 840 (1970).

[746] Lewis, B., Seaman, H., and Jones, G. W.: J. Franklin Inst. **215**, 149 (1933).

[747] Lewis, C., Ott, W. L., and Sine, N. M.: The Analysis of Nickel. Pergamon Press, Oxford 1966, p. 192.

[748] Lewis, R. R.: At. Absorption Newslett. **7**, 61 (1968).
[749] Lieber, E. R.: At. Absorption Newslett. **9**, 51 (1970).
[750] Lieberman, K. W.: Clin. Chim. Acta **46**, 217 (1973).
[751] Lind, B., Kjellström, T., Linnman, L., and Nordberg, G.: 4th ICAS, Toronto 1973, p. 104.
[752] Lindner, J.: Gesundheits-Ingenieur **95**, 39 (1974).
[753] Lindstedt, G.: Analyst **95**, 264 (1970).
[754] Lindstedt, G., and Skare, I.: 3rd CISAFA, Paris 1971, p. 581.
[755] Lindstrom, O.: Anal. Chem. **31**, 461 (1959).
[756] Ling, C.: Anal. Chem. **39**, 798 (1967).
[757] Ling, C.: Anal. Chem. **40**, 1876 (1968).
[758] Lloyd, P. D., and Lowe, R. M.: Spectrochim. Acta **27 B**, 23 (1972).
[759] Lockyer, R., and Hames, G. E.: Analyst **84**, 385 (1959).
[760] Loftin, H. P., Christian, C. M., and Robinson, J. W.: Spectrosc. Letters **3**, 161 (1970).
[761] Loken, H. F., Teal, J. S., and Eisenberg, E.: Anal. Chem. **35**, 875 (1963).
[762] Lorber, A., Cohen, R. L., Chang, C. C., and Anderson, H. E.: Arthritis and Rheumatism **11**, 170 (1968).
[763] Lowe, R. M.: Spectrochim. Acta **26 B**, 201 (1971).
[764] Luecke, W.: N. Jb. Miner. Mh. **1971**, 263.
[765] Luecke, W.: N. Jb. Miner. Mh. **1971**, 469.
[766] Luecke, W., and Emmermann, R.: Analysentechn. Berichte **19** (1970).
[767] Luecke, W., and Emmermann, R.: At. Absorption Newslett. **10**, 45 (1971).
[768] Luecke, W., Eschermann, F., Lennartz, U., and Papastamataki, A. J.: N. Jb. Miner. Abh. **120**, 178 (1974).
[769] Luecke, W., and Zielke, H-J.: Z. Anal. Chem. **253**, 20 (1971).
[770] Lundgren, G.: 4th ICAS, Toronto 1973, p. 96.
[771] Lundgren, G., Lundmark, L., and Johansson, G.: Anal. Chem. **46**, 1028 (1974).
[772] Lundy, R. G., and Watje, W. F.: At. Absorption Newslett. **8**, 124 (1969).
[773] Lurie, H. H., and Sherman, G. W.: Ind. Eng. Chem. **25**, 404 (1933).
[774] Luyten, S., Smeyers-Verbeke, J., and Massart, D. L.: At. Absorption Newslett. **12**, 131 (1973).
[775] L'vov, B. V.: Ing. Fiz. Zhur. **11**, No. 2, 44 (1959).
[776] L'vov, B. V.: Ing. Fiz. Zhur. **11**, No. 11, 56 (1959).
[777] L'vov, B. V.: Spectrochim. Acta **17**, 761 (1961).
[778] L'vov, B. V.: Spectrochim. Acta **24 B**, 53 (1969).
[779] L'vov, B. V., and Khartsyzov, A. D.: Zh. Prikl. Spektrosk. **11**, 413 (1969).
[780] L'vov, B. V., and Khartsyzov, A. D.: Zh. Analit. Khim. **24**, 799 (1969).
[781] L'vov, B. V., and Khartsyzov, A. D.: Zh. Analit. Khim. **25**, 1824 (1970).
[782] MacDonald, M. A., and Watson, L.: Clin. Chim. Acta **14**, 233 (1966).
[783] Machata, G.: Wiener Klin. Wochenschr. **85**, 216 (1973).
[784] Machata, G., and Binder, R.: Z. Rechtsmed. **73**, 29 (1973).
[785] Maes, D., Adiwinata, Y., Egglestone, D., Fyles, T., Tilgner, F., and Pate, B. D.: 4th ICAS, Toronto 1973, p. 43.
[786] Magos, L., and Cernik, A. A.: British J. Ind. Med. **26**, 144 (1969).
[787] Mahoney, J. R., Sargent, K., Greland, M., and Small, W.: Clin. Chem. **15**, 312 (1969).
[788] Makarov, D. F., Kukushkin, Y. N., and Eroshevich, T. A.: Zh. Analit. Khim. **24**, 1436 (1969).
[789] Manahan, S. E., and Jones, D. R.: Anal. Letters **6**, 745 (1973).
[790] Manahan, S. E., and Kunkel, R.: Anal. Letters **6**, 547 (1973).
[791] Manning, D. C.: At. Absorption Newslett. **3**, 84 (1964).
[792] Manning, D. C.: At. Absorption Newslett. **4**, 267 (1965).
[793] Manning, D. C.: At. Absorption Newslett. **5**, 63 (1966).
[794] Manning, D. C.: At. Absorption Newslett. **5**, 127 (1966).
[795] Manning, D. C.: At. Absorption Newslett. **6**, 35 (1967).
[796] Manning, D. C.: At. Absorption Newslett. **6**, 75 (1967).
[797] Manning, D. C.: At. Absorption Newslett. **9**, 97 (1970).
[798] Manning, D. C.: At. Absorption Newslett. **9**, 109 (1970).

[799] Manning, D. C.: At. Absorption Newslett. **10**, 123 (1971).
[800] Manning, D. C., and Capacho-Delgado, L.: Anal. Chim. Acta **36**, 312 (1966).
[801] Manning, D. C., and Fernandez, F. J.: At. Absorption Newslett. **6**, 15 (1967).
[802] Manning, D. C., and Fernandez, F. J.: At. Absorption Newslett. **9**, 65 (1970).
[803] Manning, D. C., and Fernandez, F. J.: Pittsburgh Conf. Anal. Chem. Appl. Spectrosc., Cleveland 1973.
[804] Manning, D. C., and Fernandez, F. J.: Pittsburgh Conf. Anal. Chem. Appl. Spectrosc., Cleveland 1974.
[805] Manning, D. C., and Heneage, P.: At. Absorption Newslett. **6**, 124 (1967).
[806] Manning, D. C., and Heneage, P.: At. Absorption Newslett. **7**, 80 (1968).
[807] Manning, D. C., and Kahn, H. L.: At. Absorption Newslett. **4**, 224 (1965).
[808] Manning, D. C., and Slavin, W.: At. Absorption Newslett. **1**, 39 (1962).
[809] Manning, D. C., and Slavin, S.: At. Absorption Newslett. **8**, 132 (1969).
[810] Manning, D. C., Trent, D. J., Sprague, S., and Slavin, W.: At. Absorption Newslett. **4**, 255 (1965).
[811] Manning, D. C., Trent, D. J., and Vollmer, J.: At. Absorption Newslett. **4**, 234 (1965).
[812] Manning, D. C., and Vollmer, J.: At. Absorption Newslett. **6**, 38 (1967).
[813] Manning, D. C., Vollmer, J., and Fernandez, F. J.: At. Absorption Newslett. **6**, 17 (1967).
[814] Mansell, R. E.: At. Absorption Newslett. **4**, 276 (1965).
[815] Mansell, R. E., and Emmel, H. W.: At. Absorption Newslett. **4**, 365 (1965).
[816] Mansell, R. E., Emmel, H. W., and McLaughlin, E. L.: Appl. Spectrosc. **20**, 231 (1966).
[817] Mansfield, J. M., Bratzel, M. P., Norgordon, H. O., Knapp, D. O., Zacha, K. E., and Winefordner, J. D.: Spectrochim. Acta **23 B**, 389 (1968).
[818] Mansfield, J. M., Winefordner, J. D., and Veillon, C.: Anal. Chem. **37**, 1049 (1965).
[819] Marček, M. V., Kinson, K., and Belcher, C. B.: Anal. Chim. Acta **41**, 447 (1968).
[820] Margoshes, M., and Darr, M. M.: NBS Techn. Note **272**, 18 (1965).
[821] Mariée, M., and Pinta, M.: Méth. Phys. d'Anal. (GAMS) **6**, 361 (1970).
[822] Marinkovic, M., and Vickers, T. J.: Appl. Spectrosc. **25**, 319 (1971).
[823] Marks, J. Y., and Welcher, G. G.: Anal. Chem. **42**, 1033 (1970).
[824] Marshal, G. B., and West, T. S.: Talanta **14**, 823 (1967).
[825] Marshal, G. B., and West, T. S.: Analyst **95**, 343 (1970).
[826] Martin, M.: Chem. Ind. **1971**, 514.
[827] Maruta, T., and Takeuchi, T.: Anal. Chim. Acta **62**, 253 (1972).
[828] Massmann, H.: Z. Instrumentenkunde **71**, 225 (1963).
[829] Massmann, H.: XII CSI, Exeter 1965, p. 275.
[830] Massmann, H.: 2nd Int. Sympos. Reinstoffe in Wissenschaft und Technik, Dresden 1965. II, p. 297.
[831] Massmann, H.: Z. Anal. Chem. **225**, 203 (1967).
[832] Massmann, H.: Spectrochim. Acta **23 B**, 215 (1968).
[833] Massmann, H.: Méthodes Physiques d'Analyse **4**, 193 (1968).
[834] Massmann, H.: CSI XVI, Heidelberg 1971, p. 285.
[835] Matoušek, J. P.: CSI XVII, Florence 1973, p. 57.
[836] Matoušek, J. P., and Stevens, B. J.: Clin. Chem. **17**, 363 (1971).
[837] Matoušek, J. P., and Sychra, V.: IAASC, Sheffield 1969, C 5.
[838] Matoušek, J. P., and Sychra, V.: Anal. Chem. **41**, 518 (1969).
[839] Matoušek, J. P., and Sychra, V.: Anal. Chim. Acta **49**, 175 (1970).
[840] Matsumoto, C., Taniquchi, S., Suzusho, K., and Sakaguchi, T.: J. Spect. Soc. Japan **2**, 61 (1968).
[841] Matz, A. R.: Bull. Parenteral Drug Ass. **20**, 130 (1966).
[842] Mavrodineanu, R., and Boiteux, H.: Flame Spectroscopy. John Wiley 1965.
[843] Mavrodineanu, R., and Hughes, R. C.: Spectrochim. Acta **19**, 1309 (1963).
[844] Mavrodineanu, R., and Hughes, R. C.: Appl. Optics **7**, 1281 (1968).
[845] McAuliffe, J. J.: At. Absorption Newslett. **6**, 69 (1967).
[846] McBride, C. H.: At. Absorption Newslett. **3**, 144 (1964).

[847] McBryde, W. T., and Williams, F.: US AEC Rept. Y-1178 (1957).

[848] McCrackan, M. L., Webb, H. J., Hammar, H. E., and Loadholt, C. B.: J. Ass. Offic. Agr. Chem. **50**, 5 (1967).

[849] McFarren, E. F., and Lishka, R. J.: in Trace Inorganics in Water. Am. Chem. Soc. Publ., Advan. Chem. Ser. **73**, 253 (1968).

[850] McGee, W. W., and Winefordner, J. D.: Anal. Chim. Acta **37**, 429 (1967).

[851] McPherson, G. L.: At. Absorption Newslett. **4**, 186 (1965).

[852] McPherson, G. L., Price, J. W., and Scaife, P. H.: Nature **199**, 371 (1963).

[853] Means, E. A., and Ratcliff, D.: At. Absorption Newslett. **4**, 174 (1965).

[854] Meddings, B., and Kaiser, H.: At. Absorption Newslett. **6**, 28 (1967).

[855] Medlin, J. H., Suhr, N. H., and Bodkin, J. B.: At. Absorption Newslett. **8**, 25 (1969).

[856] Meggers, W. F., Corliss, C. H., and Scribner, B. F.: Tables of Spectral Line Intensities. NBS Monograph 32-I (1961).

[857] Meggers, W. F., and Westfall, F. O.: J. Res. Nat. Bur. Stand. **44**, 447 (1950).

[858] Melton, J. R., Hoover, W. L., and Howard, P. A.: J. Ass. Offic. Agr. Chem. **52**, 950 (1969).

[859] Melton, J. R., Hoover, W. L., Howard, P. A., and Ayers, J. L.: J. Ass. Offic. Agr. Chem. **53**, 682 (1970).

[860] Menis, O., and Rains, T. C.: Anal. Chem. **32**, 1837 (1960).

[861] Mensik, J. D., and Seidemann, H. J.: At. Absorption Newslett. **13**, 8 (1974).

[862] Menzies, A. C.: Anal. Chem. **32**, 898 (1960).

[863] Mesman, B. B., and Smith, B. S.: At. Absorption Newslett. **9**, 81 (1970).

[864] Mesman, B. B., Smith, B. S., and Pierce, J. O.: Am. Ind. Hyg. Assn. J. **31**, 701 (1970).

[865] Midgett, M. R., and Fishman, M. J.: At. Absorption Newslett. **6**, 128 (1967).

[866] Miller, J. R., Helprin, J. J., and Finlayson, J. S.: J. Pharm. Sci. **58**, 455 (1969).

[867] Mitchell, A. C. G., and Zemansky, M. W.: Resonance Radiation and Excited Atoms. Cambridge Univ. Press, New York 1961.

[868] Mitchell, D. G., Jackson, K. W., and Aldous, K. M.: Anal. Chem. **45**, 1215A (1973).

[869] Mitchell, D. G., Ryan, F. J., and Aldous, K. M.: At. Absorption Newslett. **11**, 120 (1972).

[870] Mitchell, K. B.: J. Opt. Soc. Amer. **51**, 846 (1961).

[871] Moffitt, A. E., and Kupel, R. E.: At. Absorption Newslett. **9**, 113 (1970).

[872] Moldan, B.: in Atomic Absorption Spectroscopy (Editors: R. M. Dagnall, and G. F. Kirkbright). Butterworth, London 1970, p. 127.

[873] Monder, C., and Sells, N.: Anal. Biochem. **20**, 215 (1967).

[874] Monkman, J. L., Maffet, P. A., and Doherty, T. F.: Ind. Hyg. Foundation Am. Quart. **17**, 418 (1956).

[875] Montagut-Buscas, M., Obiols, J., and Rodriquez, E.: At. Absorption Newslett. **6**, 61 (1967).

[876] Montford, B., and Cribbs, S. C.: At. Absorption Newslett. **8**, 77 (1969).

[877] Moody, R. R., and Taylor, R. B.: J. Pharm. Pharmac. **24**, 848 (1972).

[878] Moore, E. J., Milner, O. I., and Glass, J. R.: Microchem. J. **10**, 148 (1966).

[879] Morgan, M. E.: At. Absorption Newslett. **3**, 43 (1964).

[880] Morrison, G. H., and Freiser, H.: Solvent Extraction in Analytical Chemistry. John Wiley & Sons, New York 1957.

[881] Morrison, G. H., and Talmi, Y.: Anal. Chem. **42**, 809 (1970).

[882] Morrow, R. W., and McElhaney, R. J.: Appl. Spectrosc. **27**, 387 (1973).

[883] Morrow, R. W., and McElhaney, R. J.: At. Absorption Newslett. **13**, 45 (1974).

[884] Mossotti, V. G., and Duggan, M.: Appl. Optics **7**, 1325 (1968).

[885] Mossotti, V. G., and Fassel, V. A.: Spectrochim. Acta **20**, 1117 (1964).

[886] Mossotti, V. G., Laqua, K., and Hagenah, W. D.: Spectrochim. Acta **23 B**, 197 (1967).

[887] Mostyn, R. A., and Cunningham, A. F.: Anal. Chem. **38**, 121 (1966).

[888] Mostyn, R. A., and Cunningham, A. F.: Anal. Chem. **39**, 433 (1967).

[889] Mostyn, R. A., and Cunningham, A. F.: J. Inst. Petrol. **53**, 101 (1967).

[890] Moten, L.: J. Ass. Offic. Agr. Chem. **53**, 916 (1970).

[891] Mrozowski, S.: Z. Physik **112**, 223 (1939).

[892] Mulford, C. E.: At. Absorption Newslett. **5**, 28 (1966).

[893] Mulford, C. E.: At. Absorption Newslett. **5,** 63 (1966).

[894] Mulford, C. E.: At. Absorption Newslett. **5,** 88 (1966).

[895] Munns, R. K., and Holland, D. C.: J. Ass. Offic. Agr. Chem. **54,** 202 (1971).

[896] Muntz, J. H.: At. Absorption Newslett. **10,** 9 (1971).

[897] Murthy, L., Menden, E. E., Eller, P. M., and Petering, H. G.: Anal. Biochem. **53,** 365 (1973).

[898] Muter, R., and Cockrell, C.: Appl. Spectrosc. **23,** 493 (1969).

[899] Muzzarelli, R. A. A., and Rocchetti, R.: Anal. Chim. Acta **64,** 371 (1973).

[900] Muzzarelli, R. A. A., and Rocchetti, R.: Anal. Chim. Acta **69,** 35 (1974).

[901] Muzzarelli, R. A. A., and Rocchetti, R.: Anal. Chim. Acta **70,** 283 (1974).

[902] Myers, D.: At. Absorption Newslett. **6,** 89 (1967).

[903] Nakahara, T., Munemori, M., and Musha, S.: Anal. Chim. Acta **62,** 267 (1972).

[904] Nakahara, T., Munemori, M., and Musha, S.: Bull. Chem. Soc. Japan **46,** 1162 (1973).

[905] Nakahara, T., Munemori, M., and Musha, S.: Bull. Chem. Soc. Japan **46,** 1172 (1973).

[906] Nesbitt, R. W.: Anal. Chim. Acta **35,** 413 (1966).

[907] Neumann, G. M.: Z. Anal. Chem. **258,** 180 (1972).

[908] Neumann, G. M.: Z. Anal. Chem. **259,** 337 (1972).

[909] Neumann, G. M.: Z. Anal. Chem. **261,** 108 (1972).

[910] Newbrun, E.: Nature **192,** 1182 (1961).

[911] Newland, B. T. N., and Mostyn, R. A.: At. Absorption Newslett. **10,** 89 (1971).

[912] Neybon, R., and Rey-Coquais, B.: At. Absorption Newslett. **6,** 92 (1967).

[913] Ng, W. K.: Anal. Chim. Acta **63,** 469 (1973).

[914] Ng, W. K.: Anal. Chim. Acta **64,** 292 (1973).

[915] Nitis, G. J., Svoboda, V., and Winefordner, J. D.: Spectrochim. Acta **27 B,** 345 (1972).

[916] Nix, J., and Goodwin, T.: At. Absorption Newslett. **9,** 119 (1970).

[917] Nixon, D. E., Fassel, V. A., and Kniseley, R. N.: Anal. Chem. **46,** 210 (1974).

[918] Nomoto, S., and Sunderman, F. W.: Clin. Chem. **16,** 477 (1970).

[919] Nonnenmacher, G., and Schleser, F. H.: Z. Anal. Chem. **209,** 284 (1965).

[920] Nord, P. J., Kadaba, M. P., and Sorenson, J. R. J.: Arch. Environ. Health. **27,** 40 (1973).

[921] Norwitz, G., and Gordon, H.: Talanta **20,** 905 (1973).

[922] Obermiller, E. L., and Freedman, R. W.: Fuel **44,** 199 (1965).

[923] Oddo, N.: Riv. Combustibili **XXV,** 153 (1971).

[924] Olivier, M.: Z. Anal. Chem. **248,** 145 (1969).

[925] Olivier, M.: Z. Anal. Chem. **257,** 135 (1971).

[926] Olivier, M.: Z. Anal. Chem. **257,** 187 (1971).

[927] Olsen, E. D., Jatlow, P. I., Fernandez, F. J., and Kahn, H. L.: Clin. Chem. **19,** 326 (1973).

[928] Olsen, R. D., and Sommerfeld, M. R.: At. Absorption Newslett. **12,** 165 (1973).

[929] Olson, A. D., and Hamlin, W. B.: At. Absorption Newslett. **7,** 69 (1968).

[930] Olson, A. D., and Hamlin, W. B.: Clin. Chem. **15,** 438 (1969).

[931] Olson, A. M.: At. Absorption Newslett. **4,** 278 (1965).

[932] Omang, S. H.: Anal. Chim. Acta **55,** 439 (1971).

[933] Omang, S. H.: Anal. Chim. Acta **56,** 470 (1971).

[934] Omang, S. H.: Anal. Chim. Acta **63,** 247 (1973).

[935] Omenetto, N., Benetti, P., Hart, L. P., Winefordner, J. D., and Alkemade, C. Th. J.: Spectrochim. Acta **28 B,** 289 (1973).

[936] Omenetto, N., Benetti, P., and Rossi, G.: Spectrochim. Acta **27 B,** 453 (1972).

[937] Omenetto, N., Hart, L. P., Benetti, P., and Winefordner, J. D.: Spectrochim. Acta **28 B,** 301 (1973).

[938] Omenetto, N., Hatch, N. N., Fraser, L. M., and Winefordner, J. D.: Spectrochim. Acta **28 B,** 65 (1973).

[939] Omenetto, N., and Rossi, G.: Spectrochim. Acta **24 B,** 95 (1969).

[940] Omenetto, N., and Rossi, G.: IAASC, Sheffield 1969, F7.

[941] Ormer, D. G., and Purdy, W. C.: Anal. Chim. Acta **64,** 93 (1973).

[942] Osborn, K. R., and Gunning, H. E.: J. Opt. Soc. Amer. **45,** 552 (1955).

[943] Osolinski, T. W., and Knight, N. H.: Appl. Spectrosc. **22,** 532 (1968).

[944] Ottaway, J. M., Coker, D. T., Rowston, W. B., and Bhattarai, D. R.: Analyst **95,** 567 (1970).
[945] Ottaway, J. M., and Pradhan, N. K.: Talanta **20,** 927 (1973).
[946] Pagliai, V., and Pozzi, F.: 3rd CISAFA, Paris 1971, p. 907.
[947] Panday, V. K., and Ganguly, A. K.: At. Absorption Newslett. **7,** 50 (1968).
[948] Pappas, E. G., and Rosenberg, L. A.: J. Ass. Offic. Agr. Chem. **49,** 782 (1966).
[949] Pappas, E. G., and Rosenberg, L. A.: J. Ass. Offic. Agr. Chem. **49,** 792 (1966).
[950] Paralusz, C. M.: Appl. Spectrosc. **22,** 520 (1968).
[951] Parker, H. E.: At. Absorption Newslett. **2,** 23 (1963).
[952] Parker, M. W., Humoller, F. L., and Mahler, D. J.: Clin. Chem. **13,** 40 (1967).
[953] Parson, M. L., McCarthy, W. J., and Winefordner, J. D.: Appl. Spectrosc. **20,** 223 (1966).
[954] Parsons, J. A., Dawson, B., Callahan, E., and Potts, J. T.: Biochem. J. **119,** 791 (1970).
[955] Paschen, A.: Physik **50,** 901 (1916).
[956] Paschen, K.: Deut. Med. Wochenschr. **95,** 2570 (1970).
[957] Paschen, K.: Private communication.
[958] Paschen, K., and Fritz, G.: Ärztl. Forsch. **24,** 202 (1970).
[959] Passmore, W. O., and Adams, P. B.: At. Absorption Newslett. **4,** 237 (1965).
[960] Passmore, W. O., and Adams, P. B.: At. Absorption Newslett. **5,** 77 (1966).
[961] Patassy, F. Z.: Plant Soil **22,** 395 (1965).
[962] Patel, B. M., and Winefordner, J. D.: Anal. Chim. Acta **64,** 135 (1973).
[963] Pau, J. C.-M., Pickett, E. E., and Koirtyohann, S. R.: Analyst **97,** 860 (1972).
[964] Paus, P. E.: At. Absorption Newslett. **10,** 44 (1971).
[965] Paus, P. E.: At. Absorption Newslett. **10,** 69 (1971).
[966] Paus, P. E.: At. Absorption Newslett. **11,** 129 (1972).
[967] Paus, P. E.: Z. Anal. Chem. **264,** 118 (1973).
[968] Pawluk, S.: At. Absorption Newslett. **6,** 53 (1967).
[969] Payne, C. E., and Combs, H. F.: Appl. Spectrosc. **22,** 786 (1968).
[970] Pekarek, R. S., and Hauer, E. C.: Fed. Proc. **31,** 700 (1972).
[971] Perkins, J.: Analyst **88,** 324 (1963).
[972] Peterson, E. A.: At. Absorption Newslett. **9,** 129 (1970).
[973] Peterson, G. E.: At. Absorption Newslett. **5,** 177 (1966).
[974] Peterson, G. E., and Kahn, H. L.: At. Absorption Newslett **9,** 71 (1970).
[975] Pichotin, B., and Chasseur, P.: Private communication.
[976] Pickett, E. E., and Koirtyohann, S. R.: Spectrochim. Acta **23 B,** 235 (1968).
[977] Pickett, E. E., and Koirtyohann, S. R.: Spectrochim. Acta **24 B,** 325 (1969).
[978] Pickett, E. E., and Koirtyohann, S. R.: Anal. Chem. **41,** 28 A (1969).
[979] Pickett, E. E., and Pau, J. C.-M.: J. Ass. Offic. Agr. Chem. **56,** 151 (1973).
[980] Pickett, E. E., Pau, J. C.-M., and Koirtyohann, S. R.: J. Ass. Offic. Agr. Chem. **54,** 796 (1971).
[981] Pickford, C. J., and Rossi, G.: Analyst **97,** 647 (1972).
[982] Pickford, C. J., and Rossi, G.: Analyst **98,** 329 (1973).
[983] Pierce, J. O., and Cholak, J.: Arch. Environ. Health **13,** 208 (1966).
[984] Pinta, M., and Riandey, C.: CSI XVII, Florence 1973, p. 71.
[985] Piper, K. G., and Higgins, G.: Proc. Assoc. Clin. Biochem. **4,** 190 (1967).
[986] Pitts, A. E., VanLoon, J. C., and Beamish, F. E.: Anal. Chim. Acta **50,** 181 (1970).
[987] Pitts, A. E., VanLoon, J. C., and Beamish, F. E.: Anal. Chim. Acta **50,** 195 (1970).
[988] Platte, J. A.: in Trace Inorganics in Water. Amer. Chem. Soc. Publ., Advan. Chem. Ser. **73,** 247 (1968).
[989] Platte, J. A., and Marcy, V. M.: At. Absorption Newslett. **4,** 289 (1965).
[990] Pollock, E. N.: At. Absorption Newslett. **9,** 47 (1970).
[991] Pollock, E. N.: At. Absorption Newslett. **10,** 77 (1971).
[992] Pollock, E. N., and Anderson, S. I.: Anal. Chim. Acta **41,** 441 (1968).
[993] Pollock, E. N., and West, S. J.: At. Absorption Newslett. **11,** 104 (1972).
[994] Pollock, E. N., and West, S. J.: At. Absorption Newslett. **12,** 6 (1973).
[995] Poluektov, N. S., and Vitkun, R. A.: Zh. Anal. Khim. **18,** 33 (1963).
[996] Poluektov, N. S., Vitkun, R. A., and Zelyukova, Y. V.: Zh. Anal. Khim. **19,** 873 (1964).

[997] Popham, R. E., and Schrenk, W. G.: Spectrochim. Acta **23 B,** 543 (1968).

[998] Porter, C.: At. Absorption Newslett. **8,** 112 (1969).

[999] Posener, D. W.: Austr. J. Physics **12,** 184 (1959).

[1000] Prasad, A. S., Oberleas, D., and Halsted, J. A.: J. Lab. Clin. Med. **66,** 508 (1965).

[1001] Prévôt, A.: At. Absorption Newslett. **5,** 13 (1966).

[1002] Price, J. P.: At. Absorption Newslett. **11,** 1 (1972).

[1003] Price, W. J., and Roos, J. T. H.: J. Sci. Fd. Agric. **20,** 437 (1969).

[1004] Price, W. J., and Roos, J. T. H.: Met. Ital. **61,** 423 (1969).

[1005] Price, W. J., Roos, J. T. H., and Clay, A. F.: Analyst **95,** 760 (1970).

[1006] Prugger, H.: Optik **21,** 320 (1964).

[1007] Prugger, H., Grosskopf, R., and Torge, R.: Spectrochim. Acta **26 B,** 191 (1971).

[1008] Purushottam, A., Naidu, P. P., and Lal, S. S.: Talanta **20,** 631 (1973).

[1009] Pybus, J., and Bowers, G. N.: Clin. Chem. **16,** 139 (1970).

[1010] Pyrih, R. S., and Bisque, R. E.: Econ. Geol. **64,** 825 (1969).

[1011] Quarrell, T. M., Powell, R. J. W., and Cluley, H. J.: Analyst **98,** 443 (1973).

[1012] Rains, T. C.: in [5], p. 349.

[1013] Rains, T. C., Epstein, M. S., and Menis, O.: Anal. Chem. **46,** 207 (1974).

[1014] Ramakrishna, T. V., Robinson, J. W., and West, P. W.: Anal. Chim. Acta **36,** 57 (1966).

[1015] Ramakrishna, T. V., Robinson, J. W., and West, P. W.: Anal. Chim. Acta **37,** 20 (1967).

[1016] Ramakrishna, T. V., Robinson, J. W., and West, P. W.: Anal. Chim. Acta **39,** 81 (1969).

[1017] Rann, C. S., and Hambly, A. N.: Anal. Chim. Acta **32,** 346 (1965).

[1018] Rann, C. S., and Hambly, A. N.: Anal. Chem. **37,** 879 (1965).

[1019] Rantala, R. T. T., and Loring, D. H.: At. Absorption Newslett. **12,** 97 (1973).

[1020] Rao, P. D.: At. Absorption Newslett. **9,** 131 (1970).

[1021] Rao, P. D.: At. Absorption Newslett. **10,** 118 (1971).

[1022] Rao, P. D.: At. Absorption Newslett. **11,** 45 (1972).

[1023] Rasmuson, J. O., Fassel, V. A., and Kniseley, R. N.: Spectrochim. Acta **28 B,** 365 (1973).

[1024] Rathje, A. O.: Amer. Ind. Hyg. Assoc. J. **30,** 126 (1969).

[1025] Rawling, B. S., Amos, M. D., and Greaves, M. C.: Aust. Inst. Mining Met. Proc. **199,** (1961).

[1026] Rawling, B. S., Greaves, M. C., and Amos, M. D.: Nature **188,** 137 (1960).

[1027] Rawson, R. A. G.: 4th ICAS, Toronto 1973, p. 28.

[1028] Reevers, J. R.: Econ. Geol. **62,** 426 (1967).

[1029] Reif, I., Fassel, V. A., and Kniseley, R. N.: Spectrochim. Acta **28 B,** 105 (1973).

[1030] Reinhold, J. G., Pascoe, E., and Kfoury, G. A.: Anal. Biochem. **25,** 557 (1968).

[1031] Renshaw, G. D.: At. Absorption Newslett. **12,** 158 (1973).

[1032] Renshaw, G. D., Pounds, C. A., and Pearson, E. F.: At. Absorption Newslett. **12,** 55 (1973).

[1033] Reversat, G.: 3rd CISAFA, Paris 1971, p. 769.

[1034] Riandey, C., and Pinta, M.: 3rd CISAFA, Paris 1971, p. 321.

[1035] Ringhardtz, I., and Welz, B.: Z. Anal. Chem. **243,** 190 (1968).

[1036] Rivalenti, G., and Sighinolfi, G. P.: Contr. Mineral. Petr. **23,** 173 (1969).

[1037] Robinson, J. W.: Anal. Chim. Acta **23,** 458 (1960).

[1038] Robinson, J. W.: Anal. Chim. Acta **24,** 451 (1961).

[1039] Robinson, J. W.: Anal. Chem. **33,** 1067 (1961).

[1040] Robles, J., and Lachica, M.: 3rd CISAFA, Paris 1971, p. 453.

[1041] Rodgerson, D. O., and Moran, I. K.: Clin. Chem. **14,** 1206 (1968).

[1042] Roe, D. A., Miller, P. S., and Lutwak, L.: Anal. Biochem. **15,** 313 (1966).

[1043] Roelandts, I.: At. Absorption Newslett. **11,** 48 (1972).

[1044] Rohleder, H. A., Dietl, F., and Sansoni, B.: Spectrochim. Acta **29 B,** 19 (1974).

[1045] Roos, J. T. H.: IAASC, Sheffield 1969, E 5.

[1046] Roos, J. T. H.: Spectrochim. Acta **24 B,** 255 (1969).

[1047] Roos, J. T. H.: Spectrochim. Acta **25 B,** 539 (1970).

[1048] Roos, J. T. H.: Spectrochim. Acta **26 B,** 285 (1971).

[1049] Roos, J. T. H.: Spectrochim. Acta **27 B,** 473 (1972).

[1050] Roos, J. T. H.: Spectrochim. Acta **28 B,** 407 (1973).

[1051] Roos, J. T. H., and Price, W. J.: Spectrochim. Acta **26 B**, 279 (1971).

[1052] Roosels, D., and Vanderkeel, J. V.: At. Absorption Newslett. **7**, 9 (1968).

[1053] Rooney, R. C., and Pratt, C. G.: IAASC, Sheffield 1969, E6.

[1054] Roschnik, R. K.: 4th ICAS, Toronto 1973, p. 106.

[1055] Roschnik, R. K.: Analyst **98**, 596 (1973).

[1056] Rose, G. A., and Willden, E. G.: Analyst **98**, 243 (1973).

[1057] Rose, S. A., and Boltz, D. F.: Anal. Chim. Acta **44**, 239 (1969).

[1058] Ross, R. T., and Gonzalez, J. G.: Anal. Chim. Acta **70**, 443 (1974).

[1059] Ross, R. T., Gonzalez, J. G., and Segar, D. A.: Anal. Chim. Acta **63**, 205 (1973).

[1060] Rossi, G., and Omenetto, N.: Appl. Spectrosc. **21**, 329 (1967).

[1061] Roth, D. J., Bohl, D. R., and Sellers, D. E.: At. Absorption Newslett. **7**, 87 (1968).

[1062] Rousselet, F., Antonetti, A., Englander, J., and Amiel, C.: 3rd CISAFA, Paris 1971, p. 175.

[1063] Rowston, W. B., and Ottaway, J. M.: Anal. Letters **3**, 411 (1970).

[1064] Rubeška, I.: Anal. Chim. Acta **40**, 187 (1968).

[1065] Rubeška, I.: in [5], p. 317.

[1066] Rubeška, I.: At. Absorption Newslett. **12**, 33 (1973).

[1067] Rubeška, I.: Spectrochim. Acta **29 B**, 263 (1974).

[1068] Rubeška, I., and Mikšovsky, M.: At. Absorption Newslett. **11**, 57 (1972).

[1069] Rubeška, I., and Moldan, B.: Anal. Chim. Acta **37**, 421 (1967).

[1070] Rubeška, I., and Moldan, B.: Appl. Optics **7**, 1341 (1968).

[1071] Rubeška, I., and Štupar, J.: At. Absorption Newslett. **5**, 69 (1966).

[1072] Rubeška, I., Šulcek, Z., and Moldan, B.: Anal. Chim. Acta **37**, 27 (1967).

[1073] Rubeška, I., and Svoboda, V.: Anal. Chim. Acta **32**, 253 (1965).

[1074] Russell, B. J., Shelton, J. P., and Walsh, A.: Spectrochim. Acta **8**, 317 (1957).

[1075] Russell, B. J., and Walsh, A.: Spectrochim. Acta **10**, 883 (1959).

[1076] Russos, G. F., and Morrow, B. H.: Appl. Spectrosc. **22**, 769 (1968).

[1077] Sachdev, S. L., Robinson, J. W., and West, P. W.: Anal. Chim. Acta **38**, 499 (1967).

[1078] Salama, C.: At. Absorption Newslett. **10**, 72 (1971).

[1079] Sattur, T. W.: At. Absorption Newslett. **5**, 37 (1966).

[1080] Savory, J., Rozel, N. O., Mushak, P., and Sunderman, F. W.: Am. J. Clin. Pathol. **50**, 505 (1968).

[1081] Savory, J., Wiggins, J. W., and Heintges, M. G.: Am. J. Clin. Pathol. **51**, 720 (1969).

[1082] Scarborough, J. M.: Anal. Chem. **41**, 250 (1969).

[1083] Scarborough, J. M., Bingham, C. D., and de Vries, P. F.: Anal. Chem. **39**, 1394 (1967).

[1084] Schaefer, C., and Vomhof, D. W.: At. Absorption Newslett. **12**, 133 (1973).

[1085] Schaller, K. H., Essing, H. G., Valentin, H., and Schäcke, G.: Z. Klin. Chem. Klin. Biochem. **10**, 434 (1972).

[1086] Schaller, K. H., Kühnert, A., and Lehnert, G.: Blut **17**, 155 (1968).

[1087] Schallis, J. E., and Kahn, H. L.: At. Absorption Newslett. **7**, 75 (1968).

[1088] Schiller, R.: At. Absorption Newslett. **9**, 111 (1970).

[1089] Schlebusch, H., and Niehoff, B.: Biochem. Anal. **74**, Munich 1974.

[1090] Schlewitz, J. H., and Shields, M. G.: At. Absorption Newslett. **10**, 39 (1971).

[1091] Schlewitz, J. H., and Shields, M. G.: At. Absorption Newslett. **10**, 43 (1971).

[1092] Schmidt, F. J., and Royer, J. L.: Anal. Letters **6**, 17 (1973).

[1093] Schmidt, W., and Sansoni, B.: Biochem. Anal. **74**, Munich 1974.

[1094] Schneider, W., and Matter, L.: Lebensmittelchem. Gerichtl. Chem. **28**, 3 (1974).

[1095] Schnepfe, M. M., and Grimaldi, F. S.: Talanta **16**, 1461 (1969).

[1096] Schramel, P.: Anal. Chim. Acta **67**, 69 (1973).

[1097] Schrenk, W. G., Lehman, D. A., and Neufeld, L.: Appl. Spectrosc. **20**, 389 (1966).

[1098] Schroeder, H. A., and Nason, A. P.: Clin. Chem. **17**, 461 (1971).

[1099] Schulz-Baldes, M.: Marine Biology **16**, 226 (1972).

[1100] Schwab, M. R., and Hembree, N. H.: At. Absorption Newslett. **10**, 15 (1971).

[1101] Schweizer, V. B.: At. Absorption Newslett. **14**, 137 (1975).

[1102] Scott, R. L.: At. Absorption Newslett. **9**, 46 (1970).

[1103] Searle, B., Chan, W., Jensen, C., and Davidow, B.: At. Absorption Newslett. **8,** 126 (1969).
[1104] Sebens, C., Vollmer, J., and Slavin, W.: At. Absorption Newslett. **3,** 165 (1964).
[1105] Sebestyen, N. A.: Spectrochim. Acta **25 B,** 261 (1970).
[1106] Segal, R. J.: Clin. Chem. **15,** 1124 (1969).
[1107] Segar, D. A.: 3rd CISAFA, Paris 1971, p. 523.
[1108] Segar, D. A.: Anal. Letters **7,** 89 (1974).
[1109] Segar, D. A., and Gonzalez, J. G.: Anal. Chim. Acta **58,** 7 (1972).
[1110] Segar, D. A., and Gonzalez, J. G.: 4th ICAS, Toronto 1973, p. 93.
[1111] Segar, D. A., Gonzalez, J. G., Gilio, J. L., and Pellenbarg, R. E.: 4th ICAS, Toronto 1973, p. 125.
[1112] Selander, S., and Cramer, K.: British J. Ind. Med. **25,** 139 (1968).
[1113] Seller, R. H., Ramirez-Muxo, O., Brest, A. N., and Moyer, J. H.: J. Amer. Med. Ass. **191,** 118 (1965).
[1114] SenGupta, J. G.: Anal. Chim. Acta **58,** 23 (1972).
[1115] SenGupta, J. G.: Anal. Chim. Acta **63,** 19 (1973).
[1116] SenGupta, J. G.: Miner. Sci. Engng. **5,** 207 (1973).
[1117] Severne, B. C., and Brooks, R. R.: Talanta **19,** 1467 (1972).
[1118] Severne, B. C., and Brooks, R. R.: Anal. Chim. Acta **58,** 216 (1972).
[1119] Shafto, R. G.: At. Absorption Newslett. **3,** 115 (1964).
[1120] Shkolnik, G. M., and Bevill, R. F.: At. Absorption Newslett. **12,** 112 (1973).
[1121] Sideman, L., Murphy, J. J., and Wilson, D. T.: Clin. Chem. **16,** 597 (1970).
[1122] Siemer, D., Lech, J. F., and Woodriff, R.: Spectrochim. Acta **28 B,** 469 (1973).
[1123] Sighinolfi, G. P.: At. Absorption Newslett. **11,** 96 (1972).
[1124] Sighinolfi, G. P.: At. Absorption Newslett. **12,** 136 (1973).
[1125] Silvester, M. D., Koop, D. J., and Barringer, A. R.: 4th ICAS, Toronto 1973, p. 92.
[1126] Simmons, E. C.: At. Absorption Newslett. **4,** 281 (1965).
[1127] Simmons, W. J.: Anal. Chem. **45,** 1947 (1973).
[1128] Simonian, J. V.: At. Absorption Newslett. **7,** 63 (1968).
[1129] Simpson, G. R., and Blay, R. A.: Food Trade Rev. **36,** No. 8, 35 (1966).
[1130] Siren, M. J.: Sci. Tools **11,** 37 (1964).
[1131] Skerry, P. J., and Chapman, W.: 4th ICAS, Toronto 1973, p. 121.
[1132] Skewes, H. R.: Aust. Inst. Mining Met. Proc. **211,** 217 (1964).
[1133] Skogerboe, R. K.: in [5], p. 381.
[1134] Skogerboe, R. K., and Woodriff, R. A.: Anal. Chem. **35,** 1977 (1963).
[1135] Slavin, S., Barnett, W. B., and Kahn, H. L.: At. Absorption Newslett. **11,** 37 (1972).
[1136] Slavin, S., Fernandez, F. J., and Manning, D. C.: 25th Pittsburgh Conf. Anal. Chem. Appl. Spectrosc., Cleveland 1974.
[1137] Slavin, S., and Sattur, T. W.: At. Absorption Newslett. **7,** 99 (1968).
[1138] Slavin, S., and Slavin, W.: At. Absorption Newslett. **5,** 106 (1966).
[1139] Slavin, W.: At. Absorption Newslett. **2,** 1 (1963).
[1140] Slavin, W.: At. Absorption Newslett. **3,** 93 (1964).
[1141] Slavin, W.: At. Absorption Newslett. **4,** 192 (1965).
[1142] Slavin, W.: At. Absorption Newslett. **4,** 243 (1965).
[1143] Slavin, W.: At. Absorption Newslett. **4,** 330 (1965).
[1144] Slavin, W.: Occupational Health Rev. **17,** 9 (1965).
[1145] Slavin, W.: At. Absorption Newslett. **6,** 9 (1967).
[1146] Slavin, W., and Manning, D. C.: Anal. Chem. **35,** 253 (1963).
[1147] Slavin, W., and Manning, D. C.: Appl. Spectrosc. **19,** 65 (1965).
[1148] Slavin, W., Sebens, C., and Sprague, S.: At. Absorption Newslett. **4,** 341 (1965).
[1149] Slavin, W., and Slavin, S.: Appl. Spectrosc. **23,** 421 (1969).
[1150] Slavin, W., and Sprague, S.: At. Absorption Newslett. **3,** 1 (1964).
[1151] Slavin, W., Trent, D. J., and Sprague, S.: At. Absorption Newslett. **4,** 180 (1965).
[1152] Slavin, W., Venghiattis, A., and Manning, D. C.: At. Absorption Newslett. **5,** 84 (1966).
[1153] Smith, D., and McLain, M. E.: Radiochem. Radioanal. Letters **16,** 89 (1974).

[1154] Smith, K. E., and Frank, C. W.: Appl. Spectrosc. **22,** 765 (1968).

[1155] Smith. R., and Winefordner, J. D.: Spectrosc. Letters **1,** 157 (1968).

[1156] Smith, R., and Winefordner, J. D.: IAASC, Sheffield 1969, C 2.

[1157] Smith, R. V.: Int. Laboratory **1973,** No. 2, 39.

[1158] Smyly, D. S., Townsend, W. P., Zeegers, P. J. Th., and Winefordner, J. D.: Spectrochim. Acta **26 B,** 531 (1971).

[1159] Snelleman, W.: in [5], p. 213.

[1160] Sobolev, N. N.: Spectrochim. Acta **11,** 310 (1956).

[1161] Spector, H., Clusman, S., Jatlow, P., and Seligson, D.: Clin. Chim. Acta **31,** 5 (1971).

[1162] Spielholtz, G. I., and Toralballa, G. C.: Analyst **94,** 1072 (1969).

[1163] Spitz, J., and Uny, G.: Appl. Optics **7,** 1345 (1968).

[1164] Spitzer, H.: Z. Erzbergbau Metallhüttenwes. **19,** 567 (1966).

[1165] Spitzer, H., and Tesik, G.: Z. Anal. Chem. **232,** 40 (1967).

[1166] Spooner, C. M., and Crassweller, P. O.: At. Absorption Newslett. **11,** 72 (1972).

[1167] Sprague, S., Manning, D. C., and Slavin, W.: At. Absorption Newslett. **3,** 27 (1964).

[1168] Sprague, S., and Slavin, W.: At. Absorption Newslett. **3,** 37 (1964).

[1169] Sprague, S., and Slavin, W.: At. Absorption Newslett. **3,** 72 (1964).

[1170] Sprague, S., and Slavin, W.: At. Absorption Newslett. **3,** 160 (1964).

[1171] Sprague, S., and Slavin, W.: At. Absorption Newslett. **4,** 228 (1965).

[1172] Sprague, S., and Slavin, W.: At. Absorption Newslett. **4,** 293 (1965).

[1173] Sprague, S., and Slavin, W.: At. Absorption Newslett. **4,** 367 (1965).

[1174] Stevens, B. J.: Clin. Chem. **18,** 1379 (1972).

[1175] Stewart, W. K., Hutchinson, F., and Fleming, L. W.: J. Lab. Clin. Med. **61,** 858 (1963).

[1176] Stone, M., and Chesher, S. E.: Analyst **94,** 1063 (1969).

[1177] Strasheim, A., and Butler, L. R. P.: Appl. Spectrosc. **16,** 109 (1962).

[1178] Strasheim, A., and Human, H. G. C.: Spectrochim. Acta **23 B,** 265 (1968).

[1179] Strasheim, A., Norval, E., and Butler, L. R. P.: J. S. Afr. Chem. Inst. **17,** 55 (1964).

[1180] Strasheim, A., Strelow, F. W. E., and Butler, L. R. P.: J. S. Afr. Chem. Inst. **13,** 73 (1960).

[1181] Strasheim, A., and Wessels, G. J.: Appl. Spectrosc. **17,** 65 (1963).

[1182] Strelow, F. W. E., Feast, E. C., Mathews, P. M., Bothma, C. J. C., and van Zyl, C. R.: Anal. Chem. **38,** 115 (1966).

[1183] Stresko, V., and Martiny, E.: At. Absorption Newslett. **11,** 4 (1972).

[1184] Strunk, D. H., and Andreasen, A. A.: At. Absorption Newslett. **6,** 111 (1967).

[1185] Stupar, J.: Z. Anal. Chem. **203,** 401 (1964).

[1186] Stupar, J.: Microchim. Acta **1966,** 722.

[1187] Stupar, J., and Dawson, J. B.: Appl. Optics **7,** 1351 (1968).

[1188] Stupar, J., and Dawson, J. B.: At. Absorption Newslett. **8,** 38 (1969).

[1189] Stupar, J., Podobnik, B., and Korosin, J.: Croat. Chem. Acta **37,** 141 (1965).

[1190] Suhr, N. H.: XIII CSI, Ottawa 1967.

[1191] Suhr, N. H., and Ingamells, C. O.: Anal. Chem. **38,** 730 (1966).

[1192] Sullivan, J. V., Parker, M., and Carson, S. B.: J. Lab. Clin. Med. **71,** 893 (1968).

[1193] Sullivan, J. V., and Walsh, A.: Spectrochim. Acta **21,** 721 (1965).

[1194] Sullivan, J. V., and Walsh, A.: Spectrochim. Acta **21,** 727 (1965).

[1195] Sullivan, J. V., and Walsh, A.: Spectrochim. Acta **22,** 1843 (1966).

[1196] Sullivan, J. V., and Walsh, A.: VI Aust. Spectrosc. Conf., Brisbane 1967.

[1197] Sullivan, J. V., and Walsh, A.: XIII CSI, Ottawa 1967.

[1198] Sullivan, J. V., and Walsh, A.: Appl. Optics **7,** 1271 (1968).

[1199] Sundberg, L. L.: Anal. Chem. **45,** 1460 (1973).

[1200] Sunderman, F. W.: Am. J. Clin. Pathol. **44,** 182 (1965).

[1201] Sunderman, F. W.: Human Pathol. **4,** 549 (1973).

[1202] Sunderman, F. W., and Carroll, J. E.: Am. J. Clin. Pathol. **43,** 302 (1965).

[1203] Sunderman, F. W., and Nechay, M. W.: Biochem. Anal. **74,** Munich 1974.

[1204] Sunderman, F. W., and Roszel, N. O.: Am. J. Clin. Pathol. **48,** 286 (1967).

[1205] Supp, G. R.: At. Absorption Newslett. **11,** 122 (1972).

[1206] Supp, G. R., Gibbs, I., and Juszli, M.: At. Absorption Newslett. **12,** 66 (1973).
[1207] Sutter, E., Platman, S. R., and Fieve, R. R.: Clin. Chem. **16,** 602 (1970).
[1208] Swider, R. T.: At Absorption Newslett. **7,** 111 (1968).
[1209] Sychra, V., and Kolihová, D.: 3rd CISAFA, Paris 1971, p. 265.
[1210] Sychra, V., and Matoušek, J.: Anal. Chim. Acta **52,** 376 (1970).
[1211] Sychra, V., and Matoušek, J.: Talanta **17,** 363 (1970).
[1212] Sychra, V., Slevin, P. J., Matoušek, J., and Bek, F.: Anal. Chim. Acta **52,** 259 (1970).
[1213] Syty, A.: Anal. Letters **4,** 531 (1971).
[1214] Syty, A.: At. Absorption Newslett. **12,** 1 (1973).
[1215] Takahashi, M., and Uruno, Y.: Bunko Kenkyu **10,** 110 (1962).
[1216] Tarlin, I. H., and Batchelder, M.: J. Pharmac. Sci. **59,** 1328 (1970).
[1217] Taulli, T. A., and Kaelble, E. F.: At. Absorption Newslett. **9,** 100 (1970).
[1218] Tavenier, P., and Hellendoorn, H. B. A.: Clin. Chim. Acta **23,** 47 (1969).
[1219] Taylor, J. H.: At. Absorption Newslett. **8,** 95 (1969).
[1220] Taylor, M. L., and Belcher, C. B.: Anal. Chim. Acta **45,** 219 (1969).
[1221] Temperli, A. T., and Misteli, H.: Anal. Biochem. **27,** 361 (1969).
[1222] Tenoutasse, N.: 3rd CISAFA, Paris 1971, p. 817.
[1223] Terashima, S.: Japan Analyst **18,** 1259 (1969).
[1224] Thilliez, G.: Anal. Chem. **39,** 427 (1967).
[1225] Thilliez, G.: Chimie Analytique **50,** 226 (1968).
[1226] Thompson, M. H.: J. Ass. Offic. Agr. Chem. **52,** 55 (1969).
[1227] Thompson, R. J., Morgan, G. B., and Purdue, L. J.: At. Absorption Newslett. **9,** 53 (1970).
[1228] Thormahlen, D. J., and Frank, E. H.: At. Absorption Newslett. **10,** 63 (1971).
[1229] Thorpe, V. A.: J. Ass. Offic. Agr. Chem. **54,** 206 (1971).
[1230] Thorpe, V. A.: J. Ass. Offic. Agr. Chem. **56,** 147 (1973).
[1231] Tindall, F. M.: At. Absorption Newslett. **4,** 339 (1965).
[1232] Tindall, F. M.: At. Absorption Newslett. **5,** 140 (1966).
[1233] Tindall, F. M.: At. Absorption Newslett. **6,** 104 (1967).
[1234] Toffoli, P., and Pannetier, G.: 3rd CISAFA, Paris 1971, p. 707.
[1235] Topping, G., and Pirie, J. M.: Anal. Chim. Acta **62,** 200 (1972).
[1236] Toshimitsu, M.: J. Physic. Soc. Japan. **17,** 1440 (1962).
[1237] Toth, S. J., and Reimer, D. N.: Indust. Water Eng. Sept. **1967,** 42.
[1238] Townsend, W. P., Smyly, D. S., Zeegers, P. J. T., Svoboda, V., and Winefordner, J. D.: Spectrochim. Acta **26 B,** 595 (1971).
[1239] Trent, D. J.: At. Absorption Newslett. **4,** 348 (1965).
[1240] Trent, D. J., Manning, D. C., and Slavin, W.: At. Absorption Newslett. **4,** 335 (1965).
[1241] Trent, D. J., and Slavin, W.: At. Absorption Newslett. **3,** 17 (1964).
[1242] Trent, D. J., and Slavin, W.: At. Absorption Newslett. **3,** 53 (1964).
[1243] Trent, D. J., and Slavin, W.: At. Absorption Newslett. **3,** 118 (1964).
[1244] Trent, D. J., and Slavin, W.: At. Absorption Newslett. **4,** 300 (1965).
[1245] Trudeau. D. L., and Freier, E. F.: Clin. Chem. **13,** 101 (1967).
[1246] Tušl, J.: J. Ass. Offic. Agr. Chem. **53,** 1190 (1970).
[1247] Tyler, J. B.: At. Absorption Newslett. **6,** 14 (1967).
[1248] Tyndall, J.: Six Lectures on Light. D. Appleton & Co, New York 1898.
[1249] Ulfvarson, U.: Acta Chem. Scand. **21,** 641 (1967).
[1250] Ure, A.M.: IAASC, Sheffield 1969, B 7.
[1251] Uthe, J. F., Armstrong, F. A. J., and Stainton, M. P.: J. Fish. Res. Bd. Canada **27,** 805 (1970).
[1252] VanLoon, J. C.: At. Absorption Newslett. **7,** 3 (1968).
[1253] VanLoon, J. C.: At. Absorption Newslett. **8,** 6 (1969).
[1254] VanLoon, J. C.: At. Absorption Newslett. **11,** 60 (1972).
[1255] VanLoon, J. C., Aarden, D., and Galbraith, J.: IAASC, Sheffield 1969, B 2.
[1256] VanLoon, J. C., and Parissis, C. M.: Anal. Letters **1,** 249 (1968).
[1257] VanLoon, J. C., and Parissis, C. M.: Analyst **94,** 1057 (1969).
[1258] Varju, M. E.: At. Absorption Newslett. **11,** 45 (1972).

[1259] Varju, M. E.: II ČS. Conf. Flame Spectrosc., Zvikov 1973.
[1260] Varju, M. E., and Elek, E.: At. Absorption Newslett. **10,** 128 (1971).
[1261] Varley, J. A., and Chin, P. Y.: Analyst **95,** 592 (1970).
[1262] Vaughn, W. W., and McCarthy, J. H.: US Geol. Survey Prof. Paper **501-D,** D-123 (1964).
[1263] Veall, N.: Medical uses of Ca-47. Int. Atom. Energ. Ag., Vienna, Techn. Report **32** (1964).
[1264] Veillon, C., Mansfield, J. M., Parsons, M. L., and Winefordner, J. D.: Anal. Chem. **38,** 204 (1966).
[1265] Veillon, C., and Margoshes, M.: Spectrochim. Acta **23 B,** 503 (1968).
[1266] Venghiattis, A. A.: Spectrochim. Acta **23 B,** 67 (1967).
[1267] Venghiattis, A. A.: Appl. Optics **7,** 1313 (1968).
[1268] Vickers, T. J., and Vaughn, R. M.: Anal. Chem. **41,** 1476 (1969).
[1269] Vidale, G. L.: General Electric T. I. S. Report R60SD330 (1961).
[1270] Vidale, G. L.: General Electric T. I. S. Report R60SD331 (1962).
[1271] Vijan, P. N., and Wood, G. R.: At. Absorption Newslett. **13,** 33 (1974).
[1272] Vogliotti, F. L.: At. Absorption Newslett. **9,** 123 (1970).
[1273] Voinovitch, I., Legrand, G., and Louvrier, J.: 3rd CISAFA, Paris 1971, p. 843.
[1274] Volhard, J.: J. Prakt. Chem. **2,** 217 (1874).
[1275] Vollmer, J.: At. Absorption Newslett. **5,** 12 (1966).
[1276] Vollmer, J.: At. Absorption Newslett. **5,** 35 (1966).
[1277] Vollmer, J., Sebens, C., and Slavin, W.: At. Absorption Newslett. **4,** 306 (1965).
[1278] Vulfson, E. K., Karyakin, A. V., and Shidlovsky, A. I.: Zh. Anal. Khim. **28,** 1253 (1973).
[1279] Wadman, B. W.: Diesel and Gas Turbine Progress, July **1971,** 16.
[1280] Walker, C. R., and Vita, O. A.: Anal. Chim. Acta **43,** 27 (1968).
[1281] Wallace, F. J.: Analyst **88,** 259 (1963).
[1282] Walsh, A.: Spectrochim. Acta **7,** 108 (1955).
[1283] Walsh, A.: X CSI, Maryland 1962.
[1284] Walsh, A.: in Atomic Absorption Spectroscopy (Editors: R. M. Dagnall, and G. F. Kirkbright). Butterworth, London 1970, p. 1.
[1285] Walsh, A.: Int. Congr. Anal. Chem., Kyoto 1972.
[1286] Walsh, A.: Appl. Spectrosc. **27,** 335 (1973).
[1287] Warren, J. M., and Spencer, H.: Clin. Chim. Acta **38,** 435 (1972).
[1288] Weger, S. J., Hossner, L. R., and Ferrara, L. W.: J. Agr. Food Chem. **17,** 1276 (1969).
[1289] Welcher, G. G., and Kriege, O. H.: At. Absorption Newslett. **8,** 97 (1969).
[1290] Welcher, G. G., and Kriege, O. H.: At. Absorption Newslett. **9,** 61 (1970).
[1291] Welcher, G. G., Kriege, O. H., and Marks, J. Y.: Anal. Chem. **46,** 1227 (1974).
[1292] Welz, B.: CZ **95,** T99 (1971).
[1293] Welz, B.: 3rd CISAFA, Paris 1971, p. 655.
[1294] Welz, B.: CZ-Chemie-Tech. **1,** 455 (1972).
[1295] Welz, B.: Fortschr. Miner. **50,** (Bh. 1), 106 (1972).
[1296] Welz, B.: Z. Werkstofftechnik **4,** 285 (1973).
[1297] Welz, B.: CSI XVII, Florence 1973, p. 67.
[1298] Welz, B.: Vom Wasser **42,** p. 119 (1974).
[1299] Welz, B., and Sebestyen, J. E.: Joint Sympos. Accurate Meth. Anal. Maj. Const., London 1970.
[1300] Welz, B., and Wiedeking, E.: Int. Sympos. Mikrochem., Vol. B. Verlag der Wiener Medizinischen Akademie, Wien 1970, p. 299.
[1301] Welz, B., and Wiedeking, E.: Z. Anal. Chem. **252,** 111 (1970).
[1302] Welz, B., and Wiedeking, E.: Z. Anal. Chem. **264,** 110 (1973).
[1303] Welz, B., and Witte, W.: 4th ICAS, Toronto 1973, p. 90.
[1304] Wendt, R. H., and Fassel, V. A.: Anal. Chem. **37,** 920 (1965).
[1305] Wendt, R. H., and Fassel, V. A.: Anal. Chem. **38,** 337 (1966).
[1306] West, A. C., Fassel, V. A., and Kniseley, R. N.: Anal. Chem. **45,** 1586 (1973).
[1307] West, C. D., and Hume, D. N.: Anal. Chem. **36,** 412 (1964).
[1308] West, F. K., West, P. W., and Ramakrishna, T. V.: Environ. Sci. Technol. **1,** 717 (1967).
[1309] West, T. S.: Endeavour **26,** 44 (1967).

[1310] West, T. S.: Int. Sympos. Mikrochem., Vol. B. Verlag der Wiener Medizinischen Akademie, Wien 1970, p. 5.
[1311] West, T. S., and Williams, X. K.: Anal. Chim. Acta **45,** 27 (1969).
[1312] Westerlund-Helmerson, U.: At. Absorption Newslett. **5,** 97 (1966).
[1313] Wheat, J. A.: XI Conf. Anal. Chem. Nucl. Techn., Gatlinburg, Tenn., 1967.
[1314] Wheat, J. A.: USAEC-Report DP-1164, Dupont Savannah River Lab., 1968.
[1315] Wheat, J. A.: Appl. Spectrosc. **25,** 3 (1971).
[1316] White, A. D.: J. Appl. Phys. **30,** 711 (1959).
[1317] White, R. A.: J. Sci. Instrum. **44,** 678 (1967).
[1318] Whittington, C. M., and Willis, J. B.: Plating **51,** 767 (1964).
[1319] Williams, A. I.: Analyst **98,** 233 (1973).
[1320] Williams, C. H., David, D. J., and Iismaa, O.: J. Agric. Sci. **59,** 381 (1962).
[1321] Williams, T. R., Wilkinson, B., Wadsworth, G. A., Barther, D. H., and Beer, W. J.: J. Sci. Fd. Agric. **17,** 344 (1967).
[1322] Willis, J. B.: Nature **184,** 186 (1959).
[1323] Willis, J. B.: Nature **186,** 249 (1960).
[1324] Willis, J. B.: Spectrochim. Acta **16,** 259 (1960).
[1325] Willis, J. B.: Spectrochim. Acta **16,** 273 (1960).
[1326] Willis, J. B.: Spectrochim. Acta **16,** 551 (1960).
[1327] Willis, J. B.: Nature **191,** 381 (1961).
[1328] Willis, J. B.: Anal. Chem. **33,** 556 (1961).
[1329] Willis, J. B.: Anal. Chem. **34,** 614 (1962).
[1330] Willis, J. B.: in Methods of Biochemical Analysis, Vol. XI. Interscience, New York/London/ Sydney 1963.
[1331] Willis, J. B.: Aust. J. Dairy Technol., June **1964,** 70.
[1332] Willis, J. B.: Clin. Chem. **11,** 251 (1965).
[1333] Willis, J. B.: Nature **207,** 715 (1965).
[1334] Willis, J. B.: Appl. Optics **7,** 1295 (1968).
[1335] Willis, J. B.: Spectrochim. Acta **26 B,** 177 (1971).
[1336] Willis, J. B., Fassel, V. A., and Fiorino, J. A.: Spectrochim. Acta **24 B,** 157 (1969).
[1337] Willis, J. B., Rasmuson J. O., Kniseley, R. N., and Fassel, V. A.: Spectrochim. Acta **23 B,** 725 (1968).
[1338] Wilson, J.: Unpublished.
[1339] Wilson, L.: Anal. Chim. Acta **30,** 377 (1964).
[1340] Wilson, L.: Anal. Chim. Acta **40,** 503 (1968).
[1341] Winefordner, J. D.: in Atomic Absorption Spectroscopy (Editors: R. M. Dagnall, and G. F. Kirkbright). Butterworth, London 1970, p. 35.
[1342] Winefordner, J. D.: CSI XVII, Florence 1973, p. 58.
[1343] Winefordner, J. D., and Mansfield, J. M.: Appl. Spectrosc. Rev. **1,** 1 (1967).
[1344] Winefordner, J. D., and Parsons, M. P.: Anal. Chem. **39,** 1593 (1966).
[1345] Winefordner, J. D., and Staab, R. A.: Anal. Chem. **36,** 165 (1964).
[1346] Winefordner, J. D., and Staab, R. A.: Anal. Chem. **36,** 1367 (1964).
[1347] Winefordner, J. D., and Vickers, T. J.: Anal. Chem. **36,** 161 (1964).
[1348] Winefordner, J. D., and Vickers, T. J.: Anal. Chem. **36,** 1947 (1964).
[1349] Witkind, I. J.: Am. Mineral. **54,** 1118 (1969).
[1350] Wollin, A.: At. Absorption Newslett. **9,** 43 (1970).
[1351] Woodriff, R., and Ramelow, G.: Spectrochim. Acta **23 B,** 665 (1968).
[1352] Woodriff, R., and Stone, R. W.: Appl. Optics 7, 1337 (1968).
[1353] Woodriff, R., Stone R. W., and Held, A. M.: Appl. Spectrosc. **22,** 408 (1968).
[1354] Woodson, T. T.: Rev. Sci. Instrum. **10,** 308 (1939).
[1355] Woodward, C.: At. Absorption Newslett. **8,** 121 (1969).
[1356] Woolley, J. F.: Spectrovision **1969,** 7.
[1357] Woolley, J. F.: 4th ICAS, Toronto 1973, p. 91.
[1358] Wu, J. Y. L., Droll, H. A., and Lott, P. F.: At. Absorption Newslett. **7,** 90 (1968).

[1359] Yamamoto, Y., Kumamaru, T., and Hayashi, Y.: Talanta **14,** 611 (1967).
[1360] Yanagisawa, M., Suzuki, M., and Takeuchi, T.: Talanta **14,** 933 (1967).
[1361] Yanagisawa, M., Suzuki, M., and Takeuchi, T.: Microchim. Acta **1973,** 475.
[1362] Yeh, Y-Y., and Zee, P.: Clin. Chem. **20,** 360 (1974).
[1363] Youden, W. J.: J. Ass. Offic. Agr. Chem. **46,** 55 (1963).
[1364] Yoza, N., and Ohashi, S.: Anal. Letters **6,** 595 (1973).
[1365] Yuan, T. L., and Breland, H. L.: Soil Sci. Soc. Amer. Proc. **33,** 868 (1969).
[1366] Yule, J. W., and Swanson, G. A.: At. Absorption Newslett. **8,** 30 (1969).
[1367] Zacha, K. E., Bratzel, M. P., Winefordner, J. D., and Mansfield, J. M.: Anal. Chem. **40,** 1733 (1968).
[1368] Zacha, K. E., and Winefordner, J. D.: Anal. Chem. **38,** 1537 (1966).
[1369] Zaidel, A. N., and Korennoi, E. P.: Opt. Spectrosc. **10,** 299 (1961).
[1370] Zaino, E. C.: At. Absorption Newslett. **6,** 93 (1967).
[1371] Zaugg, W. S.: At. Absorption Newslett. **6,** 63 (1967).
[1372] Zaugg, W. S., and Knox, R. J.: Anal. Chem. **38,** 1759 (1966).
[1373] Zaugg, W. S., and Knox, R. J.: Anal. Biochem. **20,** 282 (1967).
[1374] Zeegers, P. J. T., Smith, R., and Winefordner, J. D.: Anal. Chem. **40,** 26 (1968).
[1375] Zeegers, P. J. T., and Winefordner, J. D.: Spectrochim. Acta **26 B,** 161 (1971).
[1376] Zeeman, P. B., and Brink, J. A.: Analyst **93,** 388 (1968).
[1377] Zeeman, P. B., and Butler, L. R. P.: Tegnicon **13,** 96 (1960).
[1378] Zeeman, P. B., and Butler, L. R. P.: Appl. Spectrosc. **16,** 120 (1962).
[1379] Zettner, A.: in Adv. in Clin. Chem., Vol. 7. Academic Press, New York 1965.
[1380] Zettner, A., and Mansbach, L.: Am. J. Clin. Pathol. **44,** 517 (1965).
[1381] Zettner, A., and Mensch, A. H.: Am. J. Clin. Pathol. **48,** 225 (1967).
[1382] Zettner, A., and Mensch, A. H.: Am. J. Clin. Pathol. **49,** 196 (1968).
[1383] Zettner, A., Rafferty, K., and Jarecky, H. J.: At. Absorption Newslett. **7,** 32 (1968).
[1384] Zettner, A., and Seligson, D.: Clin. Chem. **10,** 869 (1964).
[1385] Zettner, A., Sylvia, L. C., and Capacho-Delgado, L.: Am. J. Clin. Pathol. **45,** 533 (1966).
[1386] Zinterhofer, L. J. M., Jatlow. P. I., and Fappiano, A.: J. Lab. Clin. Med. **78,** 664 (1971).
[1387] Zlalkis, A., Bruewing, W., and Bayler, E.: Anal. Chem. **41,** 1692 (1969).
[1388] Zuehlke, C. W., and Ballard, A. E.: Anal. Chem. **22,** 953 (1950).

[2000] Adams, D. B., Brown, S. S., Sunderman, F. W., and Zachariasen, H., Clin. Chem. **24,** 862 (1978).
[2001] Adams, M. J., and Kirkbright, G. F., Canad. J. Spectrosc. **21,** 127 (1976).
[2002] Adams, M. J., Kirkbright, G. F., and Rienvatana, P., At. Absorption Newslett. **14,** 105 (1975).
[2003] Ader, D., and Stoeppler, M., J. Anal. Toxicol. **1,** 252 (1977).
[2004] Agemian, H., Aspila, K. I., and Chau, A. S. Y., Anal. Chem. **47,** 1038 (1975).
[2005] Agemian, H., and Bedek, E., Anal. Chim. Acta **119,** 323 (1980).
[2006] Agemian, H., and Chau, A. S. Y., Anal. Chim. Acta **80,** 61 (1975).
[2007] Agemian, H., and Chau, A. S. Y., Anal. Chem. **50,** 13 (1978).
[2008] Agemian, H., and Cheam, V., Anal. Chim. Acta **101,** 193 (1978).
[2009] Agemian, H., and daSilva, J. A., Anal. Chim. Acta **104,** 285 (1979).
[2010] Agemian, H., Sturtevant, D. P., and Austen, K. D., Analyst **105,** 125 (1980).
[2011] Aggett, J., and Aspell, A. C., Analyst **101,** 341 (1976).
[2012] Aggett, J., and O'Brien, G., Analyst **106,** 497 (1981).
[2013] Aggett, J., and O'Brien, G., Analyst **106,** 506 (1981).
[2014] Aggett, J., and West, T. S., Anal. Chim. Acta **55,** 349 (1971).
[2015] Alder, J. F., Alger, D., Samuel, A. J., and West, T. S., Anal. Chim. Acta **87,** 301 (1976).
[2016] Alder, J. F., Baker, A. E., and West, T. S., Anal. Chim. Acta **90,** 267 (1977).
[2017] Alder, J. F., and Bucklow, P. L., At. Absorption Newslett. **18,** 123 (1979).
[2018] Alder, J. F., and Das, B. C., Anal. Chim. Acta **94,** 193 (1977).
[2019] Alder, J. F., and Das, B. C., Analyst **102,** 564 (1977).
[2020] Alder, J. F., and Das, B. C., At. Absorption Newslett. **17,** 63 (1978).

[2021] Alder, J. F., and Hickman, D. A., Anal. Chem. **49**, 336 (1977).

[2022] Alder, J. F., and Hickman, D. A., At. Absorption Newslett. **16**, 110 (1977).

[2023] Alder, J. F., Samuel, A. J., and Snook, R. D., Spectrochim. Acta **31 B**, 509 (1976).

[2024] Alder, J. F., Samuel, A. J., and West, T. S., Anal. Chim. Acta **87**, 313 (1976).

[2025] Aldous, K. M., Mitchell, D. G., and Jackson, K. W., Anal. Chem. **47**, 1034 (1975).

[2026] Aldrighetti, F., Carelli, G., Ceriati, F., Cremona, G., and Pomponi, M., At. Spectrosc. **2**, 71 (1981).

[2027] Aldrighetti, F., Carelli, G., Innaccone, A., LaBua, R., and Rimatori, V., At. Spectrosc. **2**, 13 (1981).

[2028] Alduan, J. A., Suarez, J. R. C., Polo, A. B., and del Busto, J. L., At. Spectrosc. **2**, 125 (1981).

[2029] Alkemade, C. Th. J., Snelleman, W., Boutilier, G. D., Pollard, B. D., Winefordner, J. D., Chester, T. L., and Omenetto, N., Spectrochim. Acta **33 B**, 383 (1978).

[2030] Allan, J. W., Spectrochim. Acta **24 B**, 13 (1969).

[2031] Alt, F., Z. Anal. Chem. **308**, 137 (1981).

[2032] Alt, F., and Massmann, H., Z. Anal. Chem. **279**, 100 (1976).

[2033] Alt, F., and Massmann, H., Spectrochim. Acta **33 B**, 337 (1978).

[2034] Analytical Methods Committee, Analyst **105**, 66 (1980).

[2035] Analytical Methods using the MHS Mercury/Hydride System, Bodenseewerk Perkin-Elmer, Überlingen 1979.

[2036] Andersen, I., Torjussen, W., and Zachariasen, H., Clin. Chem. **24**, 1198 (1978).

[2037] Andersson, A., At. Absorption Newslett. **15**, 71 (1976).

[2038] Andreae, M. O., Anal. Chem. **49**, 820 (1977).

[2039] Andrew, B. E., Ceramic Bull. **55**, 583 (1976).

[2040] Andrews, D. G., Aziz-Alrahman, A. M., and Headridge, J. B., Analyst **103**, 909 (1978).

[2041] Andrews, D. G., and Headridge, J. B., Analyst **102**, 436 (1977).

[2042] Armannsson, H., Anal. Chim. Acta **88**, 89 (1977).

[2043] Aurand, K., Blei und Umwelt, Verein für Wasser-, Boden- und Lufthygiene, Berlin, p. 5.

[2044] Auslitz, H. J., Arch. Lebensmittelhyg. **27**, 68 (1976).

[2045] Ayranci, B., Schweiz. mineral. petrogr. Mitt. **56**, 513 (1976).

[2046] Ayranci, B., Schweiz. mineral. petrogr. Mitt. **57**, 299 (1977).

[2047] Azad, J., Kirkbright, G. F., and Snook, R. D., Analyst **104**, 232 (1979).

[2048] Aziz-Alrahman, A. M., and Headridge, J. B., Talanta **25**, 413 (1978).

[2049] Bagliano, G., Benischek, F., and Huber, I., At. Absorption Newslett. **14**, 45 (1975).

[2050] Bagliano, G., Benischek, F., and Huber, I., Anal. Chim. Acta **123**, 45 (1981).

[2051] Baird, R. B., and Gabrielian, S. M., Appl. Spectrosc. **28**, 273 (1974).

[2052] Baker, A. A., and Headridge, J. B., Anal. Chim. Acta **125**, 93 (1981).

[2053] Baker, A. A., Headridge, J. B., and Nicholson, R. A., Anal. Chim. Acta **113**, 47 (1980).

[2054] Ball, J. W., and Gottschall, W. C., At. Absorption Newslett. **14**, 63 (1975).

[2055] Bannister, S. J., Chang, Y., Sternson, L. A., and Repta, A. J., Clin. Chem. **24**, 877 (1978).

[2056] Barfoot, R. A., and Pritchard, J. G., Analyst **105**, 551 (1980).

[2057] Barlow, P. J., and Khera, A. K., At. Absorption Newslett. **14**, 149 (1975).

[2058] Barnett, W. B., and Cooksey, M. M., At. Absorption Newslett. **18**, 61 (1979).

[2059] Barnett, W. B., and Kerber, J. D., At. Absorption Newslett. **13**, 56 (1974).

[2060] Barnett, W. B., and McLaughlin, E. A., Anal. Chim. Acta **80**, 285 (1975).

[2061] Barnett, W. B., Vollmer, J. W., and deNuzzo, S. M., At. Absorption Newslett. **15**, 33 (1976).

[2062] Batley, G. E., and Matoušek, J. P., Anal. Chem. **49**, 2031 (1977).

[2063] Batley, G. E., and Matoušek, J. P., Anal. Chem. **52**, 1570 (1980).

[2064] Baudin, G., Prog. analyt. atom. Spectrosc. **3**, 1 (1980).

[2065] Beaty, M., Barnett, W. B., and Grobenski, Z., At. Spectrosc. **1**, 72 (1980).

[2066] Beaty, R. D., and Cooksey, M. M., At. Absorption Newslett. **17**, 53 (1978).

[2067] Becker-Ross, H., and Falk, H., Spectrochim. Acta **30 B**, 253 (1975).

[2068] Bédard, M., and Kerbyson, J. D., Anal. Chem. **47**, 1441 (1975).

[2069] Bédard, M., and Kerbyson, J. D., Canad. J. Spectrosc. **21**, 64 (1976).

[2070] Begnoche, B. C., and Risby, T. H., Anal. Chem. **47**, 1041 (1975).

[2071] Behne, D., Brätter, P., Gessner, H., Hube, G., Mertz, W., and Rösik, U., Z. Anal. Chem. **278,** 269 (1976).

[2072] Behne, D., Brätter, P., and Wolters, W., Z. Anal. Chem. **277,** 355 (1975).

[2073] Belling, G. B., and Jones, G. B., Anal. Chim. Acta **80,** 279 (1975).

[2074] Benjamin, M. M., and Jenne, E. A., At. Absorption Newslett. **15,** 53 (1976).

[2075] Berndt, H., private communication (1981).

[2076] Berndt, H., and Jackwerth, E., Spectrochim. Acta **30 B,** 169 (1975).

[2077] Berndt, H., and Jackwerth, E., At. Absorption Newslett. **15,** 109 (1976).

[2078] Berndt, H., and Jackwerth, E., Z. Anal. Chem. **283,** 15 (1977).

[2079] Berndt, H., and Jackwerth, E., J. Clin. Chem. Clin. Biochem. **17,** 71 (1979).

[2080] Berndt, H., and Jackwerth, E., J. Clin. Chem. Clin. Biochem. **17,** 489 (1979).

[2081] Berndt, H., Jackwerth, E., and Kimura, M., Anal. Chim. Acta **93,** 45 (1977).

[2082] Berndt, H., and Messerschmidt, J., Spectrochim. Acta **34 B,** 241 (1979).

[2083] Berndt, H., and Slavin, W., At. Absorption Newslett. **17,** 109 (1978).

[2084] Berndt, H., Willmer, P. G., and Jackwerth, E., Z. Anal. Chem. **296,** 377 (1979).

[2085] Bettger, R. J., Ficklin, A. C., and Rees, T. F., At. Absorption Newslett. **14,** 124 (1975).

[2086] Bhattacharya, S. K., and Williams, J. C., Anal. Lett. **12,** 397 (1979).

[2087] Bisogni, J. J., and Lawrence, A. W., Environ. Sci. Technol. **8,** 850 (1974).

[2088] Bloch, L., and Bloch, E., Zeeman Verh. **1935,** 18 (1935).

[2089] Bodrov, N. V., and Nikolaev, G. I., Zh. Analit. Khim. **24,** 1314 (1969).

[2090] Boline, D. R., and Schrenk, W. G., Appl. Spectrosc. **30,** 607 (1976).

[2091] Bonilla, E., Clin. Chem. **24,** 471 (1978).

[2092] Boone, J., Hearn, T., and Lewis, S., Clin. Chem. **25,** 389 (1979).

[2093] Boriello, R., and Sciaudone, G., At. Spectrosc. **1,** 131 (1980).

[2094] Bortlisz, J., Vom Wasser **56,** 225 (1981).

[2095] Boss, C. B., and Hieftje, G. M., Anal. Chem. **49,** 2112 (1977).

[2096] Botre, C., Cacace, F., and Cozzani, R., Anal. Lett. **9,** 825 (1976).

[2097] Boutilier, G. D., Pollard, B. D., Winefordner, J. D., Chester, T. L., and Omenetto, N., Spectrochim. Acta **33 B,** 401 (1978).

[2098] Bower, N. W., and Ingle, J. D., Anal. Chem. **48,** 686 (1976).

[2099] Bower, N. W., and Ingle, J. D., Anal. Chem. **49,** 574 (1977).

[2100] Boyle, E. A., and Edmond, J. M., Anal. Chim. Acta **91,** 189 (1977).

[2101] Braman, R. S., Johnson, D. L., Foreback, C. C., Ammons, J. M., and Bricker, J. L., Anal. Chem. **49,** 621 (1977).

[2102] Braman, R. S., Justen, L. L., and Foreback, C. C., Anal. Chem. **44,** 2195 (1972).

[2103] Braman, R. S., and Tomkins, M. A., Anal. Chem. **51,** 12 (1979).

[2104] Brandvold, L. A., and Marson, S. J., At. Absorption Newslett. **13,** 125 (1974).

[2105] Bratzel, M. P., and Reed, A. J., Clin. Chem. **20,** 217 (1974).

[2106] Briese, L. A., and Giesy, J. P., At. Absorption Newslett. **14,** 133 (1975).

[2107] Brodie, K. G., and Liddell, P. R., Anal. Chem. **52,** 1059 (1980).

[2108] Brodie, K. G., and Matoušek, J. P., Anal. Chim. Acta **69,** 200 (1974).

[2109] Brooke, P. J., and Evans, W. H., Analyst **106,** 514 (1981).

[2110] Brown, J. R., Saba, C. S., Rhine, W. E., and Eisentraut, K. J., Anal. Chem. **52,** 2365 (1980).

[2111] Brown, R. M., Northway, S. J., and Fry, R. C., 31st Pittsburgh Conf. Anal. Chem. Appl. Spectrosc., Atlantic City, N. J. 1980, Paper 452.

[2112] Brown, S. S., Nomoto, S., Stoeppler, M., and Sunderman, F. W., Pure & Appl. Chem. **53,** 773 (1981).

[2113] Browner, R. F., Analyst **99,** 1183 (1974).

[2114] Burce, C. F., and Hannaford, P., Spectrochim. Acta **26 B,** 207 (1971).

[2115] Bruhn, C. F., and Navarrete, G. A., Anal. Chim. Acta **130,** 209 (1981).

[2116] Bruland, K. W., Franks, R. P., Knauer, G. A., and Martin, J. H., Anal. Chim. Acta **105,** 233 (1979).

[2117] Burba, P., Lieser, K. H., Neitzert, V., and Röber, H. M., Z. Anal. Chem. **291,** 273 (1978).

[2118] Burdo, R. A., and Wise, W. M., Anal. Chem. **47,** 2360 (1975).

[2119] Burke, K. E., Appl. Spectrosc. **28,** 234 (1974).
[2120] Butler, L. R. P., and Fulton, A., Appl. Optics **7,** 2131 (1968).
[2121] Bye, R., Paus, P. E., Solberg, R., and Thomassen, Y., At. Absorption Newslett. **17,** 131 (1978).
[2122] Callio, S., At. Spectrosc. **1,** 80 (1980).
[2123] Campbell, A. D., and Tioh, N. H., Anal. Chim. Acta **100,** 451 (1978).
[2124] Campbell, D. R., and Seitz, W. R., Anal. Lett. **9,** 543 (1976).
[2125] Campbell, J. A., Laul, J. C., Nielson, K. K., and Smith, R. D., Anal. Chem. **50,** 1032 (1978).
[2126] Campbell, W. C., and Ottaway, J. M., Talanta **21,** 837 (1974).
[2127] Campbell, W. C., and Ottaway, J. M., Talanta **22,** 729 (1975).
[2128] Campbell, W. C., and Ottaway, J. M., Analyst **102,** 495 (1977).
[2129] Campe, A., Velghe, N., and Claeys, A., At. Absorption Newslett. **17,** 100 (1978).
[2130] Campenhausen, H., and Müller-Plathe, O., Z. Klin. Chem. Klin. Biochem. **13,** 489 (1975).
[2131] Carnrick, G. R., Slavin, W., and Manning, D. C., Anal. Chem. **53,** 1866 (1981).
[2132] Carpenter, R. C., Anal. Chim. Acta **125,** 209 (1981).
[2133] Carrondo, M. J. T., Perry, R., and Lester, J. N., Anal. Chim. Acta **106,** 309 (1979).
[2134] Cary, E. E., and Olson, O. E., J. Assoc. Off. Anal. Chem. **58,** 433 (1975).
[2135] Cassidy, R. M., Hurteau, M. T., Mislan, J. P., and Ashley R. W., J. Chromatog. Sci. **14,** 444 (1976).
[2136] Castellani, F., Riccioni, R., Gusteri, M., Bartocci, V., and Cescon, P., At. Absorption Newslett. **16,** 57 (1977).
[2137] Caupeil, J. E., Hendrikse, P. W., and Bongers, J. S., Anal. Chim. Acta **81,** 53 (1976).
[2138] Cedergren, A., Frech, W., Lundberg, E., and Persson, J. A., Anal. Chim. Acta **128,** 1 (1981).
[2139] Celis, J. P., Helsen, J. A., Hermans, P., and Roos, J. R., Anal. Chim. Acta **92,** 413 (1977).
[2140] Chakrabarti, C. L., Hamed, H. A., Wan, C. C., Li, W. C., Bertels, P. C., Gregoire, D. C., and Lee, S., Anal. Chem. **52,** 167 (1980).
[2141] Chakrabarti, C. L., Wan, C. C., Hamed, H. A., and Bertels, P. C., Anal. Chem. **53,** 444 (1981).
[2142] Chakrabarti, C. L., Wan, C. C., and Li, W. C., Spectrochim. Acta **35 B,** 93 (1980).
[2143] Chakrabarti, C. L., Wan, C. C., Teskey, R. J., Chang, S. B., Hamed, H. A., and Bertels, P. C., Spectrochim. Acta **36 B,** 427 (1981).
[2144] Chakraborti, D., deJonghe, W., and Adams, F., Anal. Chim. Acta **119,** 331 (1980).
[2145] Chambers, J. C., and McClellan, B. E., Anal. Chem. **48,** 2061 (1976).
[2146] Chan, C. Y., and Vijan, P. N., Anal. Chim. Acta **101,** 33 (1978).
[2147] Chao, S. S., and Pickett, E. E., Anal. Chem. **52,** 335 (1980).
[2148] Chapman, J. F., and Dale, L. S., Anal. Chim. Acta **87,** 91 (1976).
[2149] Chapman, J. F., and Dale, L. S., Anal. Chim. Acta **89,** 363 (1977).
[2150] Chapman, J. F., Dale, L. S., and Fraser, H. J., Anal. Chim. Acta **116,** 427 (1980).
[2151] Chapman, J. F., and Leadbeatter, B. E., Anal. Lett. **13,** 439 (1980).
[2152] Chau, Y. K., Wong, P. T. S., and Goulden, P. D., Anal. Chem. **47,** 2279 (1975).
[2153] Chau, Y. K., Wong, P. T. S., and Goulden, P. D., Anal. Chim. Acta **85,** 421 (1976).
[2154] Cheam, V., and Agemian, H., Analyst **105,** 1253 (1980).
[2155] Cheng, J. T., and Agnew, W. F., At. Absorption Newslett. **13,** 123 (1974).
[2156] Chong, R. W., and Boltz, D. F., Anal. Lett. **8,** 721 (1975).
[2157] Chow, C., Analyst **104,** 154 (1979).
[2158] Chu, R. C., Barrons, G. P., and Baumgardner, P. A. W., Anal. Chem. **44,** 1476 (1972).
[2159] Clark, J. R., and Viets, J. G., Anal. Chem. **53,** 61 (1981).
[2160] Clinton, O. E., Analyst **102,** 187 (1977).
[2161] Clyburn, S. A., Kantor, T., and Veillon, C., Anal. Chem. **46,** 2214 (1974).
[2162] Cobb, W. D., Foster, W. W., and Harrison, T. S., Anal. Chim. Acta **78,** 293 (1975).
[2163] Cobb, W. D., Foster, W. W., and Harrison, T. S., Analyst **101,** 39 (1976).
[2164] Codding, E. G., Ingle, J. D., and Stratton, A. J., Anal. Chem. **52,** 2133 (1980).
[2165] Coker, D. T., Anal. Chem. **47,** 386 (1975).
[2166] Cooksey, M. M., and Barnett, W. B., At. Absorption Newslett. **18,** 1 (1979).
[2167] Corbett, J. A., and Godbeer, W. C., Anal. Chim. Acta **91,** 211 (1977).
[2168] Corbin, D. R., and Barnard, W. M., At. Absorption Newslett. **15,** 116 (1976).

[2169] Corominas, L. F., Boy, V. M., and Rojas, P., J. Assoc. Off. Anal. Chem. **64,** 704 (1981).
[2170] Coughtrey, P. J., and Martin, M. H., Chemosphere **3,** 183 (1976).
[2171] Cox, D. H., J. Analyt. Toxicol. **4,** 207 (1980).
[2172] Cranston, R. E., and Murray, J. W., Anal. Chim. Acta **99,** 275 (1978).
[2173] Cresser, M. S., Solvent Extraction in Flame Spectroscopic Analysis. Butterworths, London, Boston 1978.
[2174] Cresser, M. S., Prog. analyt. atom. Spectrosc. **4,** 219 (1981).
[2175] Crisp, P. T., Eckert, J. M., and Gibson, N. A., Anal. Chim. Acta **104,** 93 (1979).
[2176] Crisp, P. T., Eckert, J. M., Gibson, N. A., Kirkbright, G. F., and West, T. S., Anal. Chim. Acta **87,** 97 (1976).
[2177] Crow, R. F., and Connolly, J. D., J. Testing Evaluat. **1,** 382 (1973).
[2178] Cruz, R. B., Lorouso, C., George, S., Thomassen, Y., Kinrade, J. D., Butler, L. R. P., Lye, J., and VanLoon, J. C., Spectrochim. Acta **35 B,** 775 (1980).
[2179] Culver, B. R., and Surles, T., Anal. Chem. **47,** 920 (1975).
[2180] Cutter, G. A., Anal. Chim. Acta **98,** 59 (1978).
[2181] Czobik, E. J., and Matousek, J. P., Talanta **24,** 573 (1977).
[2182] Czobik, E. J., and Matousek, J. P., Anal. Chem. **50,** 2 (1978).
[2183] Czobik, E. J., and Matousek, J. P., Spectrochim. Acta **35 B,** 741 (1980).
[2184] Damiani, M., delMonte Tamba, M. G., and Bianchi, F., Analyst **100,** 643 (1975).
[2185] Dannecker, W., in: Atomspektrometrische Spurenanalytik (Editor: B. Welz). Verlag Chemie, Weinheim 1982, p. 187.
[2186] Davies, I. M., Anal. Chim. Acta **102,** 189 (1978).
[2187] Dawson, J. B., Grassan, E., Ellis, D. J., and Keir, M. J., Analyst **101,** 315 (1976).
[2188] Dědina, J., and Rubeška, I., Spectrochim. Acta **35 B,** 119 (1980).
[2189] deJong, G. J., and Brinkman, U. A. Th., Anal. Chim. Acta **98,** 243 (1978).
[2190] deJonghe, W., and Adams, F., Anal. Chim. Acta **108,** 21 (1979).
[2191] deJonghe, W., Chakraborti, D., and Adams, F., Anal. Chim. Acta **115,** 89 (1980).
[2192] deLoos-Vollebregt, M. T. C., and deGalan, L., Spectrochim. Acta **33 B,** 495 (1978).
[2193] deLoos-Vollebregt, M. T. C., and deGalan, L., Appl. Spectrosc. **33,** 616 (1979).
[2194] deLoos-Vollebregt, M. T. C., and deGalan, L., Spectrochim. Acta **35 B,** 495 (1980).
[2195] deVine, J. C., and Suhr, N. H., At. Absorption Newslett. **16,** 39 (1977).
[2196] delMonte Tamba, M. G., and Luperi, N., Analyst **102,** 489 (1977).
[2197] Delves, H. T., Analyst **102,** 403 (1977).
[2198] Delves, H. T., and Woodward, J., At. Spectrosc. **2,** 65 (1981).
[2199] Dittrich, K., Prog. analyt. atom. Spectrosc. **3,** 209 (1980).
[2200] Dittrich, K., Schneider, S., Spiwakow, B. J., Suchowejewa, L. N., and Zolotow, J. A., Spectrochim. Acta **34 B,** 257 (1979).
[2201] Dittrich, K., Vorberg, B., and Wolters, H., Talanta **26,** 747 (1979).
[2202] Dokiya, Y., Kobayashi, T., and Toda, S., J. Spectrosc. Soc. Japan. **27,** 435 (1978).
[2203] Donaldson, E. M., Talanta **27,** 499 (1980).
[2204] Doolan, K. J., and Belcher, C. B., Prog. analyt. atom. Spectrosc. **3,** 125 (1980).
[2205] Dornemann, A., and Kleist, H., Z. Anal. Chem. **305,** 379 (1981).
[2206] Drinkwater, J. E., Analyst **101,** 672 (1976).
[2207] Dudas, M. J., At. Absorption Newslett. **13,** 67 (1974).
[2208] Dudas, M. J., At. Absorption Newslett. **13,** 109 (1974).
[2209] Dulka, J. J., and Risby, T. H., Anal. Chem. **48,** 640 A (1976).
[2210] Dulski, T. R., and Bixler, R. R., Anal. Chim. Acta **91,** 199 (1977).
[2211] Dumarey, R., Heindryckx, R., and Dams, R., Anal. Chim. Acta **116,** 111 (1980).
[2212] Dunckley, J. V., Grennan, D. M., and Palmer, D. G., J. Analyt. Toxicol. **3,** 242 (1979).
[2213] Dusci, L. J., and Hackett, L. P., J. Assoc. Off. Anal. Chem. **59,** 1183 (1976).
[2214] Ebbestad, U., Gundersen, N., and Torgrimsen, T., At. Absorption Newslett. **14,** 142 (1975).
[2215] Ebdon, L., Kirkbright, G. F., and West, T. S., Anal. Chim. Acta **58,** 39 (1972).
[2216] Edgar, R. M., At. Absorption Newslett. **14,** 68 (1975).
[2217] Ediger, R. D., At. Absorption Newslett. **14,** 127 (1975).

[2218] Ediger, R. D., At. Absorption Newslett. **15**, 145 (1976).
[2219] Ediger, R. D., Knott, A. R., Peterson, G. E., and Beaty, R. D., At. Absorption Newslett. **17**, 28 (1978).
[2220] Egaas, E., and Julshamn, K., At. Absorption Newslett. **17**, 135 (1978).
[2221] Einbrodt, H. J., Rosmanith, J., and Prajsnar, D., Naturwissenschaften **63**, 148 (1976).
[2222] Eklund, R. H., and Holcombe, J. A., Anal. Chim. Acta **108**, 53 (1979).
[2223] Eklund, R. H., and Holcombe, J. A., Anal. Chim. Acta **109**, 97 (1979).
[2224] El-Defrawy, M. M. M., Posta, J., and Beck, M. T., Anal. Chim. Acta **102**, 185 (1978).
[2225] Elfving, D. C., Bache, C. A., and Lisk, D. J., J. Agric. Food Chem. **27**, 138 (1979).
[2226] Elliott, E. V., Stever, K. R., and Heady, H. H., At. Absorption Newslett. **13**, 113 (1974).
[2227] Elton-Bott, R. R., Anal. Chim. Acta **86**, 281 (1976).
[2228] Recommended Conditions for Graphite Furnace Atomic Absorption Spectrometry. Bodenseewerk Perkin-Elmer, Überlingen 1984.
[2229] Epstein, M. S., Rains, T. C., and O'Haver, T. C., Appl. Spectrosc. **30**, 324 (1976).
[2230] Erspamer, J. P., and Niemczyk, T. M., Appl. Spectrosc. **35**, 512 (1981).
[2231] Evans, W. H., Dellar, D., Lucas, B. E., Jackson, F. J., and Read, J. I., Analyst **105**, 529 (1980).
[2232] Evans, W. H., Jackson, F. J., and Dellar, D., Analyst **104**, 16 (1979).
[2233] Evans, W. H., Read, J. I., and Lucas, B. E., Analyst **103**, 580 (1978).
[2234] Evenson, M. A., and Anderson, C. T., Clin. Chem. **21**, 537 (1975).
[2235] Evenson, M. A., and Warren, B. L., Clin. Chem. **21**, 619 (1975).
[2236] Farant, J. P., Brissette, D., Moncion, L., Bigras, L., and Chartrand, A., J. Analyt. Toxicol. **5**, 47 (1981).
[2237] Favretto-Gabrielli, L., Pertoldi-Marletta, G., and Favretto, L., At. Spectrosc. **1**, 35 (1980).
[2238] Feinberg, M., and Ducauze, C., Anal. Chem. **52**, 207 (1980).
[2239] Felkel, H. L., and Pardue, H. L., Anal. Chem. **49**, 1112 (1977).
[2240] Fernandez, F. J., Clin. Chem. **21**, 558 (1975).
[2241] Fernandez, F. J., At. Absorption Newslett. **17**, 115 (1978).
[2242] Fernandez, F. J., and Hilligoss, D., At. Spectrosc. **3**, 130 (1982).
[2243] Fernandez, F. J., Beaty, M. M., and Barnett, W. B., At. Spectrosc. **2**, 16 (1981).
[2244] Fernandez, F. J., Bohler, W., Beaty, M. M., and Barnett, W. B., At. Spectrosc. **2**, 73 (1981).
[2245] Fernandez, F. J., and Iannarone, J., At. Absorption Newslett. **17**, 117 (1978).
[2246] Fernandez, F. J., Myers, S. A., and Slavin, W., Anal. Chem. **52**, 741 (1980).
[2247] Fetterolf, D. D., and Syty, A., J. Agric. Food Chem. **27**, 377 (1979).
[2248] Filipek, L. H., and Owen, R. M., Can. J. Spectrosc. **23**, 31 (1978).
[2249] Fiorino, J. A., Jones, J. W., and Capar, S. G., Anal. Chem. **48**, 120 (1976).
[2250] Fischer, H., and Weigert, P., Öff. Gesundh. Wesen **39**, 269 (1977).
[2251] Fishman, M., and Spencer, R., Anal. Chem. **49**, 1599 (1977).
[2252] Fitchett, A. W., Daughtrey, E. H., and Mushak, P., Anal. Chim. Acta **79**, 93 (1975).
[2253] Flanjak, J., J. Assoc. Off. Anal. Chem. **61**, 1299 (1978).
[2254] Fleming, H. D., and Ide, E. G., Anal. Chim. Acta **83**, 67 (1976).
[2255] Florence, T. M., and Batley, G. E., Talanta **24**, 151 (1977).
[2256] Forehand, T. J., Dupuy, A. E., and Tai, H., Anal. Chem. **48**, 999 (1976).
[2257] Forrester, J. E., Lehecka, V., Johnston, J. R., and Ott, W. L., At. Absorption Newslett. **18**, 73 (1979).
[2258] Franz, H., Görgenyi, T., Jungen, W., and Rottmann, J., Z. Anal. Chem. **292**, 353 (1978).
[2259] Frech, W., Talanta **21**, 565 (1974).
[2260] Frech, W., and Cedergren, A., Anal. Chim. Acta **82**, 83 (1976).
[2261] Frech, W., and Cedergren, A., Anal. Chim. Acta **82**, 93 (1976).
[2262] Frech, W., and Cedergren, A., Anal. Chim. Acta **88**, 57 (1977).
[2263] Frech, W., and Cedergren, A., Anal. Chim. Acta **113**, 227 (1980).
[2264] Frech, W., Lundberg, E., and Barbooti, M. M., Anal. Chim. Acta **131**, 45 (1981).
[2265] Frech, W., Lundgren, G., and Lunner, S. E., At. Absorption Newslett. **15**, 57 (1976).
[2266] Frech, W., Persson, J. A., and Cedergren, A., Prog. analyt. atom. Spectrosc. **3**, 279 (1980).
[2267] Freeman, H., Uthe, J. F., and Flemming, B., At. Absorption Newslett. **15**, 49 (1976).

[2268] Fricke, F. L., Robbins, W. B., and Caruso, J. A., Prog. analyt. atom. Spectrosc. **2**, 185 (1979).
[2269] Friend, M. T., Smith, C. A., and Wishart, D., At. Absorption Newslett. **16**, 46 (1977).
[2270] Frigieri, P., and Trucco, R., Analyst **103**, 1089 (1978).
[2271] Frigieri, P., Trucco, R., Ciaccolini, I., and Pampurini, G., Analyst **105**, 651 (1980).
[2272] Fritzsche, H., Wegscheider, W., Knapp, G., and Ortner, H. M., Talanta **26**, 219 (1979).
[2273] Fry, R. C., and Denton, M. B., Anal. Chem. **49**, 1413 (1977).
[2274] Fry, R. C., and Denton, M. B., Anal. Chem. **51**, 266 (1979).
[2275] Fryer, B. J., and Kerrich, R., At. Absorption Newslett. **17**, 4 (1978).
[2276] Fujiwara, K., Haraguchi, H., and Fuwa, K., Anal. Chem. **47**, 1670 (1975).
[2277] Fuller, C. W., Analyst **99**, 739 (1974).
[2278] Fuller, C. W., At. Absorption Newslett. **14**, 73 (1975).
[2279] Fuller, C. W., Anal. Chim. Acta **81**, 199 (1976).
[2280] Fuller, C. W., Analyst **101**, 798 (1976).
[2281] Fuller, C. W., Analyst **101**, 961 (1976).
[2282] Fuller, C. W., At. Absorption Newslett. **16**, 106 (1977).
[2283] Furr, A. K., Parkinson, T. F., Elfving, D. C., Gutemann, W. H., Pakkala, I. S., and Lisk, D. J., J. Agric. Food Chem. **27**, 135 (1979).
[2284] Fuwa, K., and Vallee, B. L., Spectrochim. Acta **35 B**, 657 (1980).
[2285] Gabrielli, L. F., Marletta, G. P., and Favretto, L., At. Absorption Newslett. **16**, 4 (1977).
[2286] Ganje, T. J., and Page, A. L., At. Absorption Newslett. **13**, 131 (1974).
[2287] Garbett, K., Goodfellow, G. I., and Marshall, G. B., Anal. Chim. Acta **126**, 135 (1981).
[2288] Garbett, K., Goodfellow, G. I., and Marshall, G. B., Anal. Chim. Acta **126**, 147 (1981).
[2289] Gardiner, P. E., and Ottaway, J. M., Talanta **26**, 841 (1979).
[2290] Gardiner, P. E., Ottaway, J. M., Fell, G. S., and Burns, R. R., Anal. Chim. Acta **124**, 281 (1981).
[2291] Gardner, D., Anal. Chim. Acta **82**, 321 (1976).
[2292] Gardner, D., Anal. Chim. Acta **93**, 291 (1977).
[2293] Gardner, M. J., and Hunt, D. T. E., Analyst **106**, 471 (1981).
[2294] Garnys, V. P., and Matousek, P., Clin. Chem. **21**, 891 (1975).
[2295] Garnys, V. P., and Smythe, L. E., Talanta **22**, 881 (1975).
[2296] Garska, K. J., At. Absorption Newslett. **15**, 38 (1976).
[2297] Geladi, P., and Adams, F., Anal. Chim. Acta **105**, 219 (1979).
[2298] Genc, Ö., Akman, S., Özdural, A. R., Ates, S., and Balkis, T., Spectrochim. Acta **36 B**, 163 (1981).
[2299] Gill, R. C. O., and Kronberg, B. I., At. Absorption Newslett. **14**, 157 (1975).
[2300] Gladfelter, W. L., and Dickerhoof, D. W., Fuel **55**, 360 (1976).
[2301] Gladney, E. S., At. Absorption Newslett. **16**, 42 (1977).
[2302] Gladney, E. S., At. Absorption Newslett. **16**, 114 (1977).
[2303] Gladney, E. S., and Apt, K. E., Anal. Chim. Acta **85**, 393 (1976).
[2304] Gladney, E. S., and Goode, W. E., Anal. Chim. Acta **91**, 411 (1977).
[2305] Gladney, E. S., and Owens, J. W., Anal. Chim. Acta **90**, 271 (1977).
[2306] Gliksman, J. E., Gibson, J. E., and Kandetzski, P. E., At. Spectrosc. **1**, 166 (1980).
[2307] Gong, H., and Suhr, N. H., Anal. Chim. Acta **81**, 297 (1976).
[2308] Gordon, D. T., J. Assoc. Off. Anal. Chem. **61**, 715 (1978).
[2309] Gorsky, J. E., and Dietz, A. A., Clin. Chem. **24**, 1485 (1978).
[2310] Gough, D. S., Anal. Chem. **48**, 1926 (1976).
[2311] Gough, D. S., and Sullivan, J. V., Anal. Chim. Acta **124**, 259 (1981).
[2312] Goulden, P. D., and Brooksbank, P., Anal. Chem. **46**, 1431 (1974).
[2313] Graef, V., J. Clin. Chem. Clin. Biochem. **14**, 181 (1976).
[2314] Graf-Harsanyi, E., and Langmyhr, F. J., Anal. Chim. Acta **116**, 105 (1980).
[2315] Greenland, L. P., and Campbell, E. Y., Anal. Chim. Acta **87**, 323 (1976).
[2316] Greig, R. A., Anal. Chem. **47**, 1682 (1975).
[2317] Grewal, D. S., and Kearns, F. X., At. Absorption Newslett. **16**, 131 (1977).
[2318] Griffin, H. R., Hocking, M. B., and Lowery, D. G., Anal. Chem. **47**, 229 (1975).

[2319] Grobenski, Z., Z. Anal. Chem. **289**, 337 (1978).

[2320] Grobenski, Z., Lehmann, R., and Welz, B., 32nd Pittsburgh Conf. Anal. Chem. Appl. Spectrosc., Atlantic City, N. J. 1981, Paper 260.

[2321] Grobenski, Z., Welz, B., Voellkopf, U., and Wolff, J., 33rd Pittsburgh Conf. Anal. Chem. Appl. Spectrosc., Atlantic City, N. J. 1982, Paper 743.

[2322] Guerra, R., At. Spectrosc. **1**, 58 (1980).

[2323] Guest, R. J., and MacPherson, D. R., Anal. Chim. Acta **78**, 299 (1975).

[2324] Guevremont, R., Anal. Chem. **52**, 1574 (1980).

[2325] Guevremont, R., Anal. Chem. **53**, 911 (1981).

[2326] Guevremont, R., Sturgeon, R. E., and Beerman, S. S., Anal. Chim. Acta **115**, 163 (1980).

[2327] Guillaumin, J. C., At. Absorption Newslett. **13**, 135 (1974).

[2328] Guimont, J., Pichette, M., and Rhéaume, N., At. Absorption Newslett. **16**, 53 (1977).

[2329] Gunčaga, J., Lentner, C., and Haas, H. G., Clin. Chim. Acta **57**, 77 (1974).

[2330] Gutsche, B., Rüdiger, K., and Herrmann, R., Z. Anal. Chem. **285**, 103 (1977).

[2331] Guttmann, W., Therapiewoche **27**, 5721 (1977).

[2332] Haddon, M. J., and Pantony, D. A., Analyst **105**, 371 (1980).

[2333] Hadeishi, T., Appl. Phys. Letts. **21**, 438 (1972).

[2334] Hadeishi, T., Church, D. A., McLaughlin, R. D., Zak, B. D., Nakamura, M., and Chang, B., Science **187**, 348 (1975).

[2335] Hadeishi, T., and McLaughlin, R. D., Science **174**, 404 (1971).

[2336] Hagedorn-Götz, H., Küppers, G., and Stoeppler, M., Arch. Toxicol. **38**, 275 (1977).

[2337] Hageman, L., Mubarak, A., and Woodriff, R., Appl. Spectrosc. **33**, 226 (1979).

[2338] Hageman, L. R., Nichols, J. A., Viswanadham, P., and Woodriff, R., Anal. Chem. **51**, 1406 (1979).

[2339] Hall, A., and Godinho, M. C., Anal. Chim. Acta **113**, 369 (1980).

[2340] Hall, S. H., At. Absorption Newslett. **18**, 126 (1979).

[2341] Halliday, M. C., Houghton, C., and Ottaway, J. M., Anal. Chim. Acta **119**, 67 (1980).

[2342] Halls, D. J., Anal. Chim. Acta **88**, 69 (1977).

[2343] Halls, D. J., Spectrochim. Acta **32 B**, 221 (1977).

[2344] Hamner, R. M., Lechak, D. L., and Greenberg, P., At. Absorption Newslett. **15**, 122 (1976).

[2345] Han, H. B., Kaiser, G., and Tölg, G., Anal. Chim. Acta **128**, 9 (1981).

[2346] Hannaford, P., and Lowe, R. M., Anal. Chem. **49**, 1852 (1977).

[2347] Hannaker, P., and Hughes, T. C., Anal. Chem. **49**, 1485 (1977).

[2348] Hansen, L. D., and Fisher, G. L., Environ. Sci. Technol. **14**, 1111 (1980).

[2349] Hansen, R. K., and Hall, R. H., Anal. Chim. Acta **92**, 307 (1977).

[2350] Haraguchi, H., and Fuwa, K., Anal. Chem. **48**, 784 (1976).

[2351] Harrington, D. E., and Bramstedt, W. R., At. Absorption Newslett. **14**, 36 (1975).

[2352] Harrington, D. E., and Bramstedt, W. R., At. Absorption Newslett. **15**, 125 (1976).

[2353] Havezov, I., Russeva, E., and Jordanov, N., Z. Anal. Chem. **296**, 125 (1979).

[2354] Havezov, I., and Tamnev, B., Z. Anal. Chem. **290**, 299 (1978).

[2355] Hawley, J. E., and Ingle, J. D., Anal. Chem. **47**, 719 (1975).

[2356] Haynes, B. W., At. Absorption Newslett. **17**, 49 (1978).

[2357] Haynes, B. W., At. Absorption Newslett. **18**, 46 (1979).

[2358] Headridge, J. B., and Thompson, R., Anal. Chim. Acta **102**, 33 (1978).

[2359] Heanes, D. L., Analyst **106**, 182 (1981).

[2360] Hegi, H. R., Hydrologie **38**, 35 (1976).

[2361] Heinemann, W., Z. Anal. Chem. **280**, 359 (1976).

[2362] Heinemann, W., Z. Anal. Chem. **281**, 291 (1976).

[2363] Heinrichs, H., Z. Anal. Chem. **295**, 355 (1979).

[2364] Helsby, C. A., Talanta **24**, 46 (1977).

[2365] Hendrikx-Jongerius, C., and de Galan, L., Anal. Chim. Acta **87**, 259 (1976).

[2366] Henn, E. L., Anal. Chem. **47**, 428 (1975).

[2367] Henn, E. L., in: Flameless Atomic Absorption Analysis: An Update. American Society for Testing and Materials, STP 618, 1977, p. 54.

[2368] Henn, K. H., Berg, R., and Hörner, L., in: Atomspektrometrische Spurenanalytik (Editor: B. Welz). Verlag Chemie, Weinheim 1982, p. 553.

[2369] Henry, C. D., At. Absorption Newslett. **16**, 128 (1977).

[2370] Hermann, R., At. Absorption Newslett. **16**, 44 (1977).

[2371] Hernandez-Mendez, J., Polo-Diez, L., and Bernal-Melchor, A., Anal. Chim. Acta **108**, 39 (1979).

[2372] Hinderberger, E. J., Kaiser, M. L., and Koirtyohann, S. R., At. Spectrosc. **2**, 1 (1981).

[2373] Hinners, T. A., Z. Anal. Chem. **277**, 377 (1975).

[2374] Hinners, T. A., Analyst **105**, 751 (1980).

[2375] Hinners, T. A., Bumgarner, J. E., and Simmons, W. S., At. Absorption Newslett. **13**, 146 (1974).

[2376] Hinrichs, G., Johannes, D., and Krause, H., Erzmetall **33**, 536 (1980).

[2377] Hirao, Y., Fukumoto, K., Sugisaki, H., and Kimura, K., Anal. Chem. **51**, 651 (1979).

[2378] Hocquellet, P., and Labeyrie, N., At. Absorption Newslett. **16**, 124 (1977).

[2379] Hodges, D. J., Analyst **102**, 66 (1977).

[2380] Hodges, D. J., and Skelding, D., Analyst **106**, 299 (1981).

[2381] Hoffmeister, W., Z. Anal. Chem. **290**, 289 (1978).

[2382] Hofsommer, H. J., and Bielig, H. J., Dtsch. Lebensm. Rdsch. **76**, 419 (1980).

[2383] Hoft, D., Oxman, J., and Gurira, R. C., J. Agric. Food Chem. **27**, 145 (1979).

[2384] Höhn, R., and Jackwerth, E., Anal. Chim. Acta **85**, 407 (1976).

[2385] Höhn, R., and Jackwerth, E., Z. Anal. Chem. **282**, 21 (1976).

[2386] Holcombe, J. A., Eklund, R. H., and Smith, J. E., Anal. Chem. **51**, 1205 (1979).

[2387] Holen, B., Bye, R., and Lund, W., Anal. Chim. Acta **130**, 257 (1981).

[2388] Holen, B., Bye, R., and Lund, W., Anal. Chim. Acta **131**, 37 (1981).

[2389] Holm, J., Fleischwirtschaft **1978**, 864.

[2390] Hon, P. K., Lau, O. W., and Mok, C. S., Analyst **105**, 919 (1980).

[2391] Hongve, D., and Holth-Larsen, B., At. Absorption Newslett. **17**, 91 (1978).

[2392] Hoover, W. L., J. Assoc. Off. Anal. Chem. **55**, 737 (1972).

[2393] Horsky, S. J., At. Spectrosc. **1**, 129 (1980).

[2394] Hoshino, Y., Utsunomiya, T., and Fukui, K., Chem. Lett. **9**, 947 (1976).

[2395] Hoshino, Y., Utsunomiya, T., and Fukui, K., J. Chem. Soc Japan. **6**, 808 (1977).

[2396] Howlett, C., and Taylor, A., Analyst **103**, 916 (1978).

[2397] Hubaux, A., and Vos, G., Anal. Chem. **42**, 849 (1970).

[2398] Huber, I., Schreinlechner, I., and Benischek, F., At. Absorption Newslett. **16**, 64 (1977).

[2399] Hubert, J., Candelaria, R. M., and Applegate, H. G., At. Spectrosc. **1**, 90 (1980).

[2400] Hull, D. A., Muhammad, N., Lanese, J. G., Reich, S. D., Finkelstein, T. T., and Fandrich, S., J. Pharmaceut. Scienc. **70**, 500 (1981).

[2401] Hurlbut, J. A., At. Absorption Newslett. **17**, 121 (1978).

[2402] Hutton, R. C., Ottaway, J. M., Epstein, M. S., and Rains, T. C., Analyst **102**, 658 (1977).

[2403] Hutton, R. C., Ottaway, J. M., Rains, T. C., and Epstein, M. S., Analyst **102**, 429 (1977).

[2404] Ihnat, M., J. Assoc. Off. Anal. Chem. **59**, 911 (1976).

[2405] Ihnat, M., and Miller, H. J., J. Assoc. Off. Anal. Chem. **60**, 1414 (1977).

[2406] Iida, C., Uchida, T., and Kojima, I., Anal. Chim. Acta **113**, 365 (1980).

[2407] Ingle, J. D., Anal. Chem. **46**, 2161 (1974).

[2408] Inglis, A. S., and Nicholls, P. W., Mikrochim. Acta (Wien) **1975 II**, 553.

[2409] International Union of Pure and Applied Chemistry. Spectrochim. Acta **33 B**, 247 (1978).

[2410] Ishizaki, M., Talanta **25**, 167 (1978).

[2411] Ishizuka, T., Uwamino, Y., and Sunahara, H., Anal. Chem. **49**, 1340 (1977).

[2412] Issaq, H. J., Anal. Chem. **51**, 657 (1979).

[2413] Issaq, H. J., and Morgenthaler, L. P., Anal. Chem. **47**, 1661 (1975).

[2414] Issaq, H. J., and Morgenthaler, L. P., Anal. Chem. **47**, 1668 (1975).

[2415] Issaq, H. J., and Morgenthaler, L. P., Anal. Chem. **47**, 1748 (1975).

[2416] Issaq, H. J., and Zielinski, W. L., Anal. Chem. **46**, 1328 (1974).

[2417] Issaq, H. J., and Zielinski, W. L., Anal. Chem. **46**, 1436 (1974).

[2418] Issaq, H. J., and Zielinski, W. L., Anal. Chem. **47**, 2281 (1975).

[2419] Iu, K. L., Pulford, I. D., and Duncan, H. J., Anal. Chim. Acta **106**, 319 (1979).

[2420] Jackson, F. J., Read, J. I., and Lucas, B. E., Analyst **105**, 359 (1980).

[2421] Jackson, K. W., Fuller, T. D., Mitchell, D. G., and Aldous, K. M., At. Absorption Newslett. **14**, 121 (1975).

[2422] Jackson, K. W., and Mitchell, D. G., Anal. Chim. Acta **80**, 39 (1975).

[2423] Jackwerth, E., Z. Anal. Chem. **271**, 120 (1974).

[2424] Jackwerth, E., in: Atomspektrometrische Spurenanalytik (Editor: B. Welz). Verlag Chemie, Weinheim 1982. p. 1.

[2425] Jackwerth, E., and Berndt, H., Anal. Chim. Acta **74**, 299 (1975).

[2426] Jackwerth, E., Lohmar, J., and Wittler, G., Z. Anal. Chem. **266**, 1 (1973).

[2427] Jackwerth, E., and Messerschmidt, J., Anal. Chim. Acta **87**, 341 (1976).

[2428] Jackwerth, E., and Willmer, P. G., Z. Anal. Chem. **279**, 23 (1976).

[2429] Jackwerth, E., and Willmer, P. G., Talanta **23**, 197 (1976).

[2430] Jackwerth, E., and Willmer, P. G., Spectrochim. Acta **33 B**, 343 (1978).

[2431] Jackwerth, E., Willmer, P. G., Höhn, R., and Berndt, H., At. Absorption Newslett. **18**, 66 (1979).

[2432] Jan, T. K., and Young, D. R., Anal. Chem. **50**, 1250 (1978).

[2433] Janoušek, I., At. Absorption Newslett. **16**, 49 (1977).

[2434] Janssen, J. R., van den Enk, J. E., Bult, R. and deGroot, D. C., Anal. Chim. Acta **84**, 319 (1976).

[2435] Jasim, F., and Barbooti, M. M., Talanta **28**, 353 (1981).

[2436] Johnson, C. A., Anal. Chim. Acta **81**, 69 (1976).

[2437] Johnson, C. A., Lewin, J. F., and Fleming, P. A., Anal. Chim. Acta **82**, 79 (1976).

[2438] Johnson, D. J., West, T. S., and Dagnall, R. M., Anal. Chim. Acta **67**, 79 (1973).

[2439] Jones, A. H., Anal. Chem. **48**, 1472 (1976).

[2440] Jones, J. W., and Boyer, K. W., J. Assoc. Off. Anal. Chem. **62**, 122 (1979).

[2441] Julshamn, K., At. Absorption Newslett. **16**, 149 (1977).

[2442] Julshamn, K., and Andersen, K. J., Anal. Biochem. **98**, 315 (1979).

[2443] Julshamn, K., and Braekkan, O. R., At. Absorption Newslett. **14**, 49 (1975).

[2444] Jungreis, E., and Ain, F., Anal. Chim. Acta **88**, 191 (1977).

[2445] Kacprzak, J. L., and Chvojka, R., J. Assoc. Off. Anal. Chem. **59**, 153 (1976).

[2446] Kahl, M., Mitchell, D. G., Kaufman, G. I., and Aldous, K. M., Anal. Chim. Acta **87**, 215 (1976).

[2447] Kaiser, G., Götz, D., Schoch, P., and Tölg, G., Talanta **22**, 889 (1975).

[2448] Kaiser, G., Götz, D., Tölg, G., Knapp, G., Maichin, B., and Spitzy, H., Z. Anal. Chem. **291**, 278 (1978).

[2449] Kalb, G. W., in: Trace Elements in Fuel (S. P. Babu, ed.). Adv. Chem. Ser. **141**, Am. Chem. Soc., Washington, D. C. 1975, p. 154.

[2450] Kallmann, S., and Blumberg, P., Talanta **27**, 827 (1980).

[2451] Kamel, H., Brown, D. H., Ottaway, J. M., and Smith, W. E., Analyst **101**, 790 (1976).

[2452] Kamel, H., Brown, D. H., Ottaway, J. M., and Smith, W. E., Analyst **102**, 645 (1977).

[2453] Kamel, H., Brown, D. H., Ottaway, J. M., and Smith, W. E., Talanta **24**, 309 (1977).

[2454] Kane, J. S., Anal. Chim. Acta **106**, 325 (1979).

[2455] Kang, H. K., and Valentine, J. L., Anal. Chem. **49**, 1829 (1977).

[2456] Kantor, T., Fodor, P., and Pungor, E., Anal. Chim. Acta **102**, 15 (1978).

[2457] Karwowska, R., Bulska, E., and Hulanicki, A., Talanta **27**, 397 (1980).

[2458] Kaszerman, R., and Theurer, K., At. Absorption Newslett. **15**, 129 (1976).

[2459] Kato, K., At. Absorption Newslett. **15**, 4 (1976).

[2460] Katz, A., and Taitel, N., Talanta **24**, 132 (1977).

[2461] Kayne, F. J., Komar, G., Laboda, H., and Vanderlinde, R. E., Clin. Chem. **24**, 2151 (1978).

[2462] Kaye, J. H., and Ballou, N. E., Anal. Chem. **50**, 2076 (1978).

[2463] Keil, R., private communication.

[2464] Keliher, P. N., and Wohlers, C. C., Anal. Chem. **48**, 333 A (1976).

[2465] Kim, C. H., Alexander, P. W., and Smythe, L. E., Talanta **23**, 229 (1976).
[2466] Kim, C. H., Alexander, P. W., and Smythe, L. E., Talanta **23**, 573 (1976).
[2467] King, A. S., Astrophys. J. **75**, 379 (1932).
[2468] Kingston, H. M., Barnes, I. L., Brady, T. J., Rains, T. C., and Champ, M. A., Anal. Chem. **50**, 2064 (1978).
[2469] Kirk, M., Perry, E. G., and Arritt, J. M., Anal. Chim. Acta **80**, 163 (1975).
[2470] Kirkbright, G. F., Hsiao-Chuan, S., and Snook, R. D., At. Spectrosc. **1**, 85 (1980).
[2471] Kirkbright, G. F., and Taddia, M., Anal. Chim. Acta **100**, 145 (1978).
[2472] Kirkbright, G. F., and Taddia, M., At. Absorption Newslett. **18**, 68 (1979).
[2473] Kirkbright, G. F., and Wilson, P. J., Anal. Chem. **46**, 1414 (1974).
[2474] Kirkbright, G. F., and Wilson, P. J., At. Absorption Newslett. **13**, 140 (1974).
[2475] Klinkhammer, G. P., Anal. Chem. **52**, 117 (1980).
[2476] Knapp, G., Z. Anal. Chem. **274**, 271 (1975).
[2477] Knapp, G., Sadjadi, B., and Spitzy, H., Z. Anal. Chem. **274**, 275 (1975).
[2478] Knauer, H. E., and Milliman, G. E., Anal. Chem. **47**, 1263 (1975).
[2479] Knezevic, G., Deut. Lebensm. Rundsch. **75**, 305 (1979).
[2480] Knudson, E. J., and Christian, G. D., Anal. Lett. **6**, 1073 (1973).
[2481] Koirtyohann, S. R., Glass, E. D., and Lichte, F. E., Appl. Spectrosc. **35**, 22 (1981).
[2482] Koirtyohann, S. R., and Khalil, M., Anal. Chem. **48**, 136 (1976).
[2483] Koizumi, H., and Yasuda, K., Anal. Chem. **47**, 1679 (1975).
[2484] Koizumi, H., and Yasuda, K., Anal. Chem. **48**, 1178 (1976).
[2485] Koizumi, H., and Yasuda, K., Spectrochim. Acta **31 B**, 237 (1976).
[2486] Koizumi, H., and Yasuda, K., Spectrochim. Acta **31 B**, 523 (1976).
[2487] Koizumi, H., Yasuda, K., and Katayama, M., Anal. Chem. **49**, 1106 (1977).
[2488] Kojima, I., Uchida, T., Nanbu, M., and Iida, C., Anal. Chim. Acta **93**, 69 (1977).
[2489] Kokot, M. L., At. Absorption Newslett. **15**, 105 (1976).
[2490] Komárek, J., Vrchlabský, M., and Sommer, L., Z. Anal. Chem. **278**, 121 (1976).
[2491] Kometani, T. Y., Anal. Chem. **49**, 2289 (1977).
[2492] König, K. H., and Neumann, P., Z. Anal. Chem. **279**, 337 (1976).
[2493] Kontas, E., At. Spectrosc. **2**, 59 (1981).
[2494] Korečková, J., Frech, W., Lundberg, E., Persson, J. A., and Cedergren, A., Anal. Chim. Acta **130**, 267 (1981).
[2495] Korenaga, T., Analyst **106**, 40 (1981).
[2496] Korkisch, J., and Gross, H., Mikrochim. Acta (Wien) **1975 II**, 413.
[2497] Korkisch, J., Hübner, H., Steffan, I., Arrhenius, G., Fisk, M., and Frazer, J., Anal. Chem. Acta **83**, 83 (1976).
[2498] Korkisch, J., and Krivanec, H., Anal. Chim. Acta **83**, 111 (1976).
[2499] Korkisch, J., and Sorio, A., Anal. Chim. Acta **82**, 311 (1976).
[2500] Korkisch, J., Steffan, I., and Gross, H., Mokrochim. Acta (Wien) **1975 II**, 569.
[2501] Korrey, J. S., and Goulden, P. D., At. Absorption Newslett. **14**, 33 (1975).
[2502] Korunová, V., and Dědina, J., Analyst **105**, 48 (1980).
[2503] Kothandaraman, P., and Dallmeyer, J. F., At. Absorption Newslett. **15**, 120 (1976).
[2504] Kovatsis, A. V., At. Absorption Newslett. **17**, 104 (1978).
[2505] Kowalska, A., and Kedziora, M., At. Spectrosc. **1**, 33 (1980).
[2506] Kraft, G., Lindenberger, D., and Beck, H., Z. Anal. Chem **282**, 119 (1976).
[2507] Kragten, J., and Reynaert, A. P., Talanta **21**, 618 (1974).
[2508] Krishnamurti, K. V., Shpirt, E., and Reddy, M. M., At. Absorption Newslett. **15**, 68 (1976).
[2509] Krishnan, S. S., Quittkat, S., and Crapper, D. R., Can. J. Spectrosc. **21**, 25 (1976).
[2510] Kujirai, O., Kobayashi, T., and Sudo, E., Z. Anal. Chem. **297**, 398 (1979).
[2511] Kuldvere, A., At. Spectrosc. **1**, 138 (1980).
[2512] Kuldvere, A., and Andreassen, B. T., At. Absorption Newslett. **18**, 106 (1979).
[2513] Kumpulainen, J., Anal. Chim. Acta **113**, 355 (1980).
[2514] Kundu, M. K., and Prévôt, A., Anal. Chem. **46**, 1591 (1974).
[2515] Kunselman, G. C., and Huff, E. A., At. Absorption Newslett. **15**, 29 (1976).

[2516] Labrecque, J. J., Appl. Spectrosc. **30**, 625 (1976).

[2517] Labrecque, J. J., Galobardes, J., and Cohen, M. E., Appl. Spectrosc. **31**, 207 (1977).

[2518] Lang, I., Šebor, G., Sychra, V., Kolihová, D., and Weisser, O., Anal. Chim. Acta **84**, 299 (1976).

[2519] Lang, I., Šebor, G., Weisser, O., and Sychra, V., Anal. Chim. Acta **88**, 313 (1977).

[2520] Langmyhr, F. J., Talanta **24**, 277 (1977).

[2521] Langmyhr, F. J., and Aadalen, U., Anal. Chim. Acta **115**, 365 (1980).

[2522] Langmyhr, F. J., and Aamodt, J., Anal. Chim. Acta **87**, 483 (1976).

[2523] Langmyhr, F. J., and Dahl, I. M., Anal. Chim. Acta **131**, 303 (1981).

[2524] Langmyhr, F. J., and Håkedal, J. T., Anal. Chim. Acta **83**, 127 (1976).

[2525] Langmyhr, F. J., and Kjuus, I., Anal. Chim. Acta **100**, 139 (1978).

[2526] Langmyhr, F. J., Lind, T., and Jonsen, J., Anal. Chim. Acta **80**, 297 (1975).

[2527] Langmyhr, F. J., and Orre, S., Anal. Chim. Acta **118**, 307 (1980).

[2528] Langmyhr, F. J., and Paus, P. E., Anal. Chim. Acta **44**, 445 (1969).

[2529] Langmyhr, F. J., and Paus, P. E., Anal. Chim. Acta **45**, 176 (1969).

[2530] Langmyhr, F. J., and Rasmussen, S., Anal. Chim. Acta **72**, 79 (1974).

[2531] Langmyhr, F. J., Solberg, R., and Thomassen, Y., Anal. Chim. Acta **92**, 105 (1977).

[2532] Langmyhr, F. J., Stubergh, J. R., Tomassen, Y., Hanssen, J. E., and Doležal, J., Anal. Chim. Acta **71**, 35 (1974).

[2533] Langmyhr, F. J., Sundli, A., and Jonsen, J., Anal. Chim. Acta **73**, 81 (1974).

[2534] Langmyhr, F. J., and Tsalev, D. L., Anal. Chim. Acta **92**, 79 (1977).

[2535] Lansford, M., McPherson, E. M., and Fishman, M. J., At. Absorption Newslett. **13**, 103 (1974).

[2536] Lau, C., Held, A., and Stephens, R., Can. J. Spectrosc. **21**, 100 (1976).

[2537] Lau, O. W., and Li, K. L., Analyst **100**, 430 (1975).

[2538] Law, S. L., and Gordon, G. E., Environ. Sci. Technol. **13**, 432 (1979).

[2539] Lawson, S. R., and Woodriff, R., Spectrochim. Acta **35 B**, 753 (1980).

[2540] leBihan, A., and Courtot-Coupez, J., Analusis **3**, 559 (1975).

[2541] Lekehal, N., and Hanocq, M., Anal. Chim. Acta **83**, 93 (1976).

[2542] Leopez-Escobar, L., and Hume, D. N., Anal. Lett. **6**, 343 (1973).

[2543] Li, R. T., and Hercules, D. M., Anal. Chem. **46**, 916 (1974).

[2544] Lichte, F. E., and Skogerboe, R. K., Anal. Chem. **44**, 1480 (1972).

[2545] Liddell, P. R., Anal. Chem. **48**, 1931 (1976).

[2546] Liddell, P. R., and Wildy, P. C., Spectrochim. Acta **35 B**, 193 (1980).

[2547] Lidums, V. V., At. Absorption Newslett. **18**, 71 (1979).

[2548] Lieu, V. T., and Woo, D. H., At. Spectrosc. **1**, 149 (1980).

[2549] Likaits, E. R., Farrell, R. F., and Mackie, A. J., At. Absorption Newslett. **18**, 53 (1979).

[2550] Lindsjo, O., in: Atomspektrometrische Spurenanalytik (Editor: B. Welz). Verlag Chemie, Weinheim 1982, p. 437.

[2551] Litman, R., Finston, H. L., and Williams, E. T., Anal. Chem. **47**, 2364 (1975).

[2552] Littlejohn, D., Fell, G. S., and Ottaway, J. M., Clin. Chem. **22**, 1719 (1976).

[2553] Littlejohn, D., and Ottaway, J. M., Analyst **102**, 553 (1977).

[2554] Littlejohn, D., and Ottaway, J. M., Anal. Chim. Acta **98**, 279 (1978).

[2555] Littlejohn, D., and Ottaway, J. M., Analyst **103**, 662 (1978).

[2556] Littlejohn, D., and Ottaway, J. M., Anal. Chim. Acta **107**, 139 (1979).

[2557] Littlejohn, D., and Ottaway, J. M., Analyst **104**, 208 (1979).

[2558] Littlejohn, D., and Ottaway, J. M., Analyst **104**, 1138 (1979).

[2559] Lo, D. B., and Christian, G. D., Can. J. Spectrosc. **22**, 45 (1977).

[2560] Lo, D. B., and Coleman, R. L., At. Absorption Newslett. **18**, 10 (1979).

[2561] Locke, J., Anal. Chim. Acta **104**, 225 (1979).

[2562] Lord, D. A., McLaren, J. W., and Wheeler, R. C., Anal. Chem. **49**, 257 (1977).

[2563] Losser, G. L., At. Absorption Newslett. **17**, 41 (1978).

[2564] Love, J. L., and Patterson, J. E., J. Assoc. Off. Anal. Chem. **61**, 627 (1978).

[2565] Luecke, W., Chem. Geol. **20**, 265 (1977).

[2566] Luecke, W., Chem. Erde **38**, 1 (1979).

[2567] Lukasiewicz, R. J., Berens, P. H., and Buell, B. E., Anal. Chem. **47**, 1046 (1975).
[2568] Lukasiewicz, R. J., and Buell, B. E., Anal. Chem. **47**, 1674 (1975).
[2569] Lund, W., Larsen, B. V., and Gundersen, N., Anal. Chim. Acta **81**, 319 (1976).
[2570] Lundberg, E., and Frech, W., Anal. Chim. Acta **104**, 67 (1979).
[2571] Lundberg, E., and Frech, W., Anal. Chim. Acta **104**, 75 (1979).
[2572] Lundberg, E., and Frech, W., Anal. Chim. Acta **108**, 75 (1979).
[2573] Lundberg, E., and Frech, W., Anal. Chem. **53**, 1437 (1981).
[2574] Lundberg, E., and Johansson, G., Anal. Chem. **48**, 1922 (1976).
[2575] L'vov, B. V., Talanta **23**, 109 (1976).
[2576] L'vov, B. V., Spectrochim. Acta **33 B**, 153 (1978).
[2577] L'vov, B. V., Keynote Lectures XXI CSI, 8th ICAS, Cambridge. Heyden, London, Philadelphia, Rheine 1979.
[2578] L'vov, B. V., Bayunov, P. A., Patrov, I. B., and Polobeiko, T. B., Zh. Anal. Khim. **35**, 1877 (1980).
[2579] L'vov, B. V., Bayunov, P. A., and Ryabchuk, G. N., Spectrochim. Acta **36 B**, 397 (1981).
[2580] L'vov, B. V., Katskov, D. A., Kruglikova, L. P., and Polzik, L. K., Spectrochim. Acta **31 B**, 49 (1976).
[2581] L'vov, B. V., Kruglikova, L. P., Polzik, L. K., and Katskov, D. A., Zh. Anal. Khim. **30**, 645 (1975).
[2582] L'vov, B. V., Kruglikova, L. P., Polzik, L. K., and Katskov, D. A., Zh. Anal. Khim. **30**, 652 (1975).
[2583] L'vov, B. V., and Orlov, N. A., Zh. Anal. Khim. **30**, 1661 (1975).
[2584] L'vov, B. V., and Pelieva, L. A., Can. J. Spectrosc. **23**, 1 (1978).
[2585] L'vov, B. V., and Pelieva, L. A., Zh. Anal. Khim. **33**, 1572 (1978).
[2586] L'vov, B. V., and Pelieva, L. A., Zh. Anal. Khim. **33**, 1695 (1978).
[2587] L'vov, B. V., and Pelieva, L. A., Zh. Anal. Khim. **34**, 1744 (1979).
[2588] L'vov, B. V., and Pelieva, L. A., Zh. Prikl. Spektrosk. **31**, 16 (1979).
[2589] L'vov, B. V., and Pelieva, L. A., Prog. analyt. atom. Spectrosc. **3**, 65 (1980).
[2590] L'vov, B. V., Pelieva, L. A., and Sharnopolsky, A. I., Zh. Prikl. Spektrosk. **27**, 395 (1977).
[2591] L'vov, B. V., Pelieva, L. A., and Sharnopolsky, A. I., Zh. Prikl. Spektrosk. **28**, 19 (1978).
[2592] L'vov, B. V., and Ribzyk, G. N., Zh. Prikl. Spektrosk. **33**, 1013 (1980).
[2593] Machata, G., Wien. klin. Wschr. **87**, 484 (1975).
[2594] Machata, G., and Binder, R., Z. Rechtsmed. **73**, 29 (1973).
[2595] Machata, G., and Binder, R., Archiv Kriminologie **155**, 87 (1975).
[2596] Machiroux, R., and Anh, D. T. K., Anal. Chim. Acta **86**, 35 (1976).
[2597] Macquet, J. P., and Theophanides, T., Spectrochim. Acta **29 B**, 241 (1974).
[2598] Macquet, J. P., and Theophanides, T., At. Absorption Newslett. **14**, 23 (1975).
[2599] Macquet, J. P., and Theophanides, T., Biochim. Biophys. Acta **442**, 142 (1976).
[2600] Maessen, F. J. M. J., Balke, J., and Massee, R., Spectrochim. Acta **33 B**, 311 (1978).
[2601] Maessen, F. J. M. J., Posma, F. D., and Balke, J., Anal. Chem. **46**, 1445 (1974).
[2602] Magill, W. A., and Svehla, G., Z. Anal. Chem. **268**, 177 (1974).
[2603] Magill, W. A., and Svehla, G., Z. Anal. Chem. **268**, 180 (1974).
[2604] Magnusson, B., and Westerlund, S., Anal. Chim. Acta **131**, 63 (1981).
[2605] Magos, L., Analyst **96**, 847 (1971).
[2606] Magyar, B., and Aeschbach, F., Spectrochim. Acta **35 B**, 839 (1980).
[2607] Maher, W. A., Anal. Chim. Acta **126**, 157 (1981).
[2608] Maier, D., Sinemus, H. W., and Wiedeking, E., Z. Anal. Chem. **296**, 114 (1979).
[2609] Majer, J. R., and Khalil, S. E. A., Anal. Chim. Acta **126**, 175 (1981).
[2610] Malloy, J. M., Keliher, P. N., and Cresser, M. S., Spectrochim. Acta **35 B**, 833 (1980).
[2611] Maloney, M. P., Moody, G. J., and Thomas, J. D. R., Analyst **105**, 1087 (1980).
[2612] Maney, J. P., and Luciano, V. J., Anal. Chim. Acta **125**, 183 (1981).
[2613] Manning, D. C., At. Absorption Newslett. **14**, 99 (1975).
[2614] Manning, D. C., At. Absorption Newslett. **17**, 107 (1978).
[2615] Manning, D. C., and Ediger, R. D., At. Absorption Newslett. **15**, 42 (1976).

[2616] Manning, D. C., Ediger, R. D., and Hoult, D. W., At. Spectrosc. **1**, 52 (1980).

[2617] Manning, D. C., and Slavin, W., Anal. Chem. **50**, 1234 (1978).

[2618] Manning, D. C., and Slavin, W., At. Absorption Newslett. **17**, 43 (1978).

[2619] Manning, D. C., and Slavin, W., Anal. Chim. Acta **118**, 301 (1980).

[2620] Manning, D. C., Slavin, W., and Carnrick, G. R., Spectrochim. Acta **37 B**, 331 (1982).

[2621] Manning, D. C., Slavin, W., and Myers, S., Anal. Chem. **51**, 2375 (1979).

[2622] Manthei, K., Ger. **2245610**, 1974.

[2623] Marcus, M., Hollander, M., Lucas, R. E., and Pfeiffer, N. C., Clin. Chem. **21**, 533 (1975).

[2624] Marks, J. Y., Spellman, R. J., and Wysocki, B., Anal. Chem. **48**, 1474 (1976).

[2625] Marks, J. Y., and Welcher, G. G., in: Flameless Atomic Absorption Analysis: An Update. American Society for Testing and Materials, STP **618**, 1977, p. 11.

[2626] Marks, J. Y., Welcher, G. G., and Spellman, R. J., Appl. Spectrosc. **31**, 9 (1977).

[2627] Markus, J. R., J. Assoc. Off. Anal. Chem. **57**, 970 (1974).

[2628] Martin, T. D., and Kopp, J. F., At. Absorption Newslett. **14**, 109 (1975).

[2629] Massmann, H., in: Ullmanns Encyklopädie der Technischen Chemie, Band 5. Verlag Chemie, Weinheim, 4th Edition, 1980, p. 423.

[2630] Massmann, H., ElGohary, Z., and Güçer, S., Spectrochim. Acta **31 B**, 399 (1976).

[2631] Massmann, H., and Güçer, S., Spectrochim. Acta **29 B**, 283 (1974).

[2632] Matoušek, J. P., Amer. Lab. **3(6)**, 45 (1971).

[2633] Matoušek, J. P., Talanta **24**, 315 (1977).

[2634] Matoušek, J. P., Prog. analyt. atom. Spectrosc. **4**, 247 (1981).

[2635] Matsunaga, K., Konishi, S., and Nishimura, M., Environ. Sci. Technol. **13**, 63 (1979).

[2636] Matsunaga, K., and Takahashi, S., Anal. Chim. Acta **87**, 487 (1976).

[2637] Matsusaki, K., Yoshino, T., and Yamamoto, Y., Talanta **26**, 377 (1979).

[2638] Matsusaki, K., Yoshino, T., and Yamamoto, Y., Anal. Chim. Acta **124**, 163 (1981).

[2639] Matter, L., and Schneider, W., Lebensmittelchem. Gerichtl. Chem. **28**, 231 (1974).

[2640] Matthes, W., Flucht, R., and Stoeppler, M., Z. Anal. Chem. **291**, 20 (1978).

[2641] Maurer, J., Z. Lebensm. Unters. Forsch. **165**, 1 (1977).

[2642] Mausbach, G., G. I. T. **23**, 898 (1979).

[2643] May, I., and Greenland, L. P., Anal. Chem. **49**, 2376 (1977).

[2644] May, K., Reisinger, K., Flucht, R., and Stoeppler, M., Vom Wasser **55**, 63 (1980).

[2645] May, K., and Stoeppler, M., Z. Anal. Chem. **293**, 127 (1978).

[2646] May, L. A., and Presley, B. J., At. Absorption Newslett. **13**, 144 (1974).

[2647] May, L. A., and Presley, B. J., Microchem. J. **21**, 119 (1976).

[2648] Mazzucotelli, A., and Frache, R., Analyst **105**, 497 (1980).

[2649] McArthur, J. M., Anal. Chim. Acta **93**, 77 (1977)

[2650] McCorriston, L. L., and Ritchie, R. K., Anal. Chem. **47**, 1137 (1975).

[2651] McDaniel, M., Shendrikar, A. D., Reiszner, K. D., and West, P. W., Anal. Chem. **48**, 2240 (1976).

[2652] McDermott, J. R., and Whitehill, I., Anal. Chim. Acta **85**, 195 (1976).

[2653] McDonald, D. C., Anal. Chem. **49**, 1336 (1977).

[2654] McElhaney, R. J., J. Radioanalyt. Chem. **32**, 99 (1976).

[2655] McHard, J. A., Winefordner, J. D., and Ataway, J. A., J. Agric. Food Chem. **24**, 41 (1976).

[2656] McHard, J. A., Winefordner, J. D., and Ting, S. V., J. Agric. Food Chem. **24**, 950 (1976).

[2657] McLaren, J. W., and Wheeler, R. C., Analyst **102**, 542 (1977).

[2658] Melcher, M., Grobenski, Z., and Welz, B., 8th Int. Microchem. Sympos. Graz 1980.

[2659] Melcher, M., and Welz, B., 29th Pittsburgh Conf. Anal. Chem. Appl. Spectrosc., Cleveland, OH. 1978.

[2660] Menden, E. E., Brockman, D., Choudhury, H., and Petering, H. G., Anal. Chem. **49**, 1644 (1977).

[2661] Menezes de Sequeira, E., Agronomia Lusit. **30**, 115 (1968).

[2662] Menz, D., and Conradi, G., in: Atomspektrometrische Spurenanalytik (Editor: B. Welz). Verlag Chemie, Weinheim 1982, p. 489.

[2663] Méranger, J. C., Hollebone, B. R., and Blanchette, G. A., J. Analyt. Toxicol. **5**, 33 (1981).

[2664] Méranger, J. C., and Subramanian, K. S., Can. J. Spectrosc. **24**, 132 (1979).
[2665] Meyer, A., Hofer, Ch., and Tölg, G., Z. Anal. Chem. **290**, 292 (1978).
[2666] Meyer, A., Hofer, Ch., Tölg, G., Raptis, S., and Knapp, G., Z. Anal. Chem. **296**, 337 (1979).
[2667] Michael, S. S., Anal. Chem. **49**, 451 (1977).
[2668] Mills, J. C., and Belcher, C. B., Prog. analyt. atom. Spectrosc. **4**, 49 (1981).
[2669] Mitcham, R. P., Analyst **105**, 43 (1980).
[2670] Mitchell, D. G., Aldous, K. M., and Ward, A. F., At. Absorption Newslett. **13**, 121 (1974).
[2671] Mojski, M., Talanta **25**, 163 (1978).
[2672] Moldan, B., Rubeška, I., Miksovsky, M., and Huka, M., Anal. Chim. Acta **52**, 91 (1970).
[2673] Montaser, A., and Crouch, S. R., Anal. Chem. **46**, 1817 (1974).
[2674] Montaser, A., and Crouch, S. R., Anal. Chem. **47**, 38 (1975).
[2675] Montaser, A., Goode, S. R., and Crouch, S. R., Anal. Chem. **46**, 599 (1974).
[2676] Müller-Vogt, G., and Wendl, W., Mat. Res. Bull. **15**, 1461 (1980).
[2677] Müller-Vogt, G., and Wendl, W., Anal. Chem. **53**, 651 (1981).
[2678] Murphy, G. F., and Stephens, R., Talanta **25**, 223 (1978).
[2679] Murphy, J., At. Absorption Newslett. **14**, 151 (1975).
[2680] Musil, J., and Doležal, J., Anal. Chim. Acta **92**, 301 (1977).
[2681] Musil, J., and Nehasilová, M., Talanta **23**, 729 (1976).
[2682] Muter, R. B., and Nice, L. L., in: Trace Elements in Fuel (S. P. Babu, ed.), Adv. Chem. Ser. 141, Am. Chem. Soc. Washington, D.C. 1975, p. 57.
[2683] Muzzarelli, R. A. A., and Rocchetti, R., Talanta **22**, 683 (1975).
[2684] Nakahara, T., and Chakrabarti, C. L., Anal. Chim. Acta **104**, 99 (1979).
[2685] Nakahara, T., and Musha, S., Anal. Chim. Acta **80**, 47 (1975).
[2686] Nakahara, T., and Musha, S., Can. J. Spectrosc. **24**, 138 (1979).
[2687] Nakahara, T., Tanaka, T., and Musha, S., Bull. Chem. Soc. Jpn. **51**, 2046 (1978).
[2688] Nakashima, S., Analyst **104**, 172 (1979).
[2689] Neumann, D. R., and Munshower, F. F., Anal. Chim. Acta **123**, 325 (1981).
[2690] Neve, J., and Hanocq, M., Anal. Chim. Acta **93**, 85 (1977).
[2691] Newton, M. P., and Davis, D. G., Anal. Chem. **47**, 2003 (1975).
[2692] Nichols, J. A., Jones, R. D., and Woodriff, R., Anal. Chem. **50**, 2071 (1978).
[2693] Nichols, J. A., and Woodriff, R., J. Assoc. Off. Anal. Chem. **63**, 500 (1980).
[2694] Noga, R. J., Anal. Chem. **47**, 332 (1975).
[2695] Noller, B. N., and Bloom, H., Anal. Chem. **49**, 346 (1977).
[2696] Noller, B. N., Bloom, H., and Arnold, A. P., Prog. analyt. atom. Spectrosc. **4**, 81 (1981).
[2697] Nomura, T., and Karasawa, I., Anal. Chim. Acta **126**, 241 (1981).
[2698] Norris, J. D., and West, T. S., Anal. Chem. **46**, 1423 (1974).
[2699] Noval, E., and Gries, W. H., Anal. Chim. Acta **83**, 393 (1976).
[2700] Nuhfer, E. B., and Romanosky, R. R., At. Absorption Newslett. **18**, 8 (1979).
[2701] O'Haver, T. C., Harnly, J. M., and Zander, A. T., Anal. Chem. **50**, 1218 (1978).
[2702] Ohta, K., and Suzuki, M., Anal. Chim. Acta **85**, 83 (1976).
[2703] Ohta, K., and Suzuki, M., Talanta **25**, 160 (1978).
[2704] Ohta, K., and Suzuki, M., Anal. Chim. Acta **104**, 293 (1979).
[2705] Ohta, K., and Suzuki, M., Talanta **26**, 207 (1979).
[2706] Okuno, I., Whitehead, J. A., and White, R. E., J. Assoc. Off. Anal. Chem. **61**, 664 (1978).
[2707] Oles, P. J., and Siggia, S., Anal. Chem. **46**, 911 (1974).
[2708] Oles, P. J., and Siggia, S., Anal. Chem. **46**, 2197 (1974).
[2709] Omenetto, N., and Winefordner, J. D., Prog. analyt. atom. Spectrosc. **2**, 1 (1979).
[2710] Ooghe, W., and Verbeek, F., Anal. Chim. Acta **73**, 87 (1974).
[2711] Orheim, R. M., and Bovee, H. H., Anal. Chem. **46**, 921 (1974).
[2712] Ortner, H. M., and Kantuscher, E., Talanta **22**, 581 (1975).
[2713] Ortner, H. M., and Lassner, E., Mikrochim. Acta, Suppl. **7**, 41 (1977).
[2714] Oster, O., Clin. Chim. Acta **114**, 53 (1981).
[2715] Oster, O., J. Clin. Chem. Clin. Biochem. **19**, 471 (1981).

[2716] Otruba, V., Jambor, J., Horák, J., and Sommer, L., Scripta Fac. Sci. Nat. Ujep. Brunensis Chemia **1**, 1 (1976).
[2717] Ottaway, J. M., Proc. Analyt. Div. Chem. Soc. **13**, 185 (1976).
[2718] Ottaway, J. M., and Shaw, F., Analyst **100**, 438 (1975).
[2719] Ottaway, J. M., and Shaw, F., Appl. Spectrosc. **31**, 12 (1977).
[2720] Owens, J. W., and Gladney, E. S., At. Absorption Newslett. **14**, 76 (1975).
[2721] Owens, J. W., and Gladney, E. S., At. Absorption Newslett. **15**, 47 (1976).
[2722] Owens, J. W., and Gladney, E. S., At. Absorption Newslett. **15**, 95 (1976).
[2723] Pabalkar, M. A., Naik, S. V., and Sanjana, N. R., Analyst **106**, 47 (1981).
[2724] Pachuta, D. G., and Love, L. J. C., Anal. Chem. **52**, 444 (1980).
[2725] Page, A. G., Godbole, S. V., Deshkar, S. B., and Joshi, B. D., Anal. Lett. **A 11**, 619 (1978).
[2726] Pakalns, P., and Farrar, Y. J., Water Research **11**, 145 (1977).
[2727] Panday, V. K., Anal. Chim. Acta **57**, 31 (1971).
[2728] Panday, V. K., and Ganguly, A. K., Spectrosc. Lett. **9**, 73 (1976).
[2729] Pandey, L. P., Ghose, A., Dasgupta, P., and Rao, A. S., Talanta **25**, 482 (1978).
[2730] Parker, C., and Pearl, A., Brit. **1385791** (1972).
[2731] Parker, R. D., Z. Anal. Chem. **292**, 362 (1978).
[2732] Parkes, A., and Murray-Smith, R., At. Absorption Newslett. **18**, 57 (1979).
[2733] Patel, B. M., Bhatt, P. M., Gupta, N., Pawar, M. M., and Joshi, B. D., Anal. Chim. Acta **104**, 113 (1979).
[2734] Paveri-Fontana, S. L., Tessari, G., and Torsi, G., Anal. Chem. **46**, 1032 (1974).
[2735] Peats, S., At. Absorption Newslett. **18**, 118 (1979).
[2736] Peck, E., Anal. Lett. **B 11**, 103 (1978).
[2737] Pedersen, B., Willems, M., and Jörgensen, S. S., Analyst **105**, 119 (1980).
[2738] Pekarek, R. S., Hauer, E. C., Wannemacher, R. W., and Beisel, W. R., Anal. Biochem. **59**, 283 (1974).
[2739] Pera, M. F., and Harder, H. C., Clin. Chem. **23**, 1245 (1977).
[2740] Perry, E. F., Koirtyohann, S. R., and Perry, H. M., Clin. Chem. **21**, 626 (1975).
[2741] Persson, J. Å., and Frech, W., Anal. Chim. Acta **119**, 75 (1980).
[2742] Persson, J. Å., Frech, W., and Cedergren, A., Anal. Chim. Acta **89**, 119 (1977).
[2743] Persson, J. Å., Frech, W., and Cedergren, A., Anal. Chim. Acta **92**, 85 (1977).
[2744] Persson, J. Å., Frech, W., and Cedergren, A., Anal. Chim. Acta **92**, 95 (1977).
[2745] Persson, J. Å., Frech, W., Pohl, G., and Lundgren, K., Analyst **105**, 1163 (1980).
[2746] Petrov, I. I., Tsalev, D. L., and Barsev, A. I., At. Spectrosc. **1**, 47 (1980).
[2747] Pickett, E. E., 5th FACSS Meeting, Boston, MA 1978.
[2748] Pickford, C. J., and Rossi, G., At. Absorption Newslett. **14**, 78 (1975)
[2749] Pierce, F. D., and Brown, H. R., Anal. Chem. **48**, 693 (1976).
[2750] Pierce, F. D., and Brown, H. R., Anal. Chem. **49**, 1417 (1977).
[2751] Pierce, F. D., Lamoreaux, T. C., Brown, H. R., and Fraser, R. S., Appl. Spectrosc. **30**, 38 (1976).
[2752] Poldoski, J. E., At. Absorption Newslett. **16**, 70 (1977).
[2753] Poldoski, J. E., Anal. Chem. **52**, 1147 (1980).
[2754] Pollock, E. N., in: Trace Elements in Fuel (S. P. Babu, ed.). Adv. Chem. Ser. 141, Am. Chem. Soc., Washington, D.C. 1975, p. 23.
[2755] Pollock E. N., At. Spectrosc. **1**, 78 (1980).
[2756] Polo-Diez, L., Hernandez-Mendez, J., and Pedraz-Penalva, F., Analyst **105**, 37 (1980).
[2757] Porter, W. K., J. Assoc. Off. Anal. Chem. **57**, 614 (1974).
[2758] Posma, F. D., Balke, J., Herber, R. F. M., and Stuik, E. J., Anal. Chem. **47**, 834 (1975).
[2759] Prager, M. J., and Graves, D., J. Assoc. Off. Anal. Chem. **60**, 609 (1977).
[2760] Prévôt, A., and Gente-Jauniaux, M., At. Absorption Newslett. **17**, 1 (1978).
[2761] Price, W. J., Dymott, T. C., and Whiteside, P. J., Spectrochim. Acta **35 B**, 3 (1980).
[2762] Price, W. J., and Roos, J. T. H., Analyst **93**, 709 (1968).
[2763] Price, W. J., and Whiteside, P. J., Analyst **102**, 664 (1977).
[2764] Pritchard, M. W., and Reeves, R. D., Anal. Chim. Acta **82**, 103 (1976).

[2765] Prugger, M., and Torge, R., Ger. **1964469** (1969).
[2766] Pruskowska, E., Carnrick, G. R., and Slavin, W., Pittsburgh Conf. Anal. Chem. Appl. Spectrosc., Atlantic City, N.J. **1982,** Paper 448.
[2767] Pyen, G., and Fishman, M., At. Absorption Newslett. **17,** 47 (1978).
[2768] Pyen, G., and Fishman, M., At. Absorption Newslett. **18,** 34 (1979).
[2769] Qureshi, M. A., Farid, M., Aziz, A., and Ejaz, M., Talanta **26,** 166 (1979).
[2770] Räde, H. S., At. Absorption Newslett. **13,** 81 (1974).
[2771] Ramelow, G., and Hornung, H., At. Absorption Newslett. **17,** 59 (1978).
[2772] Ramsey, J. M., Anal. Chem. **52,** 2141 (1980).
[2773] Rantala, R. T. T., and Loring, D. H., At. Absorption Newslett. **14,** 117 (1975).
[2774] Rantala, R. T. T., and Loring, D. H., At. Absorption Newslett. **16,** 51 (1977).
[2775] Rantala, R. T. T., and Loring, D. H., At. Spectrosc. **1,** 163 (1980).
[2776] Ranweiler, L. E., and Moyers, J. L., Environ. Sci. Technol. **8,** 152 (1974).
[2777] Rao, V. M., and Sastri, M. N., Talanta **27,** 771 (1980).
[2778] Raptis, S., Knapp, G., Meyer, A., and Tölg, G., Z. Anal. Chem. **300,** 18 (1980).
[2779] Rasmussen, L., Anal. Chim. Acta **125,** 117 (1981).
[2780] Ratcliffe, D. B., Byford, C. S., and Osman, P. B., Anal. Chim. Acta **75,** 457 (1975).
[2781] Ratzlaff, K. L., Anal. Chem. **51,** 232 (1979).
[2782] Reamer, D. C., and Veillon, C., Anal. Chem. **53,** 1192 (1981).
[2783] Regan, J. G. T., and Warren, J., Analyst **103,** 447 (1978).
[2784] Regan, J. G. T., and Warren, J., At. Absorption Newslett. **17,** 89 (1978).
[2785] Reichert, I. K., and Gruber, H., Vom Wasser **51,** 191 (1978).
[2786] Reichert, I. K., and Gruber, H., Vom Wasser **52,** 289 (1979).
[2787] Reid, R. D., and Piepmeier, E. H., Anal. Chem. **48,** 338 (1976).
[2788] Riandey, C., Gavinelli, R., and Pinta, M., Spectrochim. Acta **35 B,** 765 (1980).
[2789] Rice, T. D., Talanta **23,** 359 (1976).
[2790] Rice, T. D., Anal. Chim. Acta **91,** 221 (1977).
[2791] Riner, J. C., Wright, F. C., and McBeth, C. A., At. Absorption Newslett. **13,** 129 (1974).
[2792] Ritter, C. J., Bergman, S. C., Cothern, C. R., and Zamierowski, E. E., At. Absorption Newslett. **17,** 70 (1978).
[2793] Robbins, W. K., Anal. Chem. **46,** 2177 (1974).
[2794] Robbins, W. K., Runnels, J. H., and Merryfield, R., Anal. Chem. **47,** 2096 (1975).
[2795] Robbins, W. K., and Walker, H. H., Anal. Chem. **47,** 1269 (1975).
[2796] Robinson, J. W., Kiesel, E. L., Goodbread, J. P., Bliss, R., and Marshall, R., Anal. Chim. Acta **92,** 321 (1977).
[2797] Rombach, N., and Kock, K., Z. Anal. Chem. **292,** 365 (1978).
[2798] Rook, H. L., Gills, T. E., and LaFleur, P. D., Anal. Chem. **44,** 1114 (1972).
[2799] Rooney, R. C., Analyst **101,** 678 (1976).
[2800] Rooney, R. C., Analyst **101,** 749 (1976).
[2801] Rosales, A. T., At. Absorption Newslett. **15,** 51 (1976).
[2802] Ross, R. T., and Gonzalez, J. G., Bull. Environ. Contamin. Tech. **12,** 470 (1974).
[2803] Routh, M. W., Anal. Chem. **52,** 182 (1980).
[2804] Rowston, W. B., and Ottaway, J. M., Analyst **104,** 645 (1979).
[2805] Royal, S. J., Nat. Inst. Metallurgy Report 2063, Randburg, South Africa 1980.
[2806] Rozenblum, V., Microchem. J. **21,** 82 (1976).
[2807] Rubeška, I., Spectrochim. Acta **29 B,** 263 (1974).
[2808] Rubeška, I., Can. J. Spectrosc. **20,** 156 (1975).
[2809] Rubeška, I., and Hlavinková, V., At. Absorption Newslett. **18,** 5 (1979).
[2810] Rubeška, I., Korečková, J., and Weiss, D., At. Absorption Newslett. **16,** 1 (1977).
[2811] Rubeška, I., and Musil, J., Prog. analyt. atom. Spectrosc. **2,** 309 (1979).
[2812] Rubeška, I., and Pelikanová, M., Spectrochim. Acta **33 B,** 301 (1978).
[2813] Runnels, J. H., Merryfield, R., and Fisher, H. B., Anal. Chem. **47,** 1258 (1975).
[2814] Saba, C. S., and Eisentraut, K. J., Anal. Chem. **49,** 454 (1977).
[2815] Saeed, K., and Thomassen, Y., Anal. Chim. Acta **130,** 281 (1981).

[2816] Salmela, S., and Vuori, E., Talanta **26**, 175 (1979).
[2817] Salmon, S. G., Davis, R. H., and Holcombe, J. A., Anal. Chem. **53**, 324 (1981).
[2818] Salmon, S. G., and Holcombe, J. A., Anal. Chem. **51**, 648 (1979).
[2819] Sand, J. R., Liu, J. H., and Huber, C. O., Anal. Chim. Acta **87**, 79 (1976).
[2820] Sanzolone, R. F., Chao, T. T., and Crenshaw, G. L., Anal Chim. Acta **105**, 247 (1979).
[2821] Sastri, V. S., Chakrabarti, C. L., and Willis, D. E., Can. J. Chem. **47**, 587 (1969).
[2822] Schachter, M. M., and Boyer, K. W., Anal. Chem. **52**, 360 (1980).
[2823] Scharmann, A., and Wirz, P., in: Atomspektrometrische Spurenanalytik (Editor: B. Welz). Verlag Chemie, Weinheim 1982, p. 405.
[2824] Schattenkirchner, M., and Grobenski, Z., At. Absorption Newslett. **16**, 84 (1977).
[2825] Schierling, P., and Schaller, K. H., Arbeitsmed. Sozialmed. Präventivmed. **16**, 57 (1981).
[2826] Schierling, P., and Schaller, K. H., At. Spectrosc. **2**, 91 (1981).
[2827] Schierling, P., and Schaller, K. H., in: Atomspektrometrische Spurenanalytik (Editor: B. Welz). Verlag Chemie, Weinheim 1982, p. 97.
[2828] Schmidt, W., and Dietl, F., Z. Anal. Chem. **295**, 110 (1979).
[2829] Schneider, W., and Matter, L., Forum Städte-Hygiene **28**, 226 (1977).
[2830] Schock, M. R., and Mercer, R. B., At. Absorption Newslett. **16**, 30 (1977).
[2831] Schramel, P., Anal. Chim. Acta **72**, 414 (1974).
[2832] Schulte-Löbbert, F. J., Bohn, G., and Acker, L., Lebensmittelchem. gerichtl. Chem. **32**, 93 (1978).
[2833] Schulze, H. D., Chemie, Anlagen, Verfahren **1979** (2), 23.
[2834] Scott, K., Analyst **103**, 754 (1978).
[2835] Sebastiani, E., Ohls, K., and Riemer, G., Z. Anal. Chem. **264**, 105 (1973).
[2836] Seeger, R., At. Absorption Newslett. **15**, 45 (1976).
[2837] Seeling, W., Grünert, A., Kienle, K. H., Opferkuch, R., and Swobodnik, M., Z. Anal. Chem. **299**, 368 (1979).
[2838] Sefzik, E., Vom Wasser **50**, 285 (1978).
[2839] Seifert, B., and Drews, M., WaBoLu Berichte **1/1978**, D. Reimer Verlag, Berlin 1978.
[2840] SenGupta, J. G., Talanta **23**, 343 (1976).
[2841] SenGupta, J. G., Talanta **28**, 31 (1981).
[2842] Shabushnig, J. G., and Hieftje, G. M., Anal. Chim. Acta **126**, 167 (1981).
[2843] Shaikh, A. U., and Tallman, D. E., Anal. Chem. **49**, 1094 (1977).
[2844] Shaikh, A. U., and Tallman, D. E., Anal. Chim. Acta **98**, 251 (1978).
[2845] Shaw, F., and Ottaway, J. M., At. Absorption Newslett. **13**, 77 (1974).
[2846] Shaw, F., and Ottaway, J. M., Analyst **100**, 217 (1975).
[2847] Sheaffer, J. D., Mulvey, G., and Skogerboe, R. K., Anal. Chem. **50**, 1239 (1978).
[2848] Shearer, D. A., Cloutier, R. O., and Hidiroglou, M., J. Assoc. Off. Anal. Chem. **60**, 155 (1977).
[2849] Shendrikar, A. D., and West, P. W., Anal. Chim. Acta **89**, 403 (1977).
[2850] Shepherd, G. A., and Johnson, A. J., At. Absorption Newslett. **6**, 114 (1967).
[2851] Shum, G. T. C., Freeman, H. C., and Uthe, J. F., J. Assoc. Off. Anal. Chem. **60**, 1010 (1977).
[2852] Shum, G. T. C., Freeman, H. C., and Uthc, J. F., Anal. Chem. **51**, 414 (1979).
[2853] Siemer, D. D., and Baldwin, J. M., Anal. Chem. **52**, 295 (1980).
[2854] Siemer, D. D., and Hageman, L., Anal. Lett. **8**, 323 (1975).
[2855] Siemer, D. D., and Hageman, L., Anal. Chem. **52**, 105 (1980).
[2856] Siemer, D. D., and Koteel, P., Anal. Chem. **49**, 1096 (1977).
[2857] Siemer, D. D., Koteel, P., and Jariwala, V., Anal. Chem. **48**, 836 (1976).
[2858] Siemer, D. D., Vitek, R. K., Koteel, P., and Houser, W. C., Anal. Lett. **10**, 357 (1977).
[2859] Siemer, D. D., and Wei, H. Y., Anal. Chem. **50**, 147 (1978).
[2860] Siemer, D. D., and Woodriff, R., Spectrochim. Acta **29 B**, 269 (1974).
[2861] Siertsema, L. H., Clin. Chim. Acta **69**, 533 (1976).
[2862] Sighinolfi, G. P., Gorgoni, C., and Santos, A. M., Geostandard Newslett. **4**, 223 (1980).
[2863] Sighinolfi, G. P., and Santos, A. M., Mikrochim. Acta (Wien) **1976 II**, 33.
[2864] Sighinolfi, G. P., Santos, A. M., and Martinelli, G., Talanta **26**, 143 (1979).
[2865] Silberman, D., and Fisher, G. L., Anal. Chim. Acta **106**, 299 (1979).

[2866] Simmons, W. J., Anal. Chem. **47,** 2015 (1975).
[2867] Simmons, W. J., and Loneragan, J. F., Anal. Chem. **47,** 566 (1975).
[2868] Simon, P. J., Giessen, B. C., and Copeland, T. R., Anal. Chem. **49,** 2285 (1977).
[2869] Sinemus, H. W., Melcher, M., and Welz, B., At. Spectrosc. **2,** 81 (1981).
[2870] Sinex, S. A., Cantillo, A. Y., and Heiz, G. R., Anal. Chem. **52,** 2342 (1980).
[2871] Singh, N. P., and Joselow, M. M., At. Absorption Newslett. **14,** 42 (1975).
[2872] Sire, J., Collin, J., and Voinovitch, I. A., Spectrochim. Acta **33 B,** 31 (1978).
[2873] Sirota, G. R., and Uthe, J. F., Anal. Chem. **49,** 823 (1977).
[2874] Slavin, S., Peterson, G. E., and Lindahl, P. C., At. Absorption Newslett. **14,** 57 (1975).
[2875] Slavin, W., At. Spectrosc. **1,** 66 (1980).
[2876] Slavin, W., At. Spectrosc. **2,** 8 (1981).
[2877] Slavin, W., Carnrick, G. R., and Manning, D. C., Anal. Chem. **54,** 621 (1982).
[2878] Slavin, W., Carnrick, G. R., and Manning, D. C., Anal. Chim. Acta **138,** 103 (1982).
[2879] Slavin, W., and Manning, D. C., Anal. Chem. **51,** 261 (1979).
[2880] Slavin, W., and Manning, D. C., Spectrochim. Acta **35 B,** 701 (1980).
[2881] Slavin, W., Manning, D. C., and Carnrick, G. R., Anal. Chem. **53,** 1504 (1981).
[2882] Slavin, W., Manning, D. C., and Carnrick, G. R., At. Spectrosc. **2,** 137 (1981).
[2883] Slavin, W., Myers, S. A., and Manning, D. C., Anal. Chim. Acta **117,** 267 (1980).
[2884] Slikkerveer, F. J., Braad, A. A., and Hendrikse, P. W., At. Spectrosc. **1,** 30 (1980).
[2885] Slovák, Z., and Dočekal, B., Anal. Chim. Acta **130,** 203 (1981).
[2886] Smets, B., Spectrochim. Acta **35 B,** 33 (1980).
[2887] Smeyers-Verbeke, J., Michotte, Y., and Massart, D. L., Anal. Chem. **50,** 10 (1978).
[2888] Smeyers-Verbeke, J., Michotte, Y., van den Winkel, P., and Massart, D. L., Anal. Chem. **48,** 125 (1976).
[2889] Smeyers-Verbeke, J., Verbeelen, D., and Massart, D. L., Clin. Chim. Acta **108,** 67 (1980).
[2890] Smith, A. E., Analyst **100,** 300 (1975).
[2891] Smith, M. R., and Cochran, H. B., At. Spectrosc. **2,** 97 (1981).
[2892] Smith, R. G., Talanta **25,** 173 (1978).
[2893] Smith, R. G., VanLoon, J. C., Knechtel, J. R., Fraser, J. L., Pitts, A. E., and Hodges, A. E., Anal. Chim. Acta **93,** 61 (1977).
[2894] Smith, R. G., and Windom, H. L., Anal. Chim. Acta **113,** 39 (1980).
[2895] Sneddon, J., Ottaway, J. M., and Rowston, W. B., Analyst **103,** 776 (1978).
[2896] Sohler, A., Wolcott, P., and Pfeiffer, C. C., Clin. Chim. Acta **70,** 391 (1976).
[2897] Sommerfeld, M. R., Love, T. D., and Olsen, R. D., At. Absorption Newslett. **14,** 31 (1975).
[2898] Sperling, K. R., At. Absorption Newslett. **14,** 60 (1975).
[2899] Sperling, K. R., At. Absorption Newslett. **15,** 1 (1976).
[2900] Sperling, K. R., Z. Anal. Chem. **287,** 23 (1977).
[2901] Sperling, K. R., Z. Anal. Chem. **299,** 103 (1979).
[2902] Sperling, K. R., Vom Wasser **54,** 99 (1980).
[2903] Sperling, K. R., Z. Anal. Chem. **301,** 294 (1980).
[2904] Sperling, K. R., and Bahr, B., Z. Anal. Chem. **301,** 31 (1980).
[2905] Sperling, K. R., and Bahr, B., Z. Anal. Chem. **306,** 7 (1981).
[2906] Stafford, D. T., and Saharovici, F., Spectrochim. Acta **29 B,** 277 (1974).
[2907] Stein, V. B., Canelli, E., and Richards, A. H., At. Spectrosc. **1,** 61 (1980).
[2908] Stein, V. B., Canelli, E., and Richards, A. H., At. Spectrosc. **1,** 133 (1980).
[2909] Steinnes, E., At. Absorption Newslett. **15,** 102 (1976).
[2910] Stephens, R., Talanta **24,** 233 (1977).
[2911] Stephens, R., Talanta **25,** 435 (1978).
[2912] Stephens, R., Talanta **26,** 57 (1979).
[2913] Stephens, R., and Murphy, G. F., Talanta **25,** 441 (1978).
[2914] Stephens, R., and Ryan, D. E., Talanta **22,** 655 (1975).
[2915] Stephens, R., and Ryan, D. E., Talanta **22,** 659 (1975).
[2916] Sterritt, R. M., and Lester, J. N., Analyst **105,** 616 (1980).
[2917] Stoeppler, M., and Backhaus, F., Z. Anal. Chem. **291,** 116 (1978).

[2918] Stoeppler, M., and Brandt, K., Z. Anal. Chem. **300,** 372 (1980).
[2919] Stoeppler, M., Brandt, K., and Rains, T. C., Analyst **103,** 714 (1978).
[2920] Stoeppler, M., Kampel, M., and Welz, B., Z. Anal. Chem. **282,** 369 (1976).
[2921] Stoeppler, M., and Matthes, W., Anal. Chim. Acta **98,** 389 (1978).
[2922] Stolzenburg, T. R., and Andren, A. W., Anal. Chim. Acta **118,** 377 (1980).
[2923] Strauss, W., Air Pollution Control III, Wiby & Sons, New York 1977.
[2924] Strelow, F. W. E., Liebenberg, C. J., and Victor, A. H., Anal. Chim. Acta **46,** 1409 (1974).
[2925] Stuart, D. C., Anal. Chim. Acta **96,** 83 (1978).
[2926] Stuart, D. C., Anal. Chim. Acta **101,** 429 (1978).
[2927] Stuart, D. C., Anal. Chim. Acta **106,** 411 (1979).
[2928] Studnicki, M., Anal. Chem. **52,** 1762 (1980).
[2929] Sturgeon, R. E., Anal. Chem. **49,** 1255 A (1977).
[2930] Sturgeon, R. E., and Berman, S. S., Anal. Chem. **53,** 632 (1981).
[2931] Sturgeon, R. E., Berman, S. S., Desaulniers, J. A. H., Mykytiuk, A. P., McLaren, J. W., and Russell, D. S., Anal. Chem. **52,** 1585 (1980).
[2932] Sturgeon, R. E., and Chakrabarti, C. L., Anal. Chem. **48,** 677 (1976).
[2933] Sturgeon, R. E., and Chakrabarti, C. L., Anal. Chem. **49,** 90 (1977).
[2934] Sturgeon, R. E., and Chakrabarti, C. L., Anal. Chem. **49,** 1100 (1977).
[2935] Sturgeon, R. E., and Chakrabarti, C. L., Spectrochim. Acta **32 B,** 231 (1977).
[2936] Sturgeon, R. E., and Chakrabarti, C. L., Prog. analyt. atom. Spectrosc. **1,** 5 (1978).
[2937] Sturgeon, R. E., Chakrabarti, C. L., and Bertels, P. C., Anal. Chem. **47,** 1250 (1975).
[2938] Sturgeon, R. E., Chakrabarti, C. L., and Langford, C. H., Anal. Chem. **48,** 1792 (1976).
[2939] Sturgeon, R. E., Chakrabarti, C. L., Maines, I. S., and Bertels, P. C., Anal. Chem. **47,** 1240 (1975).
[2940] Subramanian, K. S., and Chakrabarti, C. L., Prog. analyt. atom. Spectrosc. **2,** 287 (1979).
[2941] Subramanian, K. S., and Méranger, J. C., Anal. Chim. Acta **124,** 131 (1981).
[2942] Suddendorf, R. F., Wright, S. K., and Boyer, K. W., J. Assoc. Off. Anal. Chem. **64,** 657 (1981).
[2943] Sukiman, S., Anal. Chim. Acta **84,** 419 (1976).
[2944] Sulek, A. M., Elkins, E. R., and Zink, E. W., J. Assoc. Off. Anal. Chem. **61,** 931 (1978).
[2945] Sullivan, J. V., Prog. analyt. atom. Spectrosc. **4,** 311 (1981).
[2946] Sunderman, F. W., Clin. Chem. **21,** 1873 (1975).
[2947] Sunderman, F. W., 27th Int. Congr. Pure Appl. Chem. (A. Varmavuori, ed.), Pergamon Press, Oxford, New York 1980, p. 129.
[2948] Supp, G. R., and Gibbs, I., At. Absorption Newslett. **13,** 71 (1974).
[2949] Sutcliffe, P., Analyst **101,** 949 (1976).
[2950] Suzuki, M., Hayashi, K., and Wacker, W. E. C., Anal. Chim. Acta **104,** 389 (1979).
[2951] Suzuki, M., Ohta, K., and Yamakita, T., Anal. Chim. Acta **53,** 9 (1981).
[2952] Sychra, V., Svoboda, V., and Rubeška, I., Atomic Fluorescence Spectroscopy. Van Nostrand Reinhold Co., London 1975.
[2953] Szydlowski, F. J., At. Absorption Newslett. **16,** 60 (1977).
[2954] Szydlowski, F. J., Anal. Chim. Acta **106,** 121 (1979).
[2955] Szydlowski, F. J., and Vianzon, F. R., Anal. Lett. **B 11,** 161 (1978).
[2956] Tatro, M. E., Raynolds, W. L., and Costa, F. M., At. Absorption Newslett. **16,** 143 (1977).
[2957] Tello, A., and Sepulveda, N., At. Absorption Newslett. **16,** 67 (1977).
[2958] Terashima, S., Anal. Chim. Acta **86,** 43 (1976).
[2959] Therrell, B. L., Drosche, J. M., and Dziuk, T. W., Clin. Chem. **24,** 1182 (1978).
[2960] Thiex, N., J. Assoc. Off. Anal. Chem. **63,** 496 (1980).
[2961] Thomassen, Y., Larsen, B. V., Langmyhr, F. J., and Lund, W., Anal. Chim. Acta **83,** 103 (1976).
[2962] Thomassen, Y., Solberg, R., and Hanssen, E., Anal. Chim. Acta **90,** 279 (1977).
[2963] Thompson, A. J., and Thoresby, P. A., Analyst **102,** 9 (1977).
[2964] Thompson, D. D., and Allen, R. J., At. Spectrosc. **2,** 53 (1981).
[2965] Thompson, K. C., Godden, R. G., and Thomerson, D. R., Anal. Chim. Acta **74,** 289 (1975).
[2966] Thompson, K. C., Godden, R. G., and Thomerson, D. R., Anal. Chim. Acta **74,** 389 (1975).

[2967] Thompson, K. C., and Thomerson, D. R., Analyst **99**, 595 (1974).

[2968] Thompson, K. C., and Wagstaff, K., Analyst **105**, 641 (1980).

[2969] Thompson, K. C., and Wagstaff, K., Analyst **105**, 883 (1980).

[2970] Thompson, K. C., Wagstaff, K., and Wheatstone, K. C., Analyst **102**, 310 (1977).

[2971] Thompson, M., Pahlavanpour, B., Walton, S. J., and Kirkbright, G. F., Analyst **103**, 705 (1978).

[2972] Tindall, F. M., At. Absorption Newslett. **16**, 37 (1977).

[2973] Toda, W., Lux, J., and Van Loon, J. C., Anal. Lett. **13**, 1105 (1980).

[2974] Toffaletti, J., and Savory, J., Anal. Chem. **47**, 2091 (1975).

[2975] Tölg, G., Z. Anal. Chem. **283**, 257 (1977).

[1976] Tölg, G., Nachr. Chem. Tech. Lab. **27**, 250 (1979).

[2977] Tölg, G., Z. Anal. Chem. **294**, 1 (1979).

[2978] Tölg, G., and Lorenz, I., Chemie in unserer Zeit **11**, 150 (1977).

[2979] Tominaga, M., and Umezaki, Y., Anal. Chim. Acta **110**, 55 (1979).

[2980] Tomljanovic, M., and Grobenski, Z., At. Absorption Newslett. **14**, 52 (1975).

[2981] Torsi, G., Desimoni, E., Palmisano, F., and Sabbatini, L., Anal. Chem. **53**, 1035 (1981).

[2982] Toth, J. R., and Ingle, J. D., Anal. Chim. Acta **92**, 409 (1977).

[2983] Trachman, H. L., Tyberg, A. J., and Branigan, P. D., Anal. Chem. **49**, 1090 (1977).

[2984] Treptow, H., Askar, A., and Bielig, H. J., Lebensmittelchem. Gerichtl. Chem. **32**, 63 (1978).

[2985] Tsai, W. C., Lin, C. P., Shiau, L. J., and Pan, S. D., J. Amer. Oil Chem. Soc. **55**, 695 (1978).

[2986] Tsai, W. C., and Shiau, L. J., Anal. Chem. **49**, 1641 (1977).

[2987] Tschöpel, P., Kotz, L., Schulz, W., Veber, M., and Tölg, G., Z. Anal. Chem. **302**, 1 (1980).

[2988] Tsukahara, I., and Tanaka, M., Talanta **27**, 237 (1980).

[2989] Tuell, T. M., Ullman, A. H., Pollard, B. D., Massoumi, A., Bradshaw, J. D., Bower, J. N., and Winefordner, J. D., Anal. Chim. Acta **108**, 351 (1979).

[2990] Uchida, Y., and Hattori, S., Oyo Butsuri **44**, 852 (1975).

[2991] Ullman, A. H., Prog. analyt. atom. Spectrosc. **3**, 87 (1980).

[2992] Ullucci, P. A., and Hwang, J. Y., Talanta **21**, 745 (1974).

[2993] Unvala, H. A., U. S. **4238830** (1980).

[2994] Ure, A. M., and Mitchell, M. C., Anal. Chim. Acta **87**, 283 (1976).

[2995] Vajda, F., Anal. Chim. Acta **128**, 31 (1981).

[2996] Vandeberg, J. T., Swafford, H. D., and Scott, R. W., J. Paint Technol. **47**, 84 (1975).

[2997] van den Broek, W. M. G. T., and de Galan, L., Anal. Chem. **49**, 2176 (1977).

[2998] van den Broek, W. M. G. T., de Galan, L., Matoušek, J. P., and Czobik, E. J., Anal. Chim. Acta **100**, 121 (1978).

[2999] Vanderborght, B. M., and van Grieken, R. E., Anal. Chem. **49**, 311 (1977).

[3000] van der Geugten, R. P., Z. Anal. Chem. **306**, 13 (1981).

[3001] van Eenbergen, A., and Bruninx, E., Anal. Chim. Acta **98**, 405 (1978).

[3002] Van Loon, J. C., Z. Anal. Chem. **246**, 122 (1969).

[3003] Van Loon, J. C., Radziuk, B., Kahn, N., Lichwa, J., Fernandez, F. J., and Kerber, J. D., At. Absorption Newslett. **16**, 79 (1977).

[3004] van Luipen, J., At. Absorption Newslett. **17**, 144 (1978).

[3005] Veillon, C., Guthrie, B. E., and Wolf, W. R., Anal. Chem. **52**, 457 (1980).

[3006] Veinot, D. E., and Stephens, R., Talanta **23**, 849 (1976).

[3007] Velghe, N., Campe, A., and Claeys, A., At. Absorption Newslett. **17**, 37 (1978).

[3008] Velghe, N., Campe, A., and Claeys, A., At. Absorption Newslett. **17**, 139 (1978).

[3009] Verlinden, M., and Deelstra, H., Z. Anal. Chem. **296**, 253 (1979).

[3010] Versiek, J., and Cornelis, R., Anal. Chim. Acta **116**, 217 (1980).

[3011] Vickrey, T. M., Harrison, G. V., and Ramelow, G. J., At. Spectrosc. **1**, 116 (1980).

[3012] Vickrey, T. M., Harrison, G. V., Ramelow, G. J., and Carver, J. C., Anal. Lett. **13**, 781 (1980).

[3013] Vieira, N. E., and Hansen, J. W., Clin. Chem. **27**, 73 (1981).

[3014] Vigler, M. S., and Gaylor, V. F., Appl. Spectrosc. **28**, 342 (1974).

[3015] Vijan, P. N., At. Spectrosc. **1**, 143 (1980).

[3016] Vijan, P. N., and Chan, C. Y., Anal. Chem. **48**, 1788 (1976).

[3017] Vijan, P. N., Rayner, A. C., Sturgis, D., and Wood, G. R., Anal. Chim. Acta **82,** 329 (1976).

[3018] Vijan, P. N., and Sadana, R. S., Talanta **27,** 321 (1980).

[3019] Vijan, P. N., and Wood, G. R., Talanta **23,** 89 (1976).

[3020] Volland, G., Kölblin, G., Tschöpel, P., and Tölg, G., Z. Anal. Chem. **284,** 1 (1977).

[3021] Volland, G., Tschöpel, P., and Tölg, G., Anal. Chim. Acta **90,** 15 (1977).

[3022] Völlkopf, U., Grobenski, Z., and Welz, B., At. Spectrosc. **2,** 68 (1981).

[3023] Völlkopf, U., Grobenski, Z., and Welz, B., G.I.T. Fachz. Lab. **26,** 444 (1982).

[3024] Volosin, M. T., Kubasik, N. P., and Sine, H. E., Clin. Chem. **21,** 1986 (1975).

[3025] Wagenaar, H. C., and de Galan, L., Spectrochim. Acta **30 B,** 361 (1975).

[3026] Wahab, H. S., and Chakrabarti, C. L., Spectrochim. Acta **36 B,** 475 (1981).

[3027] Wahab, H. S., and Chakrabarti, C. L., Spectrochim. Acta **36 B,** 463 (1981).

[3028] Walker, H. H., Runnels, J. H., and Merryfield, R., Anal. Chem. **48,** 2056 (1976).

[3029] Wallace, G. F., Lumas, B. K., Fernandez, F. J., and Barnett, W. B., At. Spectrosc. **2,** 130 (1981).

[3030] Walsh, J. N., Analyst **102,** 51 (1977).

[3031] Walton, G., Analyst **98,** 335 (1973).

[3032] Ward, A. F., Mitchell, D. G., and Aldous, K. M., Anal. Chem. **47,** 1656, (1975).

[3033] Ward, A. F., Mitchell, D. G., Kahl, M., and Aldous, K. M., Clin. Chem. **20,** 1199 (1974).

[3034] Warren, J., and Carter, D., Can. J. Spectrosc. **20,** 1 (1975).

[3035] Watkins, D., Corbyons, T., Bradshaw, J., and Winefordner, J. D., Anal. Chim. Acta **85,** 403 (1976).

[3036] Watling, R. J., Anal. Chim. Acta **94,** 181 (1977).

[3037] Wauchope, R. D., At. Absorpt. Newslett. **15,** 64 (1976).

[3038] Wawschinek, O., and Höfler, H., At. Absorption Newslett. **18,** 97 (1979).

[3039] Wawschinek, O., and Rainer, F., At. Absorption Newslett. **18,** 50 (1979).

[3040] Wegscheider, W., Knapp, G., and Spitzy, H., Z. Anal. Chem. **283,** 183 (1977).

[3041] Weibust, G., Langmyhr, F. J., and Thomassen, Y., Anal. Chim. Acta **128,** 23 (1981).

[3042] Weinstock, N., and Uhlemann, M., Clin. Chem. **27,** 1438 (1981).

[3043] Weiss, H. V., Guttman, M. A., Korkisch, J., and Steffan, I., Talanta **24,** 509 (1977).

[3044] Welsch, E. P., and Chao, T. T., Anal. Chim. Acta **76,** 65 (1975).

[3045] Welsch, E. P., and Chao, T. T., Anal. Chim. Acta **82,** 337 (1976).

[3046] Welz, B., Angewandte AAS **4,** Bodenseewerk Perkin-Elmer, Überlingen 1976.

[3047] Welz, B., Z. Anal. Chem. **279,** 103 (1976).

[3048] Welz, B., Proceedings Symposium III on electrothermal atomization in atomic absorption spectrometry, Chlum u Třeboně 1977, p. 126.

[3049] Welz, B., Grobenski, Z., and Melcher, M., 13. Spektrometertagung. K. H. Koch, and H. Massmann, eds., Walter de Gruyter, Berlin, New York 1981, p. 337.

[3050] Welz, B., Grobenski, Z., Melcher, M., and Weber, D., Pittsburgh Conf. Anal. Chem. Appl. Spectrosc., Cleveland, OH. 1979.

[3051] Welz, B., and Melcher, M., At. Absorption Newslett. **18,** 121 (1979).

[3052] Welz, B., and Melcher, M., At. Spectrosc. **1,** 145 (1980).

[3053] Welz, B., and Melcher, M., 19th Ann. Sympos. Analyt. Chem. Pollutants, Dortmund 1980.

[3054] Welz, B., and Melcher, M., Spectrochim. Acta **36 B,** 439 (1981).

[3055] Welz, B., and Melcher, M., Anal. Chim. Acta **131,** 17 (1981).

[3056] Welz, B., and Melcher, M., Analyst **108,** 213 (1983).

[3057] Welz, B., and Melcher, M., 9th FACSS Meeting, Philadelphia, PA 1982, Paper 356.

[3058] Welz, B., and Melcher, M., Spectrochim. Acta, in print.

[3059] Welz, B., and Melcher, M., Anal. Chim. Acta, **153,** 297 (1983).

[3060] Welz, B., and Melcher, M., Vom Wasser **59,** 407 (1982).

[3061] Welz, B., Völlkopf, U., and Grobenski, Z., Anal. Chim. Acta **136,** 201 (1982).

[3062] Welz, B., Weber, D., and Grobenski, Z., Angewandte AAS **10,** Bodenseewerk Perkin-Elmer, Überlingen 1978.

[3063] Welz, B., Weber, D., and Grobenski, Z., unpublished results.

[3064] Welz, B., Wiedeking, E., and Sigl, W., Int. Sympos. Microchem. Tech. Davos 1977.

[3065] West, A. C., Fassel, V. A., and Kniseley, R. N., Anal. Chem. **45,** 815 (1973).
[3066] West, A. C., Fassel, V. A., and Kniseley, R. N., Anal. Chem. **45,** 2420 (1973).
[3067] White, W. W., and Murphy, P. J., Anal. Chem. **49,** 255 (1977).
[3068] Whiteside, P. J., and Price, W. J., Analyst **102,** 618 (1977).
[3069] Wigfield, D. C., Croteau, S. M., and Perkins, S. L., J. Analyt. Toxicol. **5,** 52 (1981).
[3070] Willis, J. B., Anal. Chem. **47,** 1753 (1975).
[3071] Wilson, D. L., At. Absorption Newslett. **18,** 13 (1979).
[3072] Wirth, K., Geol. Mitt. **12,** 367 (1974).
[3073] Wise, W. M., and Solsky, S. D., Anal. Lett. **9,** 1047 (1976).
[3074] Wise, W. M., and Solsky, S. D., Anal. Lett. **10,** 273 (1977).
[3075] Wittmers, L. E., Alich, A., and Aufderheide, A. C., Am. J. Clin. Pathol. **75,** 80 (1981).
[3076] Wolf, W. R., Anal. Chem. **48,** 1717 (1976).
[3077] Wolf, W. R., Mertz, W., and Masironi, R., J. Agr. Food Chem. **22,** 1037 (1974).
[3078] Woodis, T. C., Hunter, G. B., and Johnson, F. J., J. Assoc. Off. Anal. Chem. **59,** 22 (1976).
[3079] Wright, F. C., and Riner, J. C., At. Absorption Newslett. **14,** 103 (1975).
[3080] Wunderlich, E., and Burghardt, M., Z. Anal. Chem. **281,** 299 (1976).
[3081] Wunderlich, E., and Hädeler, W., Z. Anal. Chem. **281,** 300 (1976).
[3082] Wuyts, L., Smeyers-Verbeke, J., and Massart, D. L., Clin. Chim. Acta **72,** 405 (1976).
[3083] Yamada, H., Uchino, K., Koizumi, H., Noda, T., and Yasuda, K., Anal. Lett. **A 11,** 855 (1978).
[3084] Yamamoto, M., Urata, K., and Yamamoto, Y., Anal. Lett. **14,** 21 (1981).
[3085] Yamamoto, Y., and Kumamaru, T., Z. Anal. Chem. **281** 353 (1976).
[3086] Yamamoto, Y., and Kumamaru, T., Z. Anal. Chem. **282,** 139 (1976).
[3087] Yamamoto, Y., Kumamaru, T., and Shiraki, A., Z. Anal. Chem. **292,** 273 (1978).
[3088] Yamasaki, S., Yoshino, A., and Kishita, A., Soil Sci. Plant Nutr. **21,** 63 (1975).
[3089] Yasuda, K., Koizumi, H., Ohishi, K., and Noda, T., Prog. analyt. atom. Spectrosc. **3,** 299 (1980).
[3090] Yasuda, S., and Kakiyama, H., Anal. Chim. Acta **84,** 291 (1976).
[3091] Yasuda, S., and Kakiyama, H., Anal. Chim. Acta **89,** 369 (1977).
[3092] Young, R. S., Talanta **28,** 25 (1981).
[3093] Yudelevich, I. G., Zelentsova, L. V., Beisel, N. F., Chanysheva, T. A., and Vechernish, L., Anal. Chim. Acta **108,** 45 (1979).
[3094] Zachariasen, H., Andersen, I., Kostøl, C., and Barton, R., Clin. Chem. **21,** 562 (1975).
[3095] Zander, A. T., O'Haver, T. C., and Keliher, P. N., Anal. Chem. **48,** 1166 (1976).
[3096] Zander, A. T., O'Haver, T. C., and Keliher, P. N., Anal. Chem. **49,** 838 (1977).
[3097] Zatka, V. J., Anal. Chem. **50,** 538 (1978).
[3098] Zeeman, P., Phil. Mag. **5,** 226 (1897).
[3099] Zielhuis, R. L., Stuik, E. J., Herber, R. F. M., Sallé, H. J. A., Verberk, M. M., Posma, F. D., and Jager, J. H., Int. Arch. Occup. Environ. Hlth. **39,** 53 (1977).
[3100] Zielke, H. J., and Luecke, W., Z. Anal. Chem. **271,** 29 (1974).
[3101] de Galan, L., and de Loos-Vollebregt, M. T. C., Keynote Lectures XXI CSI, 8th ICAS, Cambridge. Heyden, London, Philadelphia, Rheine 1979.
[3102] L'vov, B. V., Katskov, D. A., and Kruglikova, L. P., Zh. Prikl. Spektr. **14,** 784 (1971).
[3103] L'vov, B. V., and Khartsyzov, A. D., Zh. Prikl. Spektr. **11,** 9 (1969).

Index

Page numbers printed in bold type are the more important references.

Aberration 86
Absorbance 1, 13, 100
Absorptance 13, 100
Absorption, measurement of 12–14
Absorption coefficient 1, **9–11**, 12, 86, 111, **155–157**
Absorption profile of atomic lines 93
– broadening 170
Absorption spectrum 3–5, 7
Accelerators for vulcanization, analysis of 423
Accuracy 66, **109**, 115, 119, 127, 244, 262
– of background correction 137, 150–151
Acetylide formation in the burner 404
Acid digestions 123, 246–247 (*see also under individual materials for digestion techniques*)
Activated charcoal for trace enrichment 121
Additives in lubricating oils, analysis of 410–411
– – solvents for 409–410
Adsorption
– chromium on silicates 374
– mercury on containers 244–248, 372
– silicon 323
– trace elements 123–124
Aerosols, analysis of 375–376
Air, analysis of 375 *et seq.*
Air/acetylene flame **32–33**, 37, 111, 168, 171 (*refer also to the individual elements for determinations using this flame*)
Airborne dust, analysis of 375–378
Aircraft lubricating oil, analysis of 413
Air filters 377–378
Air/hydrogen flame 35, 37, 111, 169
– for the determination of
– – bismuth 274
– – cesium 279
– – magnesium 299
– – potassium 316
– – rubidium 318
– – sodium 326
– – sulfur 328
– – tellurium 329
– – tin 331–332

Air/propane flame 36, 37, 39, 132, 171
– for the determination of
– – cesium 279
– – copper 283
– – lead 294
– – lithium 296
– – magnesium 298
– – osmium 311
– – platinum 314
– – rhodium 317
– – sodium 326
Alcohol content in drinks 361
Aldehydes, indirect determination of 310
Alkali halides, molecular absorption 200–201
Alkyl mercury compounds, determination of 302–304
Alloys, analysis of 237, **399–404**, 424
Alternating current instrument 15–16, 83, 97, 129
Aluminium 267–268
– acid-soluble in steel 395
– alloys, analysis of 402–403
– analysis of 43, 122, 405, 407
– as matrix modifier 274
– atomization 169, 178, 192, 196, 197, 216
– contamination 348
– determination of 60, 211, 218
– – in aluminium alloys 403
– – in aluminium trialkyls 424
– – in brain tissue 347
– – in cement 417
– – in copper alloys 399
– – in glass 416
– – in high temperature nickel alloys 400
– – in lubricating oils 414
– – in minerals 386
– – in molybdenum 405
– – in muscle tissue 348
– – in mussels 359
– – in nitric acid 421
– – in ores 386
– – in pharmaceuticals 423
– – in polypropylene 418
– – in pulp 419
– – in rocks 382

Aluminium, determination of
- – in serum   348
- – in silicate rocks   386
- – in silicon-containing materials   417
- – in soils   365
- – in steel   393, 395–396
- – in thorium oxide fuel rods   420
- – in titanium alloys   403
- – in titanium dioxide pigments   424
- – in tungsten   405
- – in uranium   420
- – in urine   348
- – in water   369, 370
- – in wool   419
- – in zirconium alloys   403
- extraction   267, 369
- oxide catalyst supports, analysis of   422
- oxide ceramics, analysis of   417
- resonance lines   267
- trialkyls, analysis of   424
Amalgamation   **79–81,** 125, 249–250, 302, 355, 371, 379
Ammonia solution, analysis of   422
Ammonium dichromate as matrix modifier   269, 275
Ammonium nitrate as matrix modifier   204, 215, 300, 333, 373
Ammonium phosphate as matrix modifier
- for cadmium   210, 277, 380
- for chromium   210, 281, 349
- for lead   210, 215, 219, 295, 354, 380
- for nickel   210, 308
- for zinc   339, 380
Ammonium pyrrolidine dithiocarbamate (APDC)   120
Ammonium traces, determination of   340
Analyte addition technique   **117–119,** 162, 174
Analytical curve   **108,** 154–158
- ambiguity of   157
- curvature of   14, 21, 24, 28, **83** *et seq.,* 104, 131, **154–158,** 281, 299, 306–307
- – influence on accuracy   118
- – influence on precision   117
- – through ionization   183
- linearity of   14, 21, 25, 39, **83** *et seq.,* 92, **154–158,** 299
Analytical curve technique   115–116
Analytical function   **108,** 114, 115–119

Analytical measure   95, 99–101
Antagonistic elements   358
Antifoaming agent   231, 352
Antimony   268–269
- determination of   47, 71, 238
- – by the hydride technique   287, 371, 391, 397, 400
- – in catalysts   422–423
- – in chromium   405
- – in coal   408
- – in copper   240, 405
- – in copper alloys   400
- – in garbage   381
- – in iron alloys   397
- – in lead alloys   402
- – in lubricating oil additives   411
- – in nickel   405
- – in nickel alloys   400
- – in paper   419
- – in plants   367
- – in polymers   418
- – in rocks   239, 387, 391
- – in soft solders   402
- – in steel   218, 397
- – in waste waters   380
- – in water   371
- extraction of   269, 387, 418
- oxidation states   232–233, 371, 391, 397, 418
- resonance lines   268
- speciation   269
- spectral bandpass   86–87
- spectral interferences   206, 268, 398
- thermal pretreatment/atomization curves   191
- volatilization as tri-iodide   387
APDC,   *see* Ammonium pyrrolidine dithiocarbamate
Appearance temperature   189, 215
Apples, analysis of   359, 366
Argon/hydrogen diffusion flame   32, **35**
- for atomization of hydride-forming elements   47, 73, 227, 240–241, 397
- for the determination of
- – arsenic   270
- – cadmium   276
- – indium   287
- – selenium   320
- – tellurium   329
- – tin   332

Arsenic 269–272
- ascorbic acid as matrix modifier 214
- background correction 148
- determination of 7, 32, 35, 44–45, 47, 70–71, 209, 212–213, 218
- – by the hydride technique 270, 371, 391, 397
- – in biological materials 352
- – in blister copper 399
- – in body fluids 352
- – in chromium 405
- – in coal 408
- – in copper 404–405
- – in copper alloys 400
- – in estuaries 373, 382
- – in fish 357, 359
- – in flue dust 377
- – in glass 415
- – in iron alloys 397
- – in lead 238
- – in lead alloys 402
- – in minerals 391
- – in nickel 238, 405
- – in pharmaceutical products 423
- – in phosphoric acid 422
- – in plant materials 367
- – in polymers 418
- – in process waters used for crude oil recovery 415
- – in rock samples 391
- – in sea-water 373
- – in sediments 374, 382, 391
- – in soft solders 402
- – in soil extracts 365
- – in steel 397
- – in sulfidic ores 387
- – in tissue 352
- – in tungsten alloys 404
- – in urine 352
- – in waste waters 380
- – in water 271, 371
- extraction of 271, 352, 418
- influence on mercury 249
- interferences from other hydride-forming elements 240–243
- oxidation states 232, 240–243, 270, 271, 272, 371, 391, 418
- radiation sources 27

- speciation 271–272, 371, 382
- spectral interferences 206
Aryl mercury compounds, determination of 302–304
Ashing in graphite tubes 205
- of serum 350, 354
Aspiration rates 40, 42, 173
Aspiration times 96
Atom cloud density 188–189
Atom concentration in flames 41, 305
Atomic emission spectrometry 41, **251** *et seq.*
- graphite furnace 262–263, 316
- inductively coupled plasma 254 *et seq.*
Atomic fluorescence spectrometry 26, 35, 41, 50, 81, **263** *et seq.*
Atomic lines, overlapping 129–131
Atomic spectra 3–5
Atomic states, *see* Terms
Atomization
- by capacitor discharge 61, 194–195, 202, 211
- by cathodic nebulization 81
- by electron bombardment 81
- by glow discharge 81
- completeness of 34, 41
- curves 189–193
- degree of 168–170
- – in electric plasmas 254–255
- efficiency of 168–170
- in flames 165 *et seq.*
- influence of particle size on 43, 389, 413–414, 424
- in graphite furnaces 187 *et seq.*
- in heated quartz tubes 73, 227 *et seq.*
- rate of 31, 53, 61, 163, **166–168, 179–180**
- temperature in graphite furnaces 61–62, 188–189
- – influence on emission 99
- with lasers 82, 403
Atomization time 31, 49, 61, 166–168, 188
Atomizers **31** *et seq.*, 81–82, 251
Atoms
- desorption from graphite surface 187
- diffusion of 54, 55, 61
- energy states of 4–6
- excited 3, 7–8, 251 *et seq.*
- ground states of 5, 6, 31, 111, 252, 263
- lifetime in excited state 2–3, 252

Atoms
- rate of formation of 195
- rate of transport from graphite tube 195, 199
- residence time in graphite tube 53, 188
- vapor pressure in graphite tube 188
- volatilization from graphite surface 188, 214
Autoclave **123,** 247, **357**
Automation 43, **65–67,** 73–74, **105–106**
- in trace analysis 127

Babington nebulizer 43
Back extraction **121,** 205, 300, 307–308, 320, 329, 339, 359, 372, 387
Background absorption, *see* Background attenuation
Background attenuation 46–47, 50, 52, 58, 60, 73, 98, **131** *et seq.*, 171, **199** *et seq.*
- during direct solids analysis 225–226
- in the hydride technique 231
Background correction 133 *et seq.*
- accuracy of 137–139, 150
- influence on noise 98, 137, 150
- limitations 138, 205–207
- sources of error 137–139, 172, 205–207
- using the Zeeman effect **140** *et seq.*, 172, 206
- with continuum sources **135** *et seq.*, 172, 205–207, 273, 280, 349
- with hydrogen hollow cathode lamp 137, 138
Background emission 254, 262
Background radiation 21–22
Bandpass
- influence on background correction 140
- spectral 22, 28, **83–89,** 129, 136, 138, 252, 256
Band spectra in graphite furnaces 201
Barium 272–273
- determination of 34
- – in glass 415
- – in infusion solutions 423
- – in lubricating oil additives 410
- – in lubricating oils 410
- ionization 183
- salts, analysis of 422
- spectral interferences 171

Bauxite
- analysis of 390
- digestion of 383
Beef liver, analysis of 359
Beer, analysis of 361
BEER-LAMBERT law **1,** 12, 21, 100, 115
Beryllium 273–274
- atomization of 169
- determination of
- – in airborne dust 378
- – in aluminium alloys 403
- – in aluminium oxide ceramics 417
- – in biological materials 352
- – in coal 408
- – in copper alloys 399
- – in fly ash 377
- – in geological samples 389
- – in magnesium alloys 403
- – in oil 411
- extraction of 387
- thermal pretreatment/atomization curves 193
Biological materials, analysis of 341 *et seq.*, 356 *et seq.*, 361 *et seq.*
Bismuth 274–275
- analysis of 406
- determination of 44, 47, 70
- – in biological materials 239, 351, 352
- – in body fluids 352
- – in chromium 275, 405
- – in coal 408
- – in cobalt 239, 405
- – in copper 275, 404
- – in copper alloys 400
- – in iron alloys 397
- – in metallurgical samples 274–275
- – in nickel 405
- – in nickel alloys 239, 275, 400
- – in organic materials 275
- – in rocks 274, 389
- – in soft solder 402
- – in steel 275, 397, 399
- – in tissues 352
- – in urine 351
- – in water 275, 371
- extraction of 274, 389, 400
- influence on mercury determination 249
- mutual interference of hydride-forming elements 240

Bismuth
- oxidation state 275
- resonance lines 274
- spectral interferences 138
Bituminous sand 415
Biuret, indirect determination 310
- in fertilizers 310, 366
- in urea 366
Blank
- measure 130
- sample 134
- test solution 107–108
- value 119, 121, 244–248, 277, 326, 338, 378
Blaze wavelength of gratings 90
Blister copper, analysis of 399
Blood
- analysis of **45–47,** 349, **352–355**
- stability of samples 354
Boat technique 31, **44–47,** 276, 294, 302, 325, 351–352, 389
Body fluids, analysis of 341 *et seq.*
Bones, analysis of 349, 354, 355
Boron 275–276
- analysis of 421
- determination of 34
- – by ICP-AES 260
- – in apples 366
- – in glass 416
- – in plant materials 366
- – in sea-water 368, 373
- – in soils 365
- – in vegetables 362, 366
- – in water 276, 368, 370
- extraction of 276, 365
- formation of monocyanides 169
- influence of aluminium 178
- isotopes, analysis of 276
Botanical samples, analysis of 359 (*see also* Plant materials and Vegetables), 364, **366–367**
Bottom ash, element distribution 376
Bracketing technique 116–117
Brain tissue
- analysis of 347–348
- digestion of 346
Brewer's yeast, analysis of 357
Bromine, determination of 309
Buckshot, analysis of 402
Buffer action of flame gases 133, 160, **180**

Buffers *see* Spectrochemical buffers
Burner head 38 *et seq.*
Burners
- direct injection 36, **39–42,** 132, 277, 299, 316, 332, 412
- premix **38–41,** 132, 159, 171, 277
Burning velocity of flames 36, **37,** 39, 40, 161, 167–168
Butter, analysis of 361

Cadmium 276–277
- analysis of 122, 405, 407
- contamination 126, 277
- determination of 35, 44–45, 209–210
- – in aluminium alloys 206–207, 402
- – in ammonia solution 422
- – in blood 47, 205, 354–355
- – in body fluids 352, 354
- – in bones 355
- – in copper 404
- – in copper alloys 400
- – in dust deposits 377
- – in fish 357, 359–360
- – in foodstuffs 357
- – in fungicides 421
- – in glazed ceramic surfaces 417
- – in hydrofluoric acid 422
- – in iron alloys 397
- – in liver tissue 355, 359–360
- – in marine organisms 357
- – in muscle tissue 355
- – in nickel 405
- – in nickel alloys 400
- – in paper 419
- – in petroleum 411–412
- – in plant materials 359, 366
- – in rice 359
- – in rock samples 389, 390
- – in rubber seals 424
- – in sea-water 372–373
- – in seaweed 357
- – in sediments 382
- – in serum 354
- – in sewage sludge 381
- – in steel 399
- – in sulfuric acid 422
- – in suspended matter 382
- – in teeth 355

Cadmium, determination of
– – in tissues   352, 355
– – in uranium   420
– – in urine   158–159, 351, 352, 354
– – in waste waters   380
– – in water   370
– extraction of   354, 370, 372, 387
– in stack gas   376
– line profiles in Zeeman AAS   153
Calcine, analysis of   392
Calcium   277–279
– as matrix modifier   276
– atomization of   165, 170
– background attenuation   132
– determination of   38
– – in boron   421
– – in brines   278
– – in cellulose   419
– – in cement   417
– – in coalash   377
– – in dialysis solutions   343
– – in glass   415
– – in greases   411
– – in lead-calcium alloys   402
– – in lithium hydroxide   422
– – in liver tissue   343
– – in lubricating oil additives   410
– – in lubricating oils   410
– – in magnesium alloys   403
– – in molybdenum   405
– – in pharmaceutical products   423
– – in phosphate rocks   388
– – in phosphoric acid   421
– – in plant materials   277, 366
– – in plutonium   420
– – in rock samples   382
– – in sediments   374
– – in serum   277, 341 *et seq.*, 351
– – in sewage sludge   278
– – in silicate rocks   386
– – in soil samples   277
– – in tungsten   405
– – in uranium   420
– – in urine   277, 341 *et seq.*
– – in water   367
– – in zinc oxide   421
– flame profiles   176

– influence
– – of phosphate   41, 165, 277, 341, 342
– – on the determination of barium   272–273
– – on the radiation emission from graphite tubes   98
– line width   21, 131
– losses   124
Calcium hydroxide spectrum   171
Calibration   108
– techniques   115 *et seq.*
Carbide formation
– in flames   169
– in graphite furnaces   56, 63, 112, 196, 214, 273, 306, 324
Carbon disulfide, indirect determination of 310
Carbon fibers, analysis of   421
Carbon rod   50–51, 215, 216, 284
Carrier distillation (chromium)   280
Cast iron, analysis of   395–396
Catalysts, analysis of   422
Caustic soda, analysis of   422
Cellulose, analysis of   419
Cement, analysis of   417–418
Ceramics
– analysis of   415, 417
– digestion of   415
Cerium, determination of   7, 291
Cesium   279
– as ionization buffer   343
– determination of
– – in rock materials   388
– – in water   279
– radiation sources   27, 279
Characteristic concentration   108
Characteristic mass   108
Chart recorders   102–104
Chelating agents   120
– indirect determination of   310
Chelation   120
– for elimination of interferences   177
Chemicals, analysis of   421–422
Chemiluminescence   35
Chewing gum, digestion of   358
Chloric acid digestions   247, 357
Chloride, indirect determination of   309
– in liquid industrial wastes   407
– in serum   351
– in water   309

Chloroform as solvent 120
Chromium 279–281
- adsorption on silicates 374
- analysis of 405
- atomization of 169, 279
- determination of 33, 38, 46, 210
- - in aluminium alloys 403
- - in biological materials 349
- - in bismuth 406
- - in body fluids 281
- - in boron 421
- - in brewer's yeast 124, 357
- - in chromium oxide paints 424
- - in coke used for making electrodes 408
- - in environmental samples 280, 380
- - in fish 359
- - in glass 416
- - in graphite 442
- - in hair 349
- - in human milk 349
- - in leather 418
- - in lithium hydroxide 422
- - in liver tissue 349, 359–360
- - in lubricating oils 413
- - in molybdenum 405
- - in mussel tissue 359
- - in nickel alloys 400
- - in nitric acid 421
- - in oil 411
- - in paint additives 423
- - in paper 419
- - in plants 360
- - in rock materials 389
- - in sea-water 372
- - in serum 348–349
- - in sewage sludge 280–281, 381
- - in silicon tetrachloride 422
- - in sodium 406
- - in steel 394–395
- - in tellurium 406
- - in tissue 281
- - in tungsten 405
- - in uranium 420
- - in urine 204, 348–349
- - in waste waters 380
- - in water 367, 368, 370
- - in zirconium alloys 404
- extraction of 280, 369, 372

- - in iron-nickel-cobalt alloys 400–401
- - in liver 359
- - in meat 282
- influence on aluminium 178
- interference of iron 177
- line profiles in Zeeman AAS 153, 155
- losses 124
- speciation 280, 372
- volatility of 280
Chromium plating baths, analysis of 407
Citric acid as matrix modifier 373
Cleaning vessels 126, 245
Clean room 127, 244, 278–279, 326
Clinker, analysis of 417
Coal and coalash
- analysis of 377, 408
- digestion of 408
Coastal crude oil residues, analysis of 382, 414
Coating of graphite tubes **55–57,** 203, 213, 214, 215, 306, 333–334, 403, 411
- with lanthanum 56–57, 213, 274, 306, 324, 338, 411
- with molybdenum 55, 213, 324, 333
- with niobium 324
- with pyrolytic graphite **55–57,** 209, 214, 273, 276, 293, 306, 308, 335, 336, 337, 338, 350
- with tantalum 57, 213, 306, 333, 338, 400
- with tungsten 56, 213, 306, 333, 400
- with zirconium 56–57, 213, 294, 306, 324, 333, 338, 400, 411
Cobalt 281–282
- analysis of 405
- determination of 46
- - in air 377
- - in aluminium 405
- - in aluminium oxide catalysts 422
- - in apples 366
- - in blood 282, 349
- - in cement 417
- - in fish 359–360
- - in glass 415–416
- - in gold baths 407
- - in high temperature nickel alloys 400
- - in milk 361
- - in molybdenum 405
- - in paper 419

Cobalt, determination of
– – in petrochemical catalysts   423
– – in pharmaceutical products   423
– – in plant materials   282, 359, 366
– – in polymers   418
– – in proteins   423
– – in rocks   387, 389
– – in sea-water   282, 372
– – in serum   349
– – in silicon tetrachloride   422
– – in sodium   405–406
– – in soils   282
– – in tissue   349
– – in tungsten   405
– – in uranium   420
– – in urine   282
– – in vitamin B12   423
– – in water   282, 368–369, 377
– extraction   282, 349, 367–369, 387
– influence of aluminium   178
– influence on hydride-forming elements   236
– resonance lines   281
Cobalt alloys, analysis of   400–401
Cobalt-iron-nickel alloys, analysis of   400–401
Coefficient of variation   109
Coherent forward scattering   93
Coke, analysis of   408
Coke for electrodes, analysis of   408
Cold vapor technique   **75** *et seq.*, **227** *et seq.*,
   261, 302, 355, 371, 391
– chemical interferences   248 *et seq.*
– spectral interferences   244
– volume of measurement   244
Colloidal solutions   220
Columbite, analysis of   386
Combustion properties of organic solvents
   120
Complexes, stability of   300, 369
Complexing agents, indirect determination of
   309
Compound techniques for sample prepara-
   tion   121–122, 245, 358
Concentration range
– optimum   117
– usable   114
Concomitants, separation in graphite furnaces
   187
Confidence band   114

Contamination
– of pipet tips   66–67
– of the analyte element   **123–127,** 205, **244–
   247,** 277, 299, 316, 326
Continuum sources   12, **28–29,** 93, 129, 131,
   265, 291
– for background correction   58, **135** *et seq.*,
   172, 320, 321
– – noise   98, 320
Cooling water, analysis of   421
Copper   282–283
– alloys, analysis of   399
– analysis of   400, 404–405, 407
– as matrix modifier   320, 329
– atomization of   62, 195
– cyanide baths, analysis of   407
– degree of atomization   168
– determination of   47, 214
– – in aluminium alloys   402–403
– – in ammonia solution   422
– – in apples   366
– – in beef liver   382
– – in beer   361
– – in boron   421
– – in butter   361
– – in carbon fibers   421
– – in cellulose   419
– – in coal   408
– – in copper alloys   399
– – in copper cyanide baths   407
– – in dust deposits   377
– – in fertilizers   365–366
– – in fish   357, 359
– – in fish meal   283
– – in glass   415–416
– – in gold baths   407
– – in graphite   422
– – in hydrofluoric acid   422
– – in lithium hydroxide   422
– – in lithium niobate   422
– – in liver tissue   283, 346
– – in lubricating oil   413–414
– – in magnesium alloys   403
– – in meat   283
– – in milk   361
– – in muscle tissue   346
– – in mussel tissue   359
– – in oil   205, 412

Copper, determination of
– – in oil sump   409
– – in paper   419
– – in petroleum coke   408
– – in petrol (gasoline)   413
– – in pharmaceutical products   423
– – in plants   283, 359, 362, 366
– – in plating baths   407
– – in polymers   418
– – in rock materials   382, 387, 389
– – in rubber seals   424
– – in sea-water   372–373
– – in seaweed   357
– – in sediments   374
– – in serum   283, 344–346
– – in sewage sludge   283, 381
– – in silicate rocks   386
– – in silicon tetrachloride   422
– – in soft solders   402
– – in soil extracts   281
– – in steel   394
– – in sulfuric acid   422
– – in textiles   419
– – in tissue   346
– – in titanium dioxide pigments   424
– – in tungsten alloys   404
– – in uranium   420
– – in urine   283, 344–346
– – in vegetables   366
– – in zinc oxide   421
– – in zirconium alloys   404
– extraction   283, 369, 372–373, 387
– influence on the determination of mercury 249
– ironstone, analysis of   392
– line profiles   20
– resonance lines   282, 321, 329, 405
– smelting sludges, analysis of   399
Coprecipitation of
– aluminium   267
– antimony   400
– arsenic   271, 404
– bismuth   275, 400, 404
– lanthanides   338
– lead   295, 370, 400
– nickel   307
– selenium   320, 322, 404
– tellurium   330, 404
– tin   334, 400, 404

– yttrium   338
– hydride-forming elements   240
Cotton, analysis of   419
Cross-section of graphite tubes   54–55
Crude oil
– analysis of   409
– classification of   414
– residues, analysis of   414
Cyanide bands in graphite furnaces   201
Cyanides
– formation in the flame   133
– indirect determination of   309
Cyanoborohydride   71–72
Cyanogen as fuel gas   33
Cyclohexane-butyrates in oil analysis   409

Dairy products, analysis of   361, 362
Dark current   96–97
Dark noise   97
DELVES system   45–47, 276, 294, 352, 424
Deproteinization
– of milk   361
– of serum   343–344
Desorption of atoms from graphite surface
   187
Detection limit   24, 27, 44, **45, 53,** 57, 80–81,
   88, 95, **110,** 111, 113–114, 253
– in atomic emission spectrometry   252–253
– in atomic fluorescence spectrometry
   265
– in Zeeman AAS,   154
Detectors   96–97
– Vidicon   92, 97
Detergents, indirect analysis of   380
Determination limit   75, **113–114**
Deuterium lamp
– emission spectrum   138
– for background correction   **135–140,** 273,
   280
Dialysis solutions, analysis of   343
Dicarbides, formation in graphite furnaces
   198, 216
Dichromate as matrix modifier   210
Diffraction at slits   86
Diffusion flames, *see* Argon/hydrogen and
   Nitrogen/hydrogen diffusion flames
Diffusion of atoms   54, 61
Digestion bomb, *see* Autoclave

Digestions 247 (*see also* Pressure digestions
  and under *individual materials*)
– in a stream of oxygen 247
– in chloric acid 247
– independent procedures 127
– SCHÖNINGER technique 247, 328, 351, 358,
  360
– with lithium metaborate 324, 335, 377, 386–
  387, 417
Digital readout 100, 101–102
Dimethyl mercury 302
1,2-Diols, indirect determination of 310
Direct current instrument 15
Dispersion, *see* Reciprocal linear dispersion
Dissociation continuum 133, 200–201
Dissociation energy
– of boron monocyanide 169
– of halides 181
– of monoxides 168–169, 181–182
Dissociation equilibrium 170, 179–182
Dissociation in the vapor phase 194–195
– thermal 166, 170, 180, 227 *et seq.*
Distillation residues, analysis of 409
Distortion of signals 104
D-line of sodium 2
DOPPLER effect 11
Double-beam instruments **16,** 100, 137
– as a result of the Zeeman effect 16, 150
Double peaks in graphite furnaces 214–215
Drinking water, analysis of 367, 370
Drinks, analysis of 361
Droplet size in nebulizer aerosols **40–41,** 132,
  167, 171
Dry ashing
– of foodstuffs 357–358
– of garbage 381
– of plant materials 366
– of sewage sludge 381
Drying agent for mercury determinations **77,**
  248
Dual-channel instruments 16, 50, 135, 174,
  297
Dust, analysis of 375 *et seq.*
Dysprosium, resonance lines 292

Echelle monochromator 93
Edible oils, analysis of 360
Effects 159

Efficiency
– of atomization 168–170
– – in dependence on particle size 43, 389,
  413–414, 424
– of gratings 90
– of nebulizers 40, 43, 44, **173–174**
– of sample transport 159
Efficiency of atomization 168–170
Electric arc as atomizer 81
Electrodeless discharge lamps **26–28,** 99, 264,
  269–270, 279, 312, 316, 318, 320, 327
– behavior in a magnetic field 147
Electrolytic deposition 122, 205
Electron bombardment for atomization 81–82
Electronic band spectra 201
Electronic excitation spectra 138
Emission factor 63
Emission intensity 21, 26
– of nitrous oxide/acetylene flame 278
Emission lines, number of 256
Emission noise 34, 97, 111
Emission spectrum 3–5, 7
– of deuterium lamp 138
Energy level diagram 6
Energy of excitation 7–8, 251 *et seq.*
Energy states of atoms 4–5
Enrichment techniques **121** *et seq.*, 205, 221
Entrance slit of monochromator 88–89
Environmental analysis 375 *et seq.*
Erbium, resonance lines 292
Errors
– as a function of concentration 117
– caused by background correctors 138–139,
  205–207
– random 109
– relative 117
– systematic 76, 109, 119, 123–127, 221, 244 *et
  seq.*
Erythrocytes, analysis of 351
Estuaries, analysis of 372, 373, 382
Europium 293
– determination in yttrium phosphors 424
– resonance lines 292
Excitation
– equilibrium 179
– interferences 185
– potential 252
– sources 251

Excited atoms 7–8, 251 *et seq.*
Excited state 3, 7–8, 251 *et seq.*
Exit slit of monochromator 88
Extraction 387 (*see also* solvent extraction)
– of trace elements
– – from foodstuffs 359
– – from plant materials 366
– – from sea-water 372–373
– – from sewage sludge 381
– – from water 367–369

Feldspar, digestion of 383
Ferrosilicon alloys, digestion of 393
Fertilizers
– analysis of 364, 365–366
– influence on foodstuffs 356, 366
Filter monochromator 90
Fine structure of molecular bands 139, 150
Fingernails, analysis of 346
Fish, analysis of 358–359
Fish tissue, digestion of 247
Flame emission spectrometry 39, 251–254
Flame injection technique **43**, 95, 345, 351,
   422, 424
Flame technique **31** *et seq.*, **165** *et seq.*
Flames (*see also* air/acetylene, air/hydrogen,
   air/propane and nitrous oxide/acetylene flames)
– air/coal gas 36, 317
– as radiation sources 28
– atom concentration in 305
– atomization in 165 *et seq.*
– buffer action of 133, 160, 180
– burning velocity 33, **37,** 39, 40, 41, 161, 167
– carbide formation 169, 176
– diffusion *see* Argon/hydrogen and Nitro-
   gen/hydrogen diffusion flames
– dissociation equilibria in 166, 170, 179
– flow ratio 169
– geometry 166
– ionization in 182–183, 279
– mass flow pattern 161, 185
– mass flowrate 161, 173, 185
– molecular spectra in 138, 171
– nitric oxide/acetylene 34
– nitrous oxide/hydrogen 34
– observation height 34, 163, 171, **175–176,**
   182, 185, 284, 289, 298, 305, 307, 318

– observation zone 185–186
– oxide formation in 176–178, 180
– oxygen/acetylene 33, 37, 164, 167, 267, 284,
   291, 299, 316
– oxygen/cyanogen 33, 37
– oxygen/hydrogen 37, 73, 169, 227, 332
– profiles 176
– radiation absorption of **32,** 36, 97, 129, 270,
   320
– radiation emission of 32, **34,** 97, 272, 278
– residence time of sample 167
– shielded (separated) 7, 35–36, 169, 287
– solute-volatilization interferences 175–179
– spectral interferences in 170–172
– stoichiometry 32, 33, 34, 36, 99, 169, 172–
   173, 252, 298, 342
– temperature **37,** 41, 111, 163, 167, 183
– temperature profile 166
– transport interferences 172–175
– Zeeman effect 147–148
– zone structure 41, 305
Flow spoiler, in burners 38, 40
Flue gases, sample collection 376
Fluorescence spectrum 3, 5, 7, 91
Fluoride, indirect determination of 309
Fly ash
– analysis of 376
– as fertilizer 366
– distribution of elements 376
Foam formation in the hydride technique
   159, 231
Foodstuffs, analysis of 237, **356** *et seq.*
FRAUNHOFER lines 1
Fruit juice, analysis of 361
Fuel oil, analysis of 326, 411
Fuel/oxidant ratio 34
Fuming out of laboratory ware 126, 245,
   290
Fungicides, analysis of 421
Fusions 123, 386–387

Gadolinium, resonance lines 292
Gallium 283–284
– analysis of 122
– determination of
– – in aluminium 405
– – in bauxite 390
– – in ores 390

Gallium
- resonance lines 283
- spectral interferences 130
Garbage, analysis of 381
Garbage incineration plant, element distribution 376
Garbage seepage water, analysis of 380
Gas oil, analysis of 409–410
Gasoline, *see* Petrol
Gas-phase interferences in the hydride technique 240 *et seq.*
Gastric juices, analysis of 343
Gas turbine fuel oils, analysis of 411
Geological samples, analysis of 43, 45, 235, **382** *et seq.*
Geometric slit width 88
Germanium 284–285
- determination of 47, 70
- - by the hydride technique 284
- - in coal 408
- - in water 284
Glass
- analysis of 415 *et seq.*, 424
- digestion of 383, 415 *et seq.*
- for storing samples 246
- solid analysis 416
Glassy carbon, *see* vitreous carbon
Glazed ceramic surfaces 417
Glazed surfaces, analysis of 417
Glow discharge
- as excitation source 251
- for atomization 81
Glucose, indirect determination of 309
Gold 285
- analysis of 406
- baths, analysis of 407
- determination of
- - in aluminium 405
- - in blood 285, 350–351
- - in calcine 392
- - in cyanide solutions 285, 392, 406
- - in geological materials 285
- - in high purity metals 405
- - in jewelry alloys 404
- - in minerals 392–393
- - in noble metal beads 406
- - in noble metals 285, 406–407
- - in ores 285, 382, 392

- - in plating baths 285
- - in polyester 418
- - in protein fractions 285, 351
- - in serum 350–351
- - in tin/lead solders 402
- - in tissue 351
- - in urine 285, 350–351
- extraction of 285, 350, 392
- spectral interferences 138, 172
- thermal pretreatment/atomization curves 190
Grain size of samples 221–223
Graphite
- emission factor 63
- intercalation compounds 196, 208, 213
- surface oxides 215
Graphite furnace atomic emission spectrometry 262 *et seq.*, 316
Graphite furnaces
- ashing in 205, 215, 226, 350, 354
- atomization in 187 *et seq.*
- drying step 60
- emission of 97–98, 99, 273
- formation of carbides in **57,** 63, 112, 196, **213,** 273, 306, 324
- formation of halides in 216–217, 220
- formation of hydrogen in 211–212, 216
- heating rate **60** *et seq.*, 188, 269, 313
- - ultrafast **61–62,** 199, 202, 219, 262
- hydrogen as purge gas 204
- influence of thermal equilibrium on vapor-phase interferences 160, 164, **198–199, 216–217, 219**
- ionization in 216, 262
- molecular spectra in 133
- partial pressure of oxygen in 211–213, 216–217
- purge gas 49, **58** *et seq.*, 195–196, 199, 201, 278, 299, 334
- residence time of atoms in 53, 188
- stabilized temperature furnace, **64–65,** 198, **214,** 261
- - special aspects for the determination of:
    lead 295
    manganese 300
    nickel 308
    phosphorus 313
    selenium 321
    tellurium 330

Graphite furnaces
- temperature control of 62–63
- temperature gradient in 50, **64** *et seq.*, 196, 201
- temperature measurement in 62–63
- temperature programming **60** *et seq.*, 200, 202, 219, 312–313
- thermal pretreatment in **60** *et seq.*, 160, 163, **189** *et seq.*, **203**, 219, 350
- vapor-phase interferences in **215** *et seq.*, 294–295, 330
- volatilization interferences in 208 *et seq.*
Graphite furnace technique **48** *et seq.*, **186** *et seq.*, 261
Graphite surface
- influence on interferences 209, 211, 214
- sample distribution on 187–188, 214
- volatilization of atoms from 187–188, 214
Graphite tubes
- coating of **55–57**, 203, 213, 214, 215, 306, 333–334, 403, 411
- - with lanthanum 56–57, 213, 274, 306, 324, 338, 411
- - with molybdenum 55, 213, 324, 333
- - with niobium 324
- - with pyrolytic graphite **55–57**, 209, 213, 214, 273, 276, 293, 306, 308, 335, 336, 337, 338, 350
- - with tantalum 57, 213, 306, 333, 338, 400
- - with tungsten 56, 213, 306, 333, 400
- - with zirconium 56–57, 213, 294, 306, 324, 333, 338, 400, 411
- cross-section 54–55
- length 54–55
- lifetime 55, 221
- materials 55 *et seq.*
- resistance 62, 63
Gratings 89–90
Greases, analysis of 411
Ground state of atoms **5, 6**, 31, **111**, 252, 263

Hafnium 286
- atomization of 177
- oxide, analysis of 417
- resonance lines 286
Hair, analysis of 224, 346, 349, 353, 355
Half-intensity width of resonance lines **11–12**, 14, 19, **20–21**, 25, 28, 83, 131, 136
Halides, formation in graphite furnaces 216–217, 220

Halogen lamp for background correction 137, 273, 280, 349
Halogens, determination by molecular emission 309
Heating rate of graphite furnaces **60** *et seq.*, 188, 269, 313
- ultrafast **61–62**, 199, 202, 219, 262
Hemoglobin, analysis of 346
Hemoglobin-iron, removal of 343
High intensity lamps 25
High purity materials, analysis of 121
High purity metals, analysis of 237, **404–406**
Hollow cathode lamps 19 *et seq.*
- behavior in a magnetic field 147
- demountable 24
- for arsenic 269–270
- for gallium 283
- for tin 332
- multielement **24**, 92, 99, 106, 131, 281, 307
- pulsed operation 23
Holmium, resonance lines 292
Homogeneity of samples 221–224
Human milk, analysis of 349
Hydride technique 47, **69** *et seq.*, **227** *et seq.*, 261
- atomization of the hydride 73, **227** *et seq.*
- chemical interferences 233 *et seq.*
- collecting the hydride 72, 231
- competitive reactions 241
- foam formation 159, 231
- for the determination of
- - antimony 269, 371, 397
- - arsenic 271–272, 371, 397, 402
- - bismuth 275, 371, 397, 401
- - germanium 284
- - lead 295, 397
- - selenium 321–323, 371, 397, 401, 402
- - tellurium 330, 371, 397
- - tin 334, 397
- gas-phase interferences 240–243
- influence of
- - dilution 237
- - hydrogen 228–230
- - nickel and transition metals 235–240
- - oxidation state 232–233
- - oxygen 73, 228–230
- - preflush period 229–230

Hydride technique, influence of
– – the acid concentration   70, **233** *et seq.*,
   334, 397
– – the acid mixture   237 *et seq.*
– – the sample volume   **74–75,** 237
– – the surface of the quartz tube   229–231
– – the volume of measurement   **74–75,** 231–
   232
– kinetic interferences   231–232
– mutual interferences   240–243
– preflush time   229
– reductant   70–72, 73
– spectral interferences   231
– transport interferences   159
Hydrochloric acid, influence on hydride-forming elements   234–235
Hydrofluoric acid, analysis of   422
– influence on hydride-forming elements   233,
   234
Hydrogen
– influence in the hydride technique   228–230
– influence on atomization   170, 180, 228–230,
   242
– – in graphite furnaces   268, 349
– – of tin   332
Hydrogen peroxide
– analysis of   421
– as matrix modifier   210, 214, 304
Hydrogen radicals, influence on atomization
   170, 180, **227** *et seq.*, 242
Hydrolysis   124
Hydroxides, formation in the flame   133
Hyperfine structure of resonance lines   12, 14,
   21, 130, 142

ICP atomic emission spectrometry   254 *et seq.*
Impact bead, in burners   38, 40, 43
Independent analytical techniques   127,
   262
Indium   286–287
– atomization of   170
– chloride, absorption spectrum   139
– determination of
– – in aluminium   405
– – in bauxite   390
– – in ores   390
– resonance lines   286
Infusion solutions, analysis of   423

Insulin, analysis of   423
Integration of signals   100–101 (*see also* Peak
   area integration)
Intercalation compounds with graphite   196, 213
Interferences   129 *et seq.*
– chemical   34, 35, 41, 64, 159
– – for indirect determination of non-metals
   309
– – influence of organic solvents   173–174
– – in the cold vapor technique   248–250
– – in the hydride technique   233 *et seq.*
– condensed-phase   160
– excitation of analyte element   185
– gas-phase in the hydride technique   240 *et
   seq.*
– ionization   34, **161, 182–185,** 255
– kinetic   231–232, 244
– lateral diffusion   267, 305, 308, 328, 337, 340
– non-spectral   31, 35, **159** *et seq.*
– physical   43, **159,** 175
– solute-volatilization   119, **160,** 251, 255
– – in the flame technique   **175** *et seq.*, 185
– – in the graphite furnace technique   208 *et
   seq.*
– spatial distribution   **161, 185–186** (*see also*
   Lateral diffusion)
– specific   160, 162
– spectral   14, 29, 35, 83, 93, 119, **129** *et seq.*,
   252, 281, 307, 321, 404
– – in ICP-AES   257–260
– – in the cold vapor technique   244
– – in the flame technique   170 *et seq.*
– – in the graphite furnace technique   199 *et
   seq.*
– – in the hydride technique   231
– transport   67, 119, **159, 172** *et seq.*, 251, 404
– vapor-phase   64, 119, **160–161,** 255
– – in the flame technique   179 *et seq.*
– – in the graphite furnace technique   **215** *et
   seq.*, 294–295, 330
Internal standard   17, 81
Iodates, indirect determination of   309
Iodides, indirect determination of   309
Iodine   287
– determination of   7, 36
– – by means of emission bands   309
– influence on the determination of mercury
   309

Ionization 32, 34, 35, **161**, 162, **182–185**, 254, 255, 273, 279, 291, 315–316, 326
- buffer 163, 184–185
- energy 182–183
- equilibrium 182
- in graphite furnaces 216, 262
- interferences 34, **161, 182–185,** 255–256
- potential 185
Iridium 288–289
- complex 289
- determination of
- - in noble metals 288, 407
- - in rocks 288
- flame for determination 32, 289
- resonance lines 288
Iron 289–291
- analytical curve in Zeeman AAS 157
- as matrix modifier 322, 330
- as matrix modifier in the hydride technique 397
- atomization curves 190
- atomization of 169, 197
- contamination 66, 290
- determination of
- - in air 377
- - in aluminium alloys 402–403
- - in ammonia solution 422
- - in apples 366
- - in beer 361
- - in boron 421
- - in cellulose 419
- - in cement 417
- - in coal 408
- - in coalash 377
- - in coastal crude oil residues 382
- - in copper alloys 390
- - in crude oil residues 414
- - in dust deposits 377
- - in fertilizers 365
- - in fish 357
- - in glass 415
- - in gold 406
- - in gold baths 407
- - in hemoglobin 368
- - in high temperature nickel alloys 400
- - in hydrofluoric acid 422
- - in iron-nickel-cobalt alloys 400–401
- - in lithium hydroxide 422

- - in lithium niobate 422
- - in liver tissue 346
- - in lubricating oil 413
- - in milk 361
- - in molybdenum 405
- - in nitric acid 421
- - in oil sump 409
- - in paper 419
- - in pharmaceuticals 423
- - in plant materials 366
- - in plasma 345
- - in plating baths 407
- - in polymers 418
- - in rock materials 382, 384, 389
- - in rubber seals 424
- - in sea-water 368, 369, 372–373
- - in seaweed 357
- - in sediments 374
- - in serum 344–346
- - in silicate rocks 387
- - in silicon tetrachloride 422
- - in silver 406
- - in sodium 406
- - in sulfuric acid 422
- - in tissue 346
- - in titanium alloys 403–404
- - in titanium dioxide pigments 424
- - in tungsten 405
- - in tungsten alloys 404
- - in tungsten carbide 417
- - in uranium 420
- - in urine 344–346
- - in vegetables 362, 366
- - in water 368–370, 377
- - in zinc oxide 421
- - in zirconium alloys 404
- digestion of 393
- extraction of 290, 369, 395
- influence
- - of aluminium 178
- - of halogenated hydrocarbons 290
- - on hydride-forming elements 238
- oxidation states 290
- resonance lines 290
- spectral interferences 139, 171
- thermal pretreatment/atomization curves 190
- volatilization losses 208

Iron binding capacity 346
Isotope determination 81, 276, 295, 296–297, 336
Isotope shift 12, 14, 142, 145, 276, 296

Jewelry alloys, analysis of 404

Kidney tissue, analysis of 346, 354
Kinetic interferences
– in the cold vapor technique 244
– in the hydride technique 231–232
KIRCHHOFF's law 2

Lanthanides 291–293
– determination in rock samples 388, 389
Lanthanum 291, 292
– as coating for graphite tubes 56–57, 213, 214, 306, 324, 338, 411
– as matrix modifier 210, 211, 218, 312, 360, 398
– as spectrochemical buffer 343, 377
Lasers as atomizers 82, 403
Last lines 6
Lateral diffusion 267, 305, 308, 328, 337, 340
   (*see also* Interferences, spatial distribution)
– influence of particle size 185–186
Latex paints, analysis of 424
Lead 294–295
– analytical curve in Zeeman AAS 157
– atomization in graphite furnaces 196
– co-volatilization 208
– determination of, 44–47, 209–210, 214–215, 217, 222
– – in ammonia solutions 422
– – in beef liver 359
– – in beer 361
– – in biological materials 295
– – in blood 47, 158–159, 214, 353, 354
– – in body fluids 353–354
– – in bones 354
– – in columbite 386
– – in copper alloys 399
– – in dust deposits 377
– – in fish 357, 359–360
– – in foodstuffs 357, 359–360, 363
– – in garbage seepage water 380
– – in glass 415–416
– – in glazed ceramic surfaces 417
– – in gold baths 407

– – in graphite 422
– – in hair taken from beards 353
– – in hydrofluoric acid 422
– – in iron alloys 397
– – in kidney tissue 354
– – in lead alloys 402
– – in liver tissue 157–158, 354
– – in lubricating oil additives 411
– – in lubricating oils 413
– – in milk 361
– – in mussel tissue 359
– – in nickel 405
– – in nickel alloys 400–401
– – in nickel baths 407
– – in packaging materials 361
– – in paints 423–424
– – in paper 361, 419
– – in petrol (gasoline) 295, 412
– – in placenta tissue 354
– – in plant materials 360, 367
– – in polymers 418
– – in rock samples 389, 390
– – in rubber seals 424
– – in sea-water 372
– – in seaweed 357
– – in sewage sludge 381
– – in silicate rocks 387
– – in soft solders 402
– – in soils 389
– – in steel 211, 217, 397–399
– – in sulfuric acid 422
– – in suspended matter 382
– – in teeth 354
– – in tin/lead solders 402
– – in tissues 352
– – in titanium dioxide pigments 424
– – in tungsten alloys 404
– – in uranium 420
– – in urine 295, 352, 353
– – in vegetables 362, 366
– – in waste waters 308
– – in water 295, 368, 369, 370
– – in zinc oxide 421
– extraction of 351, 369, 373
– in stack gas dust 376
– isotopes determination 295
– losses 123, 209, 370
– organically bound 378

Lead
- resonance lines 294
- speciation 295, 359
Lead alkyls
- determination in air 379
- speciation in petrol 412
Lead alloys
- analysis of 402
- digestion of 402
Lead glass, analysis of 415
Leather, analysis of 418
LEIPERT amplification 287
Limiting absorbance 86
Line profile 12, 26, 150, 258–259
Line width, *see* Half-intensity width of resonance lines
Liqueurs, analysis of 361
Liquid fats, analysis of 360
Lithium 295–297
- atomization of 169
- determination of
- - in cement 417
- - in erythrocytes 373
- - in glass 415–416
- - in grease 411
- - in lubricating oil additives 411
- - in plutonium 420
- - in rock samples 296, 388
- - in serum 296, 351
- - in uranium oxide 412
- - in urine 351
- hydroxide, analysis of 422
- isotopes, determination of 297
- metaborate digestions 324, 335, 377, 386–387, 417
- niobate, analysis of 422
Liver tissue, analysis of 343, 346, 349, 355
Long tube burner 42, 182
Loop technique 47
LORENZ broadening 11
Loss of the analyte element 123–127, 244–248, 294, 323–324
Lubricating oil, analysis of 220, 410, 414
- for metals of wear 409, 413–414
Lutetium, resonance lines 292
L'VOV platform **64–65,** 199, 202, 217, 218, 219, 220 (*refer also to the individual elements for determinations in the graphite furnace using a L'VOV platform*)

Magnesium 297–299
- alloys, analysis of 403
- analysis of 407
- as reductant for hydride-forming elements 70
- atomization of 169
- determination of
- - in aluminium alloys 402
- - in apples 366
- - in boron 421
- - in cast iron 395
- - in cellulose 419
- - in cement 417
- - in coalash 377
- - in dialysis solutions 343
- - in fertilizers 365
- - in gastric juices 343
- - in glass 415
- - in lithium hydroxide 422
- - in liver tissue 343
- - in lubricating oils 413–414
- - in nickel alloys 400
- - in paper 419
- - in pharmaceutical products 423
- - in plant materials 366
- - in plutonium 420
- - in polymers 418
- - in rock samples 382–384
- - in rubber seals 424
- - in serum 341 *et seq.*
- - in silicate rocks 298, 386
- - in steel 394–395
- - in uranium 420
- - in urine 341 *et seq.*
- - in vegetables 366
- - in water 367
- flame profiles 176
- formation of spinels (MgO.Al$_2$O$_3$) 290
- losses 124
- nitrate as matrix modifier 210, 268, 274, 277, 281, 282, 291, 295, 300, 306, 308, 380
- spectral interferences 138
Magnetic field
- alternating 143 *et seq.*
- constant 143 *et seq.*
Magnetic materials, analysis of 424
Manganese 299–300
- alloys, analysis of 407

Manganese
– determination of 46
– – in airborne dust 377
– – in aluminium alloys 402–403
– – in ammonia solution 422
– – in apples 366
– – in aviation fuels 413
– – in beef liver 382
– – in bismuth 406
– – in bones 349
– – in boron 421
– – in brine 300
– – in cellulose 419
– – in cement 417
– – in copper alloys 399
– – in fertilizers 365
– – in fish 359–360
– – in glass 415–416
– – in hydrofluoric acid 422
– – in liver tissue 345
– – in magnesium alloys 403
– – in milk 361
– – in molybdenum 405
– – in muscle tissue 349
– – in nickel alloys 400
– – in oil 411
– – in paper 419
– – in petroleum products 411, 412
– – in plants 359, 366
– – in polymers 418
– – in rock materials 382, 389
– – in sea-water 372–373
– – in sediments 374
– – in serum 349
– – in silicate rocks 386
– – in silicon tetrachloride 422
– – in sodium 406
– – in steel 394–395
– – in sulfuric acid 422
– – in teeth 349
– – in tellurium 406
– – in titanium dioxide pigments 424
– – in tungsten 405
– – in uranium 420
– – in urine 349
– – in vegetables 366
– – in water 369, 370, 377
– – in zinc oxide 421

– extraction of 300, 349, 369
– nodules, analysis of 391
– stability in chelates 300, 369
Marine organisms, analysis of 357
Mass flow pattern of particles in flames 161, 185
Matrix effect 159
Matrix modification and modifiers 45, 125, 134, **163,** 203–204, **209–210,** 219
– aluminium 274
– ammonium dichromate 269, 275
– ammonium nitrate 204, 215, 300, 333, 373
– ammonium phosphate 210, 215, 219, 277, 281, 295, 308, 339, 349, 354, 380
– ascorbic acid 214
– calcium 276
– citric acid 373
– copper 209, 321, 329, 405
– dichromate 210
– EDTA 373
– hydrogen peroxide 210, 214, 304
– iron 322, 330, 397
– lanthanum 210, 211, 218, 312, 360, 398
– magnesium nitrate 210, 215, 268, 274, 277, 281, 282, 300, 306, 308, 380
– molybdenum 209, 321
– nickel 209, 213, 270–271, 275, 293, 320, 321, 350, 371, 373
– nitric acid 203, 212
– oxalic acid 209
– palladium 209, 329
– phosphoric acid 203, 210, 212, 296, 339, 370, 380
– platinum 209, 329
– potassium dichromate 304, 321
– potassium iodide 321
– sulfuric acid 296, 331
Matrix solution **107–108,** 259
Mean value 101–102
Measuring the absorption 12–14
Meat, analysis of 358
Mercury 300–304
– abundancy of 301
– amalgamation technique **79–81,** 124–125, 250, 302, 355, 371, 379
– determination of 3, 45, **75** *et seq.*, 210, **243–244**
– – in air 244, 302, 377, 379

Mercury
- – in airborne dust 377–378
- – in blood 355
- – in body fluids 352, 355
- – in caustic soda 422
- – in coal 302, 408–409
- – in coaldust 376
- – in concentrates 391
- – in fish 303, 357–358
- – in foodstuffs 357–359
- – in hair 355
- – in manganese nodules 391
- – in meat 358
- – in ores 302, 384, 391
- – in petrochemicals 302, 412
- – in pharmaceutical products 423
- – in radioactive materials 301
- – in rock samples 225, 384, 385, 390
- – in sea-water 246, 250, 302
- – in seaweed 249, 357
- – in sediments 250, 374, 382
- – in silicate materials 387, 391
- – in sludges 250, 303
- – in soil samples 244
- – in the graphite furnace 304, 390
- – in tissue 352, 355
- – in urine 351, 355
- – in water 371
- digestion of **246–247,** 358, 371, 412
- distribution function 245
- drying agents for 77, 248
- enrichment of trace elements 122, 406
- exchange (from laboratory atmosphere to aqueous phase) 245
- extraction of 303, 304
- interferences
- – through other elements 249–250
- – through water vapor 77, 244
- isotopes 304
- materials for storage 77, 125, 245, 246
- mobility of 76, 124, **244** *et seq.*, 372
- resonance lines 112
- sources 301
- speciation **303–304,** 382
- stability of solutions 244–245, 372
- systematic errors 244 *et seq.*
- total content, determination of 302–303, 371–372
- vapor discharge lamps 25
- vapor pressure 244, 301
- volatility of 244, 301
- volatilization 244
Metallurgy 393 *et seq.*
Metal salts, oil soluble 409
Metals, analysis of 404–407
Meters for display of analytical measure 100, 101
Method of least squares 155
Methyl isobutyl ketone (MIBK) 120
Methylmercury 246, 250
Methyl mercury chloride 250, **302–304**
Methyl siloxane, determination in water 324
Methyltin compounds, determination in environmental samples 380
MIBK, *see* Methyl isobutyl ketone
Mica, digestion of 383
Microanalysis 261, 339, 343, 345, 351, 372
- of solid samples 244
Microwave plasma 254
Milk, analysis of 361, 362
Minerals, analysis of 382 *et seq.*
Molecular absorption 133–134, 151, 171–172
- of alkali halides 200–201
- of indium chloride 139
Molecular emission, determination of halogens 309
Molecular spectra
- in flames 138, 171
- in graphite furnaces 133
- rotational fine structure 138, 151
Molybdenum 305–306
- analysis of 405
- as matrix modifier 209, 294, 321
- atomization of 176–177, 305–306
- determination of 33, 38, 57
- – in aluminium alloys 403
- – in aluminium oxide catalysts 422
- – in apples 366
- – in blood 349
- – in catalysts for automobile exhaust gases 423
- – in greases 411
- – in high temperature nickel alloys 400
- – in lubricating oil 414
- – in plants 367
- – in rock materials 389
- – in sea-water 306, 373

Molybdenum, determination of
– – in sediments 374
– – in serum 347
– – in slags 395
– – in steel 306
– – in uranium alloys 420
– – in urine 349
– – in vegetables 366
– extraction of 305, 369, 373
– for coating graphite tubes 55, 213, 324, 333
Monoatomic layer 187–188, 214
Monochromator 83, 88, 90, 256
– Echelle 93
– entrance slit 88
– exit slit 88
– filter 90
– resolution 83, 256
Monocyanides, formation of
– in flames 169, 180
– in graphite furnaces 60, 133, 160, **198,** 201, 216, 267, 291
Monoxides
– dissociation of 33
– formation in flames 169
Multichannel instruments 16, 92
Multielement determinations 81, 97, 260, 265
– sequential 29, 92, 257, 260
– simultaneous 29, 92–93, 257, 260
Multielement instruments 92–93, 97, 257
Muscle tissue
– analysis of 349, 355
– digestion of 346, 355
Mussels, analysis of 359

Nebulizers
– efficiency 40, 43, 44, 173–174
– pneumatic 31, **38** *et seq.*, 50, 159
Nebulizing aids 162
Neodymium, resonance lines 292
Nickel 306–308
– alloys, analysis of 222, 238, 275, 400–401
– analysis of 405
– as matrix modifier 209, 213, 270–271, 275, 320, 329, 350, 371, 373
– baths, analysis of 407
– determination of 46, 210
– – in air 377

– – in aluminium alloys 405
– – in bismuth 406
– – in blood 308
– – in boron 421
– – in coal 408
– – in coastal crude oil residues 382
– – in copper alloys 399
– – in crude oil 409
– – in edible oils 361
– – in fish 359–360
– – in gasoil 410
– – in glass 415–416
– – in gold baths 407
– – in iron-nickel-cobalt alloys 400–401
– – in lithium hydroxide 422
– – in liver tissue 308, 359–360
– – in lubricating oil 413–414
– – in molybdenum 405
– – in nitric acid 421
– – in oil 409
– – in oil residues 414
– – in oil sump 409
– – in petrochemical catalysts 423
– – in petroleum coke 408
– – in petrol (gasoline) 413
– – in plant materials 359–360
– – in rock samples 385, 387, 389
– – in sea-water 307, 372
– – in scrum 309, 349
– – in sewage sludge 308
– – in silicon tetrachloride 422
– – in sodium 406
– – in soft solders 402
– – in soil extracts 308
– – in steel 395
– – in tellurium 406
– – in tungsten 405
– – in tungsten alloys 404
– – in uranium 420
– – in urine 309, 349
– – in water 308, 367, 369, 377
– – in zinc oxide 421
– – in zirconium alloys 404
– extraction of 307, 350, 369, 372, 387
– linearity of the analytical curves 22
– oxide, analysis of 402
– reference solutions for oil analysis 409
– spectral bandpass 306–307

Nickel
- spectral interferences 307
- volatilization as carbonyl 308
Niobium 308
- alloys, analysis of 404
- atomization of 177
- determination of
- - in nickel alloys 400
- - in steel 394, 395, 397
- for coating graphite tubes 324
Nitrate, indirect determination of 310
Nitric acid
- analysis of 421
- as matrix modifier 203, 212
- influence on hydride-forming elements 234-235
Nitrogen as purge gas in graphite furnaces 60, 99, 160, 198, 216, 267, 324
Nitrogen/hydrogen diffusion flame **35,** 73
Nitrous oxide/acetylene flame **33-34,** 37, 134, 163, 168-169 (*refer also to the individual elements for determinations using this flame*)
- burner head for 39
- emission of 278
- in flame AES 252, 261
Nitrous oxide/hydrogen flame 34
Noble metals
- analysis of 406
- atomization in graphite furnaces 406
Noise **97-99,** 110-111
Non-absorbable lines 22, **83-84,** 91, 134-135
Non-ionic surfactants, analysis of 421
Non-metallic phases in steel, analysis of 396
Non-metals, determination of 7, 309-310
- by ICP-AES 260
Non-specific radiation losses, *see* Background attenuation
Nuclear fuels, analysis of 419-421
Nylon, analysis of 418

Observation height in flames
- influence on flame profile 176
- influence on interferences 34, 163, 171, 175
- influence on ionization equilibrium 182
- influence on the determination of
- - germanium 284
- - iron 289
- - magnesium 298

- - molybdenum 305
- - nickel 307
Observation zone 185-186
Oil (*see also* Lubricating oil, Edible oil)
- analysis of 205, **409-415**
- shale 415
- sump, analysis of 409
Oil shale 415
Oil sump, analysis of 409
Omnipresent concentration of elements 122
Ores, analysis of 382 *et seq.*
Organometallic compounds, analysis of 423
Orthophosphates, indirect determination of 309
Oscillator strength **9-10,** 168
Osmium 310-311
Overcompensation 138
Overlapping
- of atomic lines 129-131
- - and molecular bands 133, 138
Oxidant/fuel gas ratio 34, 169, 170
Oxides, formation in the flame 133, 176-177
8-Oxiquinoline, indirect determination of 309
Oxygen
- adsorption on graphite 212, 215
- influence on the atomization of hydride-forming elements 227-231
- partial pressure in graphite furnaces 212, 216
- use for ashing in graphite furnaces 205, 214, 226, 350, 354, 412

Packing materials, analysis of 361, 418
Paints, analysis of 423-424
Palladium 311-312
- determination of
- - in blood 351
- - in cupellation granules 406
- - in geological samples 329, 392-393
- - in noble metals 329, 406
- - in ultrapure metals 407
- - in uranium alloys 420
- - in urine 351
- extraction of 329, 392
- flames for 32
- spectral interferences 139
Paper, analysis of 361, 419
Paper pulp, analysis of 419

Partial dissolution 126, 205
Partial dissolution of the matrix 122
Partial pressure
– of atom cloud 188
– of oxygen
– – in flames 134, 255
– – in graphite furnaces 313, 398
– – in plasma 255
– of reaction products in flames 166–170
Particles as cause of radiation scattering 41,
    129, **131–134,** 171, 200
Particle size, influence on
– atomization 43, 389, 413–414, 424
– lateral diffusion (spatial distribution) 185–
    186
– oil analysis 413–414
– precision 221–223
– radiation scattering 132
Particulate matter
– analysis of 374
– elemental distribution 376
Peak area integration 195–196, 214, 219, 223,
    231, 235
Pentachlorophenol, indirect determination of
    310
Perchloric acid, interfering influences 208,
    268, 277, 291, 294–295, 327, 330–331
Petrochemical catalysts, analysis of 423
Petrochemistry 409 *et seq.*
Petrol, analysis of 412–413
– solvent 414
Petroleum products, analysis of 411–412
Petroleum/water emulsions, analysis of 409,
    413
Pharmaceutical preparations, analysis of
    423
Phenylmercury 250, **303–304**
Phosphate
– indirect determination 309
– – in body fluids 351
– interference in the determination of calci-
    um 41, 165, 277, 341
Phosphorescent compounds, analysis of
    424
Phosphoric acid
– analysis of 312, 313, 422
– as matrix modifier 203, 210, 212, 296, 339,
    370, 380

Phosphorus 312–314
– atomization in graphite furnaces 197, 313
– determination of 7, 36, 57, 211–212, 218
– – by molecular emission 309, 312–313
– – in edible oils 312–313, 360
– – in fertilizers 366
– – in fish 359–360
– – in liver tissue 351, 359–360
– – in meat extracts 312
– – in milk powder 312
– – in phosphoric acid 313
– – in plants 359–360
– – in soya oil 360
– – in steel 202, 207, 313, 398
– – in urine 351
– indirect determination of 313
– radiation sources 27
– resonance lines 112, 312
– spectral interferences 207, 398
Photodissociation 133, 150, 201
Photomultiplier 96–97
Photon noise 97
Phthalic acid, indirect determination of 310
Physical interferences 43, **159,** 175
Pigments, analysis of 43, 424
Placenta tissue, analysis of 354
PLANCK's law 2
Plant materials, analysis of 364, **366–367** (*see
    also* Botanical samples and Vegetables)
Plasma (blood plasma), analysis of 345
Plasma (electrical) 37, 254
– inductively coupled 254 *et seq.*
– microwave induced 254
– temperature 254
Plastics, analysis of 418
Platform technique, *see* L'VOV platform
Plating baths, analysis of 407, 424
Platinum 314–315
– determination of 32
– – in activated charcoal 422
– – in aluminium oxide 422
– – in biological materials 315
– – in body fluids 351
– – in catalysts 315, 422
– – in cupellation granules 406
– – in cyanide solutions 314, 406
– – in noble metals 314, 406
– – in rock samples 393

Platinum
– – in sand 315
– extraction of 315
– metals
– – atomization of 179
– – mutual influences of 179, 406
Plutonium, analysis of 420
Pneumatic nebulizer, *see* Nebulizers
Polarization of radiation in a magnetic field 140–142
Polarized components, preferential reflection in monochromators 149
Polyester, analysis of 418
Polyethylene
– analysis of 418
– as material for storage vessels 77, 125, 126, 246, 294
Polymers, analysis of **418**, 423
Polypropylene
– analysis of 418
– as material for storage vessels 77, 246
Polystyrene, analysis of 418
Polyvinyl chloride, analysis of 418
Potassium 315
– contamination 316
– determination of
– – in boron 421
– – in cellulose 419
– – in cement 417
– – in coalash 377
– – in dialysis solutions 343
– – in glass 415
– – in minerals 387
– – in pharmaceutical products 423
– – in plutonium 420
– – in rock samples 316, 382, 387
– – in rubber seals 424
– – in serum 342
– – in silicate rocks 386
– – in uranium 420
– – in urine 342–343
– – in water 367
– dichromate as matrix modifier 304, 321
– formation of monocyanides 169
– iodide as matrix modifier 321
– iodide as reductant for hydride-forming elements 232–233
– iodide as stabilizer for mercury 246

– iodide for elimination of interferences 239
Praseodymium 293
– resonance lines 293
– spectral interferences 130
Precipitation reactions for indirect determination of non-metals 309
Precision 24, 41, 66, 88, 95, **109**, 115, 221–224
Preconcentration techniques **121** *et seq.*, 205, 221
Pressure digestions 123, 247, 357
– for environmental samples 377
– for silicate materials 383, 391
Printers 102–103
Prisms 89–90
Process waters, analysis of 415
Profile of resonance lines 12, 20, 93
Promethium 291
Prospecting, geochemical 389
Proteins
– analysis of 351, 423
– indirect determination of 309
Pulsed operation of hollow cathode lamps 23
Purge gas in graphite furnaces 49, **58** *et seq.*, 195–196, 199, 201, 278, 299, 334
– influence on
– – the sensitivity **58**, 278, 299, 334
– – the temperature 60
– symmetry of 58, 59, 203
Purifying acids 124–125, 245
Pyrolytic graphite, coating of graphite tubes **55–57**, 350 (*for the significance of pyrolytic coating for the determination of individual elements by the graphite furnace technique, refer to the respective element*)
– influence on volatilization interferences 209, 213, **214**

Quantum efficiency of the photocathode 97
Quantum number 4
Quartz
– analysis of 416
– material for storage vessels 77, 125, 126, 246
Quartzite, digestion of 383
Quartz tube
– as atomizer in the hydride technique 73, 227–231
– surface characteristics 229, 231
– temperature 229–230
Quenching effect 264

Radiant energy 88, 99
Radiant flux 12, 23, 99–100
Radiant intensity 23 *et seq.*, 264
Radiant scattering 131 *et seq.*
Radiation absorption of flames 32, 36, 129, 270, 320
Radiation emission of flames 32, 34, 272
Radiation intensity, spectral 9, 11, 28–29
Radicals, participation in atomization 170, 180, 228–231, 242
Radioactive materials, analysis of 302, 419–421
Random errors 109
Rare earth elements, *see* Lanthanides
Rate of transport of atoms from graphite tube 195–196
RAYLEIGH's law of scattering 132
Reading errors 101, 102
Readout 101 *et seq.*
Reciprocal linear dispersion **87–89,** 110, 252
Reciprocal sensitivity 110
Recording of signals 102–105
Reductants
– for the cold vapor technique 76, **78–79,** 249, 250, 302
– for the hydride technique 69, **70–72**
Reduction on graphite 187, 191, **192–195,** 197
Reference element technique 134, 162, 174
Reference materials 127, **224,** 394–395
– for oil analysis 409
Reference solutions **107** *et seq.*, 159, 161
– for oil analysis 409
Relative error 117
Relative standard deviation 109
Releasers 163
Residence time
– of atoms in the flame 167
– of atoms in the radiation beam 41, 45, **53–54, 188**
Resolution of monochromators 28, **83** *et seq.*, 90, 130–131, 258–259
Resonance detector 91–92
Resonance fluorescence 2–3, 91, **263–264**
Resonance lines 6, 11–12
– broadening 11–12
– half-intensity width **11–12,** 14, 19, 20–21, 25, 28, 83, 131, 136
– hyperfine structure 12, 14, 21, 130, 142

– natural width 11
– profile 12, 20, 93
– self-absorption 14, **20–21,** 23, 25, 26
– self-reversal 20, 21, 26
Response time 96, **103–104**
– of instruments 96
– of recorders 103–104
Rhenium 316
– analysis of 406
Rhodium 317–318
– baths, analysis of 407
– determination of 32
– – in noble metals 317
– – in rock samples 393
– – in uranium alloys 420
Rock samples, analysis of 382 *et seq.*
Rollover of the analytical curve 156–158
Rubber seals, analysis of 424
Rubidium 318
– determination of
– – in blood 351
– – in rock samples 318, 388, 390
– – in serum 351
– – in urine 351
– radiation sources 27
Ruthenium 319
– determination of
– – in cyanide solutions 319
       in hydrated ruthenium chloride 422
– – in uranium alloys 420
– stability of solutions 319

Samarium 293
– resonance lines 293
Sample requirement 95–96
Sample solution 107
Sample volume 44–45, 237–238
Sampling pumps 174
Sand, digestion of 383
Scandium, resonance lines 341
SCHÖNINGER digestion technique 247, 328, 351, 358, 360
Sea-water, analysis of 204, 208, **372–374**
Seaweed, analysis of 357
Sediments, analysis of 74, **374, 381–382**
Selective amplifier **16–17,** 92, 98, 129, 256
Selective modulation 91

Selectivity (specificity) of AAS 24, 83, 131, 252, 256
Selenium 320–323
– atomization 227
– background correction 148
– collecting on filters 379
– determination of 7, 35, 45, 70, 209
– – in accelerators 423
– – in air 379
– – in apples 366
– – in biological materials 320, 321, 357
– – in bismuth 405
– – in blood 322, 350
– – in chromium 405
– – in coal 408
– – in copper 404–405
– – in copper alloys 399
– – in environmental samples 320, 380
– – in estuaries 373
– – in feedstuffs 321
– – in fish 320
– – in flue ash 366, 377, 379
– – in foodstuffs 320, 321, 322, 357
– – in glass 322, 416–417
– – in gold 405
– – in high temperature alloys 321
– – in iron 206, 397
– – in iron matrix 321
– – in lead 405
– – in metallurgical samples 322
– – in nickel 405
– – in nickel alloys 321, 401
– – in petroleum products 322, 412
– – in petrol (gasoline) 412
– – in pharmaceutical products 423
– – in plant materials 322, 366
– – in sea-water 320, 323, 373
– – in sediments 323, 374, 382
– – in semiconductor materials 406
– – in serum 350
– – in sewage sludge 381
– – in silver 405
– – in soil extracts 365
– – in steel 322, 396–398
– – in sulfuric acid 422
– – in tablets 321
– – in tissue 350
– – in vegetables 238, 366

– – in waste water 321, 380
– – in water 320–323, 371
– extraction 320, 350
– influence on the determination of mercury 249
– interference between hydride-forming elements 240–241, 322
– oxidation states 322–323, 371
– radiation sources 27
– speciation 323, 371, 382
– spectral interferences 139, 206, 321, 350
– volatilization in an oxygen stream 400, 405
Self-absorption of resonance lines 14, **20–21**, 23, 25, 26
Self-reversal of resonance lines 20, 21, 26
Semiconductor materials, analysis of 406, 424
Sensitivity 108
– for Zeeman AAS 151–154
– preselection in graphite furnaces 58
Separation of concomitants in graphite furnaces 187
Separation techniques for element enrichment **121–122**, 205
Serum
– analysis of 205, **341** *et seq.*
– ashing in graphite furnaces 350, 354
– viscosity 344
Set of calibration solutions 107
Sewage sludge
– analysis of 381
– – suspensions 222, 381
– digestion 381
Shielded flames 7, 35–36, 287, 301
Signal form in Zeeman AAS 158
Signal-to-noise ratio 24, 27, 38, 39, 84–85, 97, 100, 110, 270, 278, 294, 315, 325
Silicate materials
– analysis of 417
– digestion of 383, 386
Silicates, analysis of 324
– indirect determination of 309
Silicon 323–324
– atomization of 192, **197,** 324
– carbide, analysis of 421
– determination of
– – in aluminium alloys 403
– – in cast iron 396
– – in cement 417

Silicon, determination of
– – in glass 416
– – in lithium niobate 422
– – in niobium alloys 404
– – in paper 419
– – in rock samples 382, 388
– – in silicate materials 383
– – in silicate rocks 387, 388
– – in silicon carbide 421
– – in silicone grease 412
– – in steel 396
– – in tungsten alloys 404
– extraction 380, 419
– influence of aluminium on 178
– resonance lines 84, 323
– sensitivity 323
– sources of error 124
– speciation 324
– spectral bandpass 84–86
– stability of solutions 323–324
– tetrachloride, analysis of 422
Silicone grease, analysis of 412
Silver 324–325
– alloys, analysis of 404
– analysis of 407
– determination of 44–45
– – in air 379
– – in aluminium 405
– – in aluminium alloys 402–403
– – in beef liver 359
– – in body fluids 352
– – in chromium 405
– – in copper 404
– – in copper alloys 399
– – in copper ironstone 392
– – in cupellation granules 406
– – in gold baths 407
– – in iron 397
– – in lead alloys 402
– – in lubricating oils 413
– – in minerals 392
– – in nickel 405
– – in nickel alloys 400–401
– – in noble metals 406
– – in ores 382, 392
– – in plating baths 407
– – in rock samples 390, 393
– – in sea-water 372

– – in silicate rocks 325
– – in silver alloys 404
– – in soil samples 325, 389
– – in steel 396, 399
– – in tissue 352
– extraction 325, 396, 402, 404
– influence on the determination of mercury 249
– ores, digestion of 392
– sensitivity in Zeeman AAS 151, 156
Simple reference solution 107
Single-beam instruments 15
Singlet lines, splitting in magnetic field 142
Slags, analysis of 395
Sodium 325–326
– analysis of 406, 421
– borohydride as reductant
– – in the cold vapor technique 78, 249–250, 304
– – in the hydride technique 70, 71, 73
– bromide, molecular spectra 133
– chloride
– – interferences 204–205, 217
– – molecular spectra 133
– – separation from sea-water 373
– contamination 326
– degree of atomization 168
– determination of
– – by flame AES 325
– – in boron 421
– – in cellulose 419
– – in cement 417
– – in coalash 377
– – in dialysis solutions 343
– – in fertilizers 366
– – in fuel oil 326, 411
– – in gasoil 410
– – in glass 415
– – in greases 411
– – in minerals 326, 385, 386
– – in phosphorescent compounds 424
– – in rock samples 326, 385, 388
– – in rubber seals 424
– – in serum 342, 343
– – in silicate rocks 387
– – in ultrapure water 326
– – in uranium 420
– – in urine 342

Sodium, determination of
- – in water 367 *et seq.*
- D-line 1
- flame noise 326
- formation of monocyanides 169
- iodide, molecular spectra 133
- spectrum 4
- tungstate for coating graphite tubes 21
- vapor discharge lamps 25–26
Sodium coolant, analysis of 421
Soft solders, analysis of 402
Soil
- extracts, analysis of 365
- samples
- – analysis of 361 *et seq.*
- – storage of 244
Solid samples
- analysis of
- – alloys 401
- – coal 408
- – foodstuffs and drinks 359–360
- – geological samples 390
- – glass 416
- – in flames 31, 43, 46, 220, 389
- – in graphite furnaces 67 *et seq.*, 192, **220** *et seq.*, 261, 399
- – in miscellaneous atomizers 81
- – paper pulp 419
- background attenuation 225–226
- determination of the following metals in:
    bismuth 274
    selenium 321
    silver 325
    thallium 331
    tin 333
    zinc 339
- particle size 221–222
Solute-volatilization interferences, *see* Interferences
Solvent blank 107
Solvent effect 159
Solvent extraction **120–121,** 123, 205, 387
Solvents
- evaporation 175
- for petrochemicals 409 *et seq.*
- for polymers 418
- influence on flame temperature 180

- organic 35, 120–121, 173, 174, 180, 267, 305, 320, 332, 409–415
- – combustion properties 120
- – influence on sensitivity 173–174
Soya oil, analysis of 360
Sparks as excitation sources 251
Spatial-distribution interferences, *see* Interferences
Speciation
- of antimony 269
- of arsenic 271–272, 371, 382
- of chromium 280, 372
- of lead 295, 359, 412–413
- of mercury 302–304, 355, 358, 382
- of selenium 323, 371, 382
- of silicon 324
- of tellurium 330
- of tin 334, 380
Spectral bandpass **83** *et seq.*, 252
Spectral interferences, *see* Interferences, spectral
Spectral lines
- selection of **5–7,** 256
- splitting in a magnetic field 140–143
Spectral profile of monochromator 86
Spectral range, usable 83
Spectrochemical buffers 107, **162–163,** 342–343 (*see also* Matrix modifiers)
Spectrograph 251
Spinel formation in flames 177
Splitting of energy levels in a magnetic field 140 *et seq.*
Stability of complexes 300, 369
Stack gas, output of heavy metals 376
Standard deviation 109
Standard reference materials **127, 224,** 395
Steel
- analysis of 222, **393** *et seq.*
- – in graphite furnaces 222, **397–399**
- – interferences 397–399
- digestion 393
- extraction of iron 395
- non-metallic phases 396
- solids analysis 222, 399
Stock solution 107
Stoichiometry, *see* Flames
Strontium 326–327
- as spectrochemical buffer 341–342
- determination of 34, 326–327
- – in blood 327

Strontium, determination of
– – in cement 417
– – in coalash 377
– – in glass 415
– – in milk 361
– – in rock samples 386
– – in serum 348, 350
– – in silicate rocks 386
– – in tooth enamel 327, 350
– – in urine 350
– ionization of 326
– salts, analysis of 422
STUDENT's factor 109
Subboiling distillation 125, 245
Sulfate, indirect determination of 309, 328
– in fertilizers 366
– in rhodium baths 407
– in serum 351
– in urine 351
Sulfide, indirect determination of 309
Sulfur 327–328
– determination of 7, 36
– – by ICP-AES 260
– – by molecular absorption spectrometry
  328
– – by molecular emission 309
– – in fertilizers 328
– – in soil extracts 328
– dioxide, in air 379
– – indirect determination 328
Sulfuric acid
– analysis of 422
– as matrix modifier 296, 331
– influence on hydride-forming elements 235
Surface oxides on graphite 215
Surface tension 159
– influence on nebulization 173
Suspensions, analysis of
– by the boat technique 46
– by the flame technique 43, 220, 389
– catalysts 422
– coal slurries 408
– leaves 367
– molybdenum disulfide 411
– sediments 374
– sewage sludge 278, 280–281, 283, 308, 339,
  381
– titanium dioxide slurries 424

– wear metals in lubricating oil 413
Synthetic reference solution 107
Systematic errors 76–77, 109, **119** *et seq.*, **123**
  et seq., **244** *et seq.*

Tablets, analysis of 423
Tantalum 328
– atomization of 177
– determination of
– – by ICP-AES 260
– – in catalysts 422–423
– – in high temperature nickel alloys 400
– – in zirconium alloys 404
– foil for lining graphite tubes 48, 55, 213,
  273, 293, 338
– for coating graphite tubes 57, 213, 306, 333,
  338, 400
– influence of aluminium 328
– tantalum boat atomizer 82
Technetium 329
Teeth, analysis of 346, 349, 350, 354, 355
Tellurium 329–330
– analysis of 406
– determination of 45, 47, 70, 238
– – in accelerators 423
– – in bismuth 406
– – in body fluids 352
– – in chromium 405
   in coal 408
– – in copper 330, 404, 405
– – in copper alloys 400
– – in copper smelting sludges 399
– – in foodstuffs 330
– – in geochemical samples 329
– – in iron 397
– – in nickel 405
– – in nickel alloys 401
– – in rock samples 329, 389, 391
– – in semiconductor materials 406
– – in silicate rocks 330, 391
– – in steel 330, 397
– – in tissue 358
– – in waste water 380
– – in water 330, 371
– extraction of 329, 389, 396
– organically bound 329
– oxidation state 232–233, 330, 371
– reduction of 330

Tellurium
- speciation 330
- spectral interference 206
Temperature (*see also* Flames and Graphite
  furnaces)
- dependence of AES 252
- influence on
- - resistance of graphite tubes 62–63
- - sensitivity in the hydride technique 229–231
- - transport interferences 173
Terbium, resonance lines 293
Term diagram, *see* Energy level diagram
Terms **4–5,** 140–142
Tertra-alkyl lead
- determination of
- - in air 379
- - in foodstuffs 359
- - in petrol 412
- - in waste water 380
- - in water 370
Tetraethyl lead, determination in petrol 412
Tetramethyl lead, determination in petrol 412
Textiles, analysis of 419
Thallium 330–331
- determination of 45
- - in aluminium alloys 403
- - in biological materials 352
- - in body fluids 352
- - in cadmium 405
- - in cobalt alloys 331, 401
- - in iron 397
- - in lubricating oil 414
- - in nickel 405
- - in nickel alloys 331, 400
- - in rock samples 331, 390
- - in silicate rocks 331, 390
- - in tissue 352
- extraction of 389, 400
- vapor discharge lamps 25
- vapor-phase interferences 330–331
Thermal
- decomposition as atomization mechanism
  187, 191–194
- dissociation of hydrides 227–229
- excitation **7–8,** 262
Thermal pretreatment curves 189–193
Thermal pretreatment in graphite furnaces **60**
  et seq., 160, 163, **189** *et seq.*, **203,** 219, 350

- using air 350
- using oxygen 205, 214, 215, 226, 350, 354,
  412
Thermistors, analysis of 424
Thorium 7
Thorium oxide fuel rods, analysis of 420
Three-slot burner head 35, **38–39,** 290, 297,
  316, 326, 343, 365
Thulium, resonance lines 293
Time constant 96, 101, **103–104**
Tin 331–334
- atomization of 182, 332, 334
- determination of 46, 70, 214
- - in airborne dust 379
- - in alloys 333, 399–404
- - in aluminium alloys 403
- - in apples 366
- - in body fluids 352
- - in brass 399
- - in catalysts 422–423
- - in coal 408
- - in copper 240, 333, 334, 404
- - in copper alloys 400
- - in environmetal samples 334, 380
- - in filter materials 334
- - in foodstuffs 333
- - in iron 333, 397
- - in lead alloys 399, 402
- - in lubricating oil 413
- - in nickel 240, 400
- - in nickel alloys 333, 401
- - in ores 333
- - in packaging materials 361, 418
- - in plant materials 366
- - in polymers 418
- - in rock samples 333, 387
- - in silicate rocks 387
- - in soft solder 402
- - in steel 333, 334, 397
- - in tissue 352
- - in vegetables 366
- - in water 334, 371
- - in zirconium 404
- extraction of 333
- flames for 33, 35, 331–332
- influence of organic solvents 332
- interferences through other hydride-forming
  elements 240

Tin
- plating baths, analysis of 407
- resonance lines **112,** 331
- separation as iodide 332–333
- speciation 334, 380
- stability of solutions 333
- volatilization as tetraiodide 333, 387
Tin(II) chloride as reductant
- in the cold vapor technique **76,** 78, 248–249, 302
- in the hydride technique **70,** 73
Tissue, analysis of 341 *et seq.*
Titanium 334–335
- alloys, analysis of 403–404
- atomization of 177
- determination of 57
- - by ICP-AES 260
- - in cement 417
- - in high temperature nickel alloys 400
- - in lubricating oil 414
- - in ores 335
- - in polypropylene 418
- - in rock samples 382
- - in silicate rocks 386
- - in steel 395
- - in tetra-orthobutyltitanate 422
- influence on aluminium 267
- pigments, analysis of 43, 424
- resonance lines 133, 335
Tool steel, analysis of 394
Trace analysis **121–127**
- by AAS or AES 257, 261
- by the graphite furnace technique 200, 205
- influence of contamination 316, 326
- in solid samples 221–223
- of radioactive materials 419
Trace elements
- determination of
- - in body fluids 346 *et seq.*
- - in foodstuffs 356 *et seq.*
- - in high purity chemicals 421–422
- - in iron and steel 395, 397–399
- - in nuclear fuels 419–421
- - in oil and petroleum products 411–412
- - in rocks 383 *et seq.*
- - in tissue 346 *et seq.*
- - in water 367 *et seq.*
- essential 347 *et seq.*

- non-essential 347 *et seq.*
- normal values 347, 348
- toxic 347, 352 *et seq.*
Trace enrichment 121–122
- on activated charcoal 121, 407
- on mercury 122, 407
- on stationary phases 121
Transmission factor of monochromator 89
Transmittance 13, 100
Transport interferences, *see* Interferences
Trichloroacetic acid, interferences of 341, 343
Tungsten 335–336
- alloys, analysis of 404
- analysis of 405, 407
- carbide, analysis of 417
- coating of graphite tubes 56, 213, 306, 333, 338, 400
- determination of
- - by ICP-AES 260
- - in high temperature nickel alloys 400
- - in rock samples 388
- - in steel 336, 395, 396
- - in zirconium alloys 336
- extraction of 336, 387, 396
- foil for lining graphite tubes 338
- indirect determination 336
- influence of aluminium 178
- steels, digestion of 393
Turbulent flames, *see* Direct injection burners

Ultrasonic nebulizer 42, 50
Urea, analysis of 366
Urine, analysis of 46, 58, **341** *et seq.*

Vanadium 336–337
- determination of 57
- - in aluminium alloys 403
- - in biological materials 337, 356
- - in coal 408
- - in coastal crude oil residues 382, 414
- - in crude oil 409
- - in crude oil residues 414
- - in dust 376
- - in gasoil 410
- - in glass 416

Vanadium, determination of
– – in high temperature nickel alloys   400
– – in light fraction fuel oils   411
– – in lubricating oil   414
– – in petroleum coke   408
– – in rock samples   389
– – in sea-water   337, 372
– – in serum   348
– – in silicate rocks   387–388
– – in steel   394, 395
– – in titanium alloys   403
– – in zirconium alloys   403
– influence of aluminium   178, 180, 336
– reference solutions for oil analysis   409
– resonance lines   337
Vapor discharge lamps   **25–26**, 27, 279, 316, 318
Vapor-phase interferences   119, **160**, 251, 255
– in the flame technique   179 *et seq.*
– in the graphite furnace technique   **215** *et seq.*, 294, 330
Vapor pressure
– of atoms   188
– of mercury   244, 301
Vegetables, analysis of   362-363, 366 (*see also* plant materials)
Vidicon detector   92, 97
Viscosity   159, 162, 173
Visual display units   104–105
Vitamin B12, analysis of   423
Vitreous carbon, for storing samples   246
VOIGT profile of resonance lines   11, 21
Volatilization of atoms from graphite surface   187–188, 214
Volatilizers   162
Vulcanization accelerators, analysis of   423

Waste waters, analysis of   379–380
Water
– analysis of   45, 74, 235, **367** *et seq.*
– ultrapure   125
Watergas equilibrium   211
Wavelength, influence on radiation scattering   132
Wear metals in lubricating oil   409
– determination of   413–414
– particle size   413–414
Wet digestions
– for dust samples   376–378
– for foodstuffs   356–358
– for plant materials   366–367
– for rock samples   383–387
– for sediments   382
– for sewage sludge   381
WHITE cathode form   22–23, 25
Wire loop technique in graphite furnaces   64–65, 217, 313, 330
WOODRIFF furnace   50, 64, 217, 219
Wool, analysis of   419

Yellow cake, analysis of   420
Ytterbium
– resonance lines   293
– spectral interference   139
Yttrium   337–338
– determination
– – in hafnium oxide   417
– – in rock samples   338, 388
– – in zirconium oxide   338, 417

Zeeman effect,
– anomalous   142
– direct   143
– inverse   143
– longitudinal   142
– magnetic flux density   148
– normal   142
– on molecules   151
– transverse   142
– with flames   147–148
Zero member compensation solution   **107–108**, 259
Zinc   338–339
– analysis of   407
– as reductant in the hydride technique   69–70
– contamination   338–339
– determination of   44–45
– – in additives   409, 411
– – in aluminium alloys   402–403
– – in ammonia solution   422
– – in beer   361
– – in biological materials   339
– – in carbon fibers   421
– – in cellulose   419
– – in copper   404
– – in copper alloys   399
– – in dust deposits   377

Zinc, determination of
– – in fertilizers   364, 365
– – in fingernails   346
– – in fish   357
– – in glass   415
– – in hair   346
– – in hydrofluoric acid   422
– – in insulin   423
– – in lithium niobate   422
– – in liver tissue   346
– – in lubricating oils   409, 410
– – in magnesium alloys   403
– – in mussel tissue   359
– – in nickel alloys   400
– – in nickel baths   407
– – in nitric acid   421
– – in petrol (gasoline)   413
– – in pharmaceutical products   423
– – in plant materials   366
– – in polymers   418
– – in reference solutions for oil analysis   409
– – in rock samples   382, 387–389
– – in rubber seals   424
– – in sea-water   339, 372–373
– – in seaweed   357
– – in serum   339, 343
– – in sewage sludge   339, 381

– – in silicate rocks   386
– – in silicon tetrachloride   422
– – in soil samples   364, 389
– – in steel   339, 394, 399
– – in sulfuric acid   422
– – in teeth   339, 346
– – in tissue samples   346
– – in uranium   420
– – in urine   339, 343
– – in waste water   339, 380
– – in water   339, 367, 369
– extraction of   339, 367–369, 373, 387
– oxide, analysis of   421
– spectral interferences   404
– vapor discharge lamps   25
Zirconium   339–340
– alloys, analysis of   404
– analysis of   404, 421
– atomization of   177
– determination of   34
– – in aluminium alloys   403
– extraction of   340
– for coating graphite tubes   56, 213, 294, 306,
   324, 333, 338, 400, 411
– oxide, analysis of   417
– resonance lines   340
Zone structure in flames   41, 169, 305